## SI UNITS USED IN MECHANICS

| Quantity | Unit | SI Symbol |
|---|---|---|
| (*Base Units*) | | |
| Length | meter° | m |
| Mass | kilogram | kg |
| Time | second | s |
| (*Derived Units*) | | |
| Acceleration, linear | meter/second$^2$ | m/s$^2$ |
| Acceleration, angular | radian/second$^2$ | rad/s$^2$ |
| Area | meter$^2$ | m$^2$ |
| Density | kilogram/meter$^3$ | kg/m$^3$ |
| Force | newton | N $(= $ kg$\cdot$m/s$^2)$ |
| Frequency | hertz | Hz $(= 1/$s$)$ |
| Impulse, linear | newton-second | N$\cdot$s |
| Impulse, angular | newton-meter-second | N$\cdot$m$\cdot$s |
| Moment of force | newton-meter | N$\cdot$m |
| Moment of inertia, area | meter$^4$ | m$^4$ |
| Moment of inertia, mass | kilogram-meter$^2$ | kg$\cdot$m$^2$ |
| Momentum, linear | kilogram-meter/second | kg$\cdot$m/s $(= $ N$\cdot$s$)$ |
| Momentum, angular | kilogram-meter$^2$/second | kg$\cdot$m$^2$/s $(= $ N$\cdot$m$\cdot$s$)$ |
| Power | watt | W $(= $ J/s $= $ N$\cdot$m/s$)$ |
| Pressure, stress | pascal | Pa $(= $ N/m$^2)$ |
| Product of inertia, area | meter$^4$ | m$^4$ |
| Product of inertia, mass | kilogram-meter$^2$ | kg$\cdot$m$^2$ |
| Spring constant | newton/meter | N/m |
| Velocity, linear | meter/second | m/s |
| Velocity, angular | radian/second | rad/s |
| Volume | meter$^3$ | m$^3$ |
| Work, energy | joule | J $(= $ N$\cdot$m$)$ |
| (*Supplementary and Other Acceptable Units*) | | |
| Distance (navigation) | nautical mile | $(= 1.852$ km$)$ |
| Mass | ton (metric) | t $(= 1000$ kg$)$ |
| Plane angle | degrees (decimal) | ° |
| Plane angle | radian | — |
| Speed | knot | $(1.852$ km/h$)$ |
| Time | day | d |
| Time | hour | h |
| Time | minute | min |

° Also spelled *metre*.

## SI UNIT PREFIXES

| Multiplication Factor | Prefix | Symbol |
|---|---|---|
| 1 000 000 000 000 $= 10^{12}$ | terra | T |
| 1 000 000 000 $= 10^9$ | giga | G |
| 1 000 000 $= 10^6$ | mega | M |
| 1 000 $= 10^3$ | kilo | k |
| 100 $= 10^2$ | hecto | h |
| 10 $= 10$ | deka | da |
| 0.1 $= 10^{-1}$ | deci | d |
| 0.01 $= 10^{-2}$ | centi | c |
| 0.001 $= 10^{-3}$ | milli | m |
| 0.000 001 $= 10^{-6}$ | micro | $\mu$ |
| 0.000 000 001 $= 10^{-9}$ | nano | n |
| 0.000 000 000 001 $= 10^{-12}$ | pico | p |

## SELECTED RULES FOR WRITING METRIC QUANTITIES

1. (a) Use prefixes to keep numerical values generally between 0.1 and 1000.
   (b) Use of the prefixes hecto, decka, deci, and centi should be generally avoided except for certain areas or volumes where the numbers would be otherwise awkward.
   (c) Use prefixes only in the numerator of unit combinations. The one exception is the base unit kilogram. (*Example:* write kN/m not N/mm; J/kg not mJ/g)
   (d) Avoid double prefixes. (*Example:* write GN not kMN)
2. Unit designations
   (a) Use a dot for multiplication of units. (*Example:* write N$\cdot$m not Nm)
   (b) Avoid ambiguous double solidus. (*Example:* write N/m$^2$ not N/m/m)
   (c) Exponents refer to entire unit. (*Example:* mm$^2$ means (mm)$^2$)
3. Number grouping
   Use a space rather than a comma to separate numbers in groups of three, counting from the decimal point in both directions. (*Example:* 4 607 321.048 72) Space may be omitted for numbers of four digits. (*Example:* 4296 or 0.0476)

# ENGINEERING MECHANICS

# STATICS

# ENGINEERING MECHANICS

## VOLUME 1

# STATICS

## THIRD EDITION

### J.L. MERIAM
University of California
Santa Barbara

### L.G. KRAIGE
Virginia Polytechnic Institute and
State University

## JOHN WILEY & SONS, INC.
New York • Chichester • Brisbane • Toronto • Singapore

SI/ENGLISH VERSION

| | |
|---|---|
| *Acquisitions Editor* | Charity Robey |
| *Marketing Manager* | Susan Elbe |
| *Production Supervisor* | Nancy Prinz |
| *Cover Designer* | Pedro Noa |
| *Manufacturing Manager* | Lorraine Fumoso |
| *Copy Editing Supervisor* | Deborah Herbert |
| *Photo Researcher* | Jennifer Atkins |
| *Illustration* | John Balbalis |

This book was set in 9.5/12 Century Schoolbook by Techna Type, Inc. and printed and bound by Von Hoffman.

*Library of Congress Cataloging in Publication Data:*

Meriam, J. L. (James L.)
    Engineering mechanics / J.L. Meriam, L.G. Kraige.—3rd ed.
      p.    cm.
    "SI/English version."
    Includes indexes.
    Contents: v. 1. Statics
    ISBN 0-471-90294-2 (v. 1).
    1. Mechanics, Applied.    I. Kraige, L. G. (L. Glenn)    II. Title.
TA350.M458    1992
620.1—dc20                                    91-42953
                                                                CIP

Printed in the United States of America

10  9  8  7  6  5  4  3  2  1

# FOREWORD

The innovations and contributions of Dr. James L. Meriam to the field of engineering mechanics cannot be overstated. He has undoubtedly had as much influence on instruction in mechanics during the last forty years as any one individual. His first books on mechanics in 1951 literally reconstructed undergraduate mechanics and became the definitive textbooks for the decades that followed. His texts were logically organized, easy to read, directed to the average engineering undergraduate, and were packed with exciting examples of real-life engineering problems which were superbly illustrated. These books became the model for other engineering mechanics texts in the 1950s and beyond.

Dr. Meriam began his work in mechanical engineering at Yale University where he earned his B.E., M. Eng., and Ph.D. degrees. He had early industrial experience with Pratt and Whitney Aircraft and the General Electric Company, which stimulated his first contributions to mechanics in mathematical and experimental stress analysis. During the Second World War he served in the U.S. Coast Guard.

Dr. Meriam was a member of the faculty of the University of California, Berkeley, for twenty-one years where he served as Professor of Engineering Mechanics, Assistant Dean of Graduate Studies, and Chairman of the Division of Mechanics and Design. In 1963 he became Dean of Engineering at Duke University where he devoted his full energies to the development of its School of Engineering. In 1972 Professor Meriam followed his desire to return to full-time teaching and served as Professor of Mechanical Engineering at California Polytechnic State University, where he retired in 1980. During the following ten years he served as visiting professor at the University of California, Santa Barbara, retiring for the second time in 1990. Professor Meriam has always placed great emphasis on teaching, and this trait has been recognized by his students wherever he has taught. At Berkeley in 1963 he was the first recipient of the Outstanding Faculty Award of Tau Beta Pi, given primarily for excellence in teaching, and in 1978 he received the Distinguished Educator Award for Outstanding Service to Engineering Mechanics Education from the American Society for Engineering Education.

Professor Meriam was the first author to show clearly how the method of virtual work in statics can be employed to solve a class of problems largely neglected by previous authors. In dynamics, plane motion became understandable, and in his later editions, three-dimensional kinematics and kinetics received the same treatment. He is credited with original developments in the theory of variable-mass dynamics, which are contained in his *Dynamics, Second Edition* (1971). Professor Meriam has also been a leader in promoting the use of SI units, and his *SI Versions* of *Statics* and *Dynamics* (1975) were the first mechanics textbooks in SI units in this country.

Dr. L. Glenn Kraige, coauthor for the second time in the *Engineering Mechanics* series, has also made significant contributions to mechanics education. Dr. Kraige earned his B.S., M.S., and Ph.D. degrees at the University of Virginia, principally in aerospace engineering, and he currently serves as Professor of Engineering Science and Mechanics at Virginia Polytechnic Institute and State University.

In addition to his recognized research and publications in the field of spacecraft dynamics, Professor Kraige has devoted his attention to the teaching of mechanics at both introductory and advanced levels. His outstanding teaching has been widely recognized and has earned him several awards, including, in 1988, an AT&T award for outstanding teaching in the Southeastern Section of the American Society for Engineering Education and, also in 1988, the Outstanding Educator Award from the State Council of Higher Education for the Commonwealth of Virginia. In his teaching, Professor Kraige stresses the development of analytical capabilities along with the strengthening of physical insight and engineering judgment. In the mid 1980s, he was a leader in the development of motion simulation software for use on personal computers. More recently, he has begun a long-term effort in the area of multimedia approaches to the instruction and learning of statics and dynamics.

The third edition of *Engineering Mechanics* continues the same high standards set by previous editions and adds new features of help and interest to students. It contains a vast collection of interesting and instructive problems. Analysis and applications are the cornerstones of a successful learning experience in engineering mechanics, and J. L. Meriam and L. G. Kraige have shown again that they are the best at melding these essential characteristics.

Robert F. Steidel, Jr.
*Professor Emeritus of Mechanical Engineering*
*University of California, Berkeley*

# PREFACE
# To the Student

As you undertake the study of engineering mechanics, first statics and then dynamics, you will be building a foundation of analytical capability for the solution of a great variety of engineering problems. Modern engineering practice demands a high level of analytical capability, and you will find that your study of mechanics will help you immensely in developing this capacity.

In engineering mechanics we learn to construct and to solve mathematical models which describe the effects of force and motion on a variety of structures and machines that are of concern to engineers. In applying our principles of mechanics we formulate these models by incorporating appropriate physical assumptions and mathematical approximations. In both the formulation and solution of mechanics problems you will have frequent occasion to use your background in plane and solid geometry, scalar and vector algebra, trigonometry, analytic geometry, and calculus. Indeed, you are likely to discover new significance to these mathematical tools as you make them work for you in mechanics.

Your success in mechanics (and throughout engineering) will be highly contingent on developing a well-disciplined method of attack from hypothesis to conclusion in which the applicable principles are applied rigorously. Years of experience in teaching and engineering disclose the importance of developing the ability to represent one's work in a clear, logical, and concise manner. Mechanics is an excellent place in which to develop these habits of logical thinking and effective communication.

*Engineering Mechanics* contains a large number of sample problems in which the solutions are presented in detail. Also included in these examples are helpful observations that mention common errors and pitfalls to be avoided. In addition, the book contains a large selection of simple, introductory problems and problems of intermediate difficulty to help you gain initial confidence and understanding of each new topic. Also included are many problems that illustrate significant and contemporary engineering situations to stimulate your interest and help you to develop an appreciation for the many applications of mechanics in engineering.

We are pleased to extend our encouragement to you as a student of mechanics. We hope this book will provide both help and stimulation as you develop your background in engineering.

*J. L. Meriam*

*Santa Barbara, California*
*February 1992*

*L. Glenn Kraige*

*Blacksburg, Virginia*
*February 1992*

# PREFACE
## To the Instructor

The primary purpose of the study of engineering mechanics is to develop the capacity to predict the effects of force and motion in the course of carrying out the creative design function of engineering. Successful prediction requires more than a mere knowledge of the physical and mathematical principles of mechanics. Prediction also requires the ability to visualize physical configurations in terms of real materials, actual constraints, and the practical limitations that govern the behavior of machines and structures. One of our primary objectives in teaching mechanics is to help the student develop this ability to visualize, which is so vital to problem formulation. Indeed, the construction of a meaningful mathematical model is often a more important experience than its solution. Maximum progress is made when the principles and their limitations are learned together within the context of engineering application.

Courses in mechanics are often regarded by students as a difficult requirement and frequently as an uninteresting academic hurdle as well. The difficulty stems from the extent to which reasoning from fundamentals, as distinguished from rote learning, is required. The lack of interest that is frequently experienced is due primarily to the extent to which mechanics is presented as an academic discipline often lacking in engineering purpose and challenge. This attitude is traceable to the frequent tendency in the presentation of mechanics to use problems mainly as a vehicle to illustrate theory rather than to develop theory for the purpose of solving problems. When the first view is allowed to predominate, problems tend to become overly idealized and unrelated to engineering with the result that the exercise becomes dull, academic, and uninteresting. This approach deprives the student of much of the valuable experience in formulating problems and thus of discovering the need for and meaning of theory. The second view provides by far the stronger motive for learning theory and leads to a better balance between theory and application. The crucial role of interest and purpose in providing the strongest possible motive for learning cannot be overemphasized. Furthermore, we should stress the view that, at best, theory can only approximate the real world of mechanics rather than the view that the real world approximates the theory. This difference in philosophy is indeed basic and distinguishes the *engineering* of mechanics from the *science* of mechanics.

During the past thirty years there has been a strong trend in engineering education to increase the extent and level of theory in the engineering-science courses. Nowhere has this trend been more evident than in mechanics courses. To the extent that students are prepared to handle the accelerated treatment, the trend is beneficial. There is evidence and justifiable concern, however, that a significant disparity has more recently appeared between coverage and comprehension. Among the contributing factors we note three trends. First, emphasis on the geometric and physical meanings of prerequisite mathematics appears to have diminished. Second, there has been a significant reduction and even elimination of instruction in graphics, which in the past enhanced the visualization and representation of mechanics problems. Third, in advancing the mathematical level of our treatment of mechanics, there has been a tendency to allow the notational manipulation of vector operations to mask or replace geometric visualization. Mechanics is inherently a subject that depends on geometric and physical perception, and we should increase our efforts to develop this ability.

One of our responsibilities as teachers of mechanics is to use the mathematics that is most appropriate for the problem at hand. The use of vector notation for one-dimensional problems is usually trivial; for two-dimensional problems it is often optional; but for three-dimensional problems it is usually essential. As we introduce vector operations in two-dimensional problems, it is especially important that their geometric meaning be emphasized. A vector equation is brought to life by a sketch of the corresponding vector polygon, which often discloses through its geometry the shortest solution. There are, of course, many mechanics problems where the complexity of variable interdependence is beyond the normal powers of visualization and physical perception, and reliance on analysis is essential. Nevertheless, our students become better engineers when their abilities to perceive, visualize, and represent are developed to the fullest.

As teachers of engineering mechanics, we have the strongest obligation to the engineering profession to set reasonable standards of performance and to uphold them. In addition, we have a serious responsibility to encourage our students to think for themselves. Too much help with details that students should be reasonably able to handle from prerequisite subjects can be as bad as too little help and can easily condition them to become overly dependent on others rather than to exercise their own initiative and ability. Also, when mechanics is subdivided into an excessive number of small compartments, each with detailed and repetitious instructions, students can have difficulty seeing the "forest for the trees" and, consequently, will fail to perceive the unity of mechanics and the far-reaching applicability of its few basic principles and methods.

This third edition of *Engineering Mechanics,* as with previous editions, is written with the foregoing philosophy in mind. It is intended primarily for the first engineering course in mechanics, generally taught in the second year of study. *Engineering Mechanics* is written in a style that is both concise and friendly. The major emphasis is on basic principles and methods rather than on a multitude of special cases. Strong effort has been made to show both the cohesiveness of the relatively few fundamental ideas and the great variety of problems that these few ideas will solve. A major feature of the book is the extensive treatment of sample problems, which are presented in a single-page format for convenient self-study. In addition to presenting the solution in detail, each sample problem also contains comments and cautions keyed to salient points in the solution and printed in colored type.

*Volume 1, Statics,* contains 75 sample problems and 950 unsolved problems from which a wide choice of assignments can be made. Of these problems, more than 50 percent are new to the Third Edition. In recognition of the need for the predominant emphasis on SI units there are approximately two problems in SI units for every one in U.S. customary units. This apportionment between the two sets of units permits anywhere from a 50–50 emphasis to a 100 percent SI treatment. Many practical problems and examples of interesting engineering situations drawn from a wide range of applications are represented in the problem collection.

In a feature new to the Third Edition, most problem sets are divided into two sections entitled *Introductory Problems* and *Representative Problems.* In the first section are simple, uncomplicated problems designed to help students gain confidence with the new topic, while most of the problems in the second section are of average difficulty and length. The problems are arranged generally in order of increasing difficulty; near the end of the *Representative Problems* are more difficult exercises, which are marked with the symbol ▶. Computer-oriented problems are in a special section at the conclusion of the Review Problems at the end of each chapter. The answers to all odd-numbered problems and to all difficult problems have been provided. Simple numerical values have been used throughout so as not to complicate the solutions and divert attention from the principles. All numerical solutions have been carried out and checked with an electronic calculator without rounding intermediate values. Consequently, the final answers should be correct to within the number of significant figures cited.

A special note on the use of computers is in order. We wish to emphasize that the experience of formulating problems, where reason and judgment are developed, is vastly more important for the student than is the manipulative exercise in carrying out the solution. For this reason, we believe that computer usage must be carefully controlled. At the present time, the processes of constructing free-body diagrams and formulating governing equations are best done with pencil and paper. On the other hand, there are instances in which the *solution* to the governing equations can best be carried out and displayed via the computer. Computer-oriented problems should be genuine in the sense that there is a condition of design or criticality to be found, rather than "makework" problems in which some parameter is varied for no apparent reason other than to force artificial use of the computer. These thoughts have been in mind as we have designed the computer-oriented problems in the Third Edition. To conserve adequate time for problem formulation, it is suggested that the student be assigned only a limited number of the computer problems.

In Chapter 2 the properties of forces, moments, couples, and resultants are developed so that the student may proceed directly to the equilibrium of nonconcurrent force systems in Chapter 3 without belaboring unnecessarily the relatively trivial problem of the equilibrium of concurrent forces acting on a particle. In both Chapters 2 and 3, analysis of two-dimensional problems is presented before three-dimensional problems are treated. The vast majority of students acquire a greater physical insight and understanding of mechanics by first gaining confidence in two-dimensional analysis before coping with the third dimension.

Application of equilibrium principles to simple trusses and to frames and machines is presented in Chapter 4 with primary attention given to two-dimensional systems. A sufficient number of three-dimensional examples is included, however, to enable students to exercise more general vector tools of analysis.

The concepts and categories of distributed forces are introduced at the beginning of Chapter 5 with the balance of the chapter divided into two main sections. Section A treats centroids and mass centers where detailed examples are presented to help students master early applications of calculus to physical and geometrical problems. Section B includes the special topics of beams, flexible cables, and fluid forces, which may be omitted without loss of continuity of basic concepts.

Chapter 6 on friction is divided into Section A on the phenomenon of dry friction and Section B on selected machine applications. Although Section B may be omitted if time is limited, this material does provide a valuable experience for the student in dealing with distributed forces.

Chapter 7 presents a consolidated introduction to virtual work with application limited to single-degree-of-freedom systems. Special emphasis is placed on the advantage of the virtual-work and energy method for interconnected systems and stability determination. Virtual work provides an excellent opportunity to convince the student of the power of mathematical analysis in mechanics.

Moments and products of inertia of areas are presented in Appendix A. This topic helps to bridge the subjects of statics and solid mechanics. Appendix C contains a summary review of selected topics of elementary mathematics as well as several numerical techniques that the student should be prepared to use in computer-solved problems.

We wish to specially cite the outstanding contributions to this series of mechanics books over a period of twenty-five years by illustrator John Balbalis, who died in October 1991. His dedication to high standards of illustrative achievement greatly increased the educational potential of the *Engineering Mechanics* series by providing clarity, reality, and interest to the many thousands of students who have been challenged by his efforts.

Special recognition is due Dr. A. L. Hale of the Bell Telephone Laboratories for his continuing contribution in the form of invaluable suggestions and accurate checking of the manuscript. Dr. Hale has rendered similar service to all previous versions in this entire series of mechanics books, and his input has been a great asset. Appreciation is expressed to Professor J. M. Henderson of the University of California, Davis, for helpful comments and suggestions of selected problems. Also acknowledged are the observations and constructive comments over a period of years of Professor Alfonso Diaz-Jiménez of Bogota, Colombia. A number of members of the Department Engineering Science and Mechanics at Virginia Polytechnic Institute and State University have offered helpful suggestions, including Professors Norman E. Dowling, J. Wallace Grant, Scott L. Hendricks, Arpad A. Pap, Saad A. Ragab, and George W. Swift. The contribution by the staff of John Wiley & Sons, Inc., including editor Charity Robey, reflects a high degree of professional competence and is duly recognized. The support, in the form of a study-research leave, of VPI & SU is acknowledged. Finally, we acknowledge the patience, support, and assistance of our wives, Julia and Dale, during the many hours required to prepare this manuscript.

*J. L. Meriam*                     *L. Glenn Kraige*

*Santa Barbara, California*          *Blacksburg, Virginia*
*February 1992*                      *February 1992*

# CONTENTS

**1 INTRODUCTION TO STATICS**     **1**

1/1 Mechanics     1
1/2 Basic Concepts     2
1/3 Scalars and Vectors     3
1/4 Newton's Laws     6
1/5 Units     7
1/6 Law of Gravitation     9
1/7 Accuracy, Limits, and Approximations     11
1/8 Description of Statics Problems     13

**2 FORCE SYSTEMS**     **19**

2/1 Introduction     19
2/2 Force     19
    Section A. Two-Dimensional Force Systems     22
2/3 Rectangular Components     22
2/4 Moment     34
2/5 Couple     44
2/6 Resultants     52
    Section B. Three-Dimensional Force Systems     61
2/7 Rectangular Components     61
2/8 Moment and Couple     70
2/9 Resultants     84
2/10 Problem Formulation and Review     95

**3 EQUILIBRIUM**     **101**

3/1 Introduction     101
    Section A. Equilibrium in Two Dimensions     102
3/2 Mechanical System Isolation     102
3/3 Equilibrium Conditions     113
    Section B. Equilibrium in Three Dimensions     142
3/4 Equilibrium Conditions     142
3/5 Problem Formulation and Review     163

**4 STRUCTURES**     **173**

4/1 Introduction     173
4/2 Plane Trusses     174
4/3 Method of Joints     176

4/4  Method of Sections                                    189
4/5  Space Trusses                                         200
4/6  Frames and Machines                                   208
4/7  Problem Formulation and Review                        228

5  DISTRIBUTED FORCES                                      239

5/1  Introduction                                          239
     Section A. Centers of Mass and Centroids              241
5/2  Center of Mass                                        241
5/3  Centroids of Lines, Areas, and Volumes                243
5/4  Composite Bodies and Figures; Approximations          263
5/5  Theorems of Pappus                                    275
     Section B. Special Topics                             282
5/6  Beams—External Effects                                282
5/7  Beams—Internal Effects                                289
5/8  Flexible Cables                                       301
5/9  Fluid Statics                                         315
5/10 Problem Formulation and Review                        336

6  FRICTION                                                345

6/1  Introduction                                          345
     Section A. Frictional Phenomena                       346
6/2  Types of Friction                                     346
6/3  Dry Friction                                          347
     Section B. Application of Friction in Machines        368
6/4  Wedges                                                368
6/5  Screws                                                370
6/6  Journal Bearings                                      380
6/7  Thrust Bearings; Disk Friction                        381
6/8  Flexible Belts                                        389
6/9  Rolling Resistance                                    390
6/10 Problem Formulation and Review                        398

7  VIRTUAL WORK                                            407

7/1  Introduction                                          407
7/2  Work                                                  407
7/3  Equilibrium                                           410
7/4  Potential Energy and Stability                        428
7/5  Problem Formulation and Review                        444

APPENDIX A   AREA MOMENTS OF INERTIA                       451

A/1  Introduction                                          451
A/2  Definitions                                           452
A/3  Composite Areas                                       467
A/4  Products of Inertia and Rotation of Axes              475

**APPENDIX B    MASS MOMENTS OF INERTIA**                    **489**

**APPENDIX C    SELECTED TOPICS OF**
**                        MATHEMATICS**                               **491**

   C/1  Introduction                                           491
   C/2  Plane Geometry                                         491
   C/3  Solid Geometry                                         492
   C/4  Algebra                                                492
   C/5  Analytic Geometry                                      493
   C/6  Trigonometry                                           493
   C/7  Vector Operations                                      494
   C/8  Series                                                 497
   C/9  Derivatives                                            497
  C/10  Integrals                                                498
  C/11  Newton's Method for Solving Intractable Equations    501
  C/12  Selected Techniques for Numerical Integration        503

**APPENDIX D    USEFUL TABLES**                               **507**

  Table D/1  Physical Properties                             507
  Table D/2  Solar System Constants                          508
  Table D/3  Properties of Plane Figures                     509
  Table D/4  Properties of Homogeneous Solids                511

**PHOTO CREDITS**                                             **515**

**INDEX**                                                     **517**

Structures that support and produce large forces depend on the principles of mechanics. The ship-loading cranes of the Port of Seattle are examples of the many applications of the principles of statics.

# INTRODUCTION TO STATICS

**1**

## 1/1 MECHANICS

Mechanics is that branch of physical science which deals with the state of rest or motion of bodies under the action of forces. No one subject plays a greater role in engineering analysis than mechanics. The early history of this subject is synonymous with the very beginnings of engineering. Modern research and development in the fields of vibrations, stability and strength of structures and machines, robotics, rocket and spacecraft design, automatic control, engine performance, fluid flow, electrical machines and apparatus, and molecular, atomic, and subatomic behavior are highly dependent on the basic principles of mechanics. A thorough understanding of this subject is an essential prerequisite for work in these and many other fields.

Mechanics is the oldest of the physical sciences. The earliest recorded writings in this field are those of Archimedes (287–212 B.C.) which concern the principle of the lever and the principle of buoyancy. Substantial progress awaited the formulation of the laws of vector combination of forces by Stevinus (1548–1620), who also formulated most of the principles of statics. The first investigation of a dynamic problem is credited to Galileo (1564–1642) in connection with his experiments with falling stones. The accurate formulation of the laws of motion, as well as the law of gravitation, was made by Newton (1642–1727), who also conceived the idea of the infinitesimal in mathematical analysis. Substantial contributions to the development of mechanics were also made by da Vinci, Varignon, Euler, D'Alembert, Lagrange, Laplace, and others.

The principles of mechanics as a science embody the rigor of mathematics on which they are highly dependent. Thus, mathematics plays an important role in achieving the purpose of engineering mechanics, which is the application of these principles to the solution of practical problems. In this book we shall be concerned with both the rigorous development of principles and their application. The basic principles of mechanics are relatively few in number, but they have exceedingly wide application, and the methods employed in mechanics carry over into many fields of engineering endeavor.

The subject of mechanics is logically divided into two parts: *statics*, which concerns the equilibrium of bodies under the action of forces, and *dynamics*, which concerns the motion of bodies. *Engineering Mechanics* is divided into these two parts, *Vol. 1 Statics* and *Vol. 2 Dynamics.*

## 1/2  BASIC CONCEPTS

Certain concepts and definitions are basic to the study of mechanics, and they should be understood at the outset.

*Space* is the geometric region occupied by bodies whose positions are described by linear and angular measurements relative to a coordinate system. For three-dimensional problems three independent coordinates are needed. For two-dimensional problems only two coordinates will be required.

*Time* is the measure of the succession of events and is a basic quantity in dynamics. Time is not directly involved in the analysis of statics problems.

*Mass* is a measure of the inertia of a body, which is its resistance to a change of velocity. Mass can also be regarded as the quantity of matter in a body. Of more importance to us in statics, mass is also the property of every body by which it experiences mutual attraction to other bodies.

*Force* is the action of one body on another. A force tends to move a body in the direction of its action. The action of a force is characterized by its *magnitude*, by the *direction* of its action, and by its *point of application*. Force is a vector quantity, and its properties are discussed in detail in Chapter 2.

*Particle.* A body of negligible dimensions is called a particle. In the mathematical sense a particle is a body whose dimensions approach zero so that it may be analyzed as a point mass. Frequently a particle is chosen as a differential element of a body. Also, when the dimensions of a body are irrelevant to the description of its position or the action of forces applied to it, the body may be treated as a particle.

*Rigid body.* A body is considered rigid when the relative movements between its parts are negligible for the purpose at hand. For instance, the calculation of the tension in the cable which supports the boom of a mobile crane under load is essentially unaffected by the small internal strains (deformations) in the structural members of the boom. For the purpose, then, of determining the external forces which act on the boom, we may treat it as a rigid body. Statics deals primarily with the calculation of external forces which act on rigid bodies which are in equilibrium. To determine the internal stresses and strains, the deformation characteristics of the boom material would have to be analyzed. This type of analysis belongs in the study of the mechanics of deformable bodies, which comes after the study of statics.

## 1/3  SCALARS AND VECTORS

Mechanics deals with two kinds of quantities—scalars and vectors. Scalar quantities are those with which a magnitude alone is associated. Examples of scalar quantities in mechanics are time, volume, density, speed, energy, and mass. Vector quantities, on the other hand, possess direction as well as magnitude and must obey the parallelogram law of addition as described in this article. Examples of quantities represented by vectors are displacement, velocity, acceleration, force, moment, and momentum.

Physical quantities represented by vectors fall into one of three classifications—free, sliding, or fixed.

A *free vector* is one whose action is not confined to or associated with a unique line in space. For example, if a body moves without rotation, then the movement or displacement of any point in the body may be taken as a vector, and this vector will describe equally well the direction and magnitude of the displacement of every point in the body. Hence, we may represent the displacement of such a body by a free vector.

A *sliding vector* is one for which a unique line in space must be maintained along which the quantity acts. When we deal with the external action of a force on a rigid body, the force may be applied at any point along its line of action without changing its effect on the body as a whole* and hence may be considered a sliding vector.

A *fixed vector* is one for which a unique point of application is specified, and therefore the vector occupies a particular position in space. The action of a force on a deformable or nonrigid body must be specified by a fixed vector at the point of application of the force. In this problem the forces and deformations internal to the body will be dependent on the point of application of the force, as well as its magnitude and line of action.

A vector quantity **V** is represented by a line segment, Fig. 1/1, having the direction of the vector and having an arrowhead to indicate the sense. The length of the directed line segment represents to some convenient scale the magnitude $|\mathbf{V}|$ of the vector and is written with lightface italic type $V$. In scalar equations and frequently on diagrams where only the magnitude of a vector is labeled, the symbol will appear in lightface italic type. Boldface type is used for vector quantities whenever the directional aspect of the vector is a part of its mathematical representation. When writing vector equations, we must *always* be certain to preserve the mathematical distinction between vectors and scalars. It is recommended that in all handwritten work a distinguishing mark be used for each vector quantity, such as an underline, $\underline{V}$, or an arrow over the symbol, $\vec{V}$, to take the place of boldface type in print. The direction of the vector

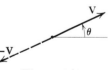

**Figure 1/1**

---

*This is the so-called *principle of transmissibility*, which is discussed in Art. 2/2.

**V** may be measured by an angle $\theta$ from some known reference direction as indicated. The negative of **V** is a vector $-$**V** directed in the sense opposite to **V** as shown in Fig. 1/1.

In addition to possessing the properties of magnitude and direction, vectors must also obey the parallelogram law of combination. This law states that two vectors $\mathbf{V}_1$ and $\mathbf{V}_2$, treated as free vectors, Fig. 1/2*a*, can be replaced by their equivalent **V**, which is the diagonal of the parallelogram formed by $\mathbf{V}_1$ and $\mathbf{V}_2$ as its two

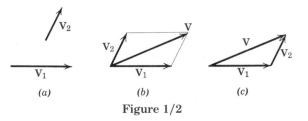

(a)        (b)        (c)

**Figure 1/2**

sides, as shown in Fig. 1/2*b*. This combination or vector sum is represented by the vector equation

$$\mathbf{V} = \mathbf{V}_1 + \mathbf{V}_2$$

where the plus sign used in conjunction with the vector quantities (boldface type) means *vector* and not *scalar* addition. The scalar sum of the magnitudes of the two vectors is written in the usual way as $V_1 + V_2$, and it is clear from the geometry of the parallelogram that $V \neq V_1 + V_2$.

The two vectors $\mathbf{V}_1$ and $\mathbf{V}_2$, again treated as free vectors, may also be added head-to-tail by the triangle law, as shown in Fig. 1/2*c*, to obtain the identical vector sum **V**. We see from the diagram that the order of addition of the vectors does not affect their sum, so that $\mathbf{V}_1 + \mathbf{V}_2 = \mathbf{V}_2 + \mathbf{V}_1$.

The difference $\mathbf{V}_1 - \mathbf{V}_2$ between the two vectors is easily obtained by adding $-\mathbf{V}_2$ to $\mathbf{V}_1$ as shown in Fig. 1/3, where either the triangle or parallelogram procedure may be used. The difference $\mathbf{V}'$ between the two vectors is expressed by the vector equation

$$\mathbf{V}' = \mathbf{V}_1 - \mathbf{V}_2$$

where the minus sign is used to denote *vector subtraction*.

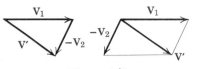

**Figure 1/3**

Any two or more vectors whose sum equals a certain vector **V** are said to be the *components* of that vector. Hence, the vectors $\mathbf{V}_1$ and $\mathbf{V}_2$ in Fig. 1/4*a* are the components of **V** in the directions 1 and

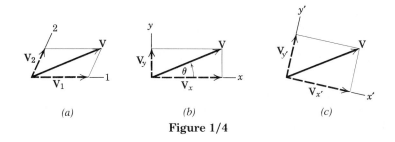

**Figure 1/4**

2, respectively. It is usually most convenient to deal with vector components that are mutually perpendicular, and these are called *rectangular components*. The vectors $\mathbf{V}_x$ and $\mathbf{V}_y$ in Fig. 1/4b are the *x*- and *y*-components, respectively, of $\mathbf{V}$. Likewise, in Fig. 1/4c, $\mathbf{V}_{x'}$ and $\mathbf{V}_{y'}$ are the *x'*- and *y'*-components of $\mathbf{V}$. When expressed in rectangular components, the direction of the vector with respect to, say, the *x*-axis is clearly specified by

$$\theta = \tan^{-1} \frac{V_y}{V_x}$$

A vector $\mathbf{V}$ may be expressed mathematically by multiplying its magnitude $V$ by a unit vector $\mathbf{n}$ whose magnitude is one and whose direction coincides with that of $\mathbf{V}$. Thus,

$$\mathbf{V} = V\mathbf{n}$$

In this way both the magnitude and direction of the vector are conveniently contained in one mathematical expression. In many problems, particularly three-dimensional ones, it is convenient to express the rectangular components of $\mathbf{V}$, Fig. 1/5, in terms of unit vectors $\mathbf{i}$, $\mathbf{j}$, and $\mathbf{k}$, which are vectors in the *x*-, *y*-, and *z*-directions, respectively, with magnitudes of unity. The vector sum of the components is written

$$\boxed{\mathbf{V} = V_x\mathbf{i} + V_y\mathbf{j} + V_z\mathbf{k}}$$

We now make use of the *direction cosines l, m,* and *n* of $\mathbf{V}$ which are defined by

$$l = \cos\theta_x \qquad m = \cos\theta_y \qquad n = \cos\theta_z$$

Thus, we may write the magnitudes of the components of $\mathbf{V}$ as

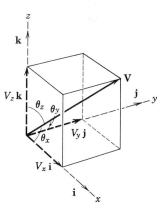

**Figure 1/5**

where

$$\boxed{\begin{array}{c} V_x = lV \qquad V_y = mV \qquad V_z = nV \\ V^2 = V_x^2 + V_y^2 + V_z^2 \end{array}}$$

Note that $l^2 + m^2 + n^2 = 1$.

## 1/4  NEWTON'S LAWS

Sir Isaac Newton was the first to state correctly the basic laws governing the motion of a particle and to demonstrate their validity.* Slightly reworded to use modern terminology, these laws are:

*Law I.*   A particle remains at rest or continues to move in a straight line with a uniform velocity if there is no unbalanced force acting on it.

*Law II.*   The acceleration of a particle is proportional to the resultant force acting on it and is in the direction of this force.

*Law III.*   The forces of action and reaction between interacting bodies are equal in magnitude, opposite in direction, and collinear.

The correctness of these laws has been verified by innumerable accurate physical measurements. Newton's second law forms the basis for most of the analysis in dynamics. As applied to a particle of mass $m$, it may be stated as

$$\boxed{\mathbf{F} = m\mathbf{a}} \tag{1/1}$$

where $\mathbf{F}$ is the resultant force acting on the particle and $\mathbf{a}$ is the resulting acceleration. This equation is a *vector* equation since the direction of $\mathbf{F}$ must agree with the direction of $\mathbf{a}$ in addition to the equality in magnitudes of $\mathbf{F}$ and $m\mathbf{a}$. Newton's first law contains the principle of the equilibrium of forces, which is the main topic of concern in statics. Actually this law is a consequence of the second law, since there is no acceleration when the force is zero, and the particle either is at rest or is moving with a constant velocity. The first law adds nothing new to the description of motion but is included here since it was a part of Newton's classical statements.

The third law is basic to our understanding of force. It states that forces always occur in pairs of equal and opposite forces. Thus, the downward force exerted on the desk by the pencil is accompanied by an upward force of equal magnitude exerted on the pencil by the desk. This principle holds for all forces, variable or constant, regardless of their source and holds at every instant of time during which the forces are applied. Lack of careful attention to this basic law is the cause of frequent error by the beginner. In analyzing bodies under the action of forces, it is absolutely necessary to be clear about which of the pair of forces is beng considered. It is necessary first of all to *isolate* the body under consideration and then to consider only the one force of the pair which acts *on* the body in question.

*Newton's original formulations may be found in the translation of his *Principia* (1687) revised by F. Cajori, University of California Press, 1934.

## 1/5  UNITS

Mechanics deals with four fundamental quantities—length, mass, force, and time. The units used to measure these quantities cannot all be chosen independently because they must be consistent with Newton's second law, Eq. 1/1. Although there are a number of different systems of units in existence, only the two systems which are in primary use in science and technology will be used in this text. The four fundamental quantities and their units and symbols in the two systems are summarized in the following table.

| QUANTITY | DIMENSIONAL SYMBOL | SI UNITS | | U.S. CUSTOMARY UNITS | |
|---|---|---|---|---|---|
| | | UNIT | SYMBOL | UNIT | SYMBOL |
| Mass | M | Base units { kilogram | kg | Base units { slug | — |
| Length | L | meter | m | foot | ft |
| Time | T | second | s | second | sec |
| Force | F | newton | N | pound | lb |

*SI units.* The International System of Units, abbreviated SI (from the French, Système International d'Unités), has been accepted in the United States and throughout the world and is a modern version of the metric system. By international agreement SI units will in time replace other systems that have been in common use. As shown in the table, in SI units mass in kilograms (kg), length in meters (m), and time in seconds (s) are selected as the base units, and force in newtons (N) is derived from the preceding three by Eq. 1/1. Thus, force (N) = mass (kg) × acceleration (m/s²) or

$$N = kg \cdot m/s^2$$

We see, then, that 1 newton is the force required to give a mass of 1 kg an acceleration of 1 m/s². 

Consider a body of mass $m$ which is allowed to fall freely near the surface of the earth. With only the force of gravitation $W$ acting on the body, it falls with an acceleration $g$ toward the center of the earth. For this gravitational experiment, the weight $W$ of the body, from Eq. 1/1, is

$$W(N) = m(kg) \times g(m/s^2)$$

*U.S. customary units.* The U.S. customary or British system of units, also called the foot-pound-second (FPS) system, has been the common system in business and industry in English-speaking countries. Although this system will in time be replaced by SI units, for many years engineers must be able to work both with SI units and with FPS units, and both systems are used freely in *Engineering*

*Mechanics.* As shown in the table, in U.S. or FPS units length in feet (ft), time in seconds (sec), and force in pounds (lb) are selected as base units, and mass in slugs is derived from Eq. 1/1. Thus, force (lb) = mass (slugs) × acceleration (ft/sec$^2$), or

$$\text{slugs} = \frac{\text{lb-sec}^2}{\text{ft}}$$

We see, then, that 1 slug is the mass which is given an acceleration of 1 ft/sec$^2$ when acted on by a force of 1 lb. From the gravitational experiment where $W$ is the gravitational force or weight and $g$ is the acceleration due to gravity, Eq. 1/1 gives

$$m \text{ (slugs)} = \frac{W(\text{lb})}{g(\text{ft/sec}^2)}$$

In U.S. units the pound is also used on occasion as a unit of mass, especially when specifying thermal properties of liquids and gases. When distinction between the two units is necessary, the force unit is frequently written as lbf and the mass unit as lbm. In this book we shall deal almost exclusively with the force unit, which is written simply as lb. Other units of force in the U.S. system which are in frequent use are the *kilopound* (kip), which equals 1000 lb, and the *ton*, which equals 2000 lb.

The International System of Units (SI) is termed an *absolute* system since the measurement of the base quantity mass is independent of its environment. On the other hand, the U.S. system (FPS) is termed a *gravitational* system since its base quantity force is defined as the gravitational attraction (weight) acting on a standard mass under specified conditions (sea level and 45° latitude). A standard pound is also the force required to give a one-pound mass an acceleration of 32.1740 ft/sec$^2$.

In SI units the kilogram is used *exclusively* as a unit of mass—*never* force. It should be pointed out that in the MKS (meter, kilogram, second) gravitational system, which has been used for many years in non-English-speaking countries, the kilogram, like the pound, has been used both as a unit of force and as a unit of mass.

***Primary standards*** for the measurements of mass, length, and time have been established by international agreement and are as follows:

***Mass.*** The kilogram is defined as the mass of a certain platinum–iridium cylinder which is kept at the International Bureau of Weights and Measures near Paris, France. An accurate copy of this cylinder is kept in the United States at the National Institute of Standards and Technology (NIST) (formerly the National Bureau of Standards) and serves as the standard of mass for this country.

***Length.*** The meter, originally defined as one ten-millionth of the distance from the pole to the equator along the meridian through Paris, was later defined as the length of a certain platinum–iridium bar kept at the International Bureau of Weights and Measures. The difficulties of accessibility and accuracy of reproduction of measurement prompted the adoption of a more accurate and reproducible standard of length for the meter, which is now defined as 1 650 763.73 wavelengths of a certain radiation of the krypton-86 atom.

***Time.*** The second was originally defined as the fraction 1/(86 400) of the mean solar day. Irregularities in the earth's rotation have led to difficulties with this definition, and a more accurate and reproducible standard has been adopted. The second is now defined as the duration of 9 192 631 770 periods of the radiation of a certain state of the cesium-133 atom.

Clearly, for most engineering work, and for our purpose in studying mechanics, the accuracy of these standards is considerably beyond our needs.

The standard value for gravitational acceleration $g$ is its value at sea level and at a 45° latitude. In the two systems these values are

$$\text{SI units} \qquad g = 9.806\ 65 \text{ m/s}^2$$
$$\text{U.S. units} \qquad g = 32.1740 \text{ ft/sec}^2$$

The approximate values of 9.81 m/s$^2$ and 32.2 ft/sec$^2$, respectively, are sufficiently accurate for the vast majority of engineering calculations.

The main characteristics of SI units are set forth inside the front cover of the book along with the numerical conversions between U.S. customary and SI units. In addition, charts which give the approximate conversions between selected quantities in the two systems appear inside the back cover of the book for convenient reference. Although these charts will be of assistance in establishing a feel for the relative size of SI and U.S. units, in time engineers will find it essential to think directly in terms of SI units rather than to rely on conversion from U.S. units. In statics we are primarily concerned with the units of length and force, with mass being involved only when we compute gravitational force, as explained in the next article.

In Fig. 1/6 are depicted examples of force, mass, and length in the two systems of units to aid in visualizing their relative magnitudes.

## 1/6  LAW OF GRAVITATION

In statics as well as dynamics we have frequent need to compute the weight of (gravitational force acting on) a body. This computation

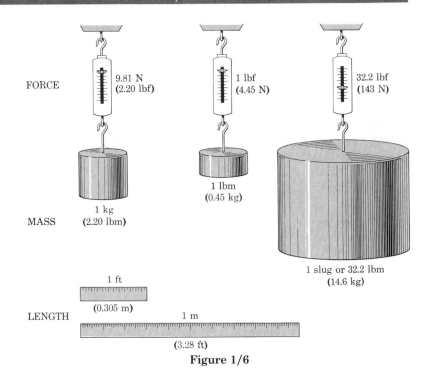

FORCE

9.81 N
(2.20 lbf)

1 lbf
(4.45 N)

32.2 lbf
(143 N)

1 lbm
(0.45 kg)

MASS

1 kg
(2.20 lbm)

1 slug or 32.2 lbm
(14.6 kg)

1 ft
(0.305 m)

LENGTH

1 m

(3.28 ft)

**Figure 1/6**

depends on the law of gravitation, which was also formulated by Newton. The law of gravitation is expressed by the equation

$$F = G \frac{m_1 m_2}{r^2}$$  **(1/2)**

where $F$ = the mutual force of attraction between two particles
  $G$ = a universal constant known as the constant of gravitation
  $m_1, m_2$ = the masses of the two particles
  $r$ = the distance between the centers of the particles

The mutual forces $F$ obey the law of action and reaction, since they are equal and opposite and are directed along the line joining the centers of the particles as shown in Fig. 1/7. By experiment the gravitational constant is found to be $G = 6.673(10^{-11})$ m³/(kg·s²).

**Figure 1/7**

Gravitational forces exist between every pair of bodies. On the surface of the earth the only gravitational force of appreciable magnitude is the force due to the earth's attraction. Thus, each of two iron spheres 100 mm in diameter is attracted to the earth with a gravitational force of 37.1 N, which is called its weight. On the other hand, the force of mutual attraction between the spheres if they are just touching is 0.000 000 095 1 N. This force is clearly negligible compared with the earth's attraction of 37.1 N, and consequently the gravitational attraction of the earth is the only gravitational force of any appreciable magnitude which need be considered for most engineering experiments conducted on the earth's surface.

The gravitational attraction of the earth on a body is known as the *weight* of the body. This force exists whether the body is at rest or in motion. Since this attraction is a force, the weight of a body should be expressed in newtons (N) in SI units and in pounds (lb) in U.S. customary units. Unfortunately in common practice the mass unit kilogram (kg) has been frequently used as a measure of weight. This usage should disappear in time because in SI units the kilogram is used exclusively for mass and the newton is used for force, including weight.

For a body of mass $m$ near the surface of the earth, the gravitational attraction on the body as specified by Eq. 1/2 may be calculated from the results of the simple gravitational experiment. If the gravitational force or weight has a magnitude $W$, then, since the body falls with an acceleration $g$, Eq. 1/1 gives

$$W = mg \qquad\qquad (1/3)$$

The weight $W$ will be in newtons (N) when the mass $m$ is in kilograms (kg) and the acceleration of gravity $g$ is in meters per second squared (m/s$^2$). In U.S. customary units, the weight $W$ will be in pounds (lb) when $m$ is in slugs and $g$ is in feet per second squared. The standard values for $g$ of 9.81 m/s$^2$ and 32.2 ft/sec$^2$ will be sufficiently accurate for our calculations in statics.

The true weight (gravitational attraction) and the apparent weight (as measured by a spring scale) are slightly different. The difference, which is due to the rotation of the earth, is quite small and will be neglected. This effect will be discussed in *Vol. 2 Dynamics*.

## 1/7 ACCURACY, LIMITS, AND APPROXIMATIONS

The number of significant figures in an answer should be no greater than the number of figures which can be justified by the accuracy of the given data. Hence, the cross-sectional area of a square bar whose 24-mm side was measured to the nearest half millimeter should be written as 580 mm$^2$ and not as 576 mm$^2$, as would be indicated if the numbers were multiplied out.

When calculations involve small differences in large quantities, greater accuracy in the data is required to achieve a given accuracy in the results. Hence, it is necessary to know the numbers 4.2503 and 4.2391 to an accuracy of five significant figures in order that their difference 0.0112 be expressed to three-figure accuracy. It is often difficult in somewhat lengthy computations to know at the outset the number of significant figures needed in the original data to ensure a certain accuracy in the answer. Accuracy to three significant figures is considered satisfactory for the majority of engineering calculations.

In this text, answers will generally be shown to three significant figures—unless the answer begins with the digit 1, in which case the answer will be shown to four significant figures. For purposes of calculation, all data which are given may be taken to be exact.

The *order* of differential quantities is the subject of frequent misunderstanding. Higher-order differentials may always be neglected compared with lower-order differentials when the mathematical limit is approached. As an example, the element of volume $\Delta V$ of a right circular cone of altitude $h$ and base radius $r$ may be taken to be a circular slice a distance $x$ from the vertex and of thickness $\Delta x$. It can be verified that the complete expression for the volume of the element may be written as

$$\Delta V = \frac{\pi r^2}{h^2} \, [x^2 \, \Delta x + x(\Delta x)^2 + \tfrac{1}{3}(\Delta x)^3]$$

It should be recognized that, when passing to the limit in going from $\Delta V$ to $dV$ and from $\Delta x$ to $dx$, the terms in $(\Delta x)^2$ and $(\Delta x)^3$ drop out, leaving merely

$$dV = \frac{\pi r^2}{h^2} \, x^2 \, dx$$

which gives an exact expression when integrated.

When dealing with small angles, we can usually make use of simplifying assumptions. Consider the right triangle of Fig. 1/8 where the angle $\theta$, expressed in radians, is relatively small. With the hypotenuse taken as unity, we see from the geometry of the figure that the arc length $1 \times \theta$ and $\sin \theta$ are very nearly the same. Also $\cos \theta$ is close to unity. Furthermore, $\sin \theta$ and $\tan \theta$ have almost the same values. Thus, for small angles we may write

$$\sin \theta \cong \tan \theta \cong \theta \qquad \cos \theta \cong 1$$

These approximations amount to retaining only the first terms in the series expansions for these three functions. As an example of these approximations, for an angle of $1°$

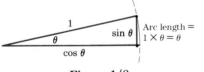

**Figure 1/8**

$$1° = 0.017\ 453 \text{ rad} \qquad \tan 1° = 0.017\ 455$$

$$\sin 1° = 0.017\ 452 \qquad\qquad \cos 1° = 0.999\ 848$$

If a closer approximation is desired, the first two terms may be retained, and they are

$$\sin \theta \cong \theta - \theta^3/6 \qquad \tan \theta \cong \theta + \theta^3/3 \qquad \cos \theta \cong 1 - \theta^2/2$$

The error in replacing the sine by the angle for 1° is only 0.005 percent. For 5° the error is 0.13 percent, and for 10° the error is still only 0.51 percent. As the angle $\theta$ approaches zero, it should now be clear that the following relations are true in the mathematical limit:

$$\sin d\theta = \tan d\theta = d\theta \qquad \cos d\theta = 1$$

The angle $d\theta$ is, of course, expressed in radian measure.

## 1/8 DESCRIPTION OF STATICS PROBLEMS

The study of statics is directed toward the quantitative description of forces that act on engineering structures in equilibrium. Mathematics establishes the relations between the various quantities involved and makes it possible for us to predict effects from these relations. A dual thought process is required in formulating this description. It is necessary to think in terms of the physical situation and in terms of the corresponding mathematical description. Analysis of every problem will require the repeated transition of thought between the physical and the mathematical. Without question, one of the most important goals for the student is to develop the ability to make this transition of thought freely. We should recognize that the mathematical formulation of a physical problem represents an ideal limiting description, or model, which approximates but never quite matches the actual physical situation.

When we construct an idealized mathematical model for a given engineering problem, certain approximations will always be involved. Some of these approximations may be mathematical, whereas others will be physical. For instance, it is often necessary for us to neglect small distances, angles, or forces compared with large distances, angles, or forces. A force which is actually distributed over a small area of the body on which it acts may be considered a concentrated force if the dimensions of the area involved are small compared with other pertinent dimensions. The weight of a steel cable per unit length may be neglected if the tension in the cable is many times greater than its total weight, whereas the cable weight may not be neglected if the problem calls for a determination of the deflection or sag of a suspended cable under the action of its weight. Thus, the degree of assumption involved depends on what information is desired and on the accuracy required. We must be constantly alert to the various assumptions called for in the formulation of real problems. The ability to understand and make use of the appropriate assumptions in the formulation and solution of engineering problems is certainly one of the most important characteristics of a successful engineer. One of the major aims of this book is to provide a maximum of opportunity to develop this ability through the formulation and analysis of many practical problems involving the principles of statics.

Graphics is an important analytical tool which serves us in three capacities. First, it makes possible the representation of a physical system on paper by means of a sketch or diagram. Geometrical representation is vital to physical interpretation and aids greatly in the visualization of the three-dimensional aspects of many problems. Second, graphics often affords a means of solving physical relations where a direct mathematical solution would be awkward or difficult. Graphical solutions not only provide us with a practical means for obtaining results, but they also aid greatly in making the transition of thought between the physical situation and the mathematical expression because both are represented simultaneously. A third use of graphics is in the display of results in charts or graphs, which become a valuable aid to representation.

An effective method of attack on statics problems, as in all engineering problems, is essential. The development of good habits in formulating problems and in representing their solutions will prove to be an invaluable asset. Each solution should proceed with a logical sequence of steps from hypothesis to conclusion, and its representation should include a clear statement of the following parts, each clearly identified:

1. Given data
2. Results desired
3. Necessary diagrams

4. Calculations
5. Answers and conclusions

In addition it is well to incorporate a series of checks on the calculations at intermediate points in the solution. We should observe reasonableness of numerical magnitudes, and the accuracy and dimensional homogeneity of terms should be frequently checked. It is also important that all work be neat and orderly. Careless solutions that cannot be easily read by others are of little or no value. The discipline involved in adherence to good form will in itself be an invaluable aid to the development of the abilities for formulation and analysis. Many problems that at first may seem difficult and complicated become clear and straightforward when begun with a logical and disciplined method of attack.

The subject of statics is based on surprisingly few fundamental concepts and involves mainly the application of these basic relations to a variety of situations. In this application the *method* of analysis is all-important. In solving a problem, it is essential that the laws which apply be carefully fixed in mind and that we apply these principles literally and exactly. In applying the principles which define the requirements for forces acting on a body, it is essential that we *isolate* the body in question from all other bodies so that a complete and accurate account of all forces which act on this body may be taken. This *isolation* should exist mentally as well as be represented on paper. The diagram of such an isolated body with the representation of *all* external forces acting on it is called a *free-body diagram*. It has long been established that the *free-body-diagram method* is the key to the understanding of mechanics. This is so because the *isolation* of a body is the tool by which *cause* and *effect* are clearly separated and by which our attention to the literal application of a principle is accurately focused. The technique of drawing free-body diagrams is covered in Chapter 3, where they are first used.

In applying the laws of statics, we may use numerical values of the quantities directly in proceeding toward the solution, or we may use algebraic symbols to represent the quantities involved and leave the answer as a formula. With numerical substitution the magnitude of each quantity expressed in its particular units is evident at each stage of the calculation. This approach offers advantage when the practical significance of the magnitude of each term is important. The symbolic solution, however, has several advantages over the numerical solution. First, the abbreviation achieved by the use of symbols aids in focusing our attention on the connection between the physical situation and its related mathematical description. Second, a symbolic solution allows us to make a dimensional check at every step, whereas dimensional homogeneity may be lost when numerical values only are used. Third, we may use a symbolic solution repeatedly for obtaining answers to the same problem when different sets and sizes of units are used. Facility with both forms of solution is essential.

The student will find that solutions to the problems of statics may be obtained in one of three ways. First, we may utilize a direct mathematical solution by hand calculation where answers appear either as algebraic symbols or as numerical results. The large majority of problems come under this category. Second, we may approximate the results of certain problems by graphical solutions. Third, solution by computer is of particular advantage when a large number of equations or when parameter variation is involved. There are a number of problems in *Vol. 1 Statics* which are designated as *computer-oriented problems*. These problems appear at the end of the Review Problem sets and are selected to illustrate the type of problem for which solution by computer offers a distinct advantage. The choice of the most expedient method of solution is an important aspect of the experience to be gained from the problem work.

## PROBLEMS

**1/1** Determine the angle made by the vector $\mathbf{V} = -5\mathbf{i} + 12\mathbf{j}$ with the positive sense of the $x$-axis.

*Ans.* $\theta_x = 112.6°$

**1/2** A vector $\mathbf{V}$ has a magnitude of 20 units and lies in the $x$-$y$ plane. If its direction cosine with respect to the $y$-axis is $-0.6$ and if its $x$-component is positive, write the vector expression for $\mathbf{V}$ using unit vectors $\mathbf{i}$ and $\mathbf{j}$. Write the unit vector $\mathbf{n}$ in the direction of $\mathbf{V}$.

**1/3** A certain force is specified by the vector $\mathbf{F} = 60\mathbf{i} - 60\mathbf{j} + 30\mathbf{k}$ N. Calculate the angles made by $\mathbf{F}$ with the $x$-, $y$-, and $z$-directions.

*Ans.* $\theta_x = 48.2°, \theta_y = 131.8°, \theta_z = 70.5°$

**1/4** The vectors $\mathbf{V}_1$ and $\mathbf{V}_2$ are the components of a single vector $\mathbf{V}$. Determine the magnitude of $\mathbf{V}$ graphically and check your result algebraically.

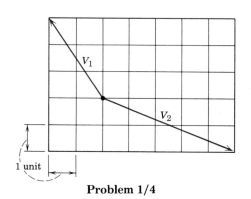

**Problem 1/4**

**1/5** Determine the weight in newtons of a man whose weight in pounds is 200. Determine your own weight in newtons.

*Ans.* 890 N

**1/6** What is the weight in both newtons and pounds of a 40-kg crate?

**1/7** A body weighs 100 lb at sea level and at a latitude of 45°. Determine its mass $m$ in both SI and U.S. units.

*Ans.* $m = 45.4$ kg, $m = 3.11$ lb-sec$^2$/ft

1/8 Calculate the weight $W$ of a body on the top of Mt. Everest (altitude 29,028 ft above sea level) if the body weighs 100 lb at sea level.

1/9 What is the percent error $n$ in replacing the sine of 20° by the value of the angle in radians? Repeat for the tangent of 20° and explain geometrically the difference in the two error percentages.

*Ans.* $n = 2.06\%$, $n = 4.09\%$

1/10 The element of volume $\Delta V$ of the right circular cone of altitude $h$ and base radius $r$ is formed by slicing the cone at a distance $x$ from the vertex. If the slice is of finite thickness $\Delta x$, show that its volume $\Delta V$ is $\pi r^2/h^2[x^2\,\Delta x + x(\Delta x)^2 + \frac{1}{3}(\Delta x)^3]$. (Recall the formula for the volume of a cone.) Explain what happens to the second and third terms when $\Delta x$ becomes the infinitesimal $dx$.

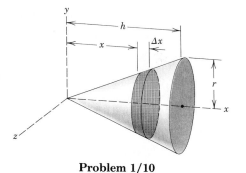

**Problem 1/10**

The properties of forces must be thoroughly mastered by engineers who analyze loaded structures and machines. The placement of this heavy panel is an example of the application and support of forces.

# FORCE SYSTEMS

# 2

## 2/1 INTRODUCTION

In this chapter and in the chapters that follow the properties and effects of various kinds of forces which act on engineering structures and mechanisms will be examined. The experience gained through this examination will prove to be of fundamental use throughout the study of mechanics and in the study of other subjects such as stress analysis, design of structures and machines, and fluid flow. The foundation for a basic understanding of not only statics but also of the entire subject of mechanics is laid in this chapter, and the student should master this material thoroughly.

## 2/2 FORCE

Before dealing with a group or *system* of forces, it is necessary to examine the properties of a single force in some detail. A force has been defined as the action of one body on another. We find that force is a *vector quantity*, since its effect depends on the direction as well as on the magnitude of the action and since forces may be combined according to the parallelogram law of vector combination. The action of the cable tension on the bracket in Fig. 2/1a is represented in Fig. 2/1b by the force vector **P** of magnitude $P$. The effect of this action on the bracket will depend on $P$, the angle $\theta$, and the location of the point of application $A$. Changing any one of these three specifications will alter the effect on the bracket, as could be detected, for instance, by the force in one of the bolts which secure the bracket to the base or the internal stress and strain in the material of the bracket at any point. Thus, the complete specification of the action of a force must include its *magnitude*, *direction*, and *point of application*, in which case it is treated as a fixed vector.

The action of a force on a body can be separated into two effects, *external* and *internal*. For the bracket of Fig. 2/1 the effects of **P** external to the bracket are the reactions or forces (not shown) exerted on the bracket by the foundation and bolts because of the action of **P**. Forces external to a body are then of two kinds, *applied* forces and *reactive* forces. The effects of **P** internal to the bracket are the resulting internal stresses and strains distributed throughout the

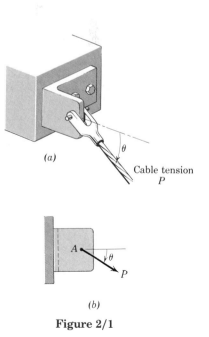

(a)

Cable tension
$P$

(b)

**Figure 2/1**

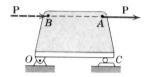

**Figure 2/2**

material of the bracket. The relation between internal forces and internal strains involves the material properties of the body and is studied in strength of materials, elasticity, and plasticity.

In dealing with the mechanics of rigid bodies, where concern is given only to the net *external* effects of forces, experience shows us that it is not necessary to restrict the action of an applied force to a given point. Hence, the force **P** acting on the rigid plate in Fig. 2/2 may be applied at *A* or at *B* or at any other point on its action line, and the net external effects of **P** on the bracket will not change. The external effects are the force exerted on the plate by the bearing support at *O* and the force exerted on the plate by the roller support at *C*. This conclusion is described by the *principle of transmissibility*, which states that a force may be applied at any point on its given line of action without altering the resultant effects of the force *external* to the *rigid* body on which it acts. When only the resultant external effects of a force are to be investigated, the force may be treated as a *sliding* vector, and it is necessary and sufficient to specify the *magnitude*, *direction*, and *line of action* of the force. Since this book deals essentially with the mechanics of rigid bodies, we will treat almost all forces as sliding vectors for the rigid body on which they act.

Forces are classified as either *contact* or *body forces*. Contact forces are generated through direct physical contact between two bodies. Body forces are those applied by remote action, such as gravitational and magnetic forces.

Forces may be either *concentrated* or *distributed*. Actually every contact force is applied over a finite area and is therefore a distributed force. When the dimensions of the area are very small compared with the other dimensions of the body, we may consider the force to be concentrated at a point with negligible loss of accuracy. Force may be distributed over an area, as in the case of mechanical contact, or it may be distributed over a volume when a body force is acting. The *weight* of a body is the force of gravitational attraction distributed over its volume and may be taken as a concentrated force acting through the center of gravity. The position of the center of gravity is frequently obvious from considerations of symmetry. If the position is not obvious, then a separate calculation, explained in Chapter 5, will be necessary to locate the center of gravity.

A force may be measured either by comparison with other known forces, using a mechanical balance, or by the calibrated movement of an elastic element. All such comparisons or calibrations have as their basis a primary standard. The standard unit of force in SI units is the newton (N) and in the U.S. customary system is the pound (lb), as defined in Art. 1/5.

The characteristic of a force expressed by Newton's third law must be carefully observed. The *action* of a force is always accompanied by an *equal* and *opposite reaction*. It is essential for us to fix clearly in mind which force of the pair is being considered. The answer is always clear when the body in question is *isolated* and the force exerted *on* that body (not *by* the body) is represented. It is very easy to make a careless mistake and consider the wrong force of the pair unless we distinguish carefully between action and reaction.

Two forces $\mathbf{F}_1$ and $\mathbf{F}_2$ that are concurrent may be added by the parallelogram law in their common plane to obtain their sum or *resultant* $\mathbf{R}$ as shown in Fig. 2/3*a*. If the two concurrent forces lie in the same plane but are applied at two different points as in Fig. 2/3*b*, by the principle of transmissibility we may move them along their lines of action and complete their vector sum $\mathbf{R}$ at the point of concurrency $A$. The resultant $\mathbf{R}$ may replace $\mathbf{F}_1$ and $\mathbf{F}_2$ without altering the external effects on the body upon which they act. The triangle law may also be used to obtain $\mathbf{R}$, but it will require moving the line of action of one of the forces as shown in Fig. 2/3*c*. In Fig. 2/3*d* the same two forces are added, and although the correct magnitude and direction of $\mathbf{R}$ are preserved, we lose the correct line of action, since $\mathbf{R}$ obtained in this way does not pass through $A$. This type of combination should be avoided. Mathematically the sum of the two forces may be written by the vector equation

$$\mathbf{R} = \mathbf{F}_1 + \mathbf{F}_2$$

In addition to the need for combining forces to obtain their resultant, we often have occasion to replace a force by its *vector components* which act in specified directions. By definition, the two or more vector components of a given vector must vectorially add to yield the given vector. Thus, the force $\mathbf{R}$ in Fig. 2/3*a* may be replaced by or *resolved* into two vector components $\mathbf{F}_1$ and $\mathbf{F}_2$ with the specified directions merely by completing the parallelogram as shown to obtain the magnitudes of $\mathbf{F}_1$ and $\mathbf{F}_2$.

The relationship between a force and its vector components along given axes must not be confused with the relationship between a force and its orthogonal projections onto the same axes. Shown in Fig. 2/3*e* are the projections $\mathbf{F}_a$ and $\mathbf{F}_b$ of the given force $\mathbf{R}$ onto axes $a$ and $b$ which are parallel to the vector components $\mathbf{F}_1$ and $\mathbf{F}_2$ of Fig. 2/3*a*. It is clear from the figure that, in general, the components of a vector are not equal to the projections of the vector onto the same axes. Furthermore, the vector sum of the projections $\mathbf{F}_a$ and $\mathbf{F}_b$ is not the force vector $\mathbf{R}$, because the parallelogram law of vector addition must be followed in forming the sum. Only when the axes $a$ and $b$ are perpendicular are the components and projections of $\mathbf{R}$ equal.

A special case of addition is presented when the two forces $\mathbf{F}_1$ and $\mathbf{F}_2$ are parallel, Fig. 2/4. They may be combined by first adding two equal, opposite, and collinear forces $\mathbf{F}$ and $-\mathbf{F}$ of convenient magnitude which taken together produce no external effect on the

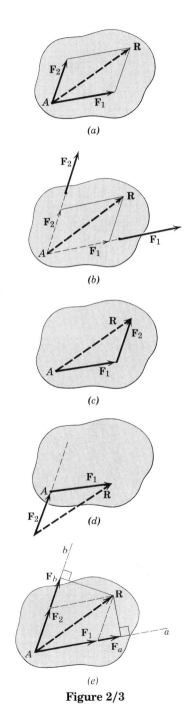

(a)

(b)

(c)

(d)

(e)

**Figure 2/3**

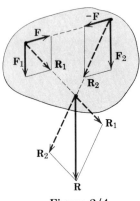

**Figure 2/4**

body. Adding $\mathbf{F}_1$ and $\mathbf{F}$ to produce $\mathbf{R}_1$ and combining with the sum $\mathbf{R}_2$ of $\mathbf{F}_2$ and $-\mathbf{F}$ yield the resultant $\mathbf{R}$ correct in magnitude, direction, and line of action. This procedure is also useful in obtaining a graphical combination of two forces that are almost parallel and hence have a point of concurrency which is remote and inconvenient.

It is usually helpful to master the analysis of force systems in two dimensions before undertaking three-dimensional analysis. To this end the remainder of Chapter 2 is subdivided into these two categories. However, for students who have a good command of vector analysis, these sections may be studied simultaneously if preferred.

# SECTION A.  TWO-DIMENSIONAL FORCE SYSTEMS

## 2/3  *RECTANGULAR COMPONENTS*

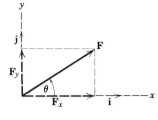

**Figure 2/5**

The most common two-dimensional resolution of a force vector is into *rectangular components*. It follows from the parallelogram rule that the vector $\mathbf{F}$ of Fig. 2/5 may be written as

$$\mathbf{F} = \mathbf{F}_x + \mathbf{F}_y \tag{2/1}$$

where $\mathbf{F}_x$ and $\mathbf{F}_y$ are *vector components* of $\mathbf{F}$. Each of the two vector components may further be written as a scalar times the appropriate unit vector. Thus, in terms of the unit vectors $\mathbf{i}$ and $\mathbf{j}$ of Fig. 2/5, we may write

$$\mathbf{F} = F_x \mathbf{i} + F_y \mathbf{j} \tag{2/2}$$

where the scalars $F_x$ and $F_y$ are the $x$ and $y$ *scalar components* of the vector $\mathbf{F}$. We observe that the scalar components may in general be positive or negative, depending on the quadrant into which $\mathbf{F}$ points. For the force vector of Fig. 2/5, the $x$ and $y$ scalar components are both positive and are related to the magnitude and direction of $\mathbf{F}$ by

$$\boxed{\begin{array}{ll} F_x = F \cos \theta & F = \sqrt{F_x^{\,2} + F_y^{\,2}} \\[2mm] F_y = F \sin \theta & \theta = \tan^{-1} \dfrac{F_y}{F_x} \end{array}} \tag{2/3}$$

It was noted in Art. 1/3 that the magnitude of a vector is written with lightface italic type; that is, $|\mathbf{F}|$ is indicated in print by $F$, a quantity which is always nonnegative. When we work with *components* of a vector, however, the scalar components, also denoted by lightface italic type, will include sign information. See Sample Problem 2/3 for further clarification.

When both a force and its vector components appear in a diagram, it is desirable to show the vector components of the force in dotted lines, as in Fig. 2/5, and the force in a full line, or vice versa. With either of these conventions it will always be clear that a force and its components are being represented and not three separate forces, as would be implied by three solid-line vectors.

Actual problems do not come with reference axes, so their assignment is a matter of arbitrary convenience, and the choice is frequently up to the student. The logical choice is usually indicated by the manner in which the geometry of the problem is specified. When the principal dimensions of a body are given in the horizontal and vertical directions, for example, then assignment of reference axes in these directions is generally convenient. However, dimensions are not always given in horizontal and vertical directions, angles need not be measured counterclockwise from the $x$-axis, and the origin of coordinates need not be on the line of action of a force. Therefore, it is essential that we be able to determine the correct components of a force no matter how the axes are oriented or how the angles are measured. Figure 2/6 suggests a few typical examples

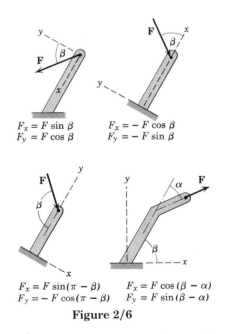

$$F_x = F \sin \beta$$
$$F_y = F \cos \beta$$

$$F_x = - F \cos \beta$$
$$F_y = - F \sin \beta$$

$$F_x = F \sin(\pi - \beta)$$
$$F_y = - F \cos(\pi - \beta)$$

$$F_x = F \cos(\beta - \alpha)$$
$$F_y = F \sin(\beta - \alpha)$$

**Figure 2/6**

of resolution situations in two dimensions, the results of which should be readily apparent. Thus, it is seen that memorization of Eqs. 2/3 is not a substitute for an understanding of the parallelogram law and for the correct projection of a vector onto a reference axis. A neatly drawn sketch always helps to clarify the geometry and avoid error.

It is often convenient to utilize rectangular components in

finding the sum or resultant **R** of two coplanar forces which are concurrent. Consider two forces **F**$_1$ and **F**$_2$ which are originally

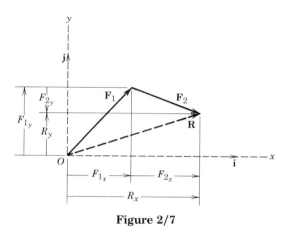

**Figure 2/7**

concurrent at a point $O$. In Fig. 2/7, the line of action of **F**$_2$ is shown shifted from $O$ to the tip of **F**$_1$ in accordance with the triangle rule of Fig. 2/3. In adding the force vectors **F**$_1$ and **F**$_2$, we may write

$$\mathbf{R} = \mathbf{F}_1 + \mathbf{F}_2 = (F_{1_x}\mathbf{i} + F_{1_y}\mathbf{j}) + (F_{2_x}\mathbf{i} + F_{2_y}\mathbf{j})$$

or

$$R_x\mathbf{i} + R_y\mathbf{j} = (F_{1_x} + F_{2_x})\mathbf{i} + (F_{1_y} + F_{2_y})\mathbf{j}$$

from which we conclude that

$$R_x = F_{1_x} + F_{2_x} = \Sigma F_x$$
$$R_y = F_{1_y} + F_{2_y} = \Sigma F_y$$

(2/4)

A term such as $\Sigma F_x$ should be read and thought of as "the algebraic sum of the $x$ scalar components". For the example shown in Fig. 2/7 note that the scalar component $F_{2_y}$ would be negative.

# Sample Problem 2/1

The forces $\mathbf{F}_1$, $\mathbf{F}_2$, and $\mathbf{F}_3$, all of which act on point $A$ of the bracket, are specified in three different ways. Determine the $x$ and $y$ scalar components of each of the three forces.

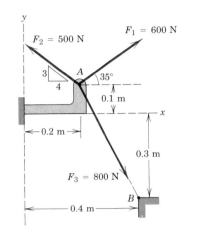

---

**Solution.** The scalar components of $\mathbf{F}_1$, from Fig. $a$, are

$$F_{1_x} = 600 \cos 35° = 491 \text{ N} \qquad \textit{Ans.}$$

$$F_{1_y} = 600 \sin 35° = 344 \text{ N} \qquad \textit{Ans.}$$

The scalar components of $\mathbf{F}_2$, from Fig. $b$, are

$$F_{2_x} = -500(\tfrac{4}{5}) = -400 \text{ N} \qquad \textit{Ans.}$$

$$F_{2_y} = 500(\tfrac{3}{5}) = 300 \text{ N} \qquad \textit{Ans.}$$

Note that the angle which orients $\mathbf{F}_2$ to the $x$-axis is never calculated. The cosine and sine of the angle are available by inspection of the 3-4-5 triangle. Also note that the $x$ scalar component of $\mathbf{F}_2$ is negative by inspection.

The scalar components of $\mathbf{F}_3$ can be obtained by first computing the angle $\alpha$ of Fig. $c$.

$$\alpha = \tan^{-1}\left[\frac{0.2}{0.4}\right] = 26.6°$$

①    Then $F_{3_x} = F_3 \sin \alpha = 800 \sin 26.6° = 358$ N    *Ans.*

$$F_{3_y} = -F_3 \cos \alpha = -800 \cos 26.6° = -716 \text{ N} \qquad \textit{Ans.}$$

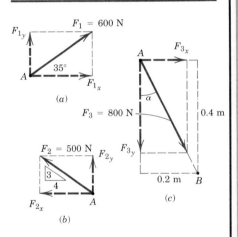

Alternatively, the scalar components of $\mathbf{F}_3$ can be obtained by writing $\mathbf{F}_3$ as a magnitude times a unit vector $\mathbf{n}_{AB}$ in the direction of the line segment $AB$. Thus,

②   
$$\mathbf{F}_3 = F_3 \mathbf{n}_{AB} = F_3 \frac{\overrightarrow{AB}}{\overline{AB}} = 800 \left[ \frac{0.2\mathbf{i} - 0.4\mathbf{j}}{\sqrt{(0.2)^2 + (-0.4)^2}} \right]$$

$$= 800[0.447\mathbf{i} - 0.894\mathbf{j}]$$

$$= 358\mathbf{i} - 716\mathbf{j} \text{ N}$$

The required scalar components are then

$$F_{3_x} = 358 \text{ N} \qquad \textit{Ans.}$$

$$F_{3_y} = -716 \text{ N} \qquad \textit{Ans.}$$

which agree with our previous results.

① The student should carefully examine the geometry of each component–determination problem and not rely on the blind use of such formulas as $F_x = F \cos \theta$ and $F_y = F \sin \theta$.

② A unit vector can be formed by dividing *any* vector, such as the geometric position vector $\overrightarrow{AB}$, by its length or magnitude. Here we use the overarrow to denote the vector which runs from $A$ to $B$ and the overbar to denote the distance between $A$ and $B$.

## Sample Problem 2/2

Combine the two forces **P** and **T**, which act on the fixed structure at $B$, into a single equivalent force **R**.

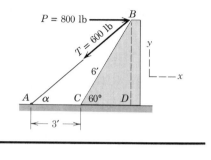

(1) **Graphical solution.** The parallelogram for the vector addition of forces **T** and **P** is constructed as shown in Fig. $a$. The scale used here is 1 in. = 1000 lb; a scale of 1 in. = 200 lb would be more suitable for regular-size paper and would give greater accuracy. Note that the angle $\alpha$ must be determined prior to construction of the parallelogram. From the given figure

$$\tan \alpha = \frac{\overline{BD}}{\overline{AD}} = \frac{6 \sin 60°}{3 + 6 \cos 60°} = 0.866 \qquad \alpha = 40.9°$$

Measurement of the length $R$ and direction $\theta$ of the resultant force **R** yields the approximate results

$$R = 525 \text{ lb} \qquad \theta = 49° \qquad\qquad Ans.$$

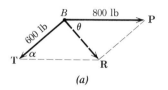

(a)

(2) **Geometric solution.** The triangle for the vector addition of **T** and **P** is shown in Fig. $b$. The angle $\alpha$ is calculated as above. The law of cosines gives

$$R^2 = (600)^2 + (800)^2 - 2(600)(800) \cos 40.9° = 274,300$$

$$R = 524 \text{ lb} \qquad\qquad Ans.$$

From the law of sines, we may determine the angle $\theta$ which orients **R**. Thus,

$$\frac{600}{\sin \theta} = \frac{524}{\sin 40.9°} \qquad \sin \theta = 0.750 \qquad \theta = 48.6° \qquad Ans.$$

(1) Note the repositioning of **P** to permit parallelogram addition at $B$.

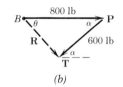

(b)

**Algebraic solution.** By using the $x$-$y$ coordinate system on the given figure, we may write

$$R_x = \Sigma F_x = 800 - 600 \cos 40.9° = 346 \text{ lb}$$

$$R_y = \Sigma F_y = -600 \sin 40.9° = -393 \text{ lb}$$

The magnitude and direction of the resultant force **R** as shown in Fig. $c$ are then

$$R = \sqrt{R_x^2 + R_y^2} = \sqrt{(346)^2 + (-393)^2} = 524 \text{ lb} \qquad Ans.$$

$$\theta = \tan^{-1} \frac{|R_y|}{|R_x|} = \tan^{-1} \frac{393}{346} = 48.6° \qquad Ans.$$

The resultant **R** may also be written in vector notation as

$$\mathbf{R} = R_x\mathbf{i} + R_y\mathbf{j} = 346\mathbf{i} - 393\mathbf{j} \text{ lb} \qquad Ans.$$

(2) Note the repositioning of **T** so as to preserve the correct line of action of the resultant **R**.

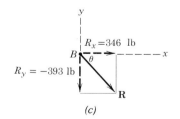

(c)

## Sample Problem 2/3

The 500-N force **F** is applied to the vertical pole as shown. (1) Write **F** in terms of the unit vectors **i** and **j** and identify both its vector and scalar components. (2) Determine the scalar components of the force vector **F** along the $x'$- and $y'$-axes. (3) Determine the scalar components of **F** along the $x$- and $y'$-axes.

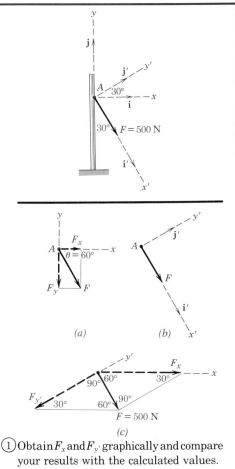

**Solution.** **Part (1)** From Fig. $a$ we may write **F** as

$$\mathbf{F} = (F \cos \theta)\mathbf{i} - (F \sin \theta)\mathbf{j}$$
$$= (500 \cos 60°)\mathbf{i} - (500 \sin 60°)\mathbf{j}$$
$$= (250\mathbf{i} - 433\mathbf{j}) \text{ N} \qquad Ans.$$

The scalar components are $F_x = 250$ N and $F_y = -433$ N. The vector components are $\mathbf{F}_x = 250\mathbf{i}$ N and $\mathbf{F}_y = -433\mathbf{j}$ N.

**Part (2).** From Fig. $b$ we may write **F** as $\mathbf{F} = 500\mathbf{i}'$ N, so that the required scalar components are

$$F_{x'} = 500 \text{ N} \qquad F_{y'} = 0 \qquad Ans.$$

**Part (3).** The components of **F** in the $x$- and $y'$-directions are nonrectangular and are obtained by completing the parallelogram as shown in Fig. $c$. The magnitudes of the components may be calculated by the law of sines. Thus,

① 
$$\frac{|F_x|}{\sin 90°} = \frac{500}{\sin 30°} \qquad |F_x| = 1000 \text{ N}$$

$$\frac{|F_{y'}|}{\sin 60°} = \frac{500}{\sin 30°} \qquad |F_{y'}| = 866 \text{ N}$$

The required scalar components are then

$$F_x = 1000 \text{ N} \qquad F_{y'} = -866 \text{ N} \qquad Ans.$$

① Obtain $F_x$ and $F_{y'}$ graphically and compare your results with the calculated values.

## Sample Problem 2/4

Forces $\mathbf{F}_1$ and $\mathbf{F}_2$ act on the bracket as shown. Determine the projection $F_b$ of their resultant **R** onto the $b$-axis.

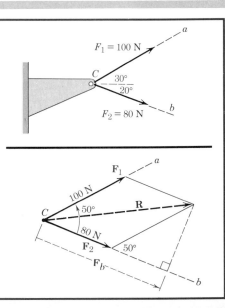

**Solution.** The parallelogram addition of $\mathbf{F}_1$ and $\mathbf{F}_2$ is shown in the figure. Using the law of cosines gives us

$$R^2 = (80)^2 + (100)^2 - 2(80)(100) \cos 130° \qquad R = 163.4 \text{ N}$$

The figure also shows the orthogonal projection $\mathbf{F}_b$ of **R** onto the $b$-axis. Its length is

$$F_b = 80 + 100 \cos 50° = 144.3 \text{ N} \qquad Ans.$$

Note that the components of a vector are in general not equal to the projections of the vector onto the same axes. If the $a$-axis had been perpendicular to the $b$-axis, then the projections and components of **R** would have been equal.

## PROBLEMS

### *Introductory problems*

**2/1** The force **F** has a magnitude of 300 N. Express **F** as a vector in terms of the unit vectors **i** and **j**. Determine the $x$ and $y$ scalar components of **F**.

*Ans.* $F_x = 260$ N, $F_y = -150$ N

$$\mathbf{F} = 260\mathbf{i} - 150\mathbf{j} \text{ N}$$

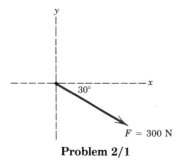

**Problem 2/1**

**2/2** The magnitude of force **F** is 500 lb. Express **F** as a vector in terms of the unit vectors **i** and **j**. Identify both the scalar and vector components of **F**.

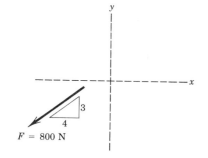

**Problem 2/2**

**2/3** The slope of the 800-N force **F** is specified as shown in the figure. Express **F** as a vector in terms of the unit vectors **i** and **j**.

*Ans.* $\mathbf{F} = -640\mathbf{i} - 480\mathbf{j}$ N

**Problem 2/3**

**2/4** The line of action of the 1600-1b force **F** runs through the points $A$ and $B$ shown in the figure. Determine the $x$ and $y$ scalar components of **F**.

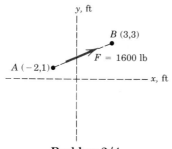

**Problem 2/4**

**2/5** Determine the resultant **R** of the two forces shown by (*a*) applying the parallelogram rule for vector addition and (*b*) summing scalar components.

$$Ans. \; \mathbf{R} = -3\mathbf{i} + 8.66\mathbf{j} \; kN$$

**2/6** Solve Prob. 2/5 graphically.

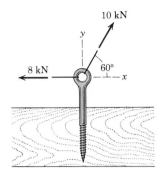

**Problem 2/5**

**2/7** The two forces shown act at point *A* of the bent bar. Determine the resultant **R** of the two forces.

$$Ans. \; \mathbf{R} = 2.35\mathbf{i} - 3.45\mathbf{j} \; kips$$

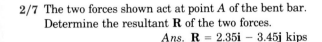

**Problem 2/7**

**2/8** Determine the components of the 2-kN force along the oblique axes *a* and *b*. Determine the projections of **F** onto the *a*- and *b*-axes.

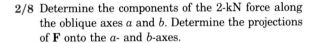

**Problem 2/8**

*Representative problems*

**2/9** If the two equal tensions *T* in the pulley cable together produce a force of 1000 lb on the pulley bearing, calculate *T*.

$$Ans. \; T = 577 \; lb$$

**2/10** If the equal tensions *T* in the pulley cable of Prob. 2/9 are 400 N, express in vector notation the force **R** exerted on the pulley by the two tensions. Determine the magnitude of **R**.

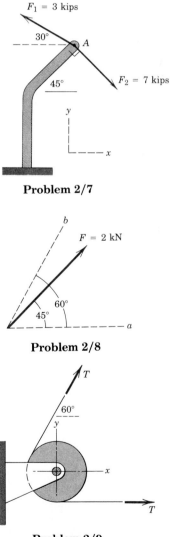

**Problem 2/9**

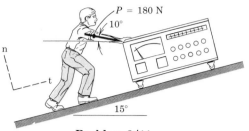

**Problem 2/11**

**2/11** While steadily pushing the machine up an incline, a person exerts a 180-N force **P** as shown. Determine the components of **P** which are parallel and perpendicular to the incline.

$$Ans. \ P_t = 163.1 \text{ N}$$
$$P_n = -76.1 \text{ N}$$

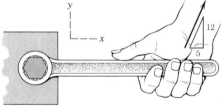

**Problem 2/12**

**2/12** Express the force **P** of magnitude 10 N as a vector in terms of the unit vectors **i** and **j**. Determine the scalar components $P_t$ and $P_n$ which are tangent and normal, respectively, to line $OA$.

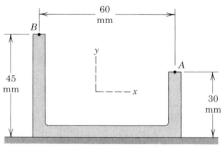

**Problem 2/13**

**2/13** The $y$-component of the force **F** which a person exerts on the handle of the box wrench is known to be 80 lb. Determine the $x$-component and the magnitude of **F**.  $Ans. \ F_x = 33.3 \text{ lb}, F = 86.7 \text{ lb}$

**Problem 2/14**

**2/14** The $x$-component of a force **P** applied to the bracket at $A$ is 160 N. If the line of action of $P$ passes through $B$, determine its magnitude and $y$-component.

**2/15** Replace the 600-lb force acting on the bracket by two forces, a 300-lb force acting in the negative $y$-direction and a force **P**, which together are equivalent to the action of the 600-lb force. Find the magnitude of **P** and the angle $\theta_y$ it makes with the $y$-axis.

$Ans.$ $P = 794$ lb, $\theta_y = 40.9°$

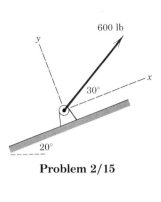

**Problem 2/15**

**2/16** Determine the scalar components $R_a$ and $R_b$ of the force **R** along the nonrectangular axes $a$ and $b$. Also determine the orthogonal projection $P_a$ of **R** onto axis $a$.

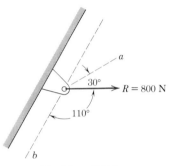

**Problem 2/16**

**2/17** The 600-N force applied to the bracket at $A$ is to be replaced by two forces, $F_a$ in the $a$-$a$ direction and $F_b$ in the $b$-$b$ direction, which together produce the same effect on the bracket as that of the 600-N force. Determine $F_a$ and $F_b$.

$Ans.$ $F_a = 693$ N, $F_b = 346$ N

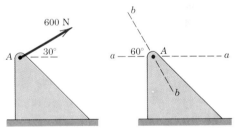

**Problem 2/17**

**2/18** The 20-kN force is to be replaced by two forces $\mathbf{F}_1$, directed along axis $a$-$a$, and $\mathbf{F}_2$, which has a magnitude of 18 kN. Determine the magnitude of $\mathbf{F}_1$ and the angle $\alpha$ which $\mathbf{F}_2$ makes with the horizontal. Note the two possible solutions.

**Problem 2/18**

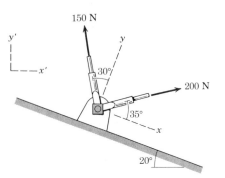

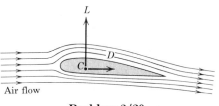

**Problem 2/19**

**2/19** Determine the resultant **R** of the two forces applied to the bracket. Write **R** in terms of unit vectors along the *x*- and *y*-axes shown.

*Ans.* **R** = 88.8**i** + 245**j** N

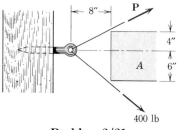

Air flow

**Problem 2/20**

**2/20** The ratio of the lift force *L* to the drag force *D* for the simple airfoil is $L/D = 10$. If the lift force on a short section of the airfoil is 50 lb, compute the magnitude of the resultant force **R** and the angle $\theta$ which it makes with the horizontal.

**Problem 2/21**

**2/21** It is desired to remove the spike from the timber by applying force along its horizontal axis. An obstruction *A* prevents direct access, so that two forces, one 400 lb and the other **P**, are applied by cables as shown. Compute the magnitude of **P** necessary to ensure axial tension *T* along the spike. Also find *T*.                                                   *Ans.* $P = 537$ lb, $T = 800$ lb

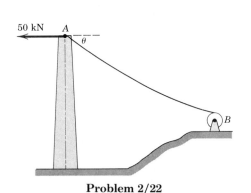

**Problem 2/22**

**2/22** The top of the fixed tower is subjected to a horizontal force of 50 kN and a tension *T* in the heavy, flexible cable, which is tightened by the power winch *B*. If the net effect of the two forces is to produce a downward compression of 30 kN on the tower at *A*, determine the tension *T* in the cable at *A* and the angle $\theta$ made by the cable with the horizontal.

**2/23** In tossing the sphere, a person exerts a 35-lb force on it for the instant shown in the figure. Determine the components, parallel and perpendicular to the forearm, of the force which the sphere exerts *on* the hand. (*Caution:* Be sure to observe Newton's third law.)

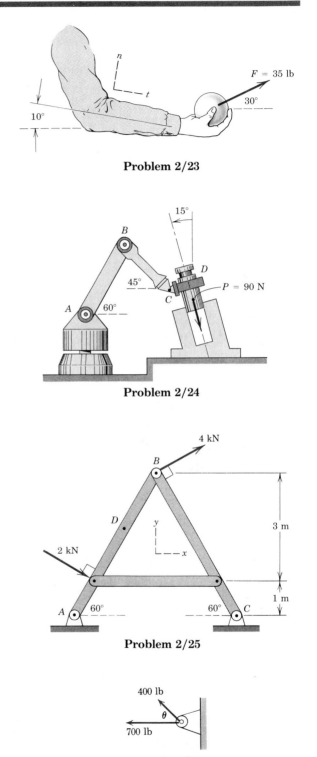

**Problem 2/23**

**2/24** As it inserts the small cylindrical part into a close-fitting circular hole, the robot arm exerts a 90-N force *P* on the part parallel to the axis of the hole as shown. Determine the components of the force which the part exerts *on* the robot along axes (*a*) parallel and perpendicular to the arm *AB*, and (*b*) parallel and perpendicular to the arm *BC*.

**Problem 2/24**

**2/25** Combine the two forces shown acting on the *A*-frame into a single force **R**. Express **R** in vector notation using unit vectors **i** and **j**, and determine its magnitude *R* and the angle $\theta$ it makes with the *x*-axis. If **R** were to be applied at a point *D* on member *AB*, find graphically the distance *s* from *A* to *D*.

*Ans.* **R** = 5.20**i** + **j** kN
$R$ = 5.29 kN
$\theta$ = 10.89°, $s$ = 2.89 m

**Problem 2/25**

**2/26** At what angle $\theta$ must the 400-lb force be applied in order that the resultant **R** of the two forces have a magnitude of 1000 lb? For this condition what will be the angle $\beta$ between **R** and the horizontal?

**Problem 2/26**

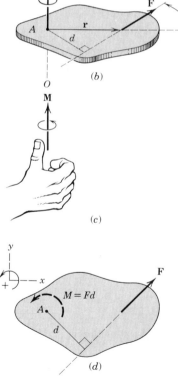

(a)

(b)

(c)

(d)

**Figure 2/8**

## 2/4  MOMENT

In addition to the tendency to move a body in the direction of its application, a force may also tend to rotate a body about an axis. The axis may be any line which neither intersects nor is parallel to the line of action of the force. This rotational tendency is known as the *moment* **M** of the force. Moment is also referred to as *torque*.

As a familiar and motivating example of the concept of moment, consider the pipe wrench of Fig. 2/8*a*. It is clear that one effect of the force applied perpendicular to the handle of the wrench is the tendency to rotate or turn the pipe about its vertical axis. The magnitude of this tendency depends on both the magnitude $F$ of the force and the effective length $d$ of the wrench handle. Common experience shows that a pull which is not perpendicular to the wrench handle is less effective than the right-angle pull shown.

Figure 2/8*b* shows a two-dimensional body acted on by a force **F** in its plane. The magnitude of the moment or tendency of the force to rotate the body about the axis *O-O* normal to the plane of the body is, clearly, proportional both to the magnitude of the force and to the *moment arm d*, which is the perpendicular distance from the axis to the line of action of the force. Therefore, the magnitude of the moment is defined as

$$\boxed{M = Fd} \qquad (2/5)$$

The moment is a vector **M** perpendicular to the plane of the body. The sense of **M** depends on the direction in which **F** tends to rotate the body. The right-hand rule, Fig. 2/8*c*, is used to identify this sense, and the moment of **F** about *O-O* may be represented as a vector pointing in the direction of the thumb, with the fingers curled in the direction of the tendency to rotate. The moment **M** obeys all the rules of vector combination and may be considered a sliding vector with a line of action coinciding with the moment axis. The basic units of moment in SI units are newton-meters (N·m) and in the U.S. customary system are pound-feet (lb-ft).

When dealing with forces all of which act in a given plane, we customarily speak of the moment about a point. Actually the moment with respect to an axis normal to the plane and passing through the point is implied. Thus, the moment of force **F** about point $A$ in Fig. 2/8*d* has the magnitude $M = Fd$ and is counterclockwise. Moment directions may be accounted for by using a stated sign convention, such as a plus sign ( + ) for counterclockwise moments and a minus sign ( − ) for clockwise moments, or vice versa. Sign-convention consistency within a given problem is essential. For the sign convention of Fig. 2/8*d*, the moment of **F** about point $A$ (or about the $z$-axis passing through point $A$) is positive. The curved arrow of the figure is a convenient way to represent moments in two-dimensional analyses.

In some two-dimensional and many three-dimensional problems to follow, it is convenient to use a vector approach for moment calculations. The moment of **F** about point $A$ of Fig. 2/8$b$ may be represented by the cross-product expression

$$\boxed{\mathbf{M} = \mathbf{r} \times \mathbf{F}} \qquad (2/6)$$

where **r** is a position vector which runs from the moment reference point $A$ to *any* point on the line of action of **F**. The magnitude of this expression is given by*

$$M = Fr \sin \alpha = Fd \qquad (2/7)$$

which agrees with the moment magnitude as given by Eq. 2/5. Note that the moment arm $d = r \sin \alpha$ does not depend on the particular point on the line of action of **F** to which the vector **r** is directed. The direction and sense of **M** are correctly established by applying the right-hand rule to the sequence $\mathbf{r} \times \mathbf{F}$. If the fingers of the right hand are curled in the direction of rotation from the positive sense of **r** to the positive sense of **F**, then the thumb points in the positive sense of **M**. We note carefully that the sequence $\mathbf{r} \times \mathbf{F}$ must be maintained, because the sequence $\mathbf{F} \times \mathbf{r}$ would produce a vector in the sense *opposite* to the correct moment. As was the case with the scalar approach, the moment **M** may be thought of as the moment about point $A$ or as the moment about the line $O\text{-}O$ which passes through point $A$ and is perpendicular to the plane containing the vectors **r** and **F**.

In evaluating the moment of a force about a given point, the choice between using the vector cross product or using the scalar expression will depend largely on how the geometry of the problem is specified. If the perpendicular distance between the line of action of the force and the moment center is given or is easily determined, then the scalar approach is generally simpler. If, however, **F** and **r** are not perpendicular and are easily expressible in vector notation, then the cross-product expression is often preferred.

In Section B of this chapter we shall see how the foregoing vector formulation of the moment of a force also applies to the determination of the moment of a force about a point in three-dimensional situations.

**Varignon's theorem.**  One of the most useful principles of mechanics is *Varignon's theorem*, which states that the moment of a force about any point is equal to the sum of the moments of the components of the force about the same point.

---

*See item 7 in Art. C/7 of Appendix C for additional information concerning the cross product.

To prove this theorem, consider the force **R** acting in the plane of the body shown in Fig. 2/9*a*. The forces **P** and **Q** represent any two nonrectangular components of **R**. The moment of **R** about point *O* is

$$\mathbf{M}_O = \mathbf{r} \times \mathbf{R}$$

Since **R** = **P** + **Q**, we may write

$$\mathbf{r} \times \mathbf{R} = \mathbf{r} \times (\mathbf{P} + \mathbf{Q})$$

Finally, using the distributive law for cross products, we have

$$\mathbf{M}_O = \mathbf{r} \times \mathbf{R} = \mathbf{r} \times \mathbf{P} + \mathbf{r} \times \mathbf{Q} \qquad (2/8)$$

which says that the moment of **R** about *O* equals the sum of the moments about *O* of its components **P** and **Q**, thus proving the principle. Varignon's theorem need not be restricted to the case of two components but applies equally well to three or more. Any number of concurrent components of **R** could have been used in the foregoing proof.*

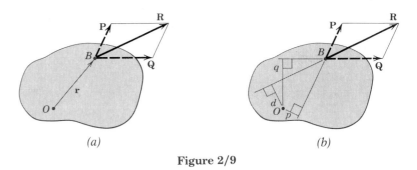

(a)                                      (b)

**Figure 2/9**

To amplify the physical meaning of the principle of moments, we may write the scalar equivalent of the vector expression of Eq. 2/8 for the configuration of Fig. 2/9*b* as

$$M_O = Rd = -pP + qQ$$

where the clockwise moment sense is taken as positive.

Sample Problem 2/5 shows how Varignon's theorem can facilitate the determination of moments.

---

*As originally stated, Varignon's theorem was limited to the case of two concurrent components of a given force. See *The Science of Mechanics*, by Ernst Mach, originally published in 1883.

## Sample Problem 2/5

Calculate the magnitude of the moment about the base point $O$ of the 600-N force in five different ways.

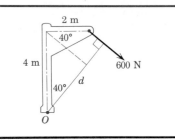

---

***Solution.*** *(I)* The moment arm to the 600-N force is

$$d = 4 \cos 40° + 2 \sin 40° = 4.35 \text{ m}$$

① By $M = Fd$ the moment is clockwise and has the magnitude

$$M_O = 600(4.35) = 2610 \text{ N·m} \qquad Ans.$$

*(II)* Replace the force by its rectangular components at $A$

$$F_1 = 600 \cos 40° = 460 \text{ N}, \qquad F_2 = 600 \sin 40° = 386 \text{ N}$$

By Varignon's theorem, the moment becomes

$$M_O = 460(4) + 386(2) = 2610 \text{ N·m} \qquad Ans.$$

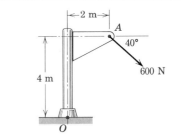

*(III)* By the principle of transmissibility move the 600-N force along its line of action to point $B$, which eliminates the moment of the component $F_2$. The moment arm of $F_1$ becomes

$$d_1 = 4 + 2 \tan 40° = 5.68 \text{ m}$$

and the moment is

$$M_O = 460(5.68) = 2610 \text{ N·m} \qquad Ans.$$

*(IV)* Moving the force to point $C$ eliminates the moment of the component $F_1$. The moment arm of $F_2$ becomes

$$d_2 = 2 + 4 \cot 40° = 6.77 \text{ m}$$

and the moment is

$$M_O = 386(6.77) = 2610 \text{ N·m} \qquad Ans.$$

*(V)* By the vector expression for a moment, and by using the coordinate system indicated on the figure together with the procedures for evaluating cross products, we have

$$M_O = \mathbf{r} \times \mathbf{F} = (2\mathbf{i} + 4\mathbf{j}) \times 600(\mathbf{i} \cos 40° - \mathbf{j} \sin 40°)$$

$$= -2610\mathbf{k} \text{ N·m}$$

The minus sign indicates that the vector is in the negative $z$-direction. The magnitude of the vector expression is

$$M_O = 2610 \text{ N·m} \qquad Ans.$$

① The required geometry here and in similar problems should not cause difficulty if the sketch is carefully drawn.

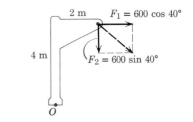

② This procedure is frequently the shortest approach.

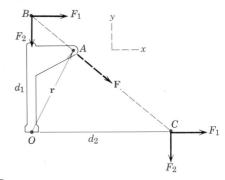

③ The fact that points $B$ and $C$ are not on the body proper should not cause concern, as the mathematical calculation of the moment of a force does not require that the force be on the body.

④ Alternative choices for the position vector $\mathbf{r}$ are $\mathbf{r} = d_1\mathbf{j} = 5.68\mathbf{j}$ m and $\mathbf{r} = d_2\mathbf{i} = 6.77\mathbf{i}$ m.

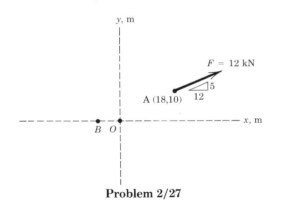

**Problem 2/27**

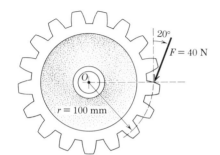

**Problem 2/28**

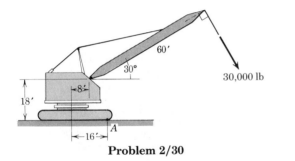

**Problem 2/29**

**Problem 2/30**

## PROBLEMS

### *Introductory problems*

**2/27** The 12-kN force **F** is applied at point *A*. Compute the moment of **F** about point *O*, expressing it both as a scalar and as a vector quantity. Specify the *x*-coordinate of the point *B* (on the *x*-axis) about which the moment of **F** is zero.

$$Ans. \ M_O = 27.7 \text{ kN·m CW}$$
$$\mathbf{M}_O = -27.7\mathbf{k} \text{ kN·m}$$
$$x_B = -6 \text{ m}$$

**2/28** A force **F** of magnitude 40 N is applied to the gear. Determine the moment of **F** about point *O*.

**2/29** Calculate the moment of the 250-N force on the handle of the monkey wrench about the center of the bolt.          *Ans. $M_O$ = 46.4 N·m CW*

**2/30** Determine the moment about point *A* caused by the 30,000-lb tension in the hoisting cable of the tractor crane.

**2/31** Compute the moment of the 0.4-lb force about the pivot $O$ of the wall-switch toggle.

                              *Ans.* $M_O = 0.268$ lb-in. CCW

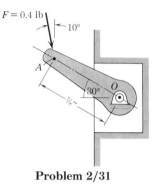

**Problem 2/31**

**2/32** The 30-N force **P** is applied perpendicular to the portion $BC$ of the bent bar. Determine the moment of **P** about point $B$ and about point $A$.

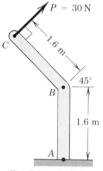

**Problem 2/32**

### Representative problems

**2/33** In order to raise the flagpole $OC$, a light frame $OAB$ is attached to the pole and a tension of 780 lb is developed in the hoisting cable by the power winch $D$. Calculate the moment $M_O$ of this tension about the hinge point $O$.

                              *Ans.* $M_O = 5010$ lb-ft CCW

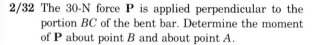

**Problem 2/33**

**2/34** A force of 200 N is applied to the end of the wrench to tighten a flange bolt which holds the wheel to the axle. Determine the moment $M$ produced by this force about the center $O$ of the wheel for the position of the wrench shown.

**Problem 2/34**

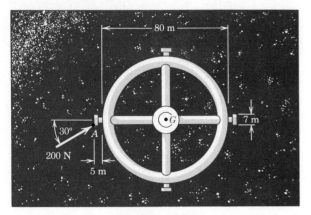

**Problem 2/35**

**2/35** In making a maneuver involving both translation and rotation, the crew of a space station fires a thruster which produces a 200-N force as shown. Determine the moment of the thruster force about the center of mass $G$ of the space station.

*Ans.* $M_G = 3890$ N·m CW

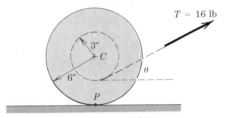

**Problem 2/36**

**2/36** A 16-lb pull $T$ is applied to a cord which is wound securely around the inner hub of the drum. Determine the moment of $T$ about the drum center $C$. At what angle $\theta$ should $T$ be applied so that the moment about the contact point $P$ is zero?

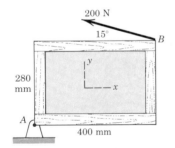

**Problem 2/37**

**2/37** Calculate the moment $M_A$ of the 200-N force about point $A$ by using three scalar methods and one vector method.

*Ans.* $M_A = 74.8$ N·m CCW

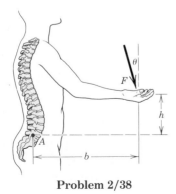

**Problem 2/38**

**2/38** The lower lumbar region $A$ of the spine is the most susceptible part of the spinal column to abuse due to its resistance to excessive bending caused by the moment about $A$ of a force $F$. For given values of $F$, $b$, and $h$, determine the angle $\theta$ which causes the most severe bending strain.

**2/39** If the combined moment about point *A* of the 50-lb force and the force **P** is zero, determine both graphically and algebraically the magnitude of **P**. The plate on which the forces act is divided into squares.

*Ans.* $P = 67.3$ lb

**Problem 2/39**

**2/40** For the angular position $\theta = 60°$ of the crank *OA*, the gas pressure on the piston induces a compressive force *P* in the connecting rod along its centerline *AB*. If this force produces a moment of 720 N·m about the crank axis *O*, calculate *P*.

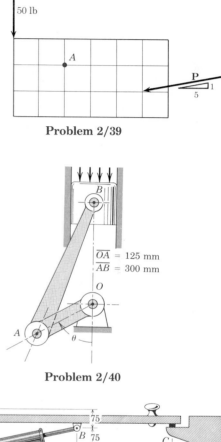

$\overline{OA} = 125$ mm
$\overline{AB} = 300$ mm

**Problem 2/40**

**2/41** The force exerted by the plunger of cylinder *AB* on the door is 40 N directed along the line *AB*, and this force tends to keep the door closed. Compute the moment of this force about the hinge *O*. What force $F_C$ normal to the plane of the door must the door stop at *C* exert on the door so that the combined moment about *O* of the two forces is zero?

*Ans.* $M_O = 7.03$ N·m CW
$F_C = 8.53$ N

Hinge

400

400

25

Dimensions in Millimeters

**Problem 2/41**

**2/42** While inserting a cylindrical part into the circular hole, the robot exerts the 90-N force on the part as shown. Determine the moment about points *A*, *B*, and *C* of the force which the part exerts on the robot.

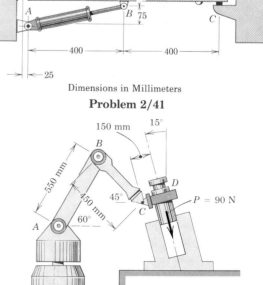

**Problem 2/42**

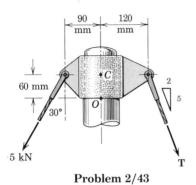

**Problem 2/43**

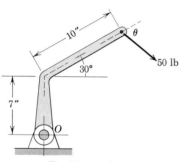

**Problem 2/44**

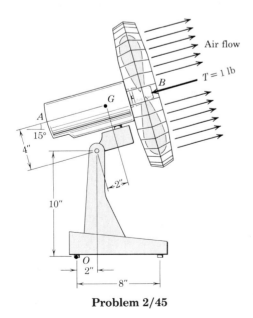

**Problem 2/45**

**2/43** The masthead fitting supports the two forces shown. Determine the magnitude of **T** which will cause no bending of the mast (zero moment) at point *O*.

*Ans.  T* = 4.04 kN

**2/44** Determine the angle $\theta$ which will maximize the moment $M_O$ of the 50-lb force about the shaft axis at *O*. Also compute $M_O$.

**2/45** The blades of the portable fan generate a 1-lb thrust **T** as shown. Compute the moment $M_O$ of this force about the rear support point *O*. For comparison, determine the moment about *O* due to the weight of the motor–fan unit *AB*, whose weight of 9 lb acts as *G*.

*Ans.  $M_O$* = 13.1 lb-in. CCW
*$M_{O_w}$* = 26.1 lb-in. CW

**2/46** The rocker arm *BD* of an automobile engine is supported by a nonrotating shaft at *C*. If the force exerted by the pushrod *AB* on the rocker arm is 80 lb, determine the force which the valve stem *DE* must exert at *D* in order for the combined moment about point *C* to be zero. Compute the resultant of these two forces exerted on the rocker arm. Note that the points *B*, *C*, and *D* form a horizontal line and that both the pushrod and valve stem exert forces along their axes.

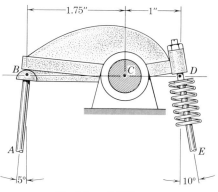

**Problem 2/46**

**2/47** If the resultant **R** of the two forces exerts a clockwise moment of 3800 lb-in. about point *A*, calculate the two values of $\theta$ and the corresponding magnitudes of **R** which satisfy this condition. Verify your results graphically with two accurately drawn diagrams.

*Ans.* $\theta = 61.0°$, $R = 93.0$ lb
$\theta = 119.0°$, $R = 155.4$ lb

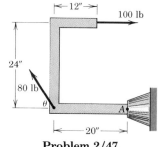

**Problem 2/47**

**2/48** If the combined moment of the two forces about point *C* is zero, determine
(a) the magnitude of the force **P**
(b) the magnitude *R* of the resultant of the two forces
(c) the coordinates *x* and *y* of the point *A* on the rim of the wheel about which the combined moment of the two forces is a maximum
(d) the combined moment $M_A$ of the two forces about *A*.

*Ans.* (a) $P = 61.6$ lb
(b) $R = 141$ lb
(c) $x = 4.90$ in., $y = 6.32$ in.
(d) $M_A = 1580$ lb-in. CW

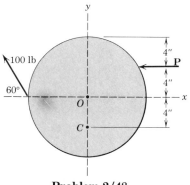

**Problem 2/48**

### 2/5 COUPLE

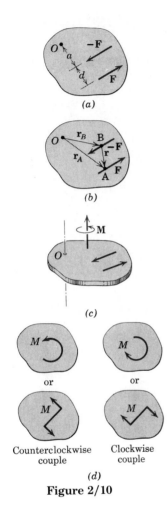

*(a)*

*(b)*

*(c)*

Counterclockwise     Clockwise
couple              couple

*(d)*

**Figure 2/10**

The moment produced by two equal and opposite and noncollinear forces is known as a *couple*. Couples have certain unique properties and have important applications in mechanics.

Consider the action of two equal and opposite forces **F** and $-$**F** a distance $d$ apart, Fig. 2/10a. These two forces cannot be combined into a single force, since their sum in every direction is zero. Their effect is entirely to produce a tendency of rotation. The combined moment of the two forces about an axis normal to their plane and passing through any point such as $O$ in their plane is the *couple* **M**. It has a magnitude

$$M = F(a + d) - Fa$$

or

$$M = Fd$$

and is counterclockwise when viewed from above for the case illustrated. Note especially that the magnitude of the couple contains no reference to the dimension $a$ which locates the forces with respect to the moment center $O$. It follows that the moment of a couple has the same value for *all* moment centers.

We may also represent the moment of a couple by vector algebra. With the cross-product notation of Eq. 2/6, the combined moment about point $O$ of the forces forming the couple of Fig. 2/10b is

$$\mathbf{M} = \mathbf{r}_A \times \mathbf{F} + \mathbf{r}_B \times (-\mathbf{F}) = (\mathbf{r}_A - \mathbf{r}_B) \times \mathbf{F}$$

where $\mathbf{r}_A$ and $\mathbf{r}_B$ are position vectors from point $O$ to arbitrary points $A$ and $B$ on the lines of action of **F** and $-$**F**, respectively. But $\mathbf{r}_A - \mathbf{r}_B = \mathbf{r}$, so that

$$\mathbf{M} = \mathbf{r} \times \mathbf{F}$$

Here again, we see that the moment expression contains no reference to the moment center $O$ and, therefore, is the same for all moment centers. Thus, we may represent **M** by a *free vector* as shown in Fig. 2/10c, where the direction of **M** is normal to the plane of the couple and the sense of **M** is established by the right-hand rule.

Since the couple vector **M** will always be perpendicular to the plane of the forces which constitute the couple, in two-dimensional analysis we can represent the sense of a couple vector as clockwise or counterclockwise by one of the conventions shown in Fig. 2/10d. Later when we deal with couple vectors in three-dimensional problems, we will make full use of the vector notation for their representation.

A couple is unchanged as long as the magnitude and direction of its vector remain constant. Consequently a given couple will not be altered by changing the values of $F$ and $d$ as long as their product remains the same. Likewise a couple is not affected by allowing the

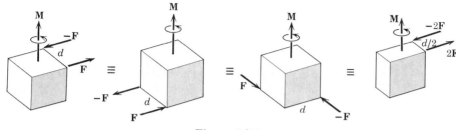

**Figure 2/11**

forces to act in any one of parallel planes. Figure 2/11 shows four different configurations of the same couple **M**. In each of the four cases, the couple is described by the same free vector that represents the identical tendencies to rotate the bodies.

*Force–couple systems.* The effect of a force acting on a body has been described in terms of the tendency to push or pull the body in the direction of the force and to rotate the body about any axis which does not intersect the line of the force. The representation of this dual effect is often facilitated by replacing the given force by an equal parallel force and a couple to compensate for the change in the moment of the force.

The replacement of a force by a force and a couple is illustrated in Fig. 2/12, where the given force **F** acting at point $A$ is replaced by an equal force **F** at some point $B$ and the counterclockwise couple $M = Fd$. The transfer is seen from the middle figure, where the

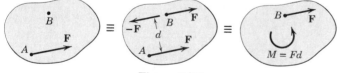

**Figure 2/12**

equal and opposite forces **F** and $-$**F** are added at point $B$ without introducing any net external effects on the body. We now see that the original force at $A$ and the equal and opposite one at $B$ constitute the couple $M = Fd$, which is counterclockwise for the sample chosen, as shown in the right-hand part of the figure. Thus, we have replaced the original force at $A$ by the same force acting at a different point $B$ and a couple without altering the external effects of the original force on the body. The combination of the force and couple in the right hand part of Fig. 2/12 is referred to as a *force–couple system.*

A given couple and a force which lies in the plane of the couple (normal to the couple vector) may be combined to produce a single force by reversing the foregoing procedure. The replacement of a force by an equivalent force-couple system, as well as the reverse procedure, are steps which find repeated application in mechanics and should be thoroughly mastered.

## Sample Problem 2/6

The rigid structural member is subjected to a couple consisting of the two 100-N forces. Replace this couple by an equivalent couple consisting of the two forces **P** and −**P**, each of which has a magnitude of 400 N. Determine the proper angle $\theta$.

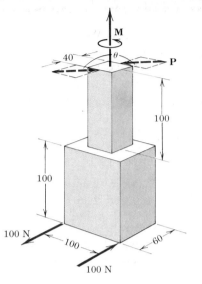

*Dimensions in Millimeters*

**Solution.** The original couple is counterclockwise when the plane of the forces is viewed from above, and its magnitude is

$$[M = Fd] \qquad M = 100(0.1) = 10 \text{ N·m}$$

The forces **P** and −**P** produce a counterclockwise couple

$$M = 400(0.040) \cos \theta$$

① Equating the two expressions gives

$$10 = 400(0.040) \cos \theta$$

$$\theta = \cos^{-1} \frac{10}{16} = 51.3° \qquad Ans.$$

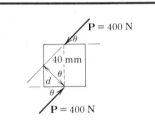

① Since the two equal couples are parallel free vectors, the only dimensions which are relevant are those which give the perpendicular distances between the forces of the couples.

## Sample Problem 2/7

Replace the horizontal 80-lb force acting on the lever by an equivalent system consisting of a force at $O$ and a couple.

**Solution.** We apply two equal and opposite 80-lb forces at $O$ and identify the counterclockwise couple

$$[M = Fd] \qquad M = 80(9 \sin 60°) = 624 \text{ lb-in.} \qquad Ans.$$

① Thus, the original force is equivalent to the 80-lb force at $O$ and the 624-lb-in. couple as shown in the third of the three equivalent figures.

① The reverse of this problem is often encountered, namely, the replacement of a force and a couple by a single force. Proceeding in reverse is the same as replacing the couple by two forces, one of which is equal and opposite to the 80-lb force at $O$. The moment arm to the second force would be $M/F = 624/80 = 7.79$ in., which is 9 sin 60°, thus determining the line of action of the single resultant force of 80 lb.

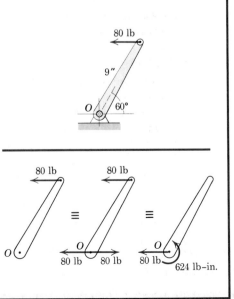

# PROBLEMS

### *Introductory problems*

**2/49** Compute the combined moment of the two 24-N forces about (*a*) point *O* and (*b*) point *A*.

Ans. $M_O = M_A = 38.4$ N·m

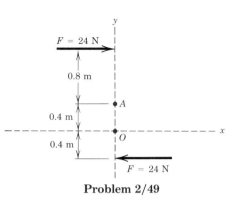

**Problem 2/49**

**2/50** Replace the 40-lb force acting at point *A* by a force–couple system at (*a*) point *O* and (*b*) point *B*.

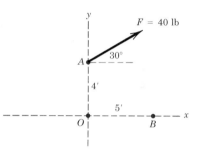

**Problem 2/50**

**2/51** Replace the force–couple system at point *O* by a single force. Specify the coordinate $x_A$ of the point on the *x*-axis through which the line of action of this resultant force acts.

Ans. $F = 1.4$ kN at $x_A = 0.6$ m

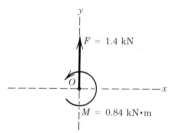

**Problem 2/51**

**2/52** When making a left turn, a driver exerts two 1.5-lb forces on a steering wheel as shown. Determine the moment associated with these forces. Discuss the effects of varying the steering-wheel diameter *d*.

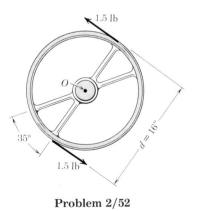

**Problem 2/52**

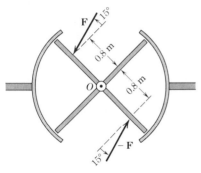

**Problem 2/53**

**2/53** The top view of a revolving entrance door is shown. Two persons simultaneously approach the door and exert forces of equal magnitude as shown. If the resulting moment about the door pivot axis at $O$ is 25 N·m, determine the force magnitude $F$.

*Ans.* $F = 16.18$ N

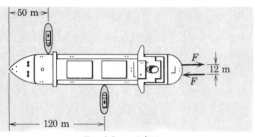

**Problem 2/54**

**2/54** Each propeller of the twin-screw ship develops a full-speed thrust of 300 kN. In maneuvering the ship, one propeller is turning full speed ahead and the other full speed in reverse. What thrust $P$ must each tug exert on the ship to counteract the turning effect of the ship's propellers?

*Representative problems*

**2/55** As part of a test, the two aircraft engines are revved up and the propeller pitches are adjusted so as to result in the fore and aft thrusts shown. What force $F$ must be exerted by the ground on each of the main braked wheels at $A$ and $B$ to counteract the turning effect of the two propeller thrusts? Neglect any effects of the nose wheel $C$, which is turned 90° and unbraked.

*Ans.* $F = 875$ lb

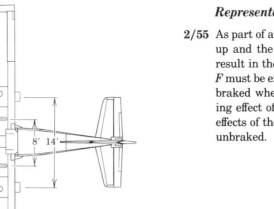

**Problem 2/55**

**2/56** In the design of the lifting hook the action of the applied force **F** at the critical section of the hook is a direct pull at $B$ and a couple. If the magnitude of the couple is 4000 lb-ft, determine the magnitude of **F**.

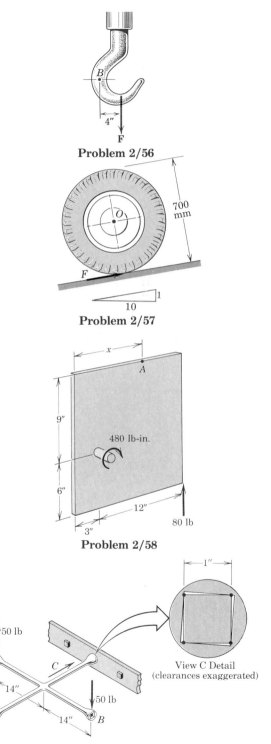

**Problem 2/56**

**2/57** During the motion of a car at constant speed up a 10 percent incline, a torque (couple) of 260 N·m is exerted on each rear wheel by the drivetrain (not shown). If the combined moment about $O$ due to the drivetrain and the friction force $F$ is to be zero, determine $F$.

*Ans.* $F = 743$ N

**Problem 2/57**

**2/58** The 480-lb-in. couple is applied to the horizontal shaft welded to the vertical square plate. If the couple and the 80-lb force are replaced by an equivalent force at point $A$, determine the distance $x$.

**Problem 2/58**

**2/59** A lug wrench is used to tighten a square-head bolt. If 50-lb forces are applied to the wrench as shown, determine the magnitude $F$ of the equal forces exerted on the four contact points on the 1-in. bolt head so that their external effect on the bolt is equivalent to that of the two 50-lb forces. Assume that the forces are perpendicular to the flats of the bolt head. *Ans.* $F = 700$ lb

**Problem 2/59**

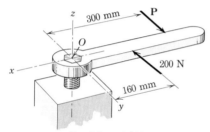

**Problem 2/60**

**2/60** The wrench is subjected to the 200-N force and the force **P** as shown. If the equivalent of the two forces is a force **R** at *O* and a couple expressed as the vector **M** = 20**k** N · m, determine the vector expressions for **P** and **R**.

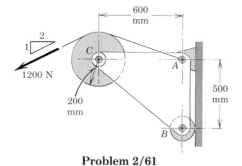

**Problem 2/61**

**2/61** Calculate the moment of the 1200-N force about pin *A* of the bracket. Begin by replacing the 1200-N force by a force–couple system at point *C*.

*Ans.* $M_A$ = 562 N·m CCW

**2/62** For the bracket of Prob. 2/61, calculate the moment of the 1200-N force about the pin at *B*.

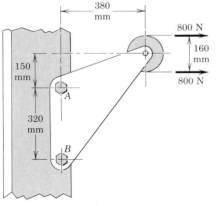

**Problem 2/63**

**2/63** The pulley bracket supports two 800-N tensions in the cable as shown and is bolted at *A* and *B* to the flange of the steel column. Replace the two forces by an equivalent force and couple, with the force midway between the bolts. Then redistribute this force and couple by replacing each by a force at *A* and a force at *B*. Combine the effects and find the force, tension or compression, supported by each bolt.

*Ans.* $F_A$ = 2.35 kN tension

$F_B$ = 0.75 kN compression

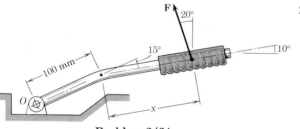

**Problem 2/64**

**2/64** A force **F** of magnitude 50 N is exerted on the automobile parking brake lever at the position *x* = 250 mm. Replace the force by an equivalent force–couple system at the pivot point *O*.

**2/65** A utility pole of mass $m$ is being slowly erected by the winch and cable arrangement shown. When $\theta = 60°$, the cable tension is 35 percent of the weight of the pole. Determine the force–couple system at $O$ which is equivalent to the tension force applied to the top of the pole. Neglect the diameter of the hoisting drum compared with $l$.

$$Ans. \ R = 0.35mg$$
$$M_O = 0.253mgl \ \text{CCW}$$

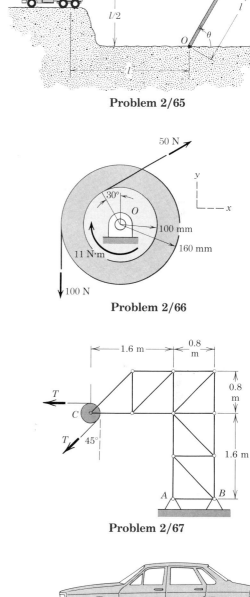

**Problem 2/65**

**2/66** The unit consisting of two rigidly connected pulleys is acted on by a couple and two tension forces, the latter exerted by belts which are securely wrapped onto the two pulley surfaces. Determine the equivalent force–couple system at the pulley axis $O$.

**Problem 2/66**

**2/67** Determine the combined moment $M_A$ about point $A$ due to the two equal tensions $T = 8$ kN in the cable acting on the pulley. Is it necessary to know the pulley diameter?

$$Ans. \ M_A = 30.9 \ \text{kN·m CCW}$$

**Problem 2/67**

**2/68** The combined drive wheels of a front-wheel-drive automobile are acted on by a 7000-N normal reaction force and a friction force $\mathbf{F}$, both of which are exerted by the road surface. If it is known that the resultant of these two forces makes a 15° angle with the vertical, determine the equivalent force–couple system at the car mass center $G$. Treat this as a two-dimensional problem.

**Problem 2/68**

## 2/6  RESULTANTS

In the previous four articles the properties of force, moment, and couple were developed. With the aid of these descriptions we are now ready to describe the resultant action of a group or *system* of forces. Most problems in mechanics deal with a system of forces, and it is usually necessary to reduce the system to its simplest form in describing its action. The resultant of a system of forces is the simplest force combination that can replace the original forces without altering the external effect of the system on the rigid body to which the forces are applied. The equilibrium of a body is the condition where the resultant of all forces that act on it is zero. When the resultant of all forces on a body is not zero, the acceleration of the body is described by equating the force resultant to the product of the mass and acceleration of the body. Thus, the determination of resultants is basic to both statics and dynamics.

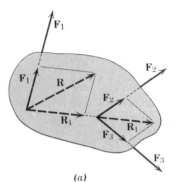

The most common type of force system occurs when the forces all act in a single plane, say, the *x-y* plane, as illustrated by the system of three forces $\mathbf{F}_1$, $\mathbf{F}_2$, and $\mathbf{F}_3$ in Fig. 2/13*a*. The resultant force $\mathbf{R}$ is obtained in magnitude and direction by forming the *force polygon* in the *b*-part of the figure where the forces are added head-to-tail in any sequence. Thus, for any system of coplanar forces we may write

$$\mathbf{R} = \mathbf{F}_1 + \mathbf{F}_2 + \mathbf{F}_3 + \cdots = \Sigma \mathbf{F}$$

$$R_x = \Sigma F_x \qquad R_y = \Sigma F_y \qquad R = \sqrt{(\Sigma F_x)^2 + (\Sigma F_y)^2}$$

$$\theta = \tan^{-1}\frac{R_y}{R_x} = \tan^{-1}\frac{\Sigma F_y}{\Sigma F_x}$$

$$(2/9)$$

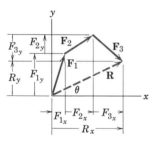

(a)

(b)

**Figure 2/13**

Graphically the correct line of action of $\mathbf{R}$ may be obtained by preserving the correct lines of action of the forces and adding them by the parallelogram law as indicated in the *a*-part of the figure for the case of three forces where the sum $\mathbf{R}_1$ of $\mathbf{F}_2$ and $\mathbf{F}_3$ is added to $\mathbf{F}_1$ to obtain $\mathbf{R}$. In this process the principle of transmissibility has been used.

Algebraically, we may locate the resultant force as follows. First, a convenient reference point is chosen and all forces are moved to that point. This process is depicted for a three-force system in Figs. 2/14*a* and *b*, where $M_1$, $M_2$, and $M_3$ are the couples resulting from the transfer of forces $\mathbf{F}_1$, $\mathbf{F}_2$, and $\mathbf{F}_3$ from their respective original lines of action to lines of action through point $O$. Then all forces at $O$ are added to form the resultant force $\mathbf{R}$, and all couples are added to form the resultant couple $M_O$. We now have the single force–couple system as shown in Fig. 2/14*c*. Finally, in Fig. 2/14*d*, the line of action of $\mathbf{R}$ is located by requiring $\mathbf{R}$ to have a moment of $M_O$ about point $O$. The reader should perceive that the force systems of Figs. 2/14*a* and 2/14*d* are equivalent, and that $\Sigma(Fd)$ in Fig. 2/14*a* is equal to $Rd$ in Fig. 2/14*d*.

This process is summarized in equation form by

$$
\begin{array}{c}
\mathbf{R} = \Sigma\mathbf{F} \\
M_O = \Sigma M = \Sigma(Fd) \\
Rd = M_O
\end{array}
\qquad (2/10)
$$

The first two of Eqs. 2/10 reduce a given system of forces to a force–couple system at an arbitrarily chosen but convenient point $O$. The last equation specifies the distance $d$ from point $O$ to the line of action of $\mathbf{R}$. This last equation, in effect, states that the moment of the resultant force about any point $O$ equals the sum of the moments of the original forces of the system about the same point. This represents an extension of Varignon's theorem to the case of *non-concurrent* force systems; this extension will be referred to hereafter as the *principle of moments*.

For a concurrent system of forces where the lines of action of all forces pass through a common point $O$, the moment sum $\Sigma M_O$ about that point will of course be zero. Thus, the line of action of the resultant $\mathbf{R} = \Sigma\mathbf{F}$, determined by the first of Eqs. 2/10, passes through point $O$. For a parallel force system, one should select a coordinate axis in the direction of the forces. If the resultant force $\mathbf{R}$ for a given force system is zero, the resultant of the system need not be zero as it may be a couple. The three forces in Fig. 2/15, for instance, have a zero resultant force but have a resultant clockwise couple $M = F_3 d$.

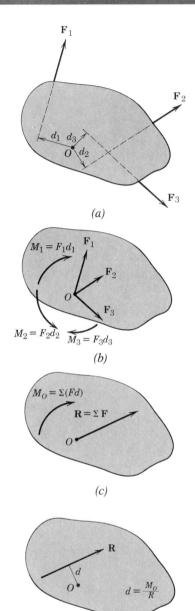

(a)

(b)

(c)

(d)

**Figure 2/14**

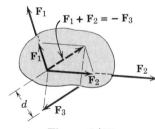

**Figure 2/15**

## Sample Problem   2/8

Determine the resultant of the four forces and one couple that act on the plate shown.

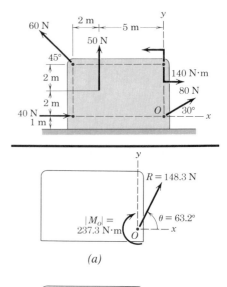

**Solution.**   Point $O$ is selected as a convenient reference point for the force–couple system that is to represent the given system.

$[R_x = \Sigma F_x]$       $R_x = 40 + 80 \cos 30° - 60 \cos 45° = 66.9$ N

$[R_y = \Sigma F_y]$       $R_y = 50 + 80 \sin 30° + 60 \cos 45° = 132.4$ N

$[R = \sqrt{R_x{}^2 + R_y{}^2}]$    $R = \sqrt{(66.9)^2 + (132.4)^2} = 148.3$ N          *Ans.*

$\left[\theta = \tan^{-1}\dfrac{R_y}{R_x}\right]$       $\theta = \tan^{-1}\dfrac{132.4}{66.9} = 63.2°$          *Ans.*

①  $[M_O = \Sigma(Fd)]$       $M_O = 140 - 50(5) + 60 \cos 45°(4) - 60 \sin 45°(7)$

$$= -237.3 \text{ N·m}$$

The force–couple system consisting of $\mathbf{R}$ and $M_O$ is shown in Fig. $a$.

We now determine the final line of action of $\mathbf{R}$ such that $\mathbf{R}$ alone represents the original system.

$[Rd = |M_O|]$          $148.3d = 237.3$     $d = 1.600$ m          *Ans.*

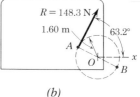

(a)

Hence, the resultant $\mathbf{R}$ may be applied at any point on the line which makes a 63.2° angle with the $x$-axis and is tangent at point $A$ to a circle of 1.6-m radius with center $O$, as shown in the $b$-part of the figure. We apply the equation $Rd = M_O$ in an absolute-value sense (ignoring any sign of $M_O$) and let the physics of the situation, as depicted in Fig. $a$, dictate the final placement of $\mathbf{R}$. Had $M_O$ been counterclockwise, the correct line of action of $\mathbf{R}$ would have been the tangent at point $B$.

The resultant $\mathbf{R}$ may also be located by determining its intercept distance $b$ to point $C$ on the $x$-axis, Fig. $c$. With $R_x$ and $R_y$ acting through point $C$, only $R_y$ exerts a moment about $O$ so that

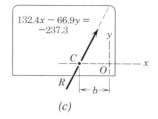

(b)

$$R_y b = |M_O| \quad \text{and} \quad b = \frac{237.3}{132.4} = 1.792 \text{ m}$$

Alternatively, the $y$-intercept could have been obtained by noting that the moment about $O$ would be due to $R_x$ only.

A more formal approach in determining the final line of action of $\mathbf{R}$ is to use the vector expression

$$\mathbf{r} \times \mathbf{R} = \mathbf{M}_O$$

where $\mathbf{r} = x\mathbf{i} + y\mathbf{j}$ is a position vector running from point $O$ to any point on the line of action of $\mathbf{R}$. Substituting the vector expressions for $\mathbf{r}$, $\mathbf{R}$, and $\mathbf{M}_O$ and carrying out the cross product result in

$$(x\mathbf{i} + y\mathbf{j}) \times (66.9\mathbf{i} + 132.4\mathbf{j}) = -237.3\mathbf{k}$$

$$(132.4x - 66.9y)\mathbf{k} = -237.3\mathbf{k}$$

Thus, the desired line of action, Fig. $c$, is given by

$$132.4x - 66.9y = -237.3$$

②  By setting $y = 0$, we obtain $x = -1.792$ m, which agrees with our earlier calculation of the distance $b$.

(c)

① We note that the choice of point $O$ as a moment center eliminates any moments due to the two forces which pass through $O$. Had the clockwise sign convention been adopted, $M_O$ would have been $+237.3$ N·m, with the plus sign indicating a sense which agrees with the sign convention. Either sign convention, of course, leads to the conclusion of a clockwise moment $M_O$.

② Note that the vector approach yields sign information automatically, whereas the scalar approach is more physically oriented. The student should master both methods.

## PROBLEMS

### *Introductory problems*

**2/69** Determine the resultant **R** of the three tension forces acting on the eye bolt. Find the magnitude of **R** and the angle $\theta_x$ which **R** makes with the positive *x*-axis.          *Ans.* $R = 17.43$ kN, $\theta_x = 26.1°$

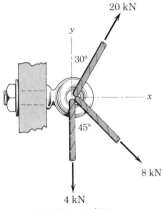

**Problem 2/69**

**2/70** Determine the equivalent force–couple system at the origin *O* for each of the three cases shown.

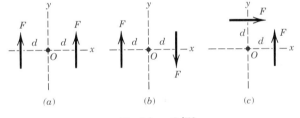

**Problem 2/70**

**2/71** Determine the equivalent force–couple system at the origin *O* for each of the three cases shown. If the resultant can be so expressed, replace this system with a standalone force.
          *Ans.* (a) **R** = 2*F***j** along *x* = −*d*
                     (b) **R** = *F***i** along *y* = 3*d*
                     (c) **R** = *F***i** along *y* = −*d*

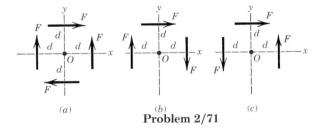

**Problem 2/71**

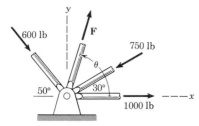

**Problem 2/72**

**2/72** Replace the two forces acting on the bent pipe by a single equivalent force **R**. Specify the distance *y* from point *A* to the line of action of **R**.

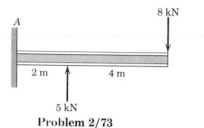

**Problem 2/73**

**2/73** Determine and locate the resultant of the two forces acting on the I-beam.

$Ans. \ R = 3$ kN down

$x = 12.67$ m (off beam)

### Representative problems

**2/74** Four slender rods are attached to the support as shown. If the forces which three of the rods exert on the support are known, determine the angle $\theta$ of the fourth rod and the magnitude $F$ of the tensile force in that rod if the resultant of all four forces is vertically upward with a magnitude of 100 lb.

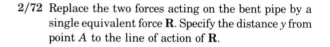

**Problem 2/74**

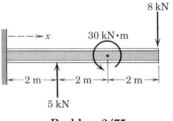

**Problem 2/75**

**2/75** Determine and locate the resultant **R** of the two forces and one couple acting on the I-beam.

$Ans. \ R = 3$ kN down at $x = 2.67$ m

**2/76** If the resultant of the two forces and couple $M$ passes through point $O$, determine $M$.

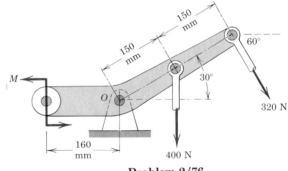

**Problem 2/76**

**2/77** A commercial airliner with four jet engines, each producing 90 kN of forward thrust, is in a steady, level cruise when engine number 3 suddenly fails. Determine and locate the resultant of the three remaining engine thrust vectors. Treat this as a two-dimensional problem.

> *Ans.* $R = 270$ kN located 4 m left of fuselage centerline

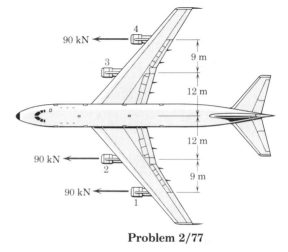

**Problem 2/77**

**2/78** The directions of the two thrust vectors of an experimental aircraft can be independently changed from the conventional forward direction within limits. For the thrust configuration shown, determine the equivalent force–couple system at point $O$. Then replace this force–couple system by a single force and specify the point on the $x$-axis through which the line of action of this resultant passes.

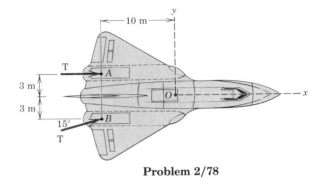

**Problem 2/78**

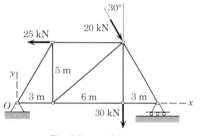

**Problem 2/79**

**2/79** Determine the resultant **R** of the three forces acting on the simple truss. Specify the points on the *x*- and *y*-axes through which **R** must pass.

$$Ans. \ \mathbf{R} = -15\mathbf{i} - 47.3\mathbf{j} \ \text{kN}$$
$$x = 7.42 \ \text{m}, \ y = -23.4 \ \text{m}$$

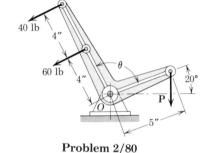

**Problem 2/80**

**2/80** In the equilibrium position shown, the resultant of the three forces acting on the bell crank passes through the bearing *O*. Determine the vertical force **P**. Does the result depend on $\theta$?

$$Ans. \ P = 119.2 \ \text{lb}, \ \text{No}$$

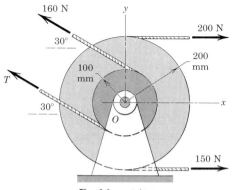

**Problem 2/81**

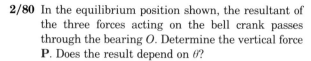

**2/81** Two integral pulleys are subjected to the belt tensions shown. If the resultant **R** of these forces passes through the center *O*, determine *T* and the magnitude of **R** and the counterclockwise angle $\theta$ it makes with the *x*-axis.

$$Ans. \ T = 60 \ \text{N}$$
$$R = 193.7 \ \text{N}$$
$$\theta = 34.6°$$

**2/82** While sliding a desk toward the doorway, three students exert the forces shown in the overhead view. Determine the equivalent force–couple system at point *A*. Then determine the equation of the line of action of the resultant force.

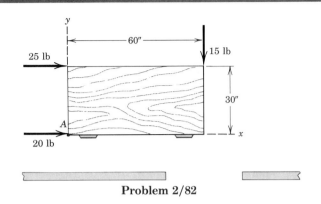

**Problem 2/82**

**2/83** Determine the point on the *y*-axis through which the line of action of the resultant of the loading system shown passes.

*Ans.* $y = 2.27$ ft

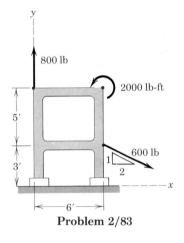

**Problem 2/83**

**2/84** The rolling rear wheel of a front-wheel-drive automobile which is accelerating to the right is subjected to the five forces and one moment shown. The forces $A_x = 60$ lb and $A_y = 500$ lb are forces transmitted from the axle to the wheel, $F = 40$ lb is the friction force exerted by the road surface on the tire, $N = 600$ lb is the normal reaction force exerted by the road surface, and $W = 100$ lb is the weight of the wheel/tire unit. The couple $M = 2$ lb-ft is the bearing friction moment. Determine and locate the resultant of the system.

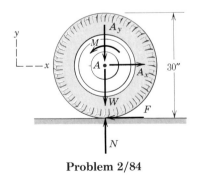

**Problem 2/84**

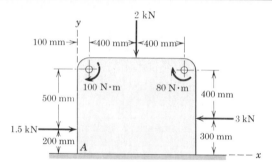

**Problem 2/85**

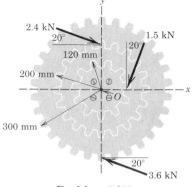

**Problem 2/86**

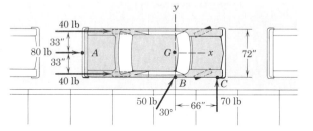

**Problem 2/87**

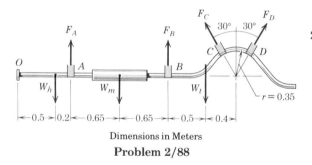

Dimensions in Meters

**Problem 2/88**

**2/85** Determine the resultant **R** of the three forces and two couples shown. Find the coordinate $x$ of the point on the $x$-axis through which **R** passes.

*Ans.* $\mathbf{R} = -1.5\mathbf{i} - 2\mathbf{j}$ kN
$x = 290$ mm

**2/86** Determine the $x$- and $y$-axis intercepts of the line of action of the resultant of the three loads applied to the gearset.

**2/87** A rear-wheel-drive car is stuck in the snow between other parked cars as shown. In an attempt to free the car, three students exert forces on the car at points $A$, $B$, and $C$ while the driver's actions result in a forward thrust of 40 lb acting parallel to the plane of rotation of each rear wheel. Treating the problem as two-dimensional, determine the equivalent force–couple system at the car center of mass $G$ and locate the position $x$ of the point on the car centerline through which the resultant passes. Neglect all forces not shown.

*Ans.* $\mathbf{R} = 185\mathbf{i} + 113.3\mathbf{j}$ lb
$M_G = 460$ lb-ft CCW
$x = 4.06$ ft

**2/88** An exhaust system for a pickup truck is shown in the figure. The weights $W_h$, $W_m$, and $W_t$ of the headpipe, muffler, and tailpipe are 10, 100, and 50 N, respectively, and act at the indicated points. If the exhaust pipe hanger at point $A$ is adjusted so that its tension $F_A$ is 50 N, determine the required forces in the hangers at points $B$, $C$, and $D$ so that the force–couple system at point $O$ is zero. Why is a zero force–couple system at $O$ desirable?

# SECTION B. THREE-DIMENSIONAL FORCE SYSTEMS

## 2/7 *RECTANGULAR COMPONENTS*

Many problems in mechanics require analysis in three dimensions, and it is often necessary to resolve a force into its three mutually perpendicular components. The force $\mathbf{F}$ acting at point $O$ in Fig. 2/16 has the *rectangular components* $F_x$, $F_y$, $F_z$, where

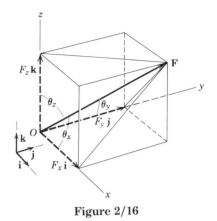

$$
\begin{aligned}
F_x &= F \cos \theta_x & F &= \sqrt{F_x^2 + F_y^2 + F_z^2} \\
F_y &= F \cos \theta_y & \mathbf{F} &= F_x\mathbf{i} + F_y\mathbf{j} + F_z\mathbf{k} \\
F_z &= F \cos \theta_z & \mathbf{F} &= F(\mathbf{i} \cos \theta_x + \mathbf{j} \cos \theta_y + \mathbf{k} \cos \theta_z)
\end{aligned}
\qquad (2/11)
$$

**Figure 2/16**

The unit vectors $\mathbf{i}, \mathbf{j}, \mathbf{k}$ are in the $x$-, $y$-, and $z$-directions, respectively. If we introduce the *direction cosines* of $\mathbf{F}$ which are $l = \cos \theta_x$, $m = \cos \theta_y$, $n = \cos \theta_z$, where $l^2 + m^2 + n^2 = 1$, we may write the force as

$$
\mathbf{F} = F(l\mathbf{i} + m\mathbf{j} + n\mathbf{k}) \qquad (2/12)
$$

We may regard the right-side expression of Eq. 2/12 as the force magnitude $F$ times a unit vector $\mathbf{n}_F$ which characterizes the direction of $\mathbf{F}$, or

$$
\mathbf{F} = F\mathbf{n}_F \qquad (2/12a)
$$

It is clear from Eqs. 2/12 and 2/12a that

$$
\mathbf{n}_F = l\mathbf{i} + m\mathbf{j} + n\mathbf{k}
$$

so that the scalar components of the unit vector $\mathbf{n}_F$ are the direction cosines of the line of action of $\mathbf{F}$.

In solving three-dimensional problems, one must usually find the $x$, $y$, and $z$ scalar components of a given or unknown force. In most cases, the direction of a force is described (*a*) by two points on the line of action of the force or (*b*) by two angles which orient the line of action.

***(a) Specification of two points on the line of action of the force.*** If the coordinates of points $A$ and $B$ of Fig. 2/17 are known, the force $\mathbf{F}$ may be written as

$$
\mathbf{F} = F\mathbf{n}_F = F \frac{\overrightarrow{AB}}{\overline{AB}} = F \frac{(x_2 - x_1)\mathbf{i} + (y_2 - y_1)\mathbf{j} + (z_2 - z_1)\mathbf{k}}{\sqrt{(x_2 - x_1)^2 + (y_2 - y_1)^2 + (z_2 - z_1)^2}}
$$

The $x$, $y$, and $z$ scalar components of $\mathbf{F}$ are thus readily available as the scalar coefficients of the unit vectors $\mathbf{i}$, $\mathbf{j}$, and $\mathbf{k}$, respectively.

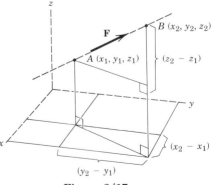

**Figure 2/17**

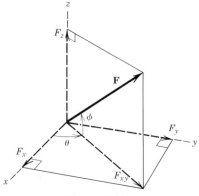

**Figure 2/18**

**(b) Specification of two angles that orient the line of action of the force.** Consider the geometry of Fig. 2/18, where the angles $\theta$ and $\phi$ are given. We may first resolve $\mathbf{F}$ into horizontal and vertical components.

$$F_{xy} = F \cos \phi$$

$$F_z = F \sin \phi$$

The horizontal component $F_{xy}$ is then further resolved into $x$- and $y$-components.

$$F_x = F_{xy} \cos \theta = F \cos \phi \cos \theta$$

$$F_y = F_{xy} \sin \theta = F \cos \phi \sin \theta$$

The quantities $F_x$, $F_y$, and $F_z$ are the desired scalar components of $\mathbf{F}$.

The choice of orientation of the coordinate system is quite arbitrary, with convenience being the primary consideration. But we *must* use a right-handed set of axes in our three-dimensional work so as to be consistent with the right-hand rule definition of the cross product. When we rotate from the $x$- to the $y$-axis through the 90° angle, the positive direction for the $z$-axis in a right-handed system is that of the advancement of a right-handed screw rotated in the same sense.

Rectangular components of a force $\mathbf{F}$ (or other vector) may be written alternatively with the aid of the vector operation known as the *dot* or *scalar product* (see item 6 in Art. C/7 of Appendix C). By definition, the dot product of two vectors $\mathbf{P}$ and $\mathbf{Q}$, Fig. 2/19a, is the

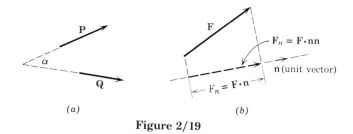

**Figure 2/19**

product of their magnitudes times the cosine of the angle $\alpha$ between them and is written

$$\mathbf{P} \cdot \mathbf{Q} = PQ \cos \alpha$$

This product may be viewed either as the orthogonal projection $P$ cos $\alpha$ of $\mathbf{P}$ in the direction of $\mathbf{Q}$ multiplied by $Q$ or as the orthogonal projection $Q$ cos $\alpha$ of $\mathbf{Q}$ in the direction of $\mathbf{P}$ multiplied by $P$. In either case the dot product of the two vectors is a scalar quantity.

Thus, the scalar component $F_x = F \cos \theta_x$ of the force $\mathbf{F}$ in Fig. 2/16, for instance, may be written as $F_x = \mathbf{F} \cdot \mathbf{i}$, where $\mathbf{i}$ is the unit vector in the $x$-direction. In more general terms, if $\mathbf{n}$ is a unit vector in a specified direction, the projection of $\mathbf{F}$ in the $\mathbf{n}$-direction, Fig. 2/19$b$, has the magnitude $F_n = \mathbf{F} \cdot \mathbf{n}$. If it is desired to write the projection in the $\mathbf{n}$-direction as a vector quantity, then its scalar component, expressed by $\mathbf{F} \cdot \mathbf{n}$, must be multiplied by the unit vector $\mathbf{n}$ to give $\mathbf{F}_n = (\mathbf{F} \cdot \mathbf{n})\mathbf{n}$, which may be written merely as $\mathbf{F}_n = \mathbf{F} \cdot \mathbf{nn}$.

If $\mathbf{n}$ has direction cosines $\alpha$, $\beta$, $\gamma$, then we may write $\mathbf{n}$ in vector component form like any other vector as

$$\mathbf{n} = \alpha\mathbf{i} + \beta\mathbf{j} + \gamma\mathbf{k}$$

where in this case its magnitude is unity. If $\mathbf{F}$ has the direction cosines $l$, $m$, $n$ with respect to reference axes $x$-$y$-$z$, then the projection of $\mathbf{F}$ in the $\mathbf{n}$-direction becomes

$$F_n = \mathbf{F} \cdot \mathbf{n} = F(l\mathbf{i} + m\mathbf{j} + n\mathbf{k}) \cdot (\alpha\mathbf{i} + \beta\mathbf{j} + \gamma\mathbf{k})$$

$$= F(l\alpha + m\beta + n\gamma)$$

since

$$\mathbf{i} \cdot \mathbf{i} = \mathbf{j} \cdot \mathbf{j} = \mathbf{k} \cdot \mathbf{k} = 1$$

and

$$\mathbf{i} \cdot \mathbf{j} = \mathbf{j} \cdot \mathbf{i} = \mathbf{i} \cdot \mathbf{k} = \mathbf{k} \cdot \mathbf{i} = \mathbf{j} \cdot \mathbf{k} = \mathbf{k} \cdot \mathbf{j} = 0$$

If the angle between the force $\mathbf{F}$ and the direction specified by the unit vector $\mathbf{n}$ is $\theta$, then by virtue of the dot-product relationship we have $\mathbf{F} \cdot \mathbf{n} = Fn \cos \theta = F \cos \theta$, where $|\mathbf{n}| = n = 1$. Thus, the angle between $\mathbf{F}$ and $\mathbf{n}$ is given by

$$\theta = \cos^{-1} \frac{\mathbf{F} \cdot \mathbf{n}}{F} \qquad (2/13)$$

or, in general, the angle between any two vectors $\mathbf{P}$ and $\mathbf{Q}$ is

$$\theta = \cos^{-1} \frac{\mathbf{P} \cdot \mathbf{Q}}{PQ} \qquad (2/13a)$$

If a force $\mathbf{F}$ is perpendicular to a line whose direction is specified by the unit vector $\mathbf{n}$, then $\cos \theta = 0$, and $\mathbf{F} \cdot \mathbf{n} = 0$. Note especially that this relationship does not mean that either $\mathbf{F}$ or $\mathbf{n}$ is zero, as would be the case with scalar multiplication where $(A)(B) = 0$ requires that either $A$ or $B$ be zero.

It should be observed that the dot-product relationship applies to nonintersecting vectors as well as to intersecting vectors. Thus, the dot product of the nonintersecting vectors $\mathbf{P}$ and $\mathbf{Q}$ in Fig. 2/20 is $Q$ times the projection of $\mathbf{P}'$ on $\mathbf{Q}$, or $P'Q \cos \alpha = PQ \cos \alpha$ since $\mathbf{P}'$ and $\mathbf{P}$ are the same free vectors.

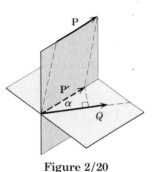

**Figure 2/20**

## Sample Problem 2/9

A force $\mathbf{F}$ with a magnitude of 100 N is applied at the origin $O$ of the axes $x$-$y$-$z$ as shown. The line of action of $\mathbf{F}$ passes through a point $A$ whose coordinates are 3 m, 4 m, and 5 m. Determine (a) the $x$, $y$, and $z$ scalar components of $\mathbf{F}$, (b) the projection $F_{xy}$ of $\mathbf{F}$ on the $x$-$y$ plane, and (c) the projection $F_{OB}$ of $\mathbf{F}$ along the line $OB$.

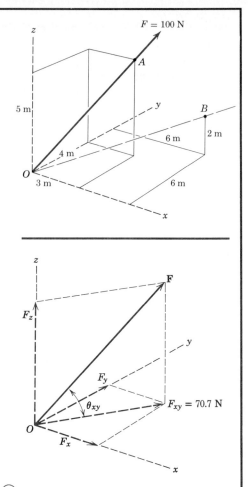

**Solution.** **Part (a).** We begin by writing the force vector $\mathbf{F}$ as its magnitude $F$ times a unit vector $\mathbf{n}_{OA}$.

$$\mathbf{F} = F\mathbf{n}_{OA} = F\frac{\overrightarrow{OA}}{OA} = 100\left[\frac{3\mathbf{i} + 4\mathbf{j} + 5\mathbf{k}}{\sqrt{3^2 + 4^2 + 5^2}}\right]$$

$$= 100[0.424\mathbf{i} + 0.566\mathbf{j} + 0.707\mathbf{k}]$$

$$= 42.4\mathbf{i} + 56.6\mathbf{j} + 70.7\mathbf{k} \text{ N}$$

The desired scalar components are thus

$$F_x = 42.4 \text{ N} \qquad F_y = 56.6 \text{ N} \qquad F_z = 70.7 \text{ N} \qquad Ans.$$

① 

**Part (b).** The cosine of the angle $\theta_{xy}$ between $\mathbf{F}$ and the $x$-$y$ plane is

$$\cos\theta_{xy} = \frac{\sqrt{3^2 + 4^2}}{\sqrt{3^2 + 4^2 + 5^2}} = 0.707$$

so that $F_{xy} = F\cos\theta_{xy} = 100\,(0.707) = 70.7$ N    *Ans.*

**Part (c).** The unit vector $\mathbf{n}_{OB}$ along $OB$ is

$$\mathbf{n}_{OB} = \frac{\overrightarrow{OB}}{OB} = \frac{6\mathbf{i} + 6\mathbf{j} + 2\mathbf{k}}{\sqrt{6^2 + 6^2 + 2^2}} = 0.688\mathbf{i} + 0.688\mathbf{j} + 0.229\mathbf{k}$$

The scalar projection of $\mathbf{F}$ on $OB$ is

② $F_{OB} = \mathbf{F}\cdot\mathbf{n}_{OB} = (42.4\mathbf{i} + 56.6\mathbf{j} + 70.7\mathbf{k})\cdot(0.688\mathbf{i} + 0.688\mathbf{j} + 0.229\mathbf{k})$

$$= (42.4)(0.688) + (56.6)(0.688) + (70.7)(0.229)$$

$$= 84.4 \text{ N} \qquad\qquad Ans.$$

If we wish to express the projection as a vector, we write

$$\mathbf{F}_{OB} = \mathbf{F}\cdot\mathbf{n}_{OB}\mathbf{n}_{OB}$$

$$= 84.4\,(0.688\mathbf{i} + 0.688\mathbf{j} + 0.229\mathbf{k})$$

$$= 58.1\mathbf{i} + 58.1\mathbf{j} + 19.35\mathbf{k} \text{ N}$$

① In this example all scalar components are positive. Be prepared for the case where a direction cosine, and hence the scalar component, is negative.

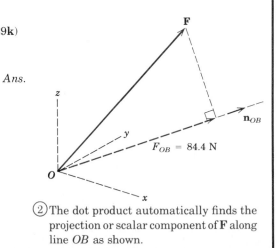

② The dot product automatically finds the projection or scalar component of $\mathbf{F}$ along line $OB$ as shown.

## PROBLEMS

### Introductory problems

**2/89** Express the 3600-lb force **F** as a vector in terms of the unit vectors **i**, **j**, and **k**. Determine the projection of **F** onto the *x*-axis.

$$\text{Ans. } \mathbf{F} = 1200\mathbf{i} + 2400\mathbf{j} + 2400\mathbf{k} \text{ lb}$$
$$F_x = 1200 \text{ lb}$$

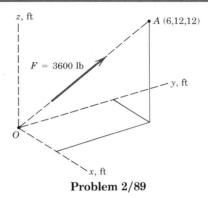

**Problem 2/89**

**2/90** Express the 400-N force **F** as a vector in terms of the unit vectors **i**, **j**, and **k**. Determine the projection of **F** onto the line *OA*.

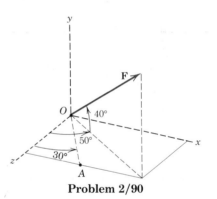

**Problem 2/90**

**2/91** The force **F** has a magnitude of 900 N and acts along the diagonal of the rectangular parallelepiped as shown. Express **F** in terms of its magnitude times the appropriate unit vector and determine its *x*-, *y*-, and *z*-components.

$$\text{Ans. } \mathbf{F} = 900(\tfrac{1}{3}\mathbf{i} - \tfrac{2}{3}\mathbf{j} - \tfrac{2}{3}\mathbf{k}) \text{ N}$$
$$F_x = 300 \text{ N}, F_y = -600 \text{ N}$$
$$F_z = -600 \text{ N}$$

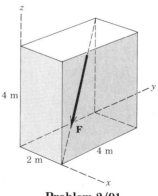

**Problem 2/91**

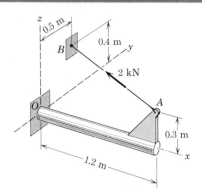

**Problem 2/92**

**2/92** The cable exerts a tension of 2 kN on the fixed bracket at $A$. Write the vector expression for the tension **T**.

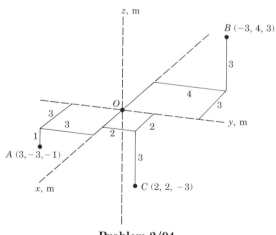

**Problem 2/93**

**2/93** The mast is restrained by the three cables shown. If the turnbuckle is tightened until the tension in $AB$ is 1.8 kN, write a vector expression for the tension **T** as a force acting on the mast at $B$.

*Ans.* $\mathbf{T} = 1.2\mathbf{i} + 0.6\mathbf{j} - 1.2\mathbf{k}$ kN

### Representative problems

**2/94** An 8-kN force **F** (not shown in the figure) acts along the line from $A$ to $B$. Express **F** in terms of the unit vectors **i**, **j**, and **k**. Determine the projection of **F** onto the line $OC$.

**Problem 2/94**

**2/95** The magnitude of force **F** is 10 kN, and its line of action passes through the points $A$ and $B$. Determine the angle made by **F** with the vertical $z$-direction and the magnitude of the projection of **F** onto the $y$-$z$ plane.

Ans. $\theta_z = 34.5°$, $F_{yz} = 8.36$ kN

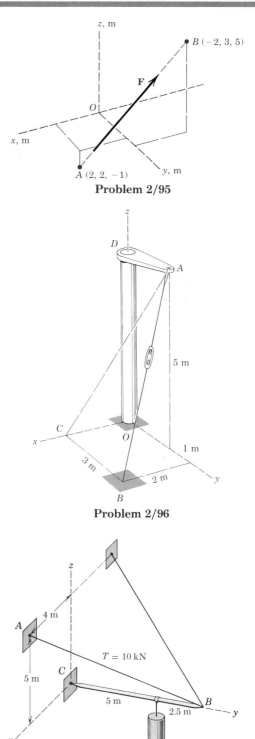

**Problem 2/95**

**2/96** The turnbuckle is tightened until the tension in the cable $AB$ equals 2.4 kN. Express the vector expression for the tension **T** as a force acting on member $AD$. Also find the magnitude of the projection of **T** along the line $AC$.

**Problem 2/96**

**2/97** The tension in the supporting cable $AB$ is 10 kN. Write the force which the cable exerts on the boom $BC$ as a vector **T**. Determine the angles $\theta_x$, $\theta_y$, and $\theta_z$ which the line of action of **T** forms with the positive $x$-, $y$-, and $z$-axes.

Ans. $\mathbf{T} = 4.06\mathbf{i} - 7.61\mathbf{j} + 5.07\mathbf{k}$ kN
$\theta_x = 66.1°$, $\theta_y = 139.5°$, $\theta_z = 59.5°$

**Problem 2/97**

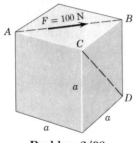

**Problem 2/98**

**2/98** Calculate the magnitude of the projection $F_{CD}$ of the 100-N force on the face diagonal $CD$ of the cube.

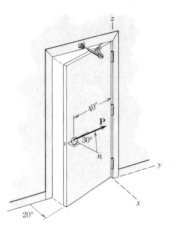

**Problem 2/99**

**2/99** In opening a door which is equipped with a heavy-duty return mechanism, a person exerts a force $\mathbf{P}$ of magnitude 8 lb as shown. Force $\mathbf{P}$ and the normal $n$ to the face of the door lie in a vertical plane. Express $\mathbf{P}$ as a vector and determine the angles $\theta_x$, $\theta_y$, and $\theta_z$ which the line of action of $\mathbf{P}$ makes with the positive $x$-, $y$-, and $z$-axes.

> *Ans.* $\mathbf{P} = 6.51\mathbf{i} + 2.37\mathbf{j} + 4.00\mathbf{k}$ lb
> $\theta_x = 35.5°$, $\theta_y = 72.8°$, $\theta_z = 60.0°$

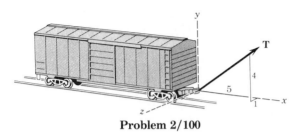

**Problem 2/100**

**2/100** An overhead crane is used to reposition the boxcar within a railroad car repair shop. If the boxcar begins to move along the rails when the $x$-component of the cable tension reaches 600 lb, calculate the necessary tension $T$ in the cable. Determine the angle $\theta_{xy}$ between the cable and the vertical $x$-$y$ plane.

**2/101** The rectangular plate is supported by hinges along its side $BC$ and by the cable $AE$. If the cable tension is 300 N, determine the projection onto line $BC$ of the force exerted on the plate by the cable. Note that $E$ is the midpoint of the horizontal upper edge of the structural support.

*Ans.* $T_{BC} = 251$ kN

**2/102** The force $\mathbf{F}$ has a magnitude of 2 kN and is directed from $A$ to $B$. Calculate the projection $F_{CD}$ of $\mathbf{F}$ onto line $CD$ and determine the angle $\theta$ between $\mathbf{F}$ and $CD$.

**2/103** The access door is held in the 30° open position by the chain $AB$. If the tension in the chain is 100 N, determine the projection of the tension force onto the diagonal axis $CD$ of the door.

*Ans.* $T_{CD} = 46.0$ N

**2/104** The power line is strung from the power-pole arm at $A$ to point $B$ on the same horizontal plane. Because of the sag of the cable in the vertical plane, the cable makes an angle of 15° with the horizontal where it attaches to $A$. If the cable tension at $A$ is 200 lb, write $\mathbf{T}$ as a vector and determine the magnitude of its projection onto the $x$-$z$ plane.

*Ans.* $\mathbf{T} = 191.0\mathbf{i} + 28.7\mathbf{j} - 51.8\mathbf{k}$ lb
$T_{xz} = 197.9$ lb

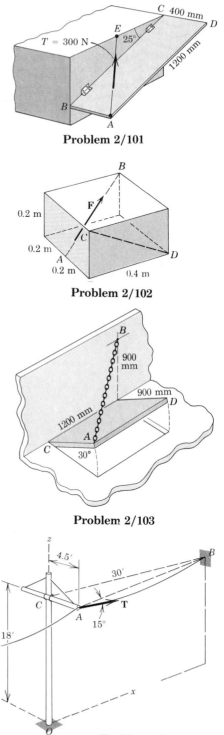

Problem 2/101

Problem 2/102

Problem 2/103

Problem 2/104

## 2/8 MOMENT AND COUPLE

The general vector definition of moment was introduced in Art. 2/4, even though in two-dimensional analyses it is often more convenient to determine a moment magnitude by scalar multiplication using the moment-arm rule. In three dimensions, however, the determination of the perpendicular distance between a point or line and the line of action of the force can be a tedious computation. The use of a vector approach using cross-product multiplication then becomes advantageous.

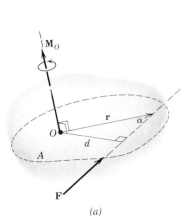

*(a)*

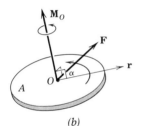

*(b)*

**Figure 2/21**

**Moment.**    Consider a force **F** with a given line of action acting on a body, Fig. 2/21*a*, and any point *O* not on this line. Point *O* and the line of **F** establish a plane *A*. The moment $\mathbf{M}_O$ of **F** about an axis through *O* normal to the plane has the magnitude $M_O = Fd$, where *d* is the perpendicular distance from *O* to the line of **F**. This moment is also referred to as the moment of **F** about the *point O*. The vector $\mathbf{M}_O$ is normal to the plane and along the axis through *O*. Both the magnitude and the direction of $\mathbf{M}_O$ may be described by the vector cross-product relation introduced in Art. 2/4. (Again refer to item 7 in Art. C/7 of Appendix C.) The vector **r** extends from *O* to *any* point on the line of action of **F**. As described in Art. 2/4, the cross product of **r** and **F** is written $\mathbf{r} \times \mathbf{F}$ and has the magnitude $(r \sin \alpha)F$, which is the same as $Fd$, the magnitude of $\mathbf{M}_O$. The correct direction and sense of the moment are established by the right-hand rule, described previously in Arts. 2/4 and 2/5. Thus, with **r** and **F** treated as free vectors, Fig. 2/21*b*, the thumb points in the direction of $\mathbf{M}_O$ if the fingers of the right hand curl in the direction of rotation from **r** to **F** through the angle $\alpha$. Therefore, we may write the moment of **F** about the axis through *O* as

$$\mathbf{M}_O = \mathbf{r} \times \mathbf{F} \qquad (2/14)$$

The order $\mathbf{r} \times \mathbf{F}$ of the vectors *must* be maintained because $\mathbf{F} \times \mathbf{r}$ would produce a vector with a sense opposite to that of $\mathbf{M}_O$; that is, $\mathbf{F} \times \mathbf{r} = -\mathbf{M}_O$.

The cross-product expression for $\mathbf{M}_O$ may be written in the determinant form

$$\mathbf{M}_O = \begin{vmatrix} \mathbf{i} & \mathbf{j} & \mathbf{k} \\ r_x & r_y & r_z \\ F_x & F_y & F_z \end{vmatrix} \qquad (2/15)$$

(Again, refer to item 7 in Art. C/7 of Appendix C if you are not already familiar with the determinant representation of the cross product.) The symmetry and order of the terms should be carefully noted, and care must be exercised to ensure the use of a *right-handed* coordinate system on which the correct evaluation of vector operations depends. Expansion of the determinant gives

$$\mathbf{M}_O = (r_y F_z - r_z F_y)\mathbf{i} + (r_z F_x - r_x F_z)\mathbf{j} + (r_x F_y - r_y F_x)\mathbf{k}$$

To gain more confidence in the cross-product relationship, let us observe the three components of the moment of a force about a point as seen from Fig. 2/22, which shows the three components of a force **F** acting at a point $A$ located from $O$ by the vector **r**. The scalar magnitudes of the moments of these forces about the positive $x$-, $y$-, and $z$-axes through $O$ are seen to be

$$M_x = r_y F_z - r_z F_y \qquad M_y = r_z F_x - r_x F_z \qquad M_z = r_x F_y - r_y F_x$$

which agree with the respective terms in the determinant expansion for the cross product **r × F**.

The moment $\mathbf{M}_\lambda$ of **F** about *any* axis $\lambda$ through $O$, Fig. 2/23, may now be written. If **n** is a unit vector in the $\lambda$-direction, then by using the dot-product expression for the component of a vector as described in Art. 2/7, the component of $\mathbf{M}_O$ in the direction of $\lambda$ is merely $\mathbf{M}_O \cdot \mathbf{n}$, which is the scalar magnitude of the moment $\mathbf{M}_\lambda$ of **F** about $\lambda$. To obtain the vector expression for the moment of **F** about $\lambda$, the magnitude must be multiplied by the directional unit vector **n** to give

$$\boxed{\mathbf{M}_\lambda = (\mathbf{r} \times \mathbf{F} \cdot \mathbf{n})\mathbf{n}} \qquad (2/16)$$

where **r × F** replaces $\mathbf{M}_O$. The expression $\mathbf{r} \times \mathbf{F} \cdot \mathbf{n}$ is known as a *triple scalar product* (see item 8 in Art. C/7, Appendix C). It need not be written $(\mathbf{r} \times \mathbf{F}) \cdot \mathbf{n}$, since the association $\mathbf{r} \times (\mathbf{F} \cdot \mathbf{n})$ would have no meaning because a cross product cannot be formed by a vector and a scalar. The triple scalar product may be represented by the determinant

$$\boxed{|\mathbf{M}_\lambda| = M_\lambda = \begin{vmatrix} r_x & r_y & r_z \\ F_x & F_y & F_z \\ \alpha & \beta & \gamma \end{vmatrix}} \qquad (2/17)$$

where $\alpha$, $\beta$, $\gamma$ are the direction cosines of the unit vector **n**.

***Varignon's theorem.*** In Art. 2/4 we introduced *Varignon's theorem* in two dimensions. The theorem is easily extended to three dimensions. In Fig. 2/24 is shown a system of concurrent forces $\mathbf{F}_1$, $\mathbf{F}_2$, $\mathbf{F}_3$, .... The sum of the moments about $O$ of these forces is

$$\mathbf{r} \times \mathbf{F}_1 + \mathbf{r} \times \mathbf{F}_2 + \mathbf{r} \times \mathbf{F}_3 + \cdots = \mathbf{r} \times (\mathbf{F}_1 + \mathbf{F}_2 + \mathbf{F}_3 + \cdots)$$

$$= \mathbf{r} \times \Sigma\mathbf{F}$$

where we have used the distributive law for cross products. Using

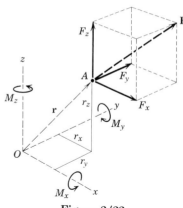

**Figure 2/22**

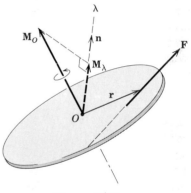

**Figure 2/23**

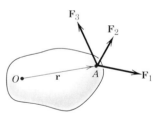

**Figure 2/24**

the symbol $\mathbf{M}_O$ to represent the sum of the moments on the left side of the above equation, we have

$$\boxed{\mathbf{M}_O = \Sigma(\mathbf{r} \times \mathbf{F}) = \mathbf{r} \times \mathbf{R}} \qquad (2/18)$$

Thus, the sum of the moments of a system of concurrent forces about a given point equals the moment of their sum about the same point. As mentioned in Art. 2/4, this principle finds repeated application in mechanics not only for the moments of force vectors but for the moments of other vectors as well.

   ***Couple.***   The concept of the couple was introduced in Art. 2/5 and is easily extended to three dimensions. Figure 2/25 shows two equal and opposite forces $\mathbf{F}$ and $-\mathbf{F}$ acting on a body. The vector $\mathbf{r}$

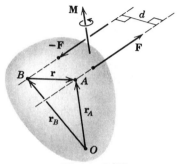

**Figure 2/25**

joins *any* point $B$ on the line of action of $-\mathbf{F}$ to *any* point $A$ on the line of action of $\mathbf{F}$. Points $A$ and $B$ are located by position vectors $\mathbf{r}_A$ and $\mathbf{r}_B$ from *any* point $O$. The combined moment of the two forces about $O$ is

$$\mathbf{M} = \mathbf{r}_A \times \mathbf{F} + \mathbf{r}_B \times (-\mathbf{F}) = (\mathbf{r}_A - \mathbf{r}_B) \times \mathbf{F}$$

But $\mathbf{r}_A - \mathbf{r}_B = \mathbf{r}$, so that all reference to the moment center $O$ disappears, and the *moment of the couple* becomes

$$\boxed{\mathbf{M} = \mathbf{r} \times \mathbf{F}} \qquad (2/19)$$

Thus, the moment of a couple is the *same about all points.* We see that the magnitude of $\mathbf{M}$ is $M = Fd$, where $d$ is the perpendicular distance between the lines of action of the two forces, as described in Art. 2/5.

   The moment of a couple is a *free vector*, whereas the moment of a force about a point (which is also the moment about a defined axis through the point) is a *sliding vector* whose direction is along the axis through the point. As in the case of two dimensions, a couple tends to produce a pure rotation of the body about an axis normal to the plane of the forces which constitute the couple.

Couple vectors obey all of the rules which govern vector quantities. Thus, in Fig. 2/26 the couple vector $\mathbf{M}_1$ due to $\mathbf{F}_1$ and $-\mathbf{F}_1$ may be added as shown to the couple vector $\mathbf{M}_2$ due to $\mathbf{F}_2$ and $-\mathbf{F}_2$ to produce the couple $\mathbf{M}$ which, in turn, can be produced by $\mathbf{F}$ and $-\mathbf{F}$.

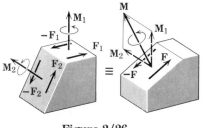

**Figure 2/26**

In Art 2/5 we learned to replace a force by its equivalent force–couple system. We should also be prepared to carry out this replacement in three dimensions. The procedure is represented in Fig. 2/27, where the force $\mathbf{F}$ acting on rigid body at point $A$ is replaced by an equal force at point $B$ and the couple $\mathbf{M} = \mathbf{r} \times \mathbf{F}$. By adding the equal and opposite forces $\mathbf{F}$ and $-\mathbf{F}$ at $B$, we obtain the couple

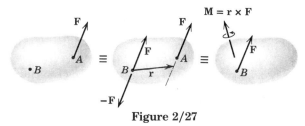

**Figure 2/27**

composed of $-\mathbf{F}$ and the original $\mathbf{F}$. Thus, we see that the couple vector is simply the moment of the original force about the point to which the force is being moved. Again we note that $\mathbf{r}$ is a vector from $B$ to *any* point on the line of action of the original force passing through $A$.

## Sample Problem 2/10

A tension $\mathbf{T}$ of magnitude 10 kN is applied to the cable attached to the top $A$ of the rigid mast and secured to the ground at $B$. Determine the moment $M_z$ of $\mathbf{T}$ about the $z$-axis passing through the base $O$.

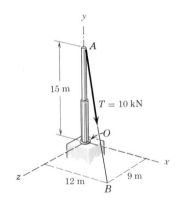

**Solution (a).** The required moment may be obtained by finding the component along the $z$-axis of the moment $\mathbf{M}_O$ of $\mathbf{T}$ about point $O$. The vector $\mathbf{M}_O$ is normal to the plane defined by $\mathbf{T}$ and point $O$ as shown in the accompanying figure. In the use of Eq. 2/14 to find $\mathbf{M}_O$, the vector $\mathbf{r}$ is any vector from point $O$ to the line of action of $\mathbf{T}$. The simplest choice

① is the vector from $O$ to $A$, which is written as $\mathbf{r} = 15\mathbf{j}$ m. The vector expression for $\mathbf{T}$ is

$$\mathbf{T} = T\mathbf{n}_{AB} = 10\left[\frac{12\mathbf{i} - 15\mathbf{j} + 9\mathbf{k}}{\sqrt{(12)^2 + (-15)^2 + (9)^2}}\right]$$

$$= 10(0.566\mathbf{i} - 0.707\mathbf{j} + 0.424\mathbf{k}) \text{ kN}$$

From Eq. 2/14,

$$[\mathbf{M}_O = \mathbf{r} \times \mathbf{F}] \qquad \mathbf{M}_O = 15\mathbf{j} \times 10(0.566\mathbf{i} - 0.707\mathbf{j} + 0.424\mathbf{k})$$

$$= 150(-0.566\mathbf{k} + 0.424\mathbf{i}) \text{ kN·m}$$

The value $M_z$ of the desired moment is the scalar component of $\mathbf{M}_O$ in the $z$-direction or $M_z = \mathbf{M}_O \cdot \mathbf{k}$. Therefore,

$$M_z = 150(-0.566\mathbf{k} + 0.424\mathbf{i}) \cdot \mathbf{k} = -84.9 \text{ kN·m} \qquad Ans.$$

② The minus sign indicates that the vector $\mathbf{M}_z$ is in the negative $z$-direction. Expressed as a vector, the moment is $\mathbf{M}_z = -84.9\mathbf{k}$ kN·m.

**Solution (b).** The force of magnitude $T$ is resolved into components $T_z$ and $T_{xy}$ in the $x$-$y$ plane. Since $T_z$ is parallel to the $z$-axis, it can exert no moment about this axis. The moment $M_z$ is, then, due only to $T_{xy}$

③ and is $M_z = T_{xy}d$, where $d$ is the perpendicular distance from $T_{xy}$ to $O$. The cosine of the angle between $T$ and $T_{xy}$ is $\sqrt{15^2 + 12^2}/\sqrt{15^2 + 12^2 + 9^2} = 0.906$, and therefore,

$$T_{xy} = 10(0.906) = 9.06 \text{ kN}$$

The moment arm $d$ equals $\overline{OA}$ multiplied by the sine of the angle between $T_{xy}$ and $OA$, or

$$d = 15\frac{12}{\sqrt{12^2 + 15^2}} = 9.37 \text{ m}$$

Hence, the moment of $\mathbf{T}$ about the $z$-axis has the magnitude

$$M_z = 9.06(9.37) = 84.9 \text{ kN·m} \qquad Ans.$$

and is clockwise when viewed in the $x$-$y$ plane.

**Solution (c).** The component $T_{xy}$ is further resolved into its components $T_x$ and $T_y$. It is clear that $T_y$ exerts no moment about the $z$-axis since it passes through it, so that the required moment is due to $T_x$ alone. The direction cosine of $\mathbf{T}$ with respect to the $x$-axis is $12/\sqrt{9^2 + 12^2 + 15^2} = 0.566$ so that $T_x = 10(0.566) = 5.66$ kN. Thus,

$$M_z = 5.66(15) = 84.9 \text{ kN·m} \qquad Ans.$$

① We could also use the vector from $O$ to $B$ for $\mathbf{r}$ and obtain the same result, but using vector $OA$ is simpler.

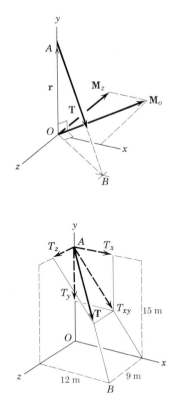

② It is always helpful to accompany your vector operations with a sketch of the vectors so as to retain a clear picture of the geometry of the problem.

③ Sketch the $x$-$y$ view of the problem and show $d$.

## Sample Problem 2/11

Determine the magnitude and direction of the couple **M** that will replace the two given couples and still produce the same external effect on the block. Specify the two forces **F** and $-$**F**, applied in the two faces of the block parallel to the $y$-$z$ plane, that may replace the four given forces. The 30-N forces act parallel to the $y$-$z$ plane.

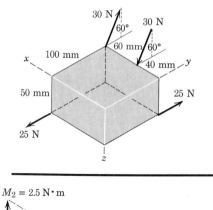

**Solution.** The couple due to the 30-N forces has the magnitude $M_1 = 30(0.06) = 1.80$ N·m. The direction of **M**$_1$ is normal to the plane defined by the two forces, and the sense, shown in the figure, is established by the right-hand convention. The couple due to the 25-N forces has the magnitude $M_2 = 25(0.10) = 2.50$ N·m with the direction and sense shown in the same figure. The two couple vectors combine to give the components

$$M_y = 1.80 \sin 60° = 1.559 \text{ N·m}$$

$$M_z = -2.50 + 1.80 \cos 60° = -1.600 \text{ N·m}$$

① Thus, $\qquad M = \sqrt{(1.559)^2 + (-1.600)^2} = 2.23$ N·m $\qquad$ *Ans.*

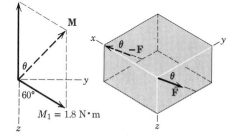

with $\qquad \theta = \tan^{-1} \dfrac{1.559}{1.600} = \tan^{-1} 0.974 = 44.3°$ $\qquad$ *Ans.*

The forces **F** and $-$**F** lie in a plane normal to the couple **M**, and their moment arm as seen from the right-hand figure is 100 mm. Thus, each force has the magnitude

$[M = Fd] \qquad F = \dfrac{2.23}{0.10} = 22.3$ N $\qquad$ *Ans.*

and the direction $\theta = 44.3°$.

① Bear in mind that the couple vectors are *free* vectors and therefore have no unique lines of action.

## Sample Problem 2/12

A force of 40 lb is applied at $A$ to the handle of the control lever that is attached to the fixed shaft $OB$. In determining the effect of the force on the shaft at a cross section such as that at $O$, we may replace the force by an equivalent force at $O$ and a couple. Describe this couple as a vector **M**.

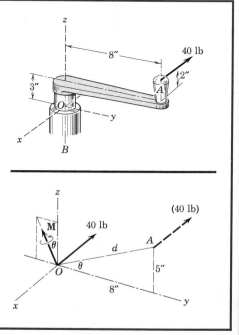

**Solution.** The couple may be expressed in vector notation as **M** = **r** × **F**, where **r** = $\vec{OA}$ = 8**j** + 5**k** in. and **F** = $-$40**i** lb. Thus,

$$\mathbf{M} = (8\mathbf{j} + 5\mathbf{k}) \times (-40\mathbf{i}) = -200\mathbf{j} + 320\mathbf{k} \text{ lb-in.}$$

Alternatively we see that moving the 40-lb force through a distance $d = \sqrt{5^2 + 8^2} = 9.43$ in. to a parallel position through $O$ requires the addition of a couple **M** whose magnitude is

$$M = Fd = 40(9.43) = 377 \text{ lb-in.} \qquad Ans.$$

The couple vector is perpendicular to the plane in which the force is shifted, and its sense is that of the moment of the given force about $O$. The direction of **M** in the $y$-$z$ plane is given by

$$\theta = \tan^{-1} \frac{5}{8} = 32.0° \qquad Ans.$$

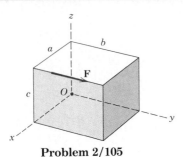

**Problem 2/105**

# PROBLEMS

## *Introductory problems*

**2/105** Determine the moment of force **F** about point $O$.

$$Ans.\ \mathbf{M}_O = F(-c\mathbf{i} + a\mathbf{k})$$

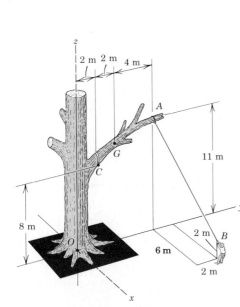

**Problem 2/106**

**2/106** Compute the moment of the force **P** about point $A$ by (*a*) using the vector cross-product relationship and (*b*) resolving **P** into components and determining the moments of the components.

**2/107** In an attempt to pull down a nearly sawn-through branch, the tree surgeon exerts a 400-N pull on the line which is looped around the branch at $A$. Determine the moment about point $C$ of the force exerted on the branch and state the magnitude of this moment.

$$Ans.\ \mathbf{M}_C = -2180\mathbf{i} + 655\mathbf{j} - 1309\mathbf{k}\ \text{N·m}$$
$$M_C = 2630\ \text{N·m}$$

**Problem 2/107**

**2/108** The figure for Prob. 2/99 is repeated here. Compute the moment of the 8-lb force **P** about the $z$-axis. Note that the force **P** and the normal $n$ to the face of the door lie in a vertical plane.

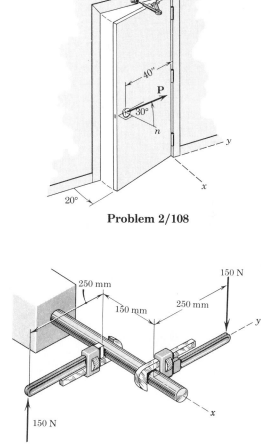

**Problem 2/108**

**2/109** The two forces acting on the handles of the pipe wrenches constitute a couple **M**. Express the couple as a vector.    *Ans.* $\mathbf{M} = -75\mathbf{i} + 22.5\mathbf{j}$ N·m

**Problem 2/109**

### Representative problems

**2/110** A steel H-beam acts as a column and supports the two vertical forces shown. Replace these forces by a single equivalent force along the vertical centerline of the column and a couple **M**.

**Problem 2/110**

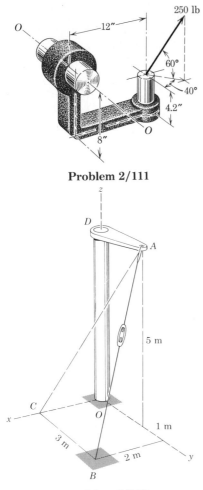

**Problem 2/111**

**2/111** Compute the moment $\mathbf{M}_O$ of the 250-lb force about the axis $O$-$O$.               *Ans.* $M_O = 2900$ lb-in.

**Problem 2/112**

**2/112** The figure for Prob. 2/96 is shown again here where the tension in the cable is 2.4 kN. Determine the moment about $O$ of the force which acts on arm $DA$. Also, explain by inspection why the moment of this force about a line from $D$ to $C$ is zero.

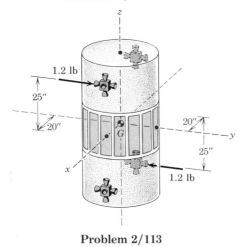

**Problem 2/113**

**2/113** Two 1.2-lb thrusters on the nonrotating satellite are simultaneously fired as shown. Compute the moment associated with this couple and state about which satellite axes rotations will begin to occur.

*Ans.* $\mathbf{M} = -60\mathbf{i} + 48\mathbf{k}$ lb-in.

**2/114** The bolt resists the torque (moment) about its axis *O-O* produced by the 100-lb force which acts on the bent bracket. Determine this torque $M_O$.

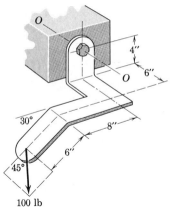

**Problem 2/114**

**2/115** A space shuttle orbiter is subjected to thrusts from five of the engines of its reaction control system. Four of the thrusts are shown in the figure; the fifth is an 850-N upward thrust at the right rear, symmetric to the 850-N thrust shown on the left rear. Compute the moment of these forces about point *G* and show that the forces have the same moment about all points.

*Ans.* $\mathbf{M} = 3400\mathbf{i} - 51\,000\mathbf{j} - 51\,000\mathbf{k}$ N·m

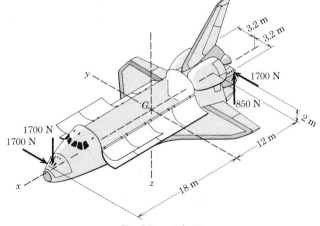

**Problem 2/115**

**2/116** The figure for Prob. 2/95 is repeated here where the force **F** has a magnitude of 10 kN and the coordinates of *A* and *B* are given in meters. Express the moment of **F** about the *x*-axis as a vector.

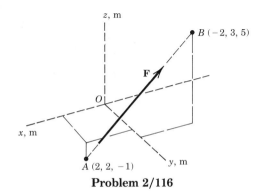

**Problem 2/116**

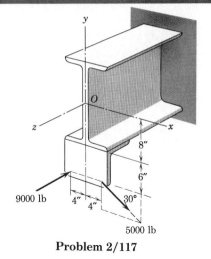

**Problem 2/117**

**2/117** A right-angle bracket is welded to the flange of the I-beam to support the 9000-lb force, applied parallel to the axis of the beam, and the 5000-lb force, applied in the end plane of the beam. In analyzing the capacity of the beam to withstand the applied loads, it is convenient to replace the forces by an equivalent force at $O$ and a corresponding couple $\mathbf{M}$. Determine the $x$-, $y$-, and $z$-components of $\mathbf{M}$.

*Ans.* $\mathbf{M} = (126\mathbf{i} - 36\mathbf{j} + 50.6\mathbf{k})10^3$ lb-in.

**Problem 2/118**

**2/118** In picking up a load from position $A$, a cable tension $\mathbf{T}$ of magnitude 21 kN is developed. Calculate the moment that $\mathbf{T}$ produces about the base $O$ of the construction crane.

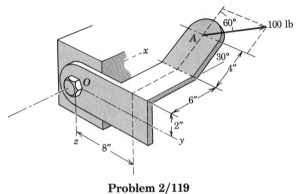

**Problem 2/119**

**2/119** Determine the vector expression for the moment $\mathbf{M}_O$ of the 100-lb force about point $O$. Also find the magnitude of the moment about the $x$-axis through the bolt at $O$.

*Ans.* $\mathbf{M}_O = 146.4\mathbf{i} + 63.4\mathbf{j} + 473\mathbf{k}$ lb-in.

$M_x = 146.4$ lb-in.

**2/120** The suspended power cable of Prob. 2/104 is shown again here where the cable exerts a tension of 200 lb on the power-pole arm at $A$. As described, due to the sag of the cable in the vertical plane, the cable makes an angle of 15° with the horizontal plane at $A$. Write expressions for the moment of $\mathbf{T}$ about the base $O$ and about the $z$-axis of the pole.

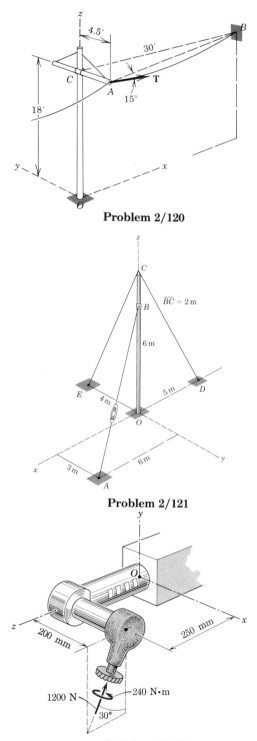

**Problem 2/120**

**2/121** The figure for Prob. 2/93 is repeated here where the tension in cable $AB$ is 1.8 kN. If the resultant of the tensions in the three supporting cables passes through the base at $O$, calculate the tensions in $CD$ and $CE$.

$Ans.\ T_{CD} = 1.698$ kN
$T_{CE} = 1.006$ kN

**Problem 2/121**

**2/122** The special-purpose milling cutter is subjected to the force of 1200 N and a couple of 240 N·m as shown. Determine the moment of this system about point $O$.

**Problem 2/122**

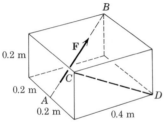

**Problem 2/123**

**2/123** The figure of Prob. 2/102 is shown again here. If the magnitude of the moment of **F** about line *CD* is 50 N·m, determine the magnitude of **F**.

*Ans. F* = 228 N

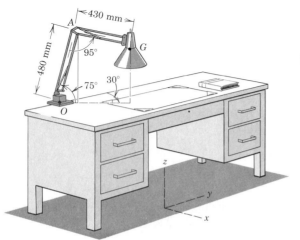

**Problem 2/124**

**2/124** The weight of the 0.8-kg light fixture acts at point *G*. Find the force–couple system at point *O* which is equivalent to this weight. Determine the moments of the weight about the *x*- and *y*-axes through *O* and explain the physical significance of these moments.

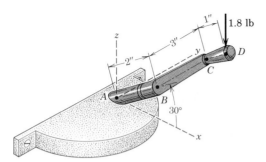

**Problem 2/125**

**2/125** A 1.8-lb vertical force is applied to the knob of the window-opener mechanism when the crank *BC* is horizontal. Determine the moment of the force about point *A* and about line *AB*.

*Ans.* $\mathbf{M}_{AB} = -4.05\mathbf{i} - 2.34\mathbf{k}$ lb-in.

**2/126** The 300-N force **P** is applied perpendicular to the handle of the box wrench and makes a 5° angle with the x'-direction (parallel to the x-axis). (Note that the plane which contains the 5° angle is *not* vertical.) Replace **P** by a force–couple system at point O.

**Problem 2/126**

**2/127** Starting from the nominal position shown, where the axis of the fixture is vertical, the track-light fixture is rotated to the position $\theta = 30°$ and $\phi = 40°$, where $\theta$ is a rotation at C about the fixed horizontal y-axis and $\phi$ is a rotation about the then-inclined axis of the pin at O (originally coincident with the horizontal x-axis). The fixture weight of 15 N acts at G. Determine the moment of this weight about point A for the specified position.

*Ans.* $\mathbf{M}_A = -1900\mathbf{i} + 462\mathbf{j}$ N·mm

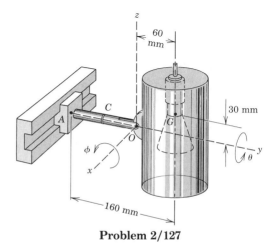

**Problem 2/127**

**2/128** The threading die is screwed onto the end of the fixed pipe which is bent through an angle of 20°. Replace the two forces by an equivalent force at O and a couple **M**. Find **M** and calculate the magnitude $M'$ of the moment which tends to screw the pipe into the fixed block about its angled axis through O.

*Ans.* $\mathbf{M} = 136\mathbf{i} - 679\mathbf{k}$ lb-in.
$M' = 685$ lb-in.

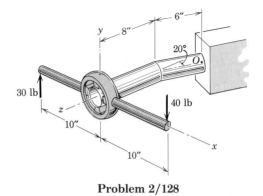

**Problem 2/128**

## 2/9 RESULTANTS

In Art. 2/6 we defined the resultant as the simplest force combination that can replace a given system of forces without altering the external effect on the rigid body on which the forces act. We found the magnitude and direction of the resultant force for the two-dimensional force system by a vector summation of forces, Eq. 2/9, and we located the line of action of the resultant force by applying the principle of moments, Eq. 2/10. These same principles may be extended to three dimensions.

In the previous article we showed that a force could be moved to a parallel position by adding a corresponding couple. Thus, for the system of forces $\mathbf{F}_1$, $\mathbf{F}_2$, $\mathbf{F}_3$ ... acting on a rigid body in Fig. 2/28a, we may move each of them in turn to the arbitrary point $O$, provided we also introduce a couple for each force transferred. Thus, for example, we may move force $\mathbf{F}_1$ to $O$, provided we introduce the

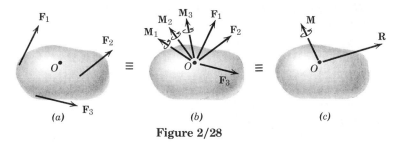

**Figure 2/28**

couple $\mathbf{M}_1 = \mathbf{r}_1 \times \mathbf{F}_1$, where $\mathbf{r}_1$ is a vector from $O$ to any point on the line of action of $\mathbf{F}_1$. When all forces are shifted to $O$ in this manner, we have a system of concurrent forces at $O$ and a system of couple vectors, as represented in the $b$-part of the figure. The concurrent forces may then be added vectorially to produce a resultant force $\mathbf{R}$, and the couples may also be added to produce a resultant couple $\mathbf{M}$, Fig. 2/28c. The general force system, then, is reduced to

$$\boxed{\begin{aligned} \mathbf{R} &= \mathbf{F}_1 + \mathbf{F}_2 + \mathbf{F}_3 + \cdots = \Sigma\mathbf{F} \\ \mathbf{M} &= \mathbf{M}_1 + \mathbf{M}_2 + \mathbf{M}_3 + \cdots = \Sigma(\mathbf{r} \times \mathbf{F}) \end{aligned}} \qquad (2/20)$$

The couple vectors are shown through point $O$, but since they are free vectors, they may be represented in any parallel positions. The magnitudes of the resultants and their components are

$$\boxed{\begin{aligned} R_x = \Sigma F_x \qquad R_y &= \Sigma F_y \qquad R_z = \Sigma F_z \\ R = \sqrt{(\Sigma F_x)^2 &+ (\Sigma F_y)^2 + (\Sigma F_z)^2} \\ M_x = \Sigma(\mathbf{r} \times \mathbf{F})_x \qquad M_y &= \Sigma(\mathbf{r} \times \mathbf{F})_y \qquad M_z = \Sigma(\mathbf{r} \times \mathbf{F})_z \\ M = \sqrt{M_x^2 &+ M_y^2 + M_z^2} \end{aligned}} \qquad (2/21)$$

The point $O$ that is selected as the point of concurrency for the forces is arbitrary, and the magnitude and direction of **M** will depend on the particular point $O$ selected. The magnitude and direction of **R**, however, are the same no matter which point is selected. In general any system of forces may be replaced by its resultant force **R** and the resultant couple **M**. In dynamics we usually select the mass center as the reference point, and the change in the linear motion of the body is determined by the resultant force and the change in the angular motion of the body is determined by the resultant couple. In statics the complete equilibrium of a body is specified when the resultant force **R** is zero and the resultant couple **M** is zero. Thus, the determination of resultants is essential in both statics and dynamics.

The resultants for several special force systems will not be noted.

*Concurrent forces.*   When forces are concurrent at a point, only the first of Eqs. 2/20 need to be used since there are no moments about the point of concurrency.

*Parallel forces.*   For a system of parallel forces not all in the same plane, the magnitude of the parallel resultant force **R** is simply the magnitude of the algebraic sum of the given forces, and the position of its line of action is obtained from the principle of moments by requiring that $\mathbf{r} \times \mathbf{R} = \mathbf{M}_O$. Here **r** is a position vector extending from the force–couple reference point $O$ to the final line of action of **R**, and $\mathbf{M}_O$ is the sum of the moments of the individual forces about $O$. See Sample Problem 2/14 for an example of parallel-force systems.

*Coplanar forces.*   Article 2/6 was devoted to this force system.

*Wrench resultant.*   When the resultant couple vector **M** is parallel to the resultant force **R**, as shown in Fig. 2/29, the resultant is said to be a *wrench*. By definition a wrench is positive if the couple and force vectors point in the same direction and negative if they point in opposite directions. A common example of a positive wrench is found with the application of a screwdriver.

Positive wrench         Negative wrench

**Figure 2/29**

Any general force system may be represented by a wrench applied along a unique line of action. This reduction is illustrated in Fig. 2/30, where the *a*-part of the figure represents for the general force system the resultant force **R** acting at some point *O* and the corresponding resultant couple **M**. Although **M** is a free vector, for convenience we represent it through *O*. In the *b*-part of the figure, **M** is resolved into components **M**$_1$ along the direction of **R** and **M**$_2$ normal to **R**. In the *c*-part of the figure, the couple **M**$_2$ is replaced by its equivalent of two forces **R** and −**R** separated a distance

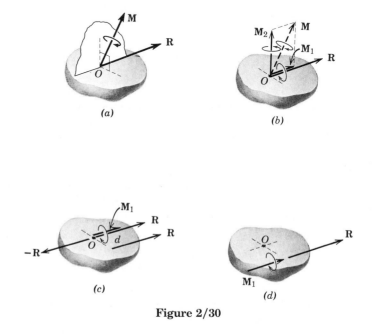

Figure 2/30

$d = M_2/R$ with −**R** applied at *O* to cancel the original **R**. This step leaves the resultant **R**, which acts along a new and unique line of action, and the parallel couple **M**$_1$, which is a free vector, as shown in the *d*-part of the figure. Thus, the resultants of the original general force system have been transformed into a wrench (positive in this illustration) with its unique axis defined by the new position of **R**. We see from Fig. 2/30 that the axis of the wrench resultant lies in a plane through *O* normal to the plane defined by **R** and **M**. The wrench is the simplest form in which the resultant of a general force system may be expressed. This form of the resultant, however, has limited application, since it is usually more convenient to use as the reference point some point *O* such as the mass center of the body or another convenient origin of coordinates not on the wrench axis.

## Sample Problem 2/13

Determine the resultant of the force and couple system which acts on the rectangular solid.

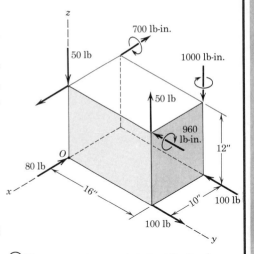

**Solution.** We choose point $O$ as a convenient reference point for the initial step of reducing the given forces to a force–couple system. The resultant force is

① $\quad \mathbf{R} = \Sigma \mathbf{F} = (80 - 80)\mathbf{i} + (100 - 100)\mathbf{j} + (50 - 50)\mathbf{k} = \mathbf{0}$ lb

The sum of the moments about $O$ is

② $\quad \mathbf{M}_O = [50(16) - 700]\mathbf{i} + [80(12) - 960]\mathbf{j} + [100(10) - 1000]\mathbf{k}$ lb-in.

$\quad = 100\mathbf{i}$ lb-in.

Hence, the resultant consists of a couple, which of course may be applied at any point on the body or the body extended.

① Since the force summation is zero, we conclude that the resultant, if it exists, must be a couple.

② The moments associated with the force pairs are easily obtained by using the $M = Fd$ rule and assigning the unit-vector direction by inspection. In many three-dimensional problems, this may be simpler than the $\mathbf{M} = \mathbf{r} \times \mathbf{F}$ approach.

## Sample Problem 2/14

Determine the resultant of the system of parallel forces which act on the plate. Solve with a vector approach.

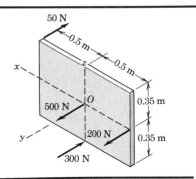

**Solution.** Transfer of all forces to point $O$ results in the force–couple system

$\quad \mathbf{R} = \Sigma \mathbf{F} = (200 + 500 - 300 - 50)\mathbf{j} = 350\mathbf{j}$ N

$\quad \mathbf{M}_O = [50(0.35) - 300(0.35)]\mathbf{i} + [-50(0.50) - 200(0.50)]\mathbf{k}$

$\quad = -87.5\mathbf{i} - 125\mathbf{k}$ N·m

The placement of $\mathbf{R}$ such that it alone represents the above force–couple system is determined by the principle of moments in vector form

$$\mathbf{r} \times \mathbf{R} = \mathbf{M}_O$$

$$(x\mathbf{i} + y\mathbf{j} + z\mathbf{k}) \times 350\mathbf{j} = -87.5\mathbf{i} - 125\mathbf{k}$$

$$350x\mathbf{k} - 350z\mathbf{i} = -87.5\mathbf{i} - 125\mathbf{k}$$

From the one vector equation we may obtain the two scalar equations

$$350x = -125 \quad \text{and} \quad -350z = -87.5$$

Hence, $x = -0.357$ m and $z = 0.250$ m are the coordinates through which the line of action of $\mathbf{R}$ must pass. The value of $y$ may, of course, be any value due to the principle of transmissibility. Thus, as expected, ① the variable $y$ drops out of the above vector analysis.

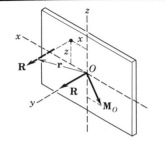

① The student should also carry out a scalar solution to this problem.

## Sample Problem 2/15

Replace the two forces and the negative wrench by a single force **R** applied at $A$ and the corresponding couple **M**.

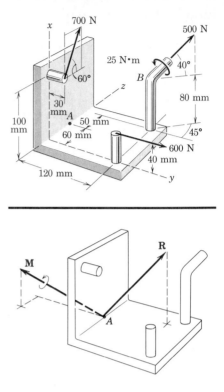

**Solution.** The resultant force has the components

$[R_x = \Sigma F_x]$  $R_x = 500 \sin 40° + 700 \sin 60° = 928$ N

$[R_y = \Sigma F_y]$  $R_y = 600 + 500 \cos 40° \cos 45° = 871$ N

$[R_z = \Sigma F_z]$  $R_z = 700 \cos 60° + 500 \cos 40° \sin 45° = 621$ N

Thus,  **R** $= 928\mathbf{i} + 871\mathbf{j} + 621\mathbf{k}$ N

and  $R = \sqrt{(928)^2 + (871)^2 + (621)^2} = 1416$ N  *Ans.*

The couple to be added as a result of moving the 500-N force is

① $[\mathbf{M} = \mathbf{r} \times \mathbf{F}]$  $\mathbf{M}_{500} = (0.08\mathbf{i} + 0.12\mathbf{j} + 0.05\mathbf{k}) \times 500(\mathbf{i} \sin 40°$

$+ \mathbf{j} \cos 40° \cos 45° + \mathbf{k} \cos 40° \sin 45°)$

where **r** is the vector from $A$ to $B$.

The term-by-term, or determinant, expansion gives

$$\mathbf{M}_{500} = 18.95\mathbf{i} - 5.59\mathbf{j} - 16.90\mathbf{k} \text{ N·m}$$

② The moment of the 600-N force about $A$ is written by inspection of its $x$- and $z$-components, which give

$$\mathbf{M}_{600} = (600)(0.060)\mathbf{i} + (600)(0.040)\mathbf{k}$$

$$= 36.0\mathbf{i} + 24.0\mathbf{k} \text{ N·m}$$

The moment of the 700-N force about $A$ is easily obtained from the moments of the $x$- and $z$-components of the force. The result becomes

$$\mathbf{M}_{700} = (700 \cos 60°)(0.030)\mathbf{i} - [(700 \sin 60°)(0.060)$$

$$+ (700 \cos 60°)(0.100)]\mathbf{j} - (700 \sin 60°)(0.030)\mathbf{k}$$

$$= 10.5\mathbf{i} - 71.4\mathbf{j} - 18.19\mathbf{k} \text{ N·m}$$

③ Also, the couple of the given wrench may be written

$$\mathbf{M}' = 25.0(-\mathbf{i} \sin 40° - \mathbf{j} \cos 40° \cos 45° - \mathbf{k} \cos 40° \sin 45°)$$

$$= -16.07\mathbf{i} - 13.54\mathbf{j} - 13.54\mathbf{k} \text{ N·m}$$

Therefore, the resultant couple on adding together the **i**-, **j**-, and **k**-terms of the four **M**'s is

④ $$\mathbf{M} = 49.4\mathbf{i} - 90.5\mathbf{j} - 24.6\mathbf{k} \text{ N·m}$$

and  $M = \sqrt{(49.4)^2 + (90.5)^2 + (24.6)^2} = 106.0$ N·m  *Ans.*

① *Suggestion*: Check the cross-product results by evaluating the moments about $A$ of the components of the 500-N force directly from the sketch.

② For the 600-N and 700-N forces it is easier to obtain the components of their moments about the coordinate directions through $A$ by inspection of the figure than it is to set up the cross-product relations.

③ The 25-N·m couple vector of the *wrench* points in the direction opposite to that of the 500-N force, and we must resolve it into its $x$-, $y$-, and $z$-components to be added to the other couple–vector components.

④ Although the resultant couple vector **M** in the sketch of the resultants is shown through $A$, we recognize that a couple vector is a free vector and therefore has no specified line of action.

## Sample Problem 2/16

Determine the wrench resultant of the three forces acting on the bracket. Calculate the coordinates of the point $P$ in the $x$-$y$ plane through which the resultant force of the wrench acts. Also find the magnitude of the couple $\mathbf{M}$ of the wrench.

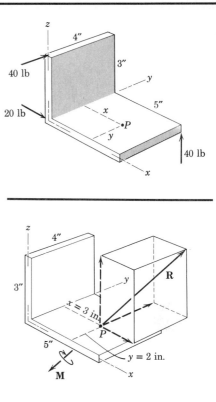

**Solution.**  The direction cosines of the couple $\mathbf{M}$ of the wrench must
① be the same as those of the resultant force $\mathbf{R}$, assuming that the wrench is positive. The resultant force is

$$\mathbf{R} = 20\mathbf{i} + 40\mathbf{j} + 40\mathbf{k} \text{ lb} \qquad R = \sqrt{(20)^2 + (40)^2 + (40)^2} = 60 \text{ lb}$$

and its direction cosines are

$$\cos \theta_x = 20/60 = 1/3 \quad \cos \theta_y = 40/60 = 2/3 \quad \cos \theta_z = 40/60 = 2/3$$

The moment of the wrench couple must equal the sum of the moments of the given forces about point $P$ through which $\mathbf{R}$ passes. The moments about $P$ of the three forces are

$$(\mathbf{M})_{R_x} = 20y\mathbf{k} \text{ lb-in.}$$

$$(\mathbf{M})_{R_y} = -40(3)\mathbf{i} - 40x\mathbf{k} \text{ lb-in.}$$

$$(\mathbf{M})_{R_z} = 40(4 - y)\mathbf{i} - 40(5 - x)\mathbf{j} \text{ lb-in.}$$

and the total moment is

$$\mathbf{M} = (40 - 40y)\mathbf{i} + (-200 + 40x)\mathbf{j} + (-40x + 20y)\mathbf{k} \text{ lb-in.}$$

The direction cosines of $\mathbf{M}$ are

$$\cos \theta_x = (40 - 40y)/M$$

$$\cos \theta_y = (-200 + 40x)/M$$

$$\cos \theta_z = (-40x + 20y)/M$$

where $M$ is the magnitude of $\mathbf{M}$. Equating the direction cosines of $\mathbf{R}$ and $\mathbf{M}$ gives

$$40 - 40y = \frac{M}{3}$$

$$-200 + 40x = \frac{2M}{3}$$

$$-40x + 20y = \frac{2M}{3}$$

Solution of the three equations gives

$$M = -120 \text{ lb-in.} \qquad x = 3 \text{ in.} \qquad y = 2 \text{ in.} \qquad Ans.$$

We see that $M$ turned out to be negative, which means that the couple vector is pointing in the direction opposite to $\mathbf{R}$, which makes the wrench negative.

① We shall assume initially that the wrench is positive. If $\mathbf{M}$ turns out to be negative, then the direction of the couple vector is opposite to that of the resultant force.

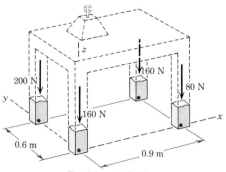

**Problem 2/129**

## PROBLEMS

### *Introductory problems*

**2/129** A table exerts the four forces shown onto the floor surface. Reduce the force system to a force–couple system at point $O$. Show that $\mathbf{R}$ is perpendicular to $\mathbf{M}_O$.

$$Ans. \ \mathbf{R} = -600\mathbf{k} \text{ N}$$
$$\mathbf{M}_O = -216\mathbf{i} + 216\mathbf{j} \text{ N·m}$$

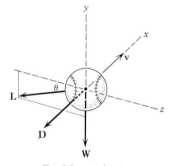

**Problem 2/130**

**2/130** A baseball is thrown with spin so that three concurrent forces act on it as shown in the figure. The weight $W$ is 5 oz, the drag $D$ is 1.7 oz, and the lift $L$ is perpendicular to the velocity $\mathbf{v}$ of the ball. If it is known that the $y$-component of the resultant is $-5.5$ oz and the $z$-component is $-0.866$ oz, determine $L$, $\theta$, and $R$.

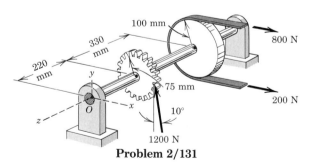

**Problem 2/131**

**2/131** The pulley and gear are subjected to the loads shown. For these forces, determine the equivalent force–couple system at point $O$.

$$Ans. \ \mathbf{R} = 792\mathbf{i} + 1182\mathbf{j} \text{ N}$$
$$\mathbf{M}_O = 260\mathbf{i} - 504\mathbf{j} + 28.6\mathbf{k} \text{ N·m}$$

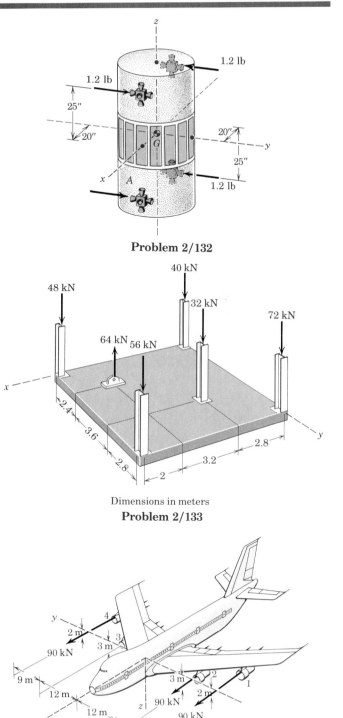

**2/132** The spacecraft of Prob 2/113 is repeated here. The plan is to fire four 1.2-lb thrusters as shown in order to spin up the spacecraft about its *z*-axis, but the thruster at *A* fails. Determine the equivalent force–couple system at *G* for the remaining three thrusters.

**Problem 2/132**

**2/133** The concrete slab supports the six vertical loads shown. Determine the *x*- and *y*-coordinates of the point on the slab through which the resultant of the loading system passes.

*Ans.* $x = 2.92$ m, $y = 6.33$ m

Dimensions in meters
**Problem 2/133**

*Representative problems*

**2/134** The commercial airliner of Prob. 2/69 is redrawn here with three-dimensional information supplied. If engine 3 suddenly fails, determine the resultant of the three remaining engine thrust vectors, each of which has a magnitude of 90 kN. Specify the *y*- and *z*-coordinates of the point through which the line of action of the resultant passes.

**Problem 2/134**

**Problem 2/135**

**2/135** Determine the angle $\theta$ so that the net downward force applied to the fixed eyebolt is 750 lb. Determine the corresponding magnitude of the resultant of the three forces and the direction cosines of the resultant.

*Ans.* $R = 1021$ lb, $\cos \theta_x = 0.525$
$\cos \theta_y = 0.430$, $\cos \theta_z = -0.734$

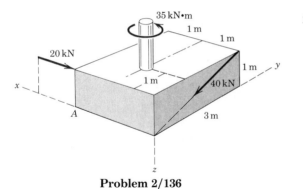

**Problem 2/136**

**2/136** Replace the two forces and single couple by an equivalent force–couple system at point $A$.

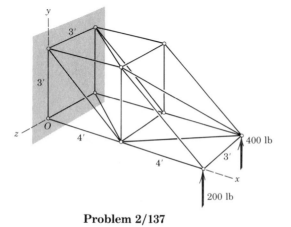

**Problem 2/137**

**2/137** Two upward loads are exerted on the small three-dimensional truss. Reduce these two loads to a single force–couple system at point $O$. Show that $\mathbf{R}$ is perpendicular to $\mathbf{M}_O$. Then determine the point in the $x$-$z$ plane through which the resultant passes.

*Ans.* $\mathbf{R} = 600\mathbf{j}$ lb, $\mathbf{M}_O = 1200\mathbf{i} + 4800\mathbf{k}$ lb-ft
$x = 8$ ft, $z = -2$ ft

**2/138** Represent the resultant of the force system acting on the pipe assembly by a single force **R** at *A* and a couple **M**.

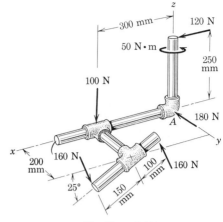

**Problem 2/138**

**2/139** The motor mounted on the bracket is acted on by its 160-N weight, and its shaft resists the 120-N thrust and 25-N·m couple applied to it. Determine the resultant of the force system shown in terms of a force **R** at *A* and a couple **M**.

$Ans.$  **R** $= -120\mathbf{i} - 160\mathbf{k}$ N
**M** $= -7\mathbf{i} + 9\mathbf{j} + 24\mathbf{k}$ N·m

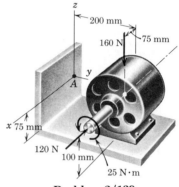

**Problem 2/139**

**2/140** The combined action of the three forces on the base at *O* may be obtained by establishing their resultant through *O*. Determine the magnitudes of **R** and the accompanying couple **M**.

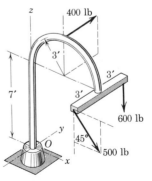

**Problem 2/140**

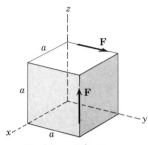

**Problem 2/141**

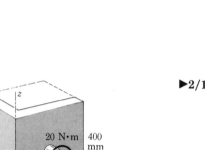

**Problem 2/142**

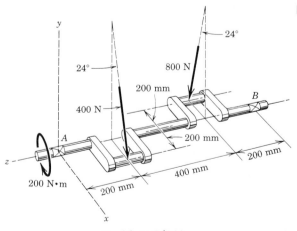

**Problem 2/144**

**2/141** Replace the two forces acting on the cube by a wrench. Write the moment **M** associated with the wrench as a vector and specify the coordinates of the point $P$ in the $x$-$z$ plane through which the line of action of the wrench passes.

$$Ans. \ \mathbf{M} = -\frac{F_a}{2}(\mathbf{j} + \mathbf{k})$$

$$x = \frac{a}{2}, z = 0$$

▶**2/142** The resultant of the two forces and couple may be represented by a wrench. Determine the vector expression for the moment **M** of the wrench and find the coordinates of the point $P$ in the $x$-$z$ plane through which the resultant force of the wrench passes.

$$Ans. \ \mathbf{M} = 10\mathbf{i} + 10\mathbf{j} \ N \cdot m$$

$$x = z = 0.1 \ m$$

▶**2/143** Replace the system of two forces and couple shown in Prob. 2/136 by a wrench. Determine the magnitude of the moment **M** of the wrench, the magnitude of the force **R** of the wrench, and the coordinates of the point $P$ in the $x$-$y$ plane through which **R** passes.

$$Ans. \ M = 26.9 \ kN \cdot m, R = 44.7 \ kN$$

$$x = 0.221 \ m, y = -0.950 \ m$$

▶**2/144** For the position shown, the crankshaft of a small two-cylinder compressor is subjected to the 400-N and 800-N forces exerted by the connecting rods and the 200-N·m couple. Replace this loading system by a force–couple system at point $A$. Show that **R** is not perpendicular to $\mathbf{M}_A$. Then replace this force–couple system by a wrench. Determine the magnitude $M$ of the moment of the wrench, the magnitude of the force **R** of the wrench, and the coordinates of the point in the $x$-$z$ plane through which the line of action of the wrench passes.

$$Ans. \ M = 85.8 \ N \cdot m, R = 1108 \ N$$

$$x = 0.1158 \ m, z = -0.478 \ m$$

## 2/10 *PROBLEM FORMULATION AND REVIEW*

In Chapter 2 we have established the properties of forces, moments, and couples and the correct procedures for representing their effects. Mastery of this material is absolutely essential for our study of equilibrium in the chapters which follow. Many of the common difficulties which occur in applying the principles of equilibrium can be traced to errors arising from failure to observe and apply correctly the procedures of Chapter 2. For this reason the student should refer back to the material of this chapter when difficulties arise to be sure that the forces, moments, and couples are correctly represented.

In order to handle forces correctly in engineering problems, their vector properties must be fully accounted for. For the cases in which the direction of a force is specified by an angle or angles or by two points on its line of action, we have developed the procedure for expressing the force in terms of its rectangular components. Also we must be able to combine two or more concurrent forces into an equivalent resultant force, as well as to solve the opposite problem of resolving a single force into components along desired directions.

The tendency of a force to rotate or turn a body about an axis has been treated with the concept of moment (or torque), which exhibits the properties of a vector. We saw that finding the moment of a force was often facilitated by combining the moments of the components of the force.

A couple was defined as the combined moment of two equal, opposite, and noncollinear forces. The unique property of a couple is to produce a pure twist or rotation regardless of where the forces are located. The couple was found to be useful in replacing a force acting at a point by a force–couple system at a different point.

With the moment and couple principles established, we treated the problem of reducing an arbitrary system of forces and couples to that of a single resultant force applied at an arbitrary point and a corresponding resultant couple. We also saw that this resultant force and couple could be combined to give a single resultant force along a unique line of action and a parallel couple vector, a combination called a wrench.

With the resultant force and resultant couple concept in mind, we can look ahead and see the essence of equilibrium. When the resultant force on a body is zero, the body will be in translational equilibrium, that is, its center of mass will be either at rest or moving in a straight line with constant velocity. In addition, if the resultant couple is zero, the body will be in rotational equilibrium, either having no rotational motion at all or rotating with a constant angular velocity. When both resultants are zero, the body is in complete equilibrium. Thus, we see at this point that the zero resultants $\Sigma \mathbf{F} = \mathbf{0}$ and $\Sigma \mathbf{M} = \mathbf{0}$ are the required conditions for equilibrium.

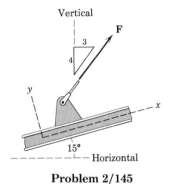

**Problem 2/145**

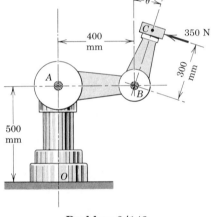

**Problem 2/146**

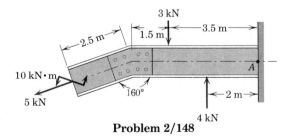

**Problem 2/148**

## REVIEW PROBLEMS

**2/145** A cable exerts a force **F** on the bracket of the structural member. If the magnitude of the $x$-component of **F** is 900 lb, calculate the $y$-component and the magnitude of **F**.

Ans. $F_y = 706$ lb
$F = 1144$ lb

**2/146** Calculate the moment $M_O$ of the 350-N force about the base point $O$ of the robot if $\theta = 20°$.

**2/147** In Sample Problem 2/5 calculate the magnitude of the smallest force **P** which can be applied at point $A$ to produce the same moment about $O$ as produced by the 600-N force. Find the corresponding angle $\theta$ between **P** and the horizontal.

Ans. $P = 584$ N, $\theta = 26.6°$

**2/148** Represent the resultant of the three forces and couple by a force–couple system located at point $A$.

**2/149**   Reduce the given loading system to a force–couple
system at point $A$. Then determine the distance $x$ to
the right of point $A$ at which the resultant of the
three forces acts.

       *Ans.* $R = 80$ lb down, $M_A = 1240$ lb-in. CW
           $x = 15.5$ in.

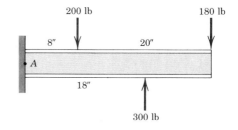

**Problem 2/149**

**2/150**   The $y$ and $z$ scalar components of a force $\mathbf{F}$ are
100 lb and 200 lb, respectively. If the direction
cosine $l = \cos \theta_x$ of the line of action of the force is
$-0.5$, write $\mathbf{F}$ as a vector.

**2/151**   The directions of rotation of the input shaft $A$ and
output shaft $B$ of the worm-gear reducer are indicated by the curved dotted arrows. An input torque
(couple) of 80 N·m is applied to shaft $A$ in the
direction of rotation. The output shaft $B$ supplies a
torque of 320 N·m to the machine that it drives (not
shown). The shaft of the driven machine exerts an
equal and opposite reacting torque on the output
shaft of the reducer. Determine the resultant $\mathbf{M}$ of
the two couples that act on the reducer unit and
calculate the direction cosine of $\mathbf{M}$ with respect to
the $x$-axis.

       *Ans.* $\mathbf{M} = -320\mathbf{i} - 80\mathbf{j}$ N·m
         $\cos \theta_x = -0.970$

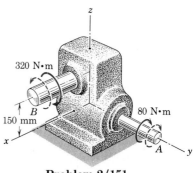

**Problem 2/151**

**2/152**   During a drilling operation, the small robotic device
is subjected to an 800-N force at point $C$ as shown.
Replace this force by an equivalent force–couple
system at point $O$.

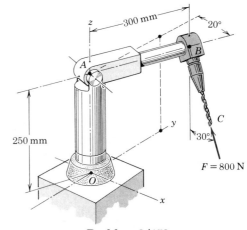

**Problem 2/152**

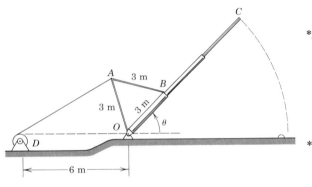

**Problem 2/153**

▶**2/153** The access door of Prob. 2/103 is shown again here. If the tension in the chain *AB* is 100 N, determine the magnitude *M* of its moment about the hinge axis. *Ans.* $M = 46.8$ N·m

### *Computer-oriented problems*

*2/154 Consider the 350-N force exerted on the robot arm of Prob. 2/146. Determine and plot the moment of this force about point *O* for the range $0 \le \theta \le 90°$. Comment on the physical significance of the maximum value of this moment. For what value(s) of $\theta$ is $M_O = 310$ N·m?

**Problem 2/155**

*2/155 A flagpole with attached light triangular frame is shown here for an arbitrary position during its raising. The 75-N tension in the erecting cable remains constant. Determine and plot the moment about the pivot *O* of the 75-N force for the range $0 \le \theta \le 90°$. Determine the maximum value of this moment and the elevation angle at which it occurs; comment on the physical significance of the latter. The effects of the diameter of the drum at *D* may be neglected.

$$Ans. \ \mathbf{M}_O = \frac{1350 \sin (\theta + 60°)}{\sqrt{45 + 36 \cos (\theta + 60°)}} \ \mathbf{k} \ \text{N·m}$$

$$M_O = 225 \ \text{N·m at} \ \theta = 60°$$

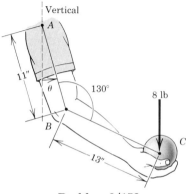

**Problem 2/156**

*2/156 A person locks his elbow so that the angle *ABC* is maintained at 130° and rotates the shoulder joint at *A* so that the arm remains in the vertical plane shown. The shoulder joint remains fixed. Determine and plot on the same axes the moments about points *A* and *B* of the weight of the 8-lb sphere as the angle $\theta$ is varied from 0° to 120°. Comment on the physical significance of the maximum value of the moment on each curve.

**\*2/157**  Assume that the hydraulic cylinder $AB$ exerts a force
$\mathbf{F}$ of constant magnitude 500 lb as the bin is elevated. Determine and plot the moment of this force
about the point $O$ for the range $0 \le \theta \le 90°$. At what
angle $\theta$ is this moment a maximum and what is the
maximum moment?

$$Ans. \ \ \mathbf{M}_O = \frac{3000(\sin \theta + \cos \theta)}{\sqrt{11 + 6(\sin \theta - \cos \theta)}} \ \mathbf{k} \text{ lb-ft}$$

$$M_O = 1414 \text{ lb-ft at } \theta = 16.9°$$

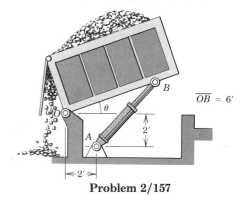

Problem 2/157

**\*2/158**  The weight of the cylinder causes a tension force $\mathbf{T}$
to be exerted by the cable on the support at $O$, and
the tension magnitude is constant at 900 N. The
small pulley at $C$ can be moved along a line which
is parallel to the $x$-axis. Determine and plot the
projection $T_{AB}$ of the tension force $\mathbf{T}$ onto the fixed
line $AB$ as a function of the pulley position $x$ for
$0 \le x \le 4$ m. Comment on the physical significance
of any sign changes in your plot.

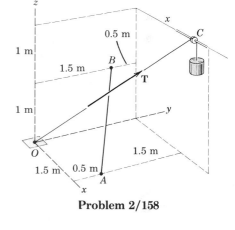

Problem 2/158

**\*2/159**  The rectangular plate is tilted about its lower edge
by a cable tensioned at a constant 600 N. Determine
and plot the moment of this tension about the lower
edge $AB$ of the plate for the range $0 \le \theta \le 90°$.

$$Ans. \ \ M_{AB} = \frac{7200 \sin \theta}{\sqrt{41 + 24 \cos \theta}} \ \text{N·m}$$

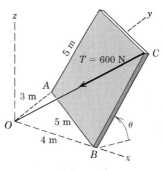

Problem 2/159

A large fraction of mechanics problems deals with bodies where the forces acting on them cancel one another, thus causing a state of equilibrium. The positioning of this loaded ship container is an example of a body in equilibrium.

# EQUILIBRIUM

# 3

## 3/1  INTRODUCTION

The subject of statics deals primarily with the description of the conditions of force that are both necessary and sufficient to maintain the state of equilibrium of engineering structures. This chapter on equilibrium, therefore, constitutes the most central part of statics and should be mastered thoroughly. We shall make continuous use of the concepts developed in Chapter 2 involving forces, moments, couples, and resultants as we apply the principles of equilibrium. The procedures which we shall develop in Chapter 3 constitute a comprehensive introduction to the approach used in the solution of countless problems in mechanics and in other engineering areas as well. This approach is basic to the successful mastery of statics, and the student is urged to read and study the following articles with special effort and attention to detail.

When a body is in equilibrium, the resultant of *all* forces acting on it is zero. Thus, the resultant force **R** and the resultant couple **M** are both zero, and we have the equilibrium equations

$$\mathbf{R} = \Sigma\mathbf{F} = 0 \qquad \mathbf{M} = \Sigma\mathbf{M} = 0 \qquad\qquad (3/1)$$

These requirements are both necessary and sufficient conditions for equilibrium.

Whereas all physical bodies are inherently three-dimensional, many of them may be treated as two-dimensional when the forces to which they are subjected act in a single plane or may be projected onto a single plane. When this simplification is not possible, the problem must be treated as three-dimensional. We shall follow the arrangement used in Chapter 2 and discuss in Section A the equilibrium of bodies subjected to two-dimensional force systems and in Section B the equilibrium of bodies subjected to three-dimensional force systems.

## SECTION A.  EQUILIBRIUM IN TWO DIMENSIONS

### 3/2  *MECHANICAL SYSTEM ISOLATION*

Before we apply Eqs. 3/1, it is essential that we define unambiguously the particular body or mechanical system to be analyzed and represent clearly and completely *all* forces which act *on* the body. The omission of a force or the inclusion of a force which does not act *on* the body in question will give erroneous results.

A mechanical system is defined as a body or group of bodies that can be isolated from all other bodies. Such a system may be a single body or a combination of connected bodies. The bodies may be rigid or nonrigid. The system may also be a defined fluid mass, liquid or gas, or the system may be a combination of fluids and solids. In statics we direct our attention primarily to a description of the forces that act on rigid bodies at rest, although consideration is also given to the statics of fluids. Once we reach a decision about which body or combination of bodies is to be analyzed, then this body or combination treated as a single body is *isolated* from all surrounding bodies. This isolation is accomplished by means of the *free-body diagram*, which is a diagrammatic representation of the isolated body or combination of bodies treated as a single body, showing all forces applied to it by mechanical contact with other bodies that are imagined to be removed. If appreciable body forces are present, such as gravitational or magnetic attraction, then these forces must also be shown on the diagram of the isolated body. Only after such a diagram has been carefully drawn should the equilibrium equations be written. Because of its critical importance we emphasize here that

> **the free-body diagram is the most important single step in the solution of problems in mechanics.**

Before we attempt to draw free-body diagrams, the mechanical characteristics of force application must be recognized. In Art. 2/2 the basic characteristics of force were described, with primary attention focused on the vector properties of force. We noted that forces are applied both by direct physical contact and by remote action and that forces may be either internal or external to the body under consideration. We further observed that the application of external forces is accompanied by reactive forces and that both applied and reactive forces may be either concentrated or distributed. Additionally the principle of transmissibility was introduced, which permits the treatment of force as a sliding vector as far as its external effects on a rigid body are concerned. We will now use these characteristics of force in developing the analytical model of an isolated mechanical system to which the equations of equilibrium will then be applied.

Figure 3/1 shows the common types of force application on mechanical systems for analysis in two dimensions. In each example the force exerted *on* the body to be isolated *by* the body to be removed is indicated. Newton's third law, which notes the existence of an equal and opposite reaction to every action, must be carefully observed. The force exerted *on* the body in question *by* a contacting or supporting member is always in the sense to oppose the movement of the body which would occur if the contacting or supporting member were removed.

In example 1 the action of a flexible cable, belt, rope, or chain on the body to which it is attached is depicted. Because of its flexibility, a rope or cable is unable to offer any resistance to bending, shear, or compression and therefore exerts a tension force in a direction tangent to the cable at its point of attachment. The force exerted *by* the cable *on* the body to which it is attached is always *away* from the body. When the tension $T$ is large compared with the weight of the cable, we may assume that the cable forms a straight line. When the cable weight is not negligible compared with its tension, the sag of the cable becomes important, and the tension in the cable changes direction and magnitude along its length. At its attachment the cable exerts a force tangent to itself.

When the smooth surfaces of two bodies are in contact, as in example 2, the force exerted by one on the other is *normal* to the tangency of the surfaces and is compressive. Although no actual surfaces are perfectly smooth, we are justified in making this assumption for practical purposes in many instances.

When mating surfaces of contacting bodies are rough, example 3, the force of contact may not necessarily be normal to the tangent to the surfaces but may be resolved into a *tangential* or *frictional component F* and a *normal component N*.

Example 4 illustrates a number of forms of mechanical support which effectively eliminate tangential friction forces, and here the net reaction is normal to the supporting surface.

Example 5 shows the action of a smooth guide on the body it supports. Resistance parallel to the guide is absent.

Example 6 illustrates the action of a pin connection. Such a connection is able to support force in any direction normal to the axis of the pin. We usually represent this action in terms of two rectangular components. The correct sense of these components in an actual problem will depend on how the member is loaded. When not otherwise initially known, the sense is arbitrarily assigned. Upon computation a positive algebraic sign for the component indicates that the assigned sense is correct. A negative sign indicates the sense is opposite to that assigned.

If the joint is free to turn about the pin, only the force $R$ can be supported. If the joint is not free to turn, a resisting couple $M$ may also be supported. Again, the sense of $M$ is arbitrarily shown

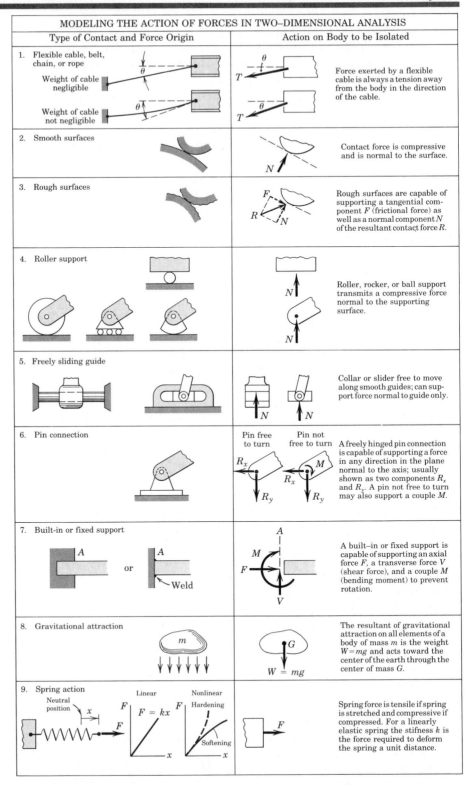

## MODELING THE ACTION OF FORCES IN TWO–DIMENSIONAL ANALYSIS

| Type of Contact and Force Origin | Action on Body to be Isolated |
|---|---|
| 1. Flexible cable, belt, chain, or rope<br><br>Weight of cable negligible<br><br>Weight of cable not negligible | Force exerted by a flexible cable is always a tension away from the body in the direction of the cable. |
| 2. Smooth surfaces | Contact force is compressive and is normal to the surface. |
| 3. Rough surfaces | Rough surfaces are capable of supporting a tangential component $F$ (frictional force) as well as a normal component $N$ of the resultant contact force $R$. |
| 4. Roller support | Roller, rocker, or ball support transmits a compressive force normal to the supporting surface. |
| 5. Freely sliding guide | Collar or slider free to move along smooth guides; can support force normal to guide only. |
| 6. Pin connection | Pin free to turn / Pin not free to turn — A freely hinged pin connection is capable of supporting a force in any direction in the plane normal to the axis; usually shown as two components $R_x$ and $R_y$. A pin not free to turn may also support a couple $M$. |
| 7. Built-in or fixed support | A built–in or fixed support is capable of supporting an axial force $F$, a transverse force $V$ (shear force), and a couple $M$ (bending moment) to prevent rotation. |
| 8. Gravitational attraction | The resultant of gravitational attraction on all elements of a body of mass $m$ is the weight $W = mg$ and acts toward the center of the earth through the center of mass $G$. |
| 9. Spring action | Spring force is tensile if spring is stretched and compressive if compressed. For a linearly elastic spring the stifness $k$ is the force required to deform the spring a unit distance. |

**Figure 3/1**

here, and in an actual problem will depend on how the member is loaded.

Example 7 shows the resultants of the rather complex distribution of force over the cross section of a slender bar or beam at a built-in or fixed support. The sense of the reactions $F$ and $V$ and the bending couple $M$ will, of course, depend in a given problem on how the member is loaded.

One of the most common forces is that due to gravitational attraction, example 8. This force affects all elements of mass of a body and is, therefore, distributed throughout it. The resultant of the gravitational forces on all elements is the weight $W = mg$ of the body, which passes through the center of mass $G$ and is directed toward the center of the earth for earthbound structures. The position of $G$ is frequently obvious from the geometry of the body, particularly where conditions of symmetry exist. When the position is not readily apparent, the location of $G$ must be calculated or determined by experiment. Similar remarks apply to the remote action of magnetic and electric forces. These forces of remote action have the same overall effect on a rigid body as forces of equal magnitude and direction applied by direct external contact.

Example 9 illustrates the action of a linear elastic spring and of a nonlinear spring with either hardening or softening characteristics. The force exerted by a linear spring, in tension or compression, is given by $F = kx$, where $k$ is the stiffness of the spring and $x$ is its deformation measured from the neutral or undeformed position.

The student is urged to study these nine conditions and to identify them in the problem work so that the correct free-body diagrams may be drawn. The representations in Fig. 3/1 are *not* free-body diagrams but are merely elements in the construction of free-body diagrams.

The full procedure for drawing a free-body diagram which accomplishes the isolation of the body or system under consideration will now be described.

*Construction of free-body diagrams.*   The following steps are involved.

*Step 1.*   A clear decision is made concerning which body or combination of bodies is to be isolated. The body chosen will usually involve one or more of the desired unknown quantities.

*Step 2.*   The body or combination chosen is next isolated by a diagram that represents its *complete external boundary*. This boundary defines the isolation of the body from *all* other contacting or attracting bodies, which are considered removed. This step is often the most crucial of all. We should always be certain that we have *completely isolated* the body before proceeding with the next step.

*Step 3.*   All forces that act *on* the isolated body as applied *by* the removed contacting and attracting bodies are next represented in their proper positions on the diagram of the isolated body. A systematic traverse of the entire boundary will disclose all contact forces. Weights, where appreciable, must be included. Known forces should be represented by vector arrows, each with its proper magnitude, direction, and sense indicated. Each unknown force should be represented by a vector arrow with the unknown magnitude or direction indicated by symbol. If the sense of the vector is also unknown, it may be arbitrarily assumed. The calculations will reveal a positive quantity if the correct sense was assumed and a negative quantity if the incorrect sense was assumed. It is necessary to be *consistent* with the assigned characteristics of unknown forces throughout all of the calculations.

*Step 4.* The choice of coordinate axes should be indicated directly on the diagram. Pertinent dimensions may also be represented for convenience. Note, however, that the free-body diagram serves the purpose of focusing accurate attention on the action of the external forces, and therefore the diagram should not be cluttered with excessive extraneous information. Force arrows should be clearly distinguished from any other arrows which may appear so that confusion will not result. For this purpose a colored pencil may be used.

When the foregoing four steps are completed, a correct free-body diagram will result, and the way will be clear for a straightforward and successful application of the governing equations, both in statics and in dynamics.

Many students are often tempted to omit from the free-body diagram certain forces that may not appear at first glance to be needed in the calculations. When we yield to this temptation, we invite serious error. It is only through *complete* isolation and a systematic representation of *all* external forces that a reliable accounting of the effects of all applied and reactive forces can be made. Very often a force that at first glance may not appear to influence a desired result does indeed have an influence. Hence, the only safe procedure is to make certain that all forces whose magnitudes are not negligible appear on the free-body diagram.

The free-body diagram has been explained in some detail because of its great importance in mechanics. The free-body method ensures an accurate definition of a mechanical system and focuses attention on the exact meaning and application of the force laws of statics and dynamics. Indeed, the free-body method is so important that students are strongly urged to reread this section several times in conjunction with their study of the sample free-body diagrams shown in Fig. 3/2 and the sample problems which appear at the end of the next article.

Figure 3/2 gives four examples of mechanisms and structures together with their correct free-body diagrams. Dimensions and magnitudes are omitted for clarity. In each case we treat the entire system as a single body, so that the internal forces are not shown. The characteristics of the various types of contact forces illustrated in Fig. 3/1 are included in the four examples as they apply.

In example 1 the truss is composed of structural elements that, taken all together, constitute a rigid framework. Thus, we may remove the entire truss from its supporting foundation and treat it as a single rigid body. In addition to the applied external load $P$, the free-body diagram must include the reactions on the truss at $A$

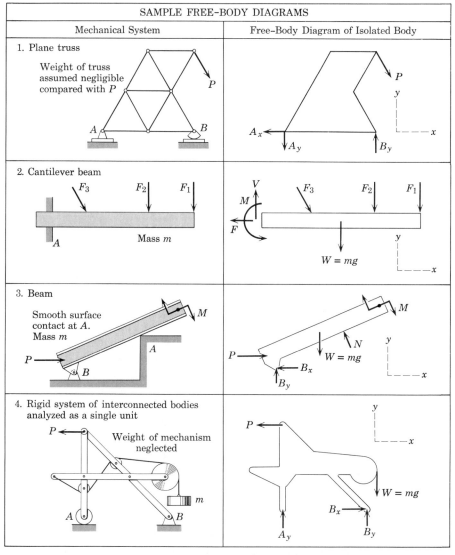

**Figure 3/2**

and $B$. The rocker at $B$ can support a vertical force only, and this force is transmitted to the structure at $B$ (example 4 of Fig. 3/1). The pin connection at $A$ (example 6 of Fig. 3/1) is capable of supplying both a horizontal and a vertical component of force to the truss. In this relatively simple example it is clear that the vertical component $A_y$ must be directed down to prevent the truss from rotating clockwise about $B$. Also, the horizontal component $A_x$ will be to the left to keep the truss from moving to the right under the influence of the horizontal component of $P$. Thus, in constructing the free-body diagram for this simple truss, the correct sense of each of the components of force exerted *on* the truss *by* the foundation at $A$ is easily perceived and may, therefore, be represented in its correct physical sense on the diagram. When the correct physical sense of a force or its component is not easily recognized by direct observation, it must be assigned arbitrarily, and the correctness of or error in the assignment is determined by the algebraic sign of its calculated value. If the total weight of the truss members is appreciable compared with $P$ and the forces at $A$ and $B$, then the weights of the members must be included on the free-body diagram as external forces.

In example 2 the cantilever beam is secured to the wall and subjected to the three applied loads. When we isolate that part of the beam to the right of the section at $A$, we must include the reactive forces applied *to* the beam *by* the wall. The resultants of these reactive forces are shown acting on the section of the beam (example 7 of Fig. 3/1). A vertical force $V$ to counteract the excess of downward applied force is shown, and a tension $F$ to balance the excess of applied force to the right must also be included. Then to prevent the beam from rotating about $A$, a counterclockwise couple $M$ is also required. The weight $mg$ of the beam must also be represented through the mass center (example 8 of Fig. 3/1). Here we have represented the somewhat complex system of forces which actually act on the cut section of the beam by the equivalent force–couple system where the force is broken down into its vertical component $V$ (shear force) and its horizontal component $F$ (tensile force). The couple $M$ is the bending moment in the beam. The free-body diagram is now complete and shows the beam in equilibrium under the action of six forces and one couple.

In example 3 the weight $W = mg$ is shown acting through the center of mass of the beam, which is assumed known (example 8 of Fig. 3/1). The force exerted by the corner $A$ on the beam is normal to the smooth surface of the beam (example 2 of Fig. 3/1). To perceive this action more clearly, visualize an enlargement of the contact point $A$, which would appear somewhat rounded, and consider the force exerted by this rounded corner on the straight surface of the beam assumed to be smooth. If the contacting surfaces at the corner were not smooth, a tangential frictional component of force could be developed. In addition to the applied force $P$ and couple $M$, there is the pin connection at $B$, which exerts both an $x$- and a $y$-component of force on the beam. The positive senses of these components are assigned arbitrarily.

In example 4 the free-body diagram of the entire isolated mechanism discloses three unknown quantities for equilibrium with the given loads $mg$ and $P$. Any one of many internal configurations for securing the cable leading from the mass $m$ would be possible without affecting the external response of the mechanism as a whole, and this fact is brought out by the free-body diagram. This hypothetical example is used to emphasize the advantage of including as much as possible in the free-body diagram and to show that the forces internal to a rigid assembly of members do not influence the values of the external reactions.

The positive senses of $B_x$ and $B_y$ in example 3 and $B_y$ in example 4 are assumed on the free-body diagrams, and the correctness of the assumptions would be proved or disproved according to whether the algebraic signs of the terms were plus or minus when the calculations were carried out in the actual problems.

The isolation of the mechanical system under consideration will be recognized as a crucial step in the formulation of the mathematical model. The most important aspect to the correct construction of the all-important free-body diagram is the clear-cut and unambiguous decision as to what is included and what is excluded. This decision becomes unambiguous only when the boundary of the free-body diagram represents a complete traverse of the body or system of bodies to be isolated, starting at some arbitrary point of the boundary and returning to that same point. The body within this closed boundary, then, is the isolated free body, and all forces transmitted to the body across the boundary by all contacting bodies which are removed must be accounted for. The student is again urged to devote special attention to this step. Before direct use is made of the free-body diagram in the application of the principles of force equilibrium in the next article, some initial practice with the drawing of free-body diagrams is helpful. The problems that follow are designed to provide this practice.

# FREE-BODY DIAGRAM EXERCISES

3/A  In each of the five following examples, the body to be isolated is shown in the left-hand
diagram, and an *incomplete* free-body diagram (FBD) of the isolated body is shown on the
right. Add whatever forces are necessary in each case to form a complete free-body
diagram. The weights of the bodies are negligible unless otherwise indicated. Dimensions
and numerical values are omitted for simplicity.

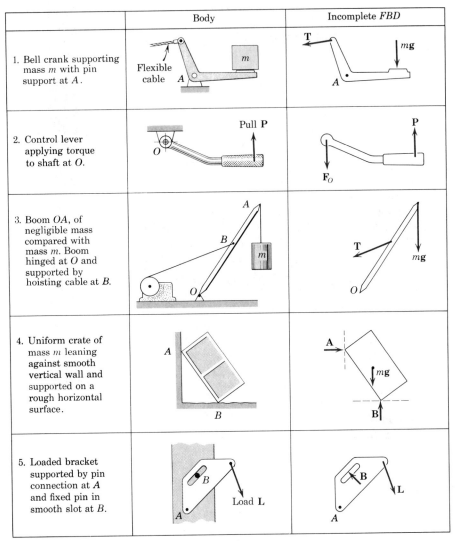

|  | Body | Incomplete *FBD* |
|---|---|---|
| 1. Bell crank supporting mass $m$ with pin support at $A$. | | |
| 2. Control lever applying torque to shaft at $O$. | | |
| 3. Boom $OA$, of negligible mass compared with mass $m$. Boom hinged at $O$ and supported by hoisting cable at $B$. | | |
| 4. Uniform crate of mass $m$ leaning against smooth vertical wall and supported on a rough horizontal surface. | | |
| 5. Loaded bracket supported by pin connection at $A$ and fixed pin in smooth slot at $B$. | | |

**Figure 3/A**

**3/B** In each of the five following examples, the body to be isolated is shown in the left-hand diagram, and either a *wrong* or an *incomplete* free-body diagram (FBD) is shown on the right. Make whatever changes or additions are necessary in each case to form a correct and complete free-body diagram. The weights of the bodies are negligible unless otherwise indicated. Dimensions and numerical values are omitted for simplicity.

| | Body | Wrong or Incomplete *FBD* |
|---|---|---|
| 1. Lawn roller of mass $m$ being pushed up incline $\theta$. | | |
| 2. Pry bar lifting body $A$ having smooth horizontal surface. Bar rests on hori–zontal rough surface. | | |
| 3. Uniform pole of mass $m$ being hoisted into position by winch. Horizontal supporting surface notched to prevent slipping of pole. | | |
| 4. Supporting angle bracket for frame. Pin joints | | |
| 5. Bent rod welded to support at $A$ and subjected to two forces and couple | | |

**Figure 3/B**

**3/C** Draw a complete and correct free-body diagram of each of the bodies designated in the statements. The weights of the bodies are significant only if the mass is stated. All forces, known and unknown, should be labeled. (*Note*: The sense of some reaction components cannot always be determined without numerical calculation.)

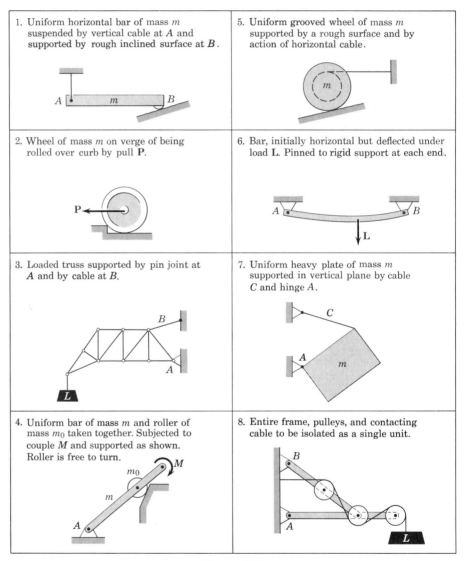

1. Uniform horizontal bar of mass $m$ suspended by vertical cable at $A$ and supported by rough inclined surface at $B$.

5. Uniform grooved wheel of mass $m$ supported by a rough surface and by action of horizontal cable.

2. Wheel of mass $m$ on verge of being rolled over curb by pull **P**.

6. Bar, initially horizontal but deflected under load **L**. Pinned to rigid support at each end.

3. Loaded truss supported by pin joint at $A$ and by cable at $B$.

7. Uniform heavy plate of mass $m$ supported in vertical plane by cable $C$ and hinge $A$.

4. Uniform bar of mass $m$ and roller of mass $m_0$ taken together. Subjected to couple $M$ and supported as shown. Roller is free to turn.

8. Entire frame, pulleys, and contacting cable to be isolated as a single unit.

**Figure 3/C**

## 3/3  *EQUILIBRIUM CONDITIONS*

In Art. 3/1 we defined equilibrium as the condition in which the resultant of all forces acting on a body is zero. Stated in another way, a body is in equilibrium if all forces and moments applied to it are in balance. These requirements are contained in the vector equations of equilibrium, Eqs. 3/1, which in two dimensions may be written in scalar form as

$$\Sigma F_x = 0 \qquad \Sigma F_y = 0 \qquad \Sigma M_O = 0 \qquad\qquad (3/2)$$

The third equation represents the zero sum of the moments of all forces about any point $O$ on or off the body. Equations 3/2 are the necessary and sufficient conditions for complete equilibrium in two dimensions. They are necessary conditions because, if they are not satisfied, there can be no force or moment balance. They are sufficient because once satisfied, there can be no unbalance, and equilibrium is assured.

The equations relating force and acceleration for rigid-body motion are developed in *Vol. 2 Dynamics* from Newton's second law of motion. These equations show that the acceleration of the mass center of a body is proportional to the resultant force $\Sigma\mathbf{F}$ acting on the body. Consequently, if a body moves with constant velocity (zero acceleration), the resultant force on it must be zero, and the body may be treated as in a state of equilibrium.

For complete equilibrium in two dimensions, all three of Eqs. 3/2 must hold. However, these conditions are independent requirements, and one may hold without another. Take, for example, a body which slides along a horizontal surface with increasing velocity under the action of applied forces. The force-equilibrium equations will be satisfied in the vertical direction where the acceleration is zero but not in the horizontal direction. Also, a body, such as a flywheel, which rotates about its fixed mass center with increasing angular speed is not in rotational equilibrium, but the two force-equilibrium equations will be satisfied.

*(a) Categories of equilibrium.*   Applications of Eqs. 3/2 fall naturally into a number of categories that are easily identified. These categories of force systems acting on bodies in two-dimensional equilibrium are summarized in Fig. 3/3 and are explained further as follows:

*Case 1,*  equilibrium of collinear forces, clearly requires only the one force equation in the direction of the forces ($x$-direction), since all other equations are automatically satisfied.

*Case 2,*  equilibrium of forces that lie in a plane ($x$-$y$ plane) and are concurrent at a point $O$, requires the two force equations

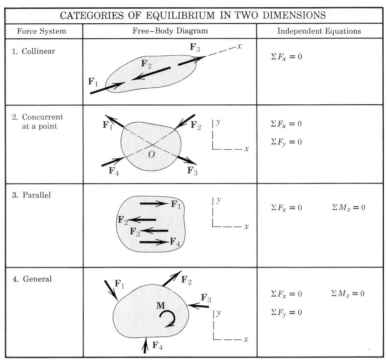

**Figure 3/3**

| CATEGORIES OF EQUILIBRIUM IN TWO DIMENSIONS | | |
|---|---|---|
| Force System | Free-Body Diagram | Independent Equations |
| 1. Collinear | | $\Sigma F_x = 0$ |
| 2. Concurrent at a point | | $\Sigma F_x = 0$ $\Sigma F_y = 0$ |
| 3. Parallel | | $\Sigma F_x = 0$     $\Sigma M_z = 0$ |
| 4. General | | $\Sigma F_x = 0$     $\Sigma M_z = 0$ $\Sigma F_y = 0$ |

only, since the moment sum about $O$, that is, about a $z$-axis through $O$, is necessarily zero. Included in this category is the case of the equilibrium of a particle.

*Case 3,* equilibrium of parallel forces in a plane, requires the one force equation in the direction of the forces ($x$-direction) and one moment equation about an axis ($z$-axis) normal to the plane of the forces.

*Case 4,* equilibrium of a general system of forces in a plane ($x$-$y$), requires the two force equations in the plane and one moment equation about an axis ($z$-axis) normal to the plane.

There are two frequently occurring equilibrium situations to which the student should be alerted. The first situation is the equilibrium of a body under the action of two forces only. Two examples are shown in Fig. 3/4, and we see that for such a *two-force member* the forces must be *equal, opposite,* and *collinear.* The shape of the member should not obscure this simple requirement. In the illustrations cited we consider the weights of the members to be negligible compared with the applied forces.

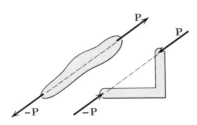

Two-force members
(a)

**Figure 3/4**

The second situation is the equilibrium of a body under the action of three forces, Fig. 3/5a. We see that the lines of action of the three forces must be *concurrent*. If they were not concurrent, then one of the forces would exert a resultant moment about the point of concurrency of the other two, which would violate the requirement of zero moment about every point. The only exception occurs when the three forces are parallel. In this case we may consider the point of concurrency to be at infinity. The principle of the concurrency of three forces in equilibrium is of considerable use in carrying out a graphical solution of the force equations. In this case the polygon of forces is drawn and made to close, as shown in Fig. 3/5b. Frequently, a body in equilibrium under the action of more than three forces may be reduced to a *three-force member* by a combination of two or more of the known forces.

**(b) Alternative equilibrium equations.** There are two additional ways in which we may express the general conditions for the equilibrium of forces in two dimensions. For the body shown in Fig. 3/6a, if $\Sigma M_A = 0$, then the resultant, if it still exists, cannot be a couple but must be a force **R** passing through $A$. If now the equation $\Sigma F_x = 0$ holds, where the $x$-direction is perfectly arbitrary, it follows from Fig. 3/6b that the resultant force **R**, if it still exists, not only must pass through $A$, but also must be perpendicular to the $x$-direction as shown. Now, if $\Sigma M_B = 0$, where $B$ is any point such that the line $AB$ is not perpendicular to the $x$-direction, we see that

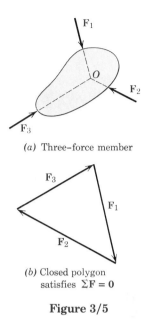

(a) Three-force member

(b) Closed polygon
satisfies $\Sigma \mathbf{F} = 0$

**Figure 3/5**

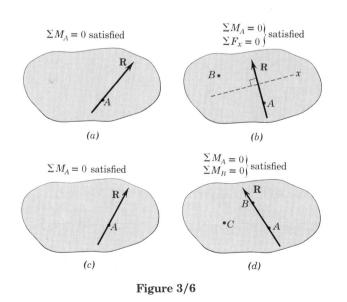

$\Sigma M_A = 0$ satisfied

(a)

$\left.\begin{array}{l} \Sigma M_A = 0 \\ \Sigma F_x = 0 \end{array}\right\}$ satisfied

(b)

$\Sigma M_A = 0$ satisfied

(c)

$\left.\begin{array}{l} \Sigma M_A = 0 \\ \Sigma M_B = 0 \end{array}\right\}$ satisfied

(d)

**Figure 3/6**

**R** must be zero, and hence the body is in equilibrium. Therefore, an alternative set of equilibrium equations is

$$\Sigma F_x = 0 \qquad \Sigma M_A = 0 \qquad \Sigma M_B = 0$$

where the two points $A$ and $B$ must not lie on a line perpendicular to the $x$-direction.

A third formulation of the conditions of equilibrium may be made for a coplanar force system. Again, if $\Sigma M_A = 0$ for any body such as that shown in Fig. 3/6c, the resultant, if any, must be a force **R** through $A$. In addition if $\Sigma M_B = 0$, the resultant, if one still exists, must pass through $B$ as shown in Fig. 3/6d. Such a force cannot exist, however, if $\Sigma M_C = 0$, where $C$ is not collinear with $A$ and $B$. Hence, we may write the equations of equilibrium as

$$\Sigma M_A = 0 \qquad \Sigma M_B = 0 \qquad \Sigma M_C = 0$$

where $A$, $B$, and $C$ are any three points not on the same straight line.

When equilibrium equations are written which are not independent, redundant information is obtained, and solution of the equations will yield $0 = 0$. For example, for a general problem in two dimensions with three unknowns, three moment equations written about three points which lie on the same straight line are not independent. Such equations will contain duplicated information, and solution of two of them can at best determine two of the unknowns, with the third equation merely verifying the identity $0 = 0$.

*(c) Constraints and statical determinacy.*  The equilibrium equations developed in this article, once satisfied, are both necessary and sufficient conditions to establish the equilibrium of a body. However, they do not necessarily provide all the information that is required to calculate all the unknown forces that may act on a body in equilibrium. The question of adequacy lies in the characteristics of the constraints against possible movement of the body provided by its supports. By constraint we mean the restriction of movement. In example 4 of Fig. 3/1 the roller, ball, and rocker provide constraint normal to the surface of contact but none tangent to the surface. Hence, a tangential force cannot be supported. For the collar and slider of example 5 constraint exists only normal to the guide. In example 6 the fixed-pin connection provides constraint in both directions, but offers no resistance to rotation about the pin unless the pin is not free to turn. The fixed support of example 7, however, offers constraint against rotation as well as lateral movement.

If the rocker that supports the truss of example 1 in Fig. 3/2 were replaced by a pin joint, as at $A$, there would be one additional constraint beyond that required to support an equilibrium configuration without collapse. The three scalar conditions of equilibrium, Eqs. 3/2, would not provide sufficient information to determine all four unknowns, since $A_x$ and $B_x$ could not be separated. These two components of force would be dependent on the deformation of the members of the truss as influenced by their corresponding stiffness properties. The horizontal reactions $A_x$ and $B_x$ would also be dependent on any initial deformation required to fit the dimensions of the structure to those of the foundation between $A$ and $B$. Again referring to Fig. 3/2, we see that if the pin $B$ in example 3 were not free to turn, the support could transmit a couple to the beam through the pin. Therefore, there would be four unknown supporting reactions acting on the beam, namely, the force at $A$, the two components of force at $B$, and the couple at $B$. Consequently the three independent scalar equations of equilibrium would not provide enough information to compute all four unknowns.

A body, or rigid combination of elements treated as a single body, that possesses more external supports or constraints than are necessary to maintain an equilibrium position is called *statically indeterminate*. Supports that can be removed without destroying the equilibrium condition of the body are said to be *redundant*. The number of redundant supporting elements present corresponds to the degree of statical indeterminacy and equals the total number of unknown external forces, minus the number of available independent equations of equilibrium. On the other hand, bodies that are supported by the minimum number of constraints necessary to ensure an equilibrium configuration are called *statically determinate*, and for such bodies the equilibrium equations are sufficient to determine the unknown external forces.

The problems on equilibrium in this article and throughout *Vol. 1 Statics* are generally restricted to statically determinate bodies where the constraints are just sufficient to ensure a stable position and where the unknown supporting forces can be completely determined by the available independent equations of equilibrium. The student is alerted at this point by this brief discussion, however, to the fact that we must be aware of the nature of the constraints before we attempt to solve an equilibrium problem. A body will be recognized as statically indeterminate when there are more unknown external reactions than there are available independent equilibrium equations for the force system involved. It is always well to count the number of unknown forces on a given body and to be certain that an equal number of independent equations may be written; otherwise, effort may be wasted in attempting an impossible solution with the aid of the equilibrium equations only. Unknowns may be forces, couples, distances, or angles.

In discussing the relationship between constraints and equilibrium, we should look further at the question of the adequacy of constraints. The existence of three constraints for a two-dimensional

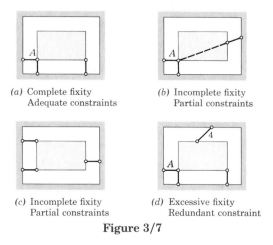

(a) Complete fixity
Adequate constraints

(b) Incomplete fixity
Partial constraints

(c) Incomplete fixity
Partial constraints

(d) Excessive fixity
Redundant constraint

**Figure 3/7**

problem does not always guarantee a stable configuration. Figure 3/7 shows four different types of constraints. In the *a*-part of the figure point *A* of the rigid body is fixed by the two links and cannot move, and the third link prevents any rotation about *A*. Thus, this body is completely fixed with three adequate (proper) constraints. In the *b*-part of the figure the third link is positioned so that the force transmitted by it passes through point *A* where the other two constraint forces act. Thus, this configuration of constraints can offer no initial resistance to rotation about *A*, which would occur when external loads were applied to the body. We conclude, therefore, that this body is incompletely fixed under partial constraints. The configuration in the *c*-part of the figure gives us a similar condition of incomplete fixity since the three parallel links could offer no initial resistance to a small vertical movement of the body as a result of external loads applied to it in this direction. The constraints in these two examples are often termed *improper*. In the *d*-part of Fig. 3/7 we have a condition of complete fixity, with link 4 acting as an unnecessary fourth constraint to maintain a fixed position. Link 4, then, is a *redundant* constraint, and the body is statically indeterminate.

As in the four examples of Fig. 3/7, it is generally possible by direct observation to conclude whether the constraints on a body in two-dimensional equilibrium are adequate (proper), partial (improper), or redundant. As indicated previously, the vast majority of problems in this book are statically determinate with adequate (proper) constraints.

*(d) Problem solution.* The sample problems at the end of the article illustrate the application of free-body diagrams and the equations of equilibrium to typical statics problems. These solutions should be studied thoroughly. In the problem work of this chapter

and throughout mechanics it is important to develop a logical and systematic approach which includes the following steps:

1. Identify clearly the quantities which are known and unknown.

2. Make an unambiguous choice of the body (or group of connecting bodies treated as a single body) to be isolated and draw its complete free-body diagram, labeling all external known and unknown forces and couples which act on it.

3. Designate a convenient set of reference axes, always using right-handed axes when vector cross products are employed. Choose moment centers with a view to simplifying the calculations. Generally the best choice is one through which as many unknown forces pass as possible. Simultaneous solutions of equilibrium equations are frequently necessary but can be minimized or avoided by a careful choice of reference axes and moment centers.

4. Identify and state the applicable force and moment principles or equations which govern the equilibrium conditions of the problem. In the sample problems which are included, these relations are set forth in brackets and precede each major calculation.

5. Match the number of independent equations with the number of unknowns in each problem.

6. Carry out the solution and check the results. In many problems engineering judgment can be developed by first making a reasonable guess or estimate of the result prior to the calculation and then comparing the estimate with the calculated value.

## Sample Problem 3/1

Determine the magnitudes of the forces **C** and **T** which, along with the other three forces shown, act on the bridge-truss joint.

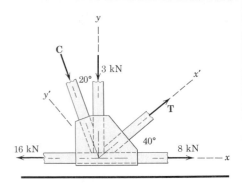

_**Solution.**_ The given sketch constitutes the free-body diagram of the isolated section of the joint in question and shows the five forces which ① are in equilibrium.

_**Solution I (scalar algebra).**_ For the $x$-$y$ axes as shown we have

$[\Sigma F_x = 0]$ $\qquad 8 + T \cos 40° + C \sin 20° - 16 = 0$

$[\Sigma F_y = 0]$ $\qquad\qquad 0.766T + 0.342C = 8$ $\qquad (a)$

$\qquad\qquad T \sin 40° - C \cos 20° - 3 = 0$

$\qquad\qquad 0.643T - 0.940C = 3$ $\qquad (b)$

Simultaneous solution of Eqs. ($a$) and ($b$) produces

$$T = 9.09 \text{ kN} \qquad C = 3.03 \text{ kN} \qquad\qquad Ans.$$

_**Solution II (scalar algebra).**_ To avoid a simultaneous solution, we ② may use axes $x'$-$y'$ with the first summation in the $y'$-direction to eliminate reference to $T$. Thus,

$[\Sigma F_{y'} = 0]$ $\quad -C \cos 20° - 3 \cos 40° - 8 \sin 40° + 16 \sin 40° = 0$

$\qquad\qquad C = 3.03 \text{ kN} \qquad\qquad Ans.$

$[\Sigma F_{x'} = 0]$ $\quad T + 8 \cos 40° - 16 \cos 40° - 3 \sin 40° - 3.03 \sin 20° = 0$

$\qquad\qquad T = 9.09 \text{ kN} \qquad\qquad Ans.$

_**Solution III (vector algebra).**_ With unit vectors **i** and **j** in the $x$- and $y$-directions, the zero summation of forces for equilibrium yields the vector equation

$[\Sigma \mathbf{F} = 0] \quad 8\mathbf{i} + (T \cos 40°)\mathbf{i} + (T \sin 40°)\mathbf{j} - 3\mathbf{j} + (C \sin 20°)\mathbf{i}$
$$- (C \cos 20°)\mathbf{j} - 16\mathbf{i} = \mathbf{0}$$

Equating the coefficients of the **i**- and **j**-terms to zero gives

$$8 + T \cos 40° + C \sin 20° - 16 = 0$$
$$T \sin 40° - 3 - C \cos 20° = 0$$

which are the same, of course, as Eqs. ($a$) and ($b$), which we solved above.

_**Solution IV (geometric).**_ The polygon representing the zero vector sum of the five forces is shown. Equations ($a$) and ($b$) are seen immediately to give the projections of the vectors onto the $x$- and $y$-directions. Similarly, projections onto the $x'$- and $y'$-directions give the alternative equations in Solution II.

A graphical solution is easily obtained. The known vectors are laid ③ off head-to-tail to some convenient scale, and the directions of **T** and **C** are then drawn to close the polygon. The resulting intersection at point $P$ completes the solution, thus enabling us to measure the magnitudes of **T** and **C** directly from the drawing to whatever degree of accuracy we incorporate in the construction.

① Since this is a problem of concurrent forces, no moment equation is necessary.

② The selection of reference axes to facilitate computation is always an important consideration. Alternatively in this example we could take a set of axes along and normal to the direction of **C** and employ a force summation normal to **C** to eliminate it.

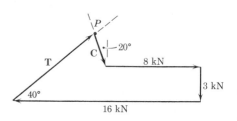

③ The known vectors may be added in any order desired, but they must be added before the unknown vectors.

## Sample Problem 3/2

Calculate the tension $T$ in the cable that supports the 1000-lb load with the pulley arrangement shown. Each pulley is free to rotate about its bearing, and the weights of all parts are small compared with the load. Find the magnitude of the total force on the bearing of pulley $C$.

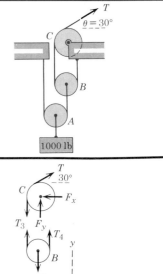

**Solution.** The free-body diagram of each pulley is drawn in its ① relative position to the others. We begin with pulley $A$, which includes the only known force. With the unspecified pulley radius designated by $r$, the equilibrium of moments about its center $O$ and the equilibrium of forces in the vertical direction require

② $[\Sigma M_O = 0]$ $\qquad T_1 r - T_2 r = 0 \qquad T_1 = T_2$

$[\Sigma F_y = 0] \qquad T_1 + T_2 - 1000 = 0 \qquad 2T_1 = 1000 \qquad T_1 = T_2 = 500$ lb

From the example of pulley $A$ we may write the equilibrium of forces on pulley $B$ by inspection as

$$T_3 = T_4 = T_2/2 = 250 \text{ lb}$$

For pulley $C$ the angle $\theta = 30°$ in no way affects the moment of $T$ about the center of the pulley, so that moment equilibrium requires

$$T = T_3 \qquad \text{or} \qquad T = 250 \text{ lb} \qquad\qquad Ans.$$

Equilibrium of the pulley in the $x$- and $y$-directions requires

$[\Sigma F_x = 0] \qquad 250 \cos 30° - F_x = 0 \qquad\qquad F_x = 217 \text{ lb}$

$[\Sigma F_y = 0] \qquad F_y + 250 \sin 30° - 250 = 0 \qquad F_y = 125 \text{ lb}$

$[F = \sqrt{F_x^2 + F_y^2}] \qquad F = \sqrt{(217)^2 + (125)^2} = 250 \text{ lb} \qquad\qquad Ans.$

① Note that we are careful to observe Newton's third law of action and equal and opposite reaction.

② Clearly the radius $r$ does not influence the results. Once having analyzed a simple pulley, the results should be perfectly clear by inspection.

---

## Sample Problem 3/3

The uniform 100-kg I-beam is supported initially by its end rollers on the horizontal surface at $A$ and $B$. By means of the cable at $C$ it is desired to elevate end $B$ to a position 3 m above end $A$. Determine the required tension $P$, the reaction at $A$, and the angle $\theta$ made by the beam with the horizontal in the elevated position.

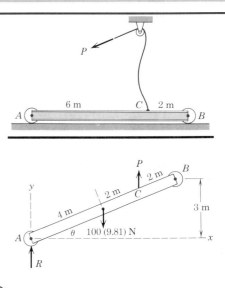

**Solution.** In constructing the free-body diagram, we note that the reaction on the roller at $A$ and the weight are vertical forces. Consequently, in the absence of other horizontal forces, $P$ must also be vertical. From Sample Problem 3/2 we see immediately that the tension $P$ in the cable equals the tension $P$ applied to the beam at $C$.

Moment equilibrium about $A$ eliminates force $R$ and gives

① $[\Sigma M_A = 0] \qquad P(6 \cos \theta) - 981(4 \cos \theta) = 0 \qquad P = 654 \text{ N} \qquad Ans.$

Equilibrium of vertical forces requires

$[\Sigma F_y = 0] \qquad 654 + R - 981 = 0 \qquad R = 327 \text{ N} \qquad Ans.$

The angle $\theta$ depends only on the specified geometry and is

$$\sin \theta = 3/8 \qquad \theta = 22.0° \qquad\qquad Ans.$$

① Clearly the equilibrium of this parallel force system is independent of $\theta$.

## Sample Problem 3/4

Determine the magnitude $T$ of the tension in the supporting cable and the magnitude of the force on the pin at $A$ for the jib crane shown. The beam $AB$ is a standard 0.5-m I-beam with a mass of 95 kg per meter of length.

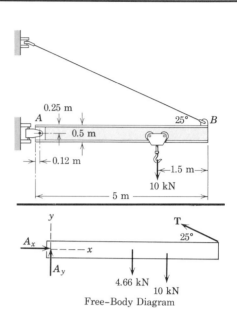

---

**Algebraic solution.** The system is symmetrical about the vertical $x$-$y$ plane through the center of the beam, so the problem may be analyzed as the equilibrium of a coplanar force system. The free-body diagram of the beam is shown in the figure with the pin reaction at $A$ represented in terms of its two rectangular components. The weight of the beam is $95(10^{-3})(5)9.81 = 4.66$ kN and acts through its center. Note that there are three unknowns $A_x$, $A_y$, and $T$ which may be found from the three equations of equilibrium. We begin with a moment equation about $A$ which eliminates two of the three unknowns from the equation. In applying the moment equation about $A$, it is simpler to consider the moments of the $x$- and $y$-components of $\mathbf{T}$ than it is to compute the perpendicular distance from $\mathbf{T}$ to $A$. Hence, with the counterclockwise sense as positive we write

① The justification for this step is Varignon's theorem, explained in Art. 2/4. Be prepared to take full advantage of this principle frequently.

$[\Sigma M_A = 0]$    $(T \cos 25°)0.25 + (T \sin 25°)(5 - 0.12)$
$$- 10(5 - 1.5 - 0.12) - 4.66(2.5 - 0.12) = 0$$

from which    $T = 19.61$ kN    *Ans.*

Equating the sum forces in the $x$- and $y$-directions to zero gives

② The calculation of moments in two-dimensional problems is generally handled more simply by scalar algebra than by the vector cross product $\mathbf{r} \times \mathbf{F}$. In three dimensions, as we shall see later, the reverse is often the case.

$[\Sigma F_x = 0]$        $A_x - 19.61 \cos 25° = 0$    $A_x = 17.77$ kN
$[\Sigma F_y = 0]$    $A_y + 19.61 \sin 25° - 4.66 - 10 = 0$    $A_y = 6.37$ kN
$[A = \sqrt{A_x^2 + A_y^2}]$    $A = \sqrt{(17.77)^2 + (6.37)^2}$    $A = 18.88$ kN    *Ans.*

**Graphical solution.** The principle that three forces in equilibrium must be concurrent is utilized for a graphical solution by combining the two known vertical forces of 4.66 and 10 kN into a single 14.66-kN force, located as shown on the modified free-body diagram of the beam in the lower figure. The position of this resultant load may easily be determined graphically or algebraically. The intersection of the 14.66-kN force with the line of action of the unknown tension $\mathbf{T}$ defines the point of concurrency $O$ through which the pin reaction $\mathbf{A}$ must pass. The unknown magnitudes of $\mathbf{T}$ and $\mathbf{A}$ may now be found by adding the forces head-to-tail to form the closed equilibrium polygon of forces, thus satisfying their zero vector sum. After the known vertical load is laid off to a convenient scale, as shown in the lower part of the figure, a line representing the given direction of the tension $\mathbf{T}$ is drawn through the tip of the 14.7-kN vector. Likewise a line representing the direction of the pin reaction $\mathbf{A}$, determined from the concurrency established with the free-body diagram, is drawn through the tail of the 14.7-kN vector. The intersection of the lines representing vectors $\mathbf{T}$ and $\mathbf{A}$ establishes the magnitudes $T$ and $A$ necessary to make the vector sum of the forces equal to zero. These magnitudes are scaled from the diagram. The $x$- and $y$-components of $\mathbf{A}$ may be constructed on the force polygon if desired.

③ The direction of the force at $A$ could be easily calculated if desired. However, in designing the pin $A$ or in checking its strength, it is only the magnitude of the force that matters.

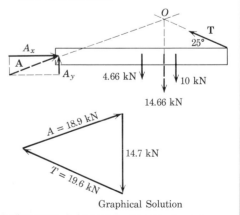

Graphical Solution

# PROBLEMS

## *Introductory problems*

**3/1** Determine the tensions in cables $CA$ and $CB$.
$$Ans. \ T_{CA} = 586 \ lb, \ T_{CB} = 717 \ lb$$

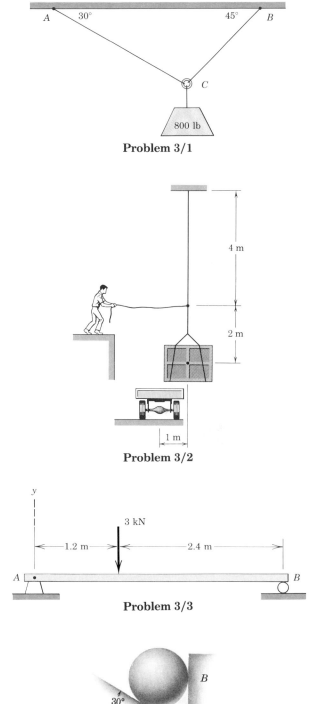

**Problem 3/1**

**3/2** What horizontal force $P$ must a worker exert on the rope to position the 50-kg crate directly over the transport vehicle?

**Problem 3/2**

**3/3** The uniform beam has a mass per unit length of 60 kg/m. Determine the reactions at the supports.
$$Ans. \ A_y = 3060 \ N, \ B_y = 2060 \ N$$

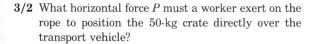

**Problem 3/3**

**3/4** The 50-kg homogeneous smooth sphere rests on the 30-deg incline $A$ and bears against the smooth vertical wall $B$. Calculate the contact forces at $A$ and $B$.

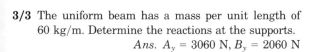

**Problem 3/4**

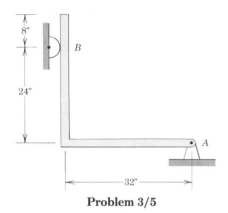

**Problem 3/5**

**3/5** The uniform angle bar with equal legs weighs 40 lb and is supported in the vertical plane as shown. Calculate the force $F_A$ supported by the pin at $A$.

*Ans.* $F_A = 56.6$ lb

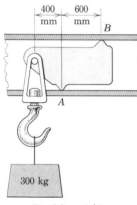

**Problem 3/6**

**3/6** To facilitate shifting the position of a lifting hook when it is not under load, the sliding hanger shown is used. The projections at $A$ and $B$ engage the flanges of a box beam when a load is supported, and the hook projects through a horizontal slot in the beam. Compute the forces at $A$ and $B$ when the hook supports a 300-kg mass.

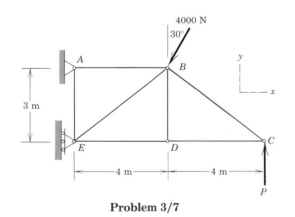

**Problem 3/7**

**3/7** Determine the reactions at $A$ and $E$ if $P = 500$ N. What is the maximum value that $P$ may have for static equilibrium? Neglect the weight of the structure compared with the applied loads.

*Ans.* $A_x = -1290$ N, $A_y = 2960$ N
$E_x = 3290$ N, $P = 1732$ N

**3/8** The pin at *A* can support a maximum force of 3.2 kN. What is the corresponding maximum load *L* which can be supported by the bracket?

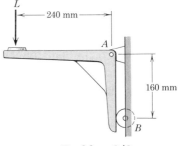

**Problem 3/8**

**3/9** The uniform bar *AB* weighs 100 lb and supports the 400-lb load at *A*. Calculate the tension *T* in the supporting cable and the magnitude $F_B$ of the force supported by the pin at *B*.

*Ans.* $T = 1559$ lb, $F_B = 1153$ lb

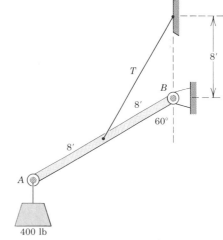

**3/10** Solve Prob. 3/9 graphically. (*Suggestion:* Combine the 100-lb weight and the 400-lb force into a single force.)

**Problem 3/9**

**3/11** A man pushes the 40-kg machine with mass center at *G* up an incline at a steady speed. Determine the required force magnitude *P* and the normal reaction forces at *A* and *B*. Neglect the small effects of friction.

*Ans.* $P = 112.1$ N, $N_B = 219$ N, $N_A = 207$ N

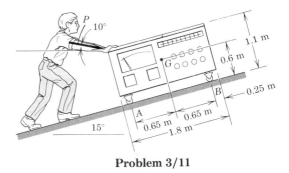

**Problem 3/11**

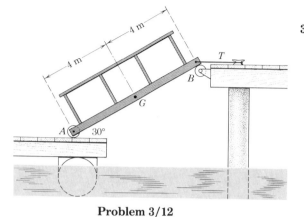

**Problem 3/12**

**3/12** To accommodate the rise and fall of the tide, a walkway from a pier to a float is supported by two rollers as shown. If the mass center of the 300-kg walkway is at $G$, calculate the tension $T$ in the horizontal cable which is attached to the cleat and find the force under the roller at $A$.

**Problem 3/13**

**3/13** During an engine test, a propeller thrust $T = 3000$ N is generated on the 1800-kg airplane with mass center at $G$. The main wheels at $B$ are locked and do not skid; the small tail wheel at $A$ has no brake. Compute the percent change $n$ in the normal forces at $A$ and $B$ as compared with their "engine-off" values.          *Ans.* $n_A = -32.6\%$, $n_B = 2.28\%$

**Problem 3/14**

**3/14** The 100-kg wheel rests on a rough surface and bears against the roller $A$ when the couple $M$ is applied. If $M = 60$ N·m and the wheel does not slip, compute the reaction on the roller $A$.

### *Representative problems*

**3/15** What fraction $n$ of the weight $W$ of a jet airplane is the net thrust (nozzle thrust $T$ minus air resistance $R$) in order for the airplane to climb with a constant speed at an angle $\theta$ with the horizontal?          *Ans.* $n = \sin \theta$

**Problem 3/15**

**3/16** Determine the force $P$ required to maintain the 200-kg engine in the position for which $\theta = 30°$. The diameter of the pulley at $B$ is negligible.

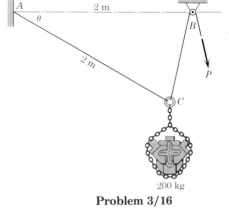

Problem 3/16

**3/17** The three cables are secured to a ring at $B$, and the turnbuckle at $C$ is tightened until it supports a tension of 1.6 kN. Calculate the moment $M$ produced by the tension in cable $AB$ about the base of the mast at $D$.          *Ans.* $M = 8$ kN·m

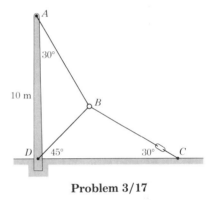

Problem 3/17

**3/18** Calculate the tensions in cables 1 and 2 of the pulley system in terms of the mass $m$ of the supported cylinder. Neglect the mass of the small pulleys. The angle $\alpha$ is 30°.

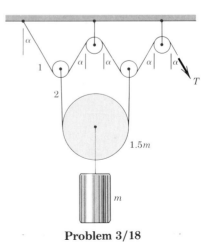

Problem 3/18

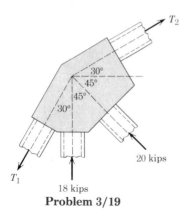

**Problem 3/19**

**3/19** The forces acting on four intersecting girders in a bridge truss are shown. Calculate $T_1$ and $T_2$. Solve in two ways, first, by the simultaneous solution of the two equilibrium equations in the two unknowns and, second, by choosing reference axes so as to solve for each unknown independently of the other. (1 kip equals 1000 lb.)

*Ans.* $T_1 = 69.8$ kips, $T_2 = 56.6$ kips

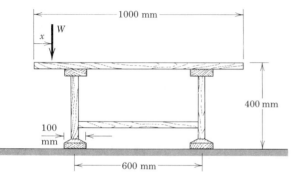

**Problem 3/20**

**3/20** The symmetrical wooden bench has a mass of 15 kg. Determine the minimum distance $x$ from the end of the benchtop at which a 90-kg person must place his weight when sitting in order to avoid tipping the bench.

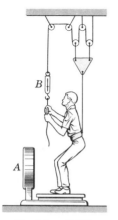

**Problem 3/21**

**3/21** A former student of mechanics wishes to weigh himself but has access only to a scale $A$ with capacity limited to 100 lb and a small 20-lb spring dynamometer $B$. With the rig shown he discovers that when he exerts a pull on the rope so that $B$ registers 19 lb, the scale $A$ reads 67 lb. What is his correct weight? *Ans.* $W = 162$ lb

**3/22** Determine the tension $T$ in the turnbuckle for the pulley–cable system in terms of the mass $m$ of the body which it supports. Neglect the mass of the pulleys and cable.

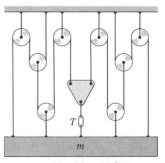

**Problem 3/22**

**3/23** The cable shown weighs 1.2 lb per foot of length and supports the pulley and lifting hook, which together weigh 12 lb. Determine the force $P$ necessary to maintain equilibrium. *Ans.* $P = 11.74$ lb

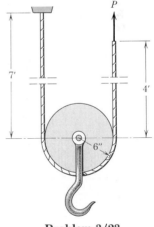

**Problem 3/23**

**3/24** The 200-N force produces a torque (moment) of 40 N·m about the axis of the bolt in order to tighten the hexagonal nut. Find the forces between the smooth jaws of the wrench and the nut if contact occurs at the corners $A$ and $B$ of the hexagon.

**Problem 3/24**

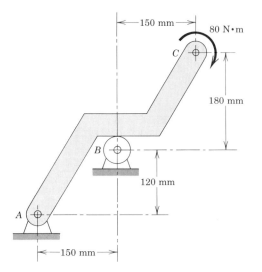

**Problem 3/25**

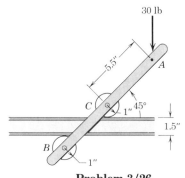

**Problem 3/26**

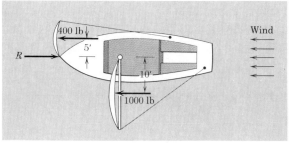

**Problem 3/27**

**3/25** Calculate the magnitude $F_A$ of the force supported by the pin at $A$ due to the action of the 80 N·m couple applied to the end $C$ of the bar.

*Ans.* $F_A = 533$ N

**3/26** The device shown is used to apply pressure when bonding laminate to each side of a countertop near an edge. If a 30-lb force is applied to the handle, determine the force which each roller exerts on its corresponding surface.

**3/27** In sailing at a constant speed with the wind, the sailboat is driven by a 1000-lb force against its mainsail and a 400-lb force against its staysail as shown. The total resistance due to fluid friction through the water is the force $R$. Determine the resultant of the lateral forces perpendicular to motion applied to the hull by the water.

*Ans.* $M = 8000$ lb-ft

**3/28** The uniform 15-m pole has a mass of 150 kg and is supported by its smooth ends against the vertical walls and by the tension $T$ in the vertical cable. Compute the reactions at $A$ and $B$.

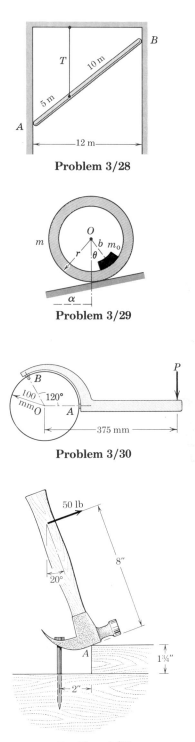

**Problem 3/28**

**3/29** A uniform ring of mass $m$ and radius $r$ carries an eccentric mass $m_0$ at a radius $b$ and is in an equilibrium position on the incline, which makes an angle $\alpha$ with the horizontal. If the contacting surfaces are rough enough to prevent slipping, write the expression for the angle $\theta$ which defines the equilibrium position.

$$Ans. \quad \theta = \sin^{-1}\left[\frac{r}{b}\left(1 + \frac{m}{m_0}\right)\sin \alpha\right]$$

**Problem 3/29**

**3/30** The hook wrench or pin spanner is used to turn shafts and collars. If a moment of 80 N·m is required to turn the 200-mm-diameter collar about its center $O$ under the action of the applied force $P$, determine the contact force $R$ on the smooth surface at $A$. Engagement of the pin at $B$ may be considered to occur at the periphery of the collar.

**Problem 3/30**

**3/31** A block placed under the head of the claw hammer as shown greatly facilitates the extraction of the nail. If the 50-lb pull on the handle is required to pull the nail, calculate the tension $T$ in the nail and the magnitude $A$ of the force exerted by the hammer head on the block. The contacting surfaces at $A$ are sufficiently rough to prevent slipping.

$$Ans. \quad T = 200 \text{ lb}, A = 188.8 \text{ lb}$$

**Problem 3/31**

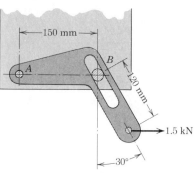

**Problem 3/32**

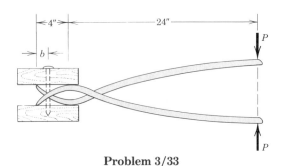

**Problem 3/33**

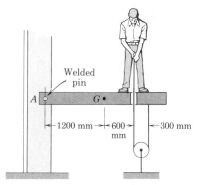

**Problem 3/34**

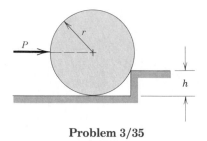

**Problem 3/35**

**3/32** Calculate the magnitude of the force supported by the pin at $A$ under the action of the 1.5-kN load applied to the bracket. Neglect friction in the slot.

**3/33** The two planks are connected by a large spike. If a force $P = 30$ lb is required on the handle of each crowbar to loosen the spike, calculate the corresponding tension $T$ in the spike. Also find the value of $b$ which will eliminate any tendency to bend the spike. State any assumptions which you make.

*Ans.* $T = 390$ lb, $b = 2.15$ in.

**3/34** The pin $A$, which connects the 200-kg steel beam with center of gravity at $G$ to the vertical column, is welded both to the beam and to the column. To test the weld, the 80-kg man loads the beam by exerting a 300-N force on the rope which passes through a hole in the beam as shown. Calculate the torque (couple) $M$ supported by the pin.

**3/35** Determine the force $P$ required to begin rolling the uniform cylinder of mass $m$ over the obstruction of height $h$.

*Ans.* $P = \dfrac{mg\sqrt{2rh - h^2}}{r - h}$

**3/36** The uniform beam $OA$ of mass $m$ is supported by the cable of length $c$. Prove from the geometry of the force polygon and the figure that the cable tension is $T = mgc/(2h)$.

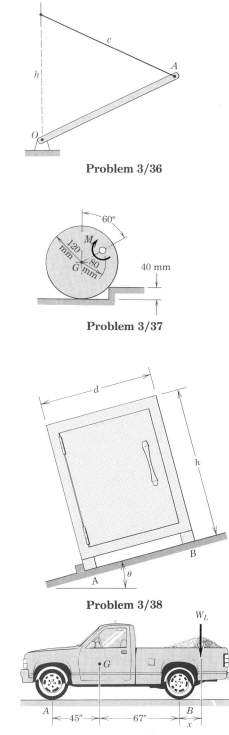

**Problem 3/36**

**3/37** Determine the couple $M$ applied to the 30-kg wheel with mass center at $G$ required to roll the wheel over the 40-mm curb. The contacting surfaces of the wheel and curb are sufficiently rough to prevent slipping.                                   *Ans.* $M = 26.3$ N·m

**Problem 3/37**

**3/38** A small obstruction near foot $A$ prevents sliding of the uniform cabinet of mass $m$. If the feet are small and friction is negligible, determine the normal forces at $A$ and $B$ as functions of the incline angle $\theta$. For what value of $\theta$ does the normal force at $B$ vanish, and what is the physical significance of this condition?

**3/39** The indicated location of the center of gravity of the 3600-lb pickup truck is for the unladen condition. If a load whose center of gravity is $x = 16$ in. behind the rear axle is added to the truck, determine the load weight $W_L$ for which the normal forces under the front and rear wheels are equal.

*Ans.* $W_L = 550$ lb

**Problem 3/38**

**3/40** The indicated location of the center of gravity of the 3600-lb pickup truck of Prob. 3/39 is for the unladen condition. A load $W_L = 800$ lb is placed in the truckbed and the weight distribution is measured to be 46 percent front and 54 percent rear. Determine the location $x$ of the center of gravity of the load.

**Problem 3/39**

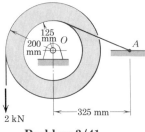

**Problem 3/41**

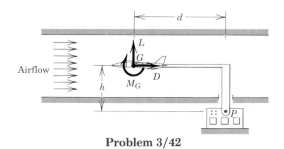

**Problem 3/42**

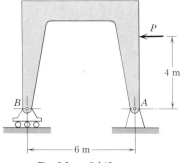

**Problem 3/43**

**3/41** Two light pulleys are fastened together and form an integral unit. They are prevented from turning about their bearing at $O$ by a cable wound securely around the smaller pulley and fastened to point $A$. Calculate the magnitude $R$ of the force supported by the bearing $O$ for the applied 2-kN load.

*Ans.* $R = 4.38$ kN

**3/42** The aircraft model is being tested in a wind tunnel. Its support bracket is connected to a force and moment balance, which is zeroed when there is no airflow. Under test conditions, the lift $L$, drag $D$, and pitching moment $M_G$ act as shown. The force balance records the lift, drag, and a moment $M_P$. Determine $M_G$ in terms of $L$, $D$, and $M_P$.

**3/43** The symmetrical frame has a mass of 1200 kg and is supported and loaded as shown. If the total force supported by the pin at $A$ is limited to 20 kN, determine the maximum permissible side load $P$.

*Ans.* $P = 18.9$ kN

**3/44** Because of long-term loading from the 900-N tension in the power line, the uniform light pole has developed a 5-deg lean. If the mass per unit length of the 9-m pole is 24 kg/m and the mass of the lamp fixture is negligible, determine the reactions at the ground support $A$.

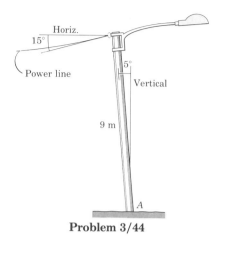

**Problem 3/44**

**3/45** The exercise machine consists of a lightweight cart which is mounted on small rollers so that it is free to move along the inclined ramp. Two cables are attached to the cart—one for each hand. If the hands are together so that the cables are parallel and if each cable lies essentially in a vertical plane, determine the force $P$ which each hand must exert on its cable in order to maintain an equilibrium position. The mass of the person is 70 kg, the ramp angle is 15°, and the angle $\beta$ is 18°. In addition, calculate the force $R$ which the ramp exerts on the cart.

*Ans.* $P = 45.5$ N, $R = 691$ N

**Problem 3/45**

**3/46** Pulley $A$ delivers a steady torque (moment) of 900 lb-in. to a pump through its shaft at $C$. The tension in the lower side of the belt is 150 lb. The driving motor $B$ weighs 200 lb and rotates clockwise. Determine the magnitude $R$ of the force on the supporting pin at $O$.

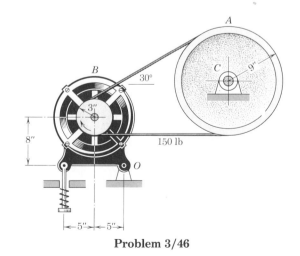

**Problem 3/46**

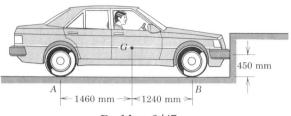

**Problem 3/47**

**3/47** The rear-wheel-drive car has a mass of 1400 kg with mass center at $G$. First calculate the normal forces under the front and rear wheel pairs when the car is at normal rest. Then repeat your calculations for the case shown where the front bumper is tested by causing the car to push, via its rear wheels, against the fixed barrier with a 2500-N force. Neglect friction at the bumper-barrier interface, but not at the tire-ground interface.

> *Ans.* Normally, $N_A = 6310$ N, $N_B = 7430$ N
>
> Under test, $N_A = 6720$ N, $N_B = 7010$ N

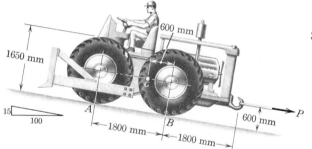

**Problem 3/48**

**3/48** The rubber-tired tractor shown has a mass of 13.5 Mg with center of mass at $G$ and is used for pushing or pulling heavy loads. Determine the load $P$ which the tractor can pull at a constant speed of 5 km/h up the 15-percent grade if the driving force exerted by the ground on each of its four wheels is 80 percent of the normal force under that wheel. Also find the total normal reaction $N_B$ under the rear pair of wheels at $B$.

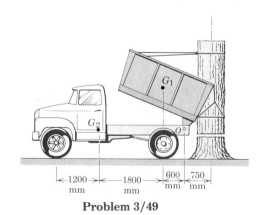

**Problem 3/49**

**3/49** The dump truck is used to lift a cut section of trunk from the stump of a large tree. The section of trunk is measured and calculated to have a mass of 600 kg. The body of the rotatable dump has a mass of 300 kg, and its center of mass $G_1$ is directly over the rear wheels in the position for lifting. Calculate the necessary torque $M$ applied to the dump through its shaft at $O$ to make the lift. Also calculate the corresponding forces under the front and rear pairs of wheels of the truck. The truck has a mass of 3100 kg exclusive of the dump, and its center of mass is at $G_2$.          *Ans.* $M = 2650$ N·m

> $R_F = 15.60$ kN, $R_R = 23.6$ kN

**3/50** The cargo door for an airplane of circular fuselage section consists of the uniform semicircular cowling $AB$ of mass $m$. Determine the compression $C$ in the horizontal strut at $B$ to hold the door open in the position shown. Also find an expression for the total force supported by the hinge at $A$. (Consult Table D/3 of Appendix D for the position of the centroid or mass center of the cowling.)

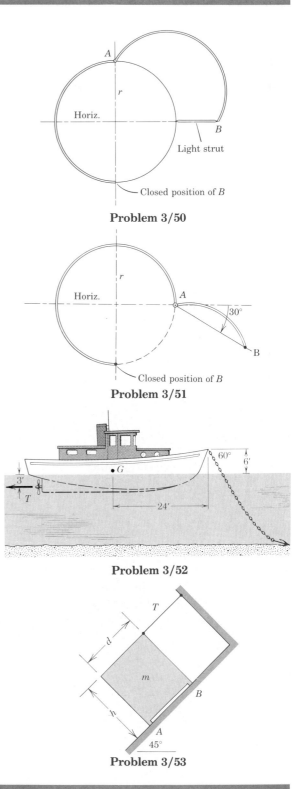

**Problem 3/50**

**3/51** The cargo door for an airplane of circular fuselage section consists of the uniform quarter-circular segment $AB$ of mass $m$. A detent in the hinge at $A$ holds the door open in the position shown. Determine the moment exerted by the hinge on the door.

                 *Ans.* $M_A = 0.709mgr$

**Problem 3/51**

**3/52** When setting the anchor so that it will dig into the sandy bottom, the engine of the 80,000-lb cruiser with center of gravity at $G$ is run in reverse to produce a horizontal thrust $T$ of 500 lb. If the anchor chain makes an angle of 60° with the horizontal, determine the forward shift $b$ of the center of buoyancy from its position when the boat is floating free. The center of buoyancy is the point through which the resultant of the buoyant forces passes.

**Problem 3/52**

**3/53** Determine the tension $T$ in the cable and the reactions at $A$ and $B$ for the cases (*a*) $h = 0.75d$ and (*b*) $h = 1.25d$. The object of mass $m$ is homogeneous and the feet are small. Friction is negligible. (*Caution:* Ensure that equilibrium is possible in each case.)

     *Ans.* (*a*) $T = \dfrac{\sqrt{2}}{2}\, mg,\ N_A = \dfrac{\sqrt{2}}{16}\, mg,\ N_B = \dfrac{7\sqrt{2}}{16}\, mg$

         (*b*) No equilibrium

**Problem 3/53**

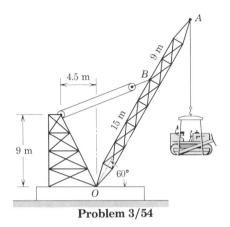

**Problem 3/54**

**3/54** The crane is hoisting a 4.20-Mg bulldozer. The mass center of the 2-Mg boom *OA* is at its midlength. Calculate the tension *T* in the cable where it attaches at *B* and the magnitude of the force supported by the boom at its hinge *O* for equilibrium in the 60° position. Neglect the width of the boom.

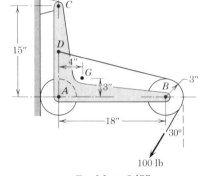

**Problem 3/55**

**3/55** The bracket and pulley assembly weighs 85 lb with combined center of gravity at *G*. Calculate the magnitude of the force supported by the pin at *C* when a tension of 100 lb is applied in the vertical plane to the cable. (*Suggestion:* Make use of the principle of replacement of a force by a force–couple system to simplify your calculation. Why is the result independent of the location of the attachment point *D* of the cable on the bracket?)     *Ans. C* = 226 lb

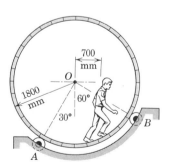

**Problem 3/56**

**3/56** The uniform 400-kg drum is mounted on a line of rollers at *A* and a line of rollers at *B*. An 80-kg man moves slowly a distance of 700 mm from the vertical centerline before the drum begins to rotate. All rollers are perfectly free to rotate, except one of them at *B* which must overcome appreciable friction in its bearing. Calculate the friction force *F* exerted by that one roller tangent to the drum and find the magnitude *R* of the force exerted by all rollers at *A* on the drum for this condition.

**3/57** The 20-lb dolly, with center of gravity at $G_1$, is carrying a 100-lb waste container, with center of gravity at $G_2$, down the 10-percent incline. Determine the force $P$ and the angle $\theta$ required of the operator to maintain a constant speed.

*Ans.* $P = 31.2$ lb, $\theta = 61.8°$

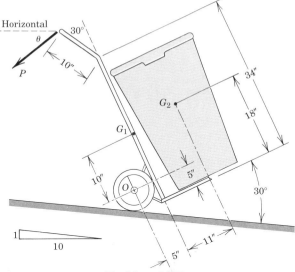

**Problem 3/57**

**3/58** The figure shows a series of rectangular plates and their constraints, all confined to the plane of representation. The plates could be subjected to various known loads applied in the plane of the plates. Identify the plates that belong to each of the following categories:

(a) Complete fixity with minimum number of adequate constraints

(b) Partial fixity with inadequate constraints

(c) Complete fixity with redundant constraints

(d) Partial fixity with redundant constraints

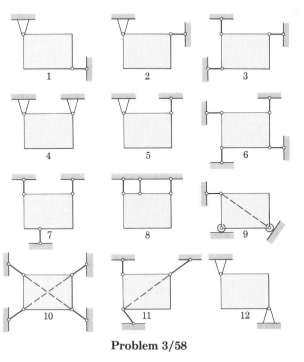

**Problem 3/58**

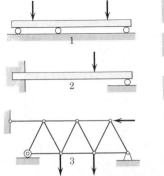

Dimensions in meters
**Problem 3/59**

**3/59** If the load on the crane is increased by 4 Mg, compute the corresponding increment $\Delta A$ in the force supported by the pin at $A$.        *Ans.* $\Delta A = 56.6$ kN

**3/60** Each of the structures is statically indeterminate. Describe at least one modification in the supports for each case which would make the structure statically determinate.

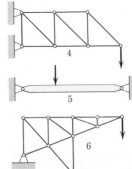

**Problem 3/60**

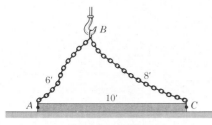

**Problem 3/61**

**3/61** The 10-ft uniform steel beam weighing 800 lb is to be lifted from the ring at $B$ with the two chains, $AB$ of length 6 ft and $CB$ of length 8 ft. Determine the tension $T$ in chain $AB$ when the beam is clear of the platform.        *Ans.* $T = 480$ lb

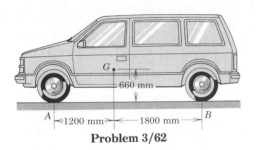

**Problem 3/62**

**3/62** Calculate the normal forces associated with the front and rear wheel pairs of the 1600-kg front-wheel-drive van. Then repeat the calculations when the van (*a*) climbs a 10-percent grade and (*b*) descends a 10-percent grade, both at constant speed. Compute the percent changes $n_A$ and $n_B$ in the normal forces compared with the nominal values. Be sure to recognize that propulsive and braking forces are present for cases (*a*) and (*b*).

▶**3/63** A special jig for turning large concrete pipe sections (shown dotted) consists of an 80-Mg sector mounted on a line of rollers at *A* and a line of rollers at *B*. One of the rollers at *B* is a gear which meshes with a ring of gear teeth on the sector so as to turn the sector about its geometric center *O*. When $\alpha = 0$, a counterclockwise torque of 2460 N·m must be applied to the gear at *B* to keep the assembly from rotating. When $\alpha = 30°$, a clockwise torque of 4680 N·m is required to prevent rotation. Locate the mass center *G* of the jig by calculating $\bar{r}$ and $\theta$.

Ans. $\bar{r} = 367$ mm, $\theta = 79.8°$

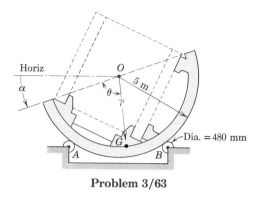

**Problem 3/63**

▶**3/64** The base plate of a special industrial caster is mounted to the carriage, with soft elastic washers on either side of the plate in order to absorb shock during operation. Bolts *A* and *B* are initially tightened to a tension of 4 kN with no load on the caster. Determine the force in each bolt when the caster supports a load of 3 kN. The forces acting at each bolt location may be modeled by replacing the washers by identical stiff springs as shown in the separate view. Consider the bolts and base plate to be perfectly rigid. (*Caution:* Draw your free-body diagrams of each part carefully.)

Ans. $F_A = 1.964$ kN tension
$F_B = 4.54$ kN tension

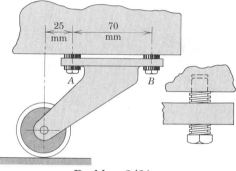

**Problem 3/64**

## SECTION B.  EQUILIBRIUM IN THREE DIMENSIONS

### 3/4  *EQUILIBRIUM CONDITIONS*

We now extend our principles and methods developed for two-dimensional equilibrium to the case of three-dimensional equilibrium. In Art. 3/1 the general conditions for the equilibrium of a body were stated in Eqs. 3/1, which require that the resultant force and resultant couple on a body in equilibrium must be zero. These two vector equations of equilibrium and their scalar components may be written as

$$
\begin{array}{lll}
\Sigma \mathbf{F} = \mathbf{0} & \text{or} & \left\{ \begin{array}{l} \Sigma F_x = 0 \\ \Sigma F_y = 0 \\ \Sigma F_z = 0 \end{array} \right. \\[4em]
\Sigma \mathbf{M} = \mathbf{0} & \text{or} & \left\{ \begin{array}{l} \Sigma M_x = 0 \\ \Sigma M_y = 0 \\ \Sigma M_z = 0 \end{array} \right.
\end{array}
\tag{3/3}
$$

The first three scalar equations state that there is no resultant force acting on a body in equilibrium in any of the three coordinate directions. The second three scalar equations express the further equilibrium requirement that there is no resultant moment acting on the body about any of the coordinate axes or about axes parallel to the coordinate axes. These six equations are both necessary and sufficient conditions for complete equilibrium. The reference axes may be chosen arbitrarily as a matter of convenience, the only restriction being that a right-handed coordinate system must be used with vector notation.

The six scalar relationships of Eqs. 3/3 are independent conditions since any of them may be valid without the others. For example, for a car which accelerates on a straight and level road in the $x$-direction, Newton's second law tells us that the resultant force on the car equals its mass times its acceleration. Hence, $\Sigma F_x \neq 0$ but the remaining two force-equilibrium equations are satisfied since all other acceleration components are zero. Similarly, if the flywheel of the engine of the accelerating car is rotating with increasing angular speed about the $x$-axis, it is not in rotational equilibrium about this axis. Hence, for the flywheel alone, $\Sigma M_x \neq 0$ along with $\Sigma F_x \neq 0$, but the remaining four equilibrium equations for the flywheel would be satisfied for its mass-center axes.

In applying the vector form of Eqs. 3/3, we first express each of the forces in terms of the coordinate unit vectors $\mathbf{i}$, $\mathbf{j}$, and $\mathbf{k}$. For the first equation, $\Sigma \mathbf{F} = \mathbf{0}$, the vector sum will be zero only if the coefficients of $\mathbf{i}$, $\mathbf{j}$, and $\mathbf{k}$ in the expression are, respectively, zero. These three sums when each is set equal to zero yield precisely the three scalar equations of equilibrium, $\Sigma F_x = 0$, $\Sigma F_y = 0$, and $\Sigma F_z = 0$.

For the second equation, $\Sigma \mathbf{M} = \mathbf{0}$, where the moment sum may be taken about any convenient point $O$, we express the moment of each force as the cross product $\mathbf{r} \times \mathbf{F}$, where $\mathbf{r}$ is the position vector from $O$ to any point on the line of action of the force $\mathbf{F}$. Thus, $\Sigma \mathbf{M} = \Sigma(\mathbf{r} \times \mathbf{F}) = \mathbf{0}$. The coefficients of $\mathbf{i}$, $\mathbf{j}$, and $\mathbf{k}$ in the resulting moment equation when set equal to zero, respectively, produce the three scalar moment equations $\Sigma M_x = 0$, $\Sigma M_y = 0$, and $\Sigma M_z = 0$.

*(a) Free-body diagrams.*   The summations in Eqs. 3/3 include the effects of *all* forces on the body under consideration. We learned in the previous article that the free-body diagram is the only reliable method for disclosing all forces and moments which should be included in our equilibrium equations. In three dimensions the free-body diagram serves the same essential purpose as it does in two dimensions and should *always* be drawn. We have our choice either of drawing a pictorial view of the isolated body with all external forces represented or of drawing the orthogonal projections of the free-body diagram. Both representations will be illustrated in the sample problems at the end of this article.

The correct representation of forces on the free-body diagram requires a knowledge of the characteristics of contacting surfaces. These characteristics were set forth in Fig. 3/1 for two-dimensional problems, and their extension to three-dimensional problems is represented in Fig. 3/8 for the most common situations of force transmission. The representations in both Figs. 3/1 and 3/8 will be used in three-dimensional analysis.

Inasmuch as the essential purpose of the free-body diagram is to develop a reliable picture of the physical action of all forces (including couples if any) acting on a body, it is helpful to represent the forces in their correct physical sense whenever possible. In this way the free-body diagram becomes a closer model to the actual physical problem than would be the case if the forces were arbitrarily assigned or always assigned in the same mathematical sense as that of the assigned coordinate axis. For example, in part 4 of Fig. 3/8, the correct sense of the unknowns $R_x$ and $R_y$ may be known or perceived to be in the sense opposite to those of the assigned coordinate axes. Similar conditions apply to the sense of couple vectors, parts 5 and 6, where their sense by the right-hand rule may be assigned opposite to that of the respective coordinate direction. By this time the student should recognize that a negative answer for an unknown force or couple vector merely indicates that its physical action is in the sense opposite to that assigned on the free-body diagram. There are frequent occasions, of course, when the correct physical sense is not known initially, so that an arbitrary assignment on the free-body diagram becomes necessary.

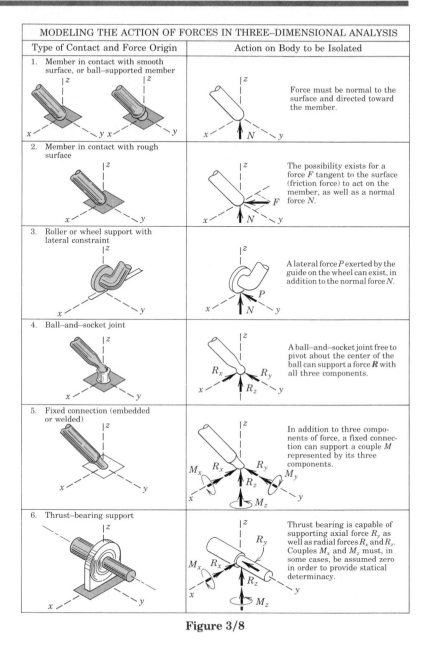

**Figure 3/8**

*(b) Categories of equilibrium.* Application of Eqs. 3/3 falls into four categories which we identify with the aid of Fig. 3/9.

*Case 1,* equilibrium of forces all concurrent at a point $O$, requires all three force equations but no moment equations since the moment of the forces about any axis through $O$ is zero.

| CATEGORIES OF EQUILIBRIUM IN THREE DIMENSIONS | | |
|---|---|---|
| Force System | Free–Body Diagram | Independent Equations |
| 1. Concurrent at a point |  | $\Sigma F_x = 0$<br>$\Sigma F_y = 0$<br>$\Sigma F_z = 0$ |
| 2. Concurrent with a line | | $\Sigma F_x = 0 \qquad \Sigma M_y = 0$<br>$\Sigma F_y = 0 \qquad \Sigma M_z = 0$<br>$\Sigma F_z = 0$ |
| 3. Parallel | | $\Sigma F_x = 0 \qquad \Sigma M_y = 0$<br>$\qquad\qquad\quad \Sigma M_z = 0$ |
| 4. General | | $\Sigma F_x = 0 \qquad \Sigma M_x = 0$<br>$\Sigma F_y = 0 \qquad \Sigma M_y = 0$<br>$\Sigma F_z = 0 \qquad \Sigma M_z = 0$ |

**Figure 3/9**

*Case 2,* equilibrium of forces that are concurrent with a line, requires all equations except the moment equation about that line, which is automatically satisfied.

*Case 3,* equilibrium of parallel forces, requires only one force equation in the direction of the forces (*x*-direction as shown) and two moment equations about axes (*y* and *z*) that are normal to the direction of the forces.

*Case 4,* equilibrium of a general system of forces, requires all three force equations and all three moment equations.

The observations contained in these statements are generally quite evident when a given problem is being solved.

*(c) Constraints and statical determinacy.* The six scalar relations of Eqs. 3/3, although necessary and sufficient conditions to
establish equilibrium, do not necessarily provide all of the information required to calculate the unknown forces acting in a three-
dimensional equilibrium situation. Again, as we found with two
dimensions, the question of adequacy of information lies in the
characteristics of the constraints provided by the supports. An
analytical criterion for determining the adequacy of constraints is
available but is beyond the scope of this treatment.* In Fig. 3/10,
however, we cite four examples of constraint conditions to alert the
reader to the problem. In the *a*-part of the figure is shown a rigid
body whose corner point *A* is completely fixed by the links 1, 2, and
3. Links 4, 5, and 6 prevent rotations about the axes of links 1, 2,
and 3, respectively, so that the body is completely fixed and the

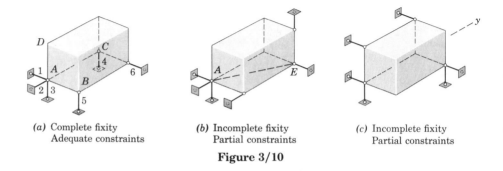

(*a*) Complete fixity          (*b*) Incomplete fixity          (*c*) Incomplete fixity
    Adequate constraints            Partial constraints              Partial constraints

**Figure 3/10**

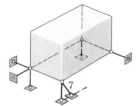

(*d*) Excessive fixity
    Redundant constraint

constraints are said to be adequate. The *b*-part of Fig. 3/10 shows
the same number of constraints, but we see that they provide no
resistance to a moment which might be applied about axis *AE*. Here
the body is incompletely fixed and only partially constrained. Similarly, in Fig. 3/10*c* the constraints provide no resistance to an
unbalanced force in the *y*-direction, so here also is a case of incomplete
fixity with partial constraints. In Fig. 3/10*d*, if a seventh constraining
link were imposed on a system of six constraints placed properly for
complete fixity, more supports would be provided than would be
necessary to establish the equilibrium position, and link 7 would be
*redundant*. The body would then be *statically indeterminate* with
such a seventh link in place. With only a few exceptions, the
supporting constraints for rigid bodies in equilibrium in this book
are adequate and the bodies are statically determinate.

*See the senior author's *Statics, 2nd Edition SI Version 1975,* Art. 16.

## Sample Problem 3/5

The uniform 7-m steel shaft has a mass of 200 kg and is supported by a ball-and-socket joint at $A$ in the horizontal floor. The ball end $B$ rests against the smooth vertical walls as shown. Compute the forces exerted by the walls and the floor on the ends of the shaft.

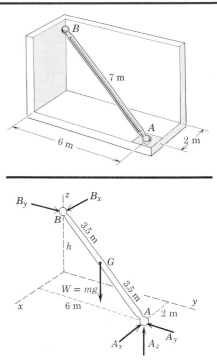

**Solution.** The free-body diagram of the shaft is first drawn where the contact forces acting on the shaft at $B$ are shown normal to the wall surfaces. In addition to the weight $W = mg = 200(9.81) = 1962$ N, the force exerted by the floor on the ball joint at $A$ is represented by its $x$-, $y$-, and $z$-components. These components are shown in their correct physical sense, as should be evident from the requirement that $A$ is held in place. The vertical position of $B$ is found from $7 = \sqrt{2^2 + 6^2 + h^2}$, $h = 3$ m. Right-handed coordinate axes are conveniently assigned as shown.

**Vector solution.** We will use $A$ as a moment center to eliminate reference to the forces at $A$. The position vectors needed to compute the moments about $A$ are

$$\mathbf{r}_{AG} = -1\mathbf{i} - 3\mathbf{j} + 1.5\mathbf{k} \text{ m} \qquad \text{and} \qquad \mathbf{r}_{AB} = -2\mathbf{i} - 6\mathbf{j} + 3\mathbf{k} \text{ m}$$

where the mass center $G$ is located halfway between $A$ and $B$.

The vector moment equation gives

$$[\Sigma \mathbf{M}_A = 0] \qquad \mathbf{r}_{AB} \times (\mathbf{B}_x + \mathbf{B}_y) + \mathbf{r}_{AG} \times \mathbf{W} = 0$$

$$(-2\mathbf{i} - 6\mathbf{j} + 3\mathbf{k}) \times (B_x\mathbf{i} + B_y\mathbf{j}) + (-\mathbf{i} - 3\mathbf{j} + 1.5\mathbf{k}) \times (-1962\mathbf{k}) = 0$$

$$\begin{vmatrix} \mathbf{i} & \mathbf{j} & \mathbf{k} \\ -2 & -6 & 3 \\ B_x & B_y & 0 \end{vmatrix} + \begin{vmatrix} \mathbf{i} & \mathbf{j} & \mathbf{k} \\ -1 & -3 & 1.5 \\ 0 & 0 & -1962 \end{vmatrix} = 0$$

$$(-3B_y + 5886)\mathbf{i} + (3B_x - 1962)\mathbf{j} + (-2B_y + 6B_x)\mathbf{k} = 0$$

Equating the coefficients of $\mathbf{i}$, $\mathbf{j}$, and $\mathbf{k}$ to zero and solving give

$$B_x = 654 \text{ N} \qquad \text{and} \qquad B_y = 1962 \text{ N} \qquad Ans.$$

The forces at $A$ are easily determined by

$$[\Sigma \mathbf{F} = 0] \quad (654 - A_x)\mathbf{i} + (1962 - A_y)\mathbf{j} + (-1962 + A_z)\mathbf{k} = 0$$

and $\qquad A_x = 654$ N $\qquad A_y = 1962$ N $\qquad A_z = 1962$ N

Finally $\qquad A = \sqrt{A_x{}^2 + A_y{}^2 + A_z{}^2}$

$$= \sqrt{(654)^2 + (1962)^2 + (1962)^2} = 2850 \text{ N} \qquad Ans.$$

**Scalar solution.** Evaluating the scalar moment equations about axes through $A$ parallel, respectively, to the $x$- and $y$-axes, gives

$$[\Sigma M_{A_x} = 0] \qquad 1962(3) - 3B_y = 0 \qquad B_y = 1962 \text{ N}$$
$$[\Sigma M_{A_y} = 0] \qquad -1962(1) + 3B_x = 0 \qquad B_x = 654 \text{ N}$$

The force equations give, simply,

$$[\Sigma F_x = 0] \qquad -A_x + 654 = 0 \qquad A_x = 654 \text{ N}$$
$$[\Sigma F_y = 0] \qquad -A_y + 1962 = 0 \qquad A_y = 1962 \text{ N}$$
$$[\Sigma F_z = 0] \qquad A_z - 1962 = 0 \qquad A_z = 1962 \text{ N}$$

① We could, of course, assign all of the unknown components of force in the positive mathematical sense, in which case $A_x$ and $A_y$ would turn out to be negative upon computation. The free-body diagram describes the physical situation, so it is generally preferable to show the forces in their correct physical senses wherever possible.

② Note that the third equation $-2B_y + 6B_x = 0$ merely checks the results of the first two equations. This result could be anticipated from the fact that an equilibrium system of forces concurrent with a line requires only two moment equations (Case 2 under *Categories of equilibrium*).

③ We observe that a moment sum about an axis through $A$ parallel to the $z$-axis merely gives us $6B_x - 2B_y = 0$, which serves only as a check as noted previously. Alternatively we could have first obtained $A_z$ from $\Sigma F_z = 0$ and then taken our moment equations about axes through $B$ to obtain $A_x$ and $A_y$.

# Sample Problem 3/6

A 200-N force is applied to the handle of the hoist in the direction shown. The bearing $A$ supports the thrust (force in the direction of the shaft axis), while bearing $B$ supports only radial load (load normal to the shaft axis). Determine the mass $m$ which can be supported and the total radial force exerted on the shaft by each bearing. Assume neither bearing to be capable of supporting a moment about a line normal to the shaft axis.

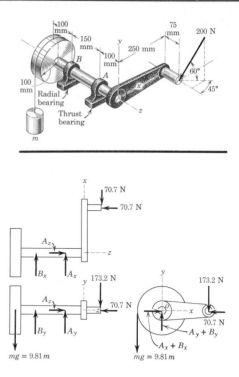

**Solution.** The system is clearly three-dimensional with no lines or planes of symmetry, and therefore the problem must be analyzed as a general space system of forces. A scalar solution is used here to illustrate this approach, although a solution using vector notation would also be satisfactory. The free-body diagram of the shaft, lever, and drum considered a single body could be shown by a space view if desired, but ① is represented here by its three orthogonal projections.

The 200-N force is resolved into its three components, and each of the three views shows two of these components. The correct directions of $A_x$ and $B_x$ may be seen by inspection by observing that the line of action of the resultant of the two 70.7-N forces passes between $A$ and $B$. The correct sense of the forces $A_y$ and $B_y$ cannot be determined until the magnitudes of the moments are obtained, so they are arbitrarily assigned. The $x$-$y$ projection of the bearing forces is shown in terms of the sums of the unknown $x$- and $y$-components. The addition of $A_z$ and the weight $W = mg$ completes the free-body diagrams. It should be noted that the three views represent three two-dimensional problems related by the corresponding components of the forces.

② From the $x$-$y$ projection

$[\Sigma M_O = 0]$    $100(9.81m) - 250(173.2) = 0$    $m = 44.1$ kg    *Ans.*

From the $x$-$z$ projection

$[\Sigma M_A = 0]$    $150B_x + 175(70.7) - 250(70.7) = 0$    $B_x = 35.4$ N

$[\Sigma F_x = 0]$    $A_x + 35.4 - 70.7 = 0$    $A_x = 35.3$ N

③ The $y$-$z$ view gives

$[\Sigma M_A = 0]$   $150B_y + 175(173.2) - 250(44.1)(9.81) = 0$   $B_y = 520$ N

$[\Sigma F_y = 0]$   $A_y + 520 - 173.2 - (44.1)(9.81) = 0$   $A_y = 86.8$ N

$[\Sigma F_z = 0]$   $A_z = 70.7$ N

The total radial forces on the bearings become

$[A_r = \sqrt{A_x^2 + A_y^2}]$    $A_r = \sqrt{(35.3)^2 + (86.8)^2} = 93.5$ N    *Ans.*

$[B = \sqrt{B_x^2 + B_y^2}]$    $B = \sqrt{(35.4)^2 + (520)^2} = 521$ N    *Ans.*

① If the standard three views of orthographic projection are not entirely familiar, then review and practice them. Visualize the three views as the images of the body projected onto the front, top, and end surfaces of a clear plastic box placed over and aligned with the body.

② We could have started with the $x$-$z$ projection rather than with the $x$-$y$ projection.

③ The $y$-$z$ view could have followed immediately after the $x$-$y$ view since the determination of $A_y$ and $B_y$ may be made after $m$ is found.

④ Without the assumption of zero moment supported by each bearing about a line normal to the shaft axis, the problem would be statically indeterminate.

# Sample Problem 3/7

The welded tubular frame is secured to the horizontal $x$-$y$ plane by a ball-and-socket joint at $A$ and receives support from the loose-fitting ring at $B$. Under the action of the 2-kN load, rotation about a line from $A$ to $B$ is prevented by the cable $CD$, and the frame is stable in the position shown. Neglect the weight of the frame compared with the applied load and determine the tension $T$ in the cable, the reaction at the ring, and the reaction components at $A$.

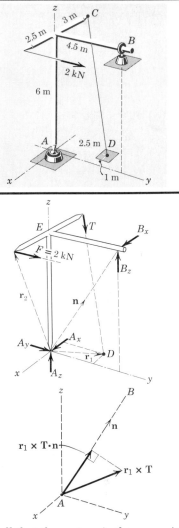

**Solution.** The system is clearly three-dimensional with no lines or planes of symmetry, and therefore the problem must be analyzed as a general space system of forces. The free-body diagram is drawn, where the ring reaction is shown in terms of its two components. All unknowns except $\mathbf{T}$ may be eliminated by a moment sum about the line $AB$.

The direction of $AB$ is specified by the unit vector $\mathbf{n} = \dfrac{1}{\sqrt{6^2 + 4.5^2}}$
$(4.5\mathbf{j} + 6\mathbf{k}) = \frac{1}{5}(3\mathbf{j} + 4\mathbf{k})$. The moment of $\mathbf{T}$ about $AB$ is the component in the direction of $AB$ of the vector moment about the point $A$ and equals $\mathbf{r}_1 \times \mathbf{T} \cdot \mathbf{n}$. Similarly the moment of the applied load $F$ about $AB$ is $\mathbf{r}_2 \times \mathbf{F} \cdot \mathbf{n}$. With $\overline{CD} = \sqrt{46.25}$ m, the vector expressions for $\mathbf{T}$, $\mathbf{F}$, $\mathbf{r}_1$, and $\mathbf{r}_2$ are

$$\mathbf{T} = \frac{T}{\sqrt{46.25}}\,(2\mathbf{i} + 2.5\mathbf{j} - 6\mathbf{k}) \qquad \mathbf{F} = 2\mathbf{j} \text{ kN}$$

$$\mathbf{r}_1 = -\mathbf{i} + 2.5\mathbf{j} \text{ m} \qquad \mathbf{r}_2 = 2.5\mathbf{i} + 6\mathbf{k} \text{ m}$$

The moment equation now becomes

$$[\Sigma M_{AB} = 0] \quad (-\mathbf{i} + 2.5\mathbf{j}) \times \frac{T}{\sqrt{46.25}}\,(2\mathbf{i} + 2.5\mathbf{j} - 6\mathbf{k}) \cdot \tfrac{1}{5}(3\mathbf{j} + 4\mathbf{k})$$

$$+ (2.5\mathbf{i} + 6\mathbf{k}) \times (2\mathbf{j}) \cdot \tfrac{1}{5}(3\mathbf{j} + 4\mathbf{k}) = 0$$

Completion of the vector operations gives

$$-\frac{48T}{\sqrt{46.25}} + 20 = 0 \qquad T = 2.83 \text{ kN} \qquad\qquad Ans.$$

and the components of $T$ become

$$T_x = 0.833 \text{ kN} \qquad T_y = 1.042 \text{ kN} \qquad T_z = -2.50 \text{ kN}$$

We may find the remaining unknowns by moment and force summations as follows:

$[\Sigma M_z = 0] \quad 2(2.5) - 4.5B_x - 1.042(3) = 0 \quad B_x = 0.417 \text{ kN} \qquad Ans.$

$[\Sigma M_x = 0] \quad 4.5B_z - 2(6) - 1.042(6) = 0 \quad B_z = 4.06 \text{ kN} \qquad Ans.$

$[\Sigma F_x = 0] \quad A_x + 0.417 + 0.833 = 0 \quad A_x = -1.250 \text{ kN} \qquad Ans.$

$[\Sigma F_y = 0] \quad A_y + 2 + 1.042 = 0 \quad A_y = -3.04 \text{ kN} \qquad Ans.$

$[\Sigma F_z = 0] \quad A_z + 4.06 - 2.50 = 0 \quad A_z = -1.56 = \text{kN} \qquad Ans.$

① Recall that the vector $\mathbf{r}$ in the expression $\mathbf{r} \times \mathbf{F}$ for the moment of a force is a vector from the moment center to *any* point on the line of action of the force. Instead of $\mathbf{r}_1$, an equally simple choice would be the vector $\overrightarrow{AC}$.

② The advantage of using vector notation in this problem is the freedom to take moments directly about any axis. In this problem this freedom permits the choice of an axis that eliminates five of the unknowns.

③ The negative signs associated with the $A$-components indicate that they are in the opposite direction to those shown on the free-body diagram.

## PROBLEMS

### Introductory problems

**3/65** A scale is placed under each of the three supports $A$, $B$, and $C$ of the nonhomogeneous slab, and the readings are 150 lb, 120 lb, and 170 lb, respectively. Determine the weight of the slab and the $x$- and $y$-coordinates of its center of mass.

*Ans.* $W = 440$ lb, $x = 1.873$ ft, $y = 0.545$ ft

**3/66** Two steel I-beams, each with a mass of 100 kg, are welded together at right angles and lifted by vertical cables so that the beams remain in a horizontal plane. Compute the tension in each of the cables $A$, $B$, and $C$.

**3/67** Determine the tensions in the three cables which support the uniform 180-lb plate.

*Ans.* $T_A = T_B = 47.9$ lb, $T_C = 90$ lb

**3/68** The turnbuckle at $F$ is tightened until the tension in cable $AE$ is 5 kN. Determine the tensions in cables $AB$, $AC$, and $AD$.

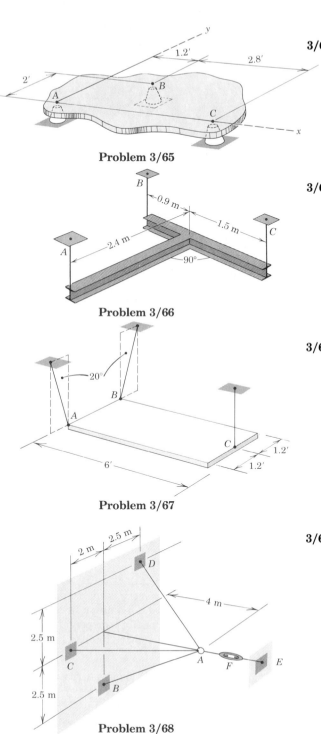

Problem 3/65

Problem 3/66

Problem 3/67

Problem 3/68

**3/69** Determine the tensions in wires $AB$, $AC$, and $AD$.

*Ans.* $T_{AB} = 0.501mg$, $T_{AC} = 0.290mg$
$T_{AD} = 0.412mg$

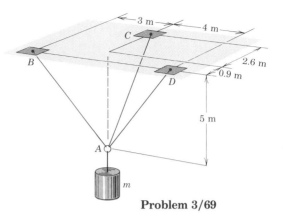

**Problem 3/69**

**3/70** The mass per unit length of the uniform beam is 60 kg/m. Determine the reactions at the built-in support at $A$ during application of the 5-kN force at $B$.

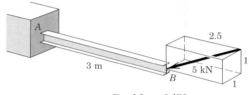

**Problem 3/70**

**3/71** The light right-angle boom which supports the 400-kg cylinder is supported by three cables and a ball-and-socket joint at $O$ attached to the vertical $x$-$y$ surface. Determine the reactions at $O$ and the cable tensions.

*Ans.* $O_x = 1962$ N, $O_y = 0$, $O_z = 6540$ N
$T_{AC} = 4810$ N, $T_{BD} = 2770$ N, $T_{BE} = 654$ N

**Problem 3/71**

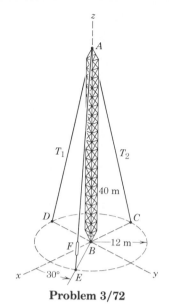

**Problem 3/72**

3/72 The radio tower has a mass of 1.2 Mg and is supported in the vertical position by the three cables from its top $A$ to the ground positions shown. A tensioning device at $F$ permits adjustment of the cable tension. If a tension of 4.8 kN is developed in cable $AE$, determine the tension in each of the other two cables.

**Problem 3/73**

3/73 Determine the reactions of the supports $A$, $B$, and $C$ on the rigid prismatic truss loaded as shown. The cross section of the truss is an equilateral triangle of side length 4 m.

$Ans.$ $A_x = -3.90$ kN, $A_y = 3$ kN
$C_x = 28.5$ kN
$B_x = -19.65$ kN, $B_y = 6.06$ kN
$B_z = -3.5$ kN

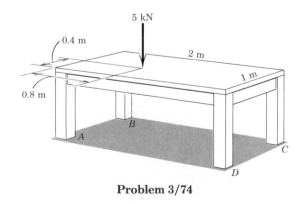

**Problem 3/74**

*Representative problems*

3/74 A 60-kg heavy-duty table is tested by application of a 5-kN vertical force as shown. If instruments reveal the normal reaction force under leg $A$ to be 2000 N, determine the normal forces at $B$, $C$, and $D$. Assume that the normal forces act at the outer corners of the legs.

**3/75** The mass center of the 30-kg door is in the center of the panel. If the weight of the door is supported entirely by the lower hinge $A$, calculate the magnitude of the total force supported by the hinge at $B$.          *Ans.* $B = 190.2$ N

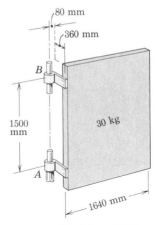

**Problem 3/75**

**3/76** The young tree, originally bent, has been brought into the vertical position by adjusting the three guy-wire tensions to $AB = 0$, $AC = 10$ lb, and $AD = 15$ lb. Determine the force and moment reactions at the trunk base point $O$. Neglect the weight of the tree.

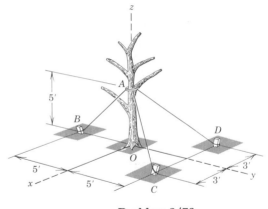

**Problem 3/76**

**3/77** Because of a combination of soil support conditions and the tension in the single power cable, the utility pole has developed the indicated 5° lean. The 9-m uniform pole has a mass per unit length of 25 kg/m, and the tension in the power cable is 900 N. Determine the reactions at the base $O$. Note that the power cable lies in a vertical plane parallel to the $x$-$z$ plane.

   *Ans.* $O_x = 869$ N, $O_y = 0$, $O_z = 2440$ N
   $M_x = -524$ N·m, $M_y = 8700$ N·m
   $M_z = 341$ N·m

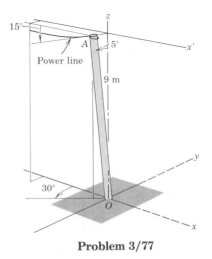

**Problem 3/77**

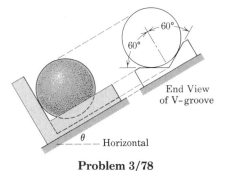

End View
of V–groove

$\theta$ —— Horizontal

**Problem 3/78**

**3/78** The smooth homogeneous sphere rests in the 120° groove and bears against the end plate which is normal to the direction of the groove. Determine the angle $\theta$, measured from the horizontal, for which the reaction on each side of the groove equals the force supported by the end plate.

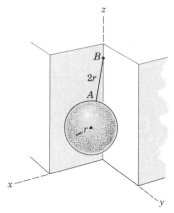

**Problem 3/79**

**3/79** A smooth homogeneous sphere of mass $m$ and radius $r$ is suspended by a wire $AB$ of length $2r$ from point $B$ on the line of intersection of the two vertical walls at right angles to one another. Determine the reaction $R$ of each wall against the sphere.
*Ans.* $R = mg/\sqrt{7}$

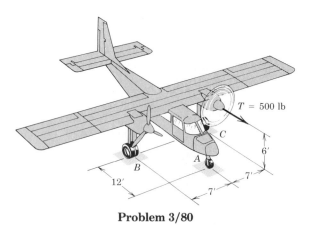

$T = 500$ lb

**Problem 3/80**

**3/80** During a test, the left engine of the twin-engine airplane is revved up and a 500-lb thrust is generated. The main wheels at $B$ and $C$ are braked in order to prevent motion. Determine the change (compared with the nominal values with both engines off) in the normal reaction forces at $A$, $B$, and $C$.

**3/81** The square steel plate has a mass of 1800 kg with mass center at its center $G$. Calculate the tension in each of the three cables with which the plate is lifted while remaining horizontal.

$$Ans. \ T_A = T_B = 5.41 \ \text{kN}, \ T_C = 9.87 \ \text{kN}$$

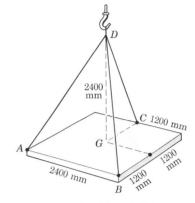

**Problem 3/81**

**3/82** The uniform panel door weighs 60 lb and is prevented from opening by the strut $C$, which is a light two-force member whose upper end is secured under the door knob and whose lower end is attached to a rubber cup which does not slip on the floor. Of the door hinges $A$ and $B$, only $B$ can support force in the vertical $z$-direction. Calculate the compression $C$ in the strut and the horizontal components of the forces supported by hinges $A$ and $B$ when a horizontal force $P = 50$ lb is applied normal to the plane of the door as shown.

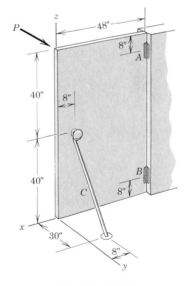

**Problem 3/82**

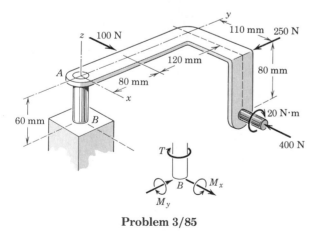

Dimensions in Millimeters

**Problem 3/83**

**Problem 3/84**

**Problem 3/85**

**3/83** One of the three landing pads for the Mars Viking lander is shown in the figure with its approximate dimensions. The mass of the lander is 600 kg. Compute the force in each leg when the lander is resting on a horizontal surface on Mars. (Assume equal support by the pads and consult Table D/2 in Appendix D as needed.)

$Ans.$ $F_{AC} = F_{CB} = 240$ N tension
$F_{CD} = 1046$ N compression

**3/84** Determine the mechanical effect of the tire wrench on the wheel nut. Then insert the vertical support shown in dotted lines and repeat your calculations. Assume that the support carries all of the vertical loading. Explain the difference in the two cases.

**3/85** The bracket of negligible weight is welded to the shaft at $A$, and the shaft, in turn, is welded to the rigid support at $B$. Compute the torsional moment $T$ (moment about the $z$-axis) and the bending moment $M$ (moment about an axis normal to the shaft) at $B$ as a result of the three forces and one couple applied to the bracket.

$Ans.$ $T = 44.5$ N·m, $M = 14.87$ N·m

**3/86** Circular rods are embedded along two edges of the solid structure and are supported as shown. If the weight of the entire structure is small compared with the loads shown, determine the reactions at the four supports. All supporting surfaces are smooth; those at $A$ and $D$ are vertical and that at $B$ is horizontal. The support at $C$ is thrust-bearing.

**3/87** In an attempt to pull down the nearly sawn-through branch, the tree surgeon exerts a 400-N tension in the line which is looped around the branch at $A$. The portion $CA$ of the branch has a mass of 300 kg and its center of mass is at point $G$. Determine the magnitudes of the force and moment reactions at point $C$.        *Ans.* $F_C = 3280$ N, $M_C = 8200$ N·m

**3/88** The uniform 15-kg plate is welded to the vertical shaft, which is supported by bearings $A$ and $B$. Calculate the magnitude of the force supported by bearing $B$ during application of the 120-N·m couple to the shaft. The cable from $C$ to $D$ prevents the plate and shaft from turning, and the weight of the assembly is carried entirely by bearing $A$.

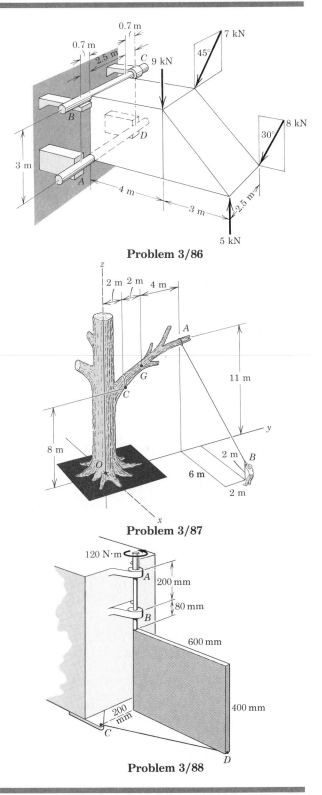

Problem 3/86

Problem 3/87

Problem 3/88

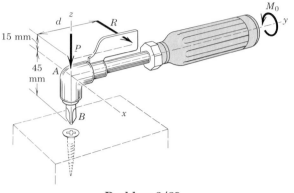

**Problem 3/89**

**3/89** A specialized screwdriver for use where space is limited consists of a geared head $A$ which transmits rotation of the handle about the $y$-axis to rotation of the Phillips-head bit about the $z$-axis of the screw. The bit is held down by an axial force $P$. One turn of the handle produces one turn of the screw, so that an applied torque (couple) $M_O$ to the handle results in an equal torque applied to the screw about the $z$-axis. If $M_O = 10$ N·m, determine the necessary distance $d$ and the force $R$ applied to the lever, which is secured to the gear housing $A$, in order for the screwdriver to operate in the position shown. Note that constraint in the $x$-$y$ plane is provided to the bit when engaged.

*Ans.* $R = 166.7$ N, $d = 60$ mm

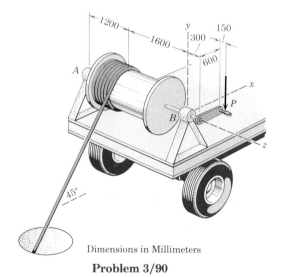

Dimensions in Millimeters

**Problem 3/90**

**3/90** A force $P$ of 200 N on the handle of the cable reel is required to wind up the underground cable as it comes from the manhole. The drum diameter is 1000 mm. For the horizontal position of the crank handle shown, calculate the magnitudes of the bearing forces at $A$ and $B$. Neglect the weight of the drum.

**3/91** The unit *ABCD* of the radial-arm saw weighs 40 lb with center of gravity at *G*. If a horizontal 10-lb force is applied to the control handle in sawing the board, calculate the corresponding bending moment acting on the column at *A* (total moment about a horizontal axis through *A*). The reaction of the wood on the saw teeth is 15 lb in the direction shown, and to a close approximation, its point of application may be taken as *E*. What justification exists for treating the saw as being in equilibrium?

*Ans.* $M = 711$ lb-in.

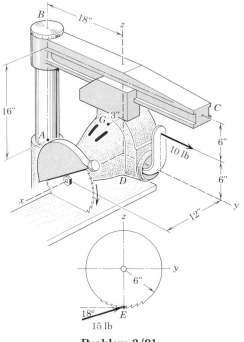

**Problem 3/91**

**3/92** The upper ends of the vertical coil springs in the stock racecar can be moved up and down by means of a screw mechanism not shown. This adjustment permits a change in the downward force at each wheel as an optimum handling setup is sought. Initially, scales indicate the normal forces to be 800 lb, 800 lb, 1000 lb, and 1000 lb at *A*, *B*, *C*, and *D*, respectively. If the top of the right rear spring at *A* is lowered so that the scale at *A* reads an additional 100 lb, determine the corresponding changes in the normal forces at *B*, *C*, and *D*. Neglect the effects of the small attitude changes (pitch and roll angles) caused by the spring adjustment. The front wheels are the same distance apart as the rear wheels.

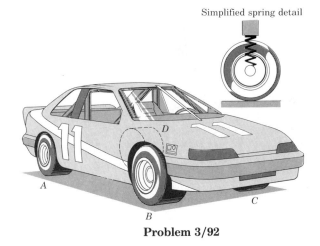

Simplified spring detail

**Problem 3/92**

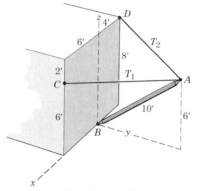

**Problem 3/93**

**3/93** The uniform steel boom $AB$ weighs 500 lb and is supported by the cables $AC$ and $AD$ attached to the vertical wall and by a ball-and-socket joint at $B$. Calculate the tension $T_1$ in cable $AC$ and the magnitude of the force supported at $B$.

*Ans.* $T_1 = 138.9$ lb, $B = 536$ lb

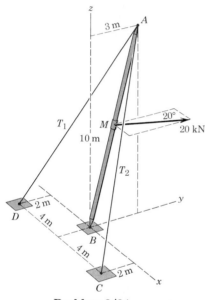

**Problem 3/94**

**3/94** The boom $AB$ lies in the vertical $y$-$z$ plane and is supported by a ball-and-socket joint at $B$ and by the two cables at $A$. Calculate the tension in each cable resulting from the 20-kN force acting in the horizontal plane and applied at the midpoint $M$ of the boom. Neglect the weight of the boom.

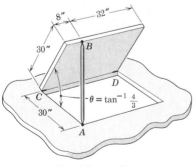

**Problem 3/95**

**3/95** The uniform 30- by 40-in. trap door weighs 200 lb and is propped open by the light strut $AB$ at the angle $\theta = \tan^{-1}(4/3)$. Calculate the compression $F_B$ in the strut and the force supported by the hinge $D$ normal to the hinge axis. Assume that the hinges act at the extreme ends of the lower edge.

*Ans.* $F_B = 70$ lb, $D_n = 101$ lb

**3/96** A portable electric grinder with a mass of 3.2 kg and mass center at $G$ is held in position ($z$-axis horizontal) by gripping the housing at $B$ and $C$. The grip at $B$ applies a force only in the $x$-$y$ plane, whereas the grip at $C$ must apply a couple $M_z$ about the $z$-axis as well as a force. If the tangential friction force $F$ is 60 percent of the normal force $R$ under the wheel, calculate the couple $M_z$ and the $x$- and $y$-components of the gripping forces at $B$ and $C$ if the normal force $R$ is maintained at 30 N.

**Problem 3/96**

**▶3/97** A rectangular sign over a store has a mass of 100 kg, with the center of mass in the center of the rectangle. The support against the wall at point $C$ may be treated as a ball-and-socket joint. At corner $D$ support is provided in the $y$-direction only. Calculate the tensions $T_1$ and $T_2$ in the supporting wires, the total force supported at $C$, and the lateral force $R$ supported at $D$.     *Ans.* $T_1 = 347$ N, $T_2 = 431$ N
$R = 63.1$ N, $C = 768$ N

**Problem 3/97**

**▶3/98** Under the action of the 40-N·m torque (couple) applied to the vertical shaft, the restraining cable $AC$ limits the rotation of the arm $OA$ and attached shaft to an angle of 60° measured from the $y$-axis. The collar $D$ fastened to the shaft prevents downward motion of the shaft in its bearing. Calculate the bending moment $M$, the compression $P$, and the shear force $V$ in the shaft at section $B$. (*Note*: Bending moment, expressed as a vector, is normal to the shaft axis, and shear force is also normal to the shaft axis.)
    *Ans.* $M = 47.7$ N·m, $P = 320$ N, $V = 274$ N

**Problem 3/98**

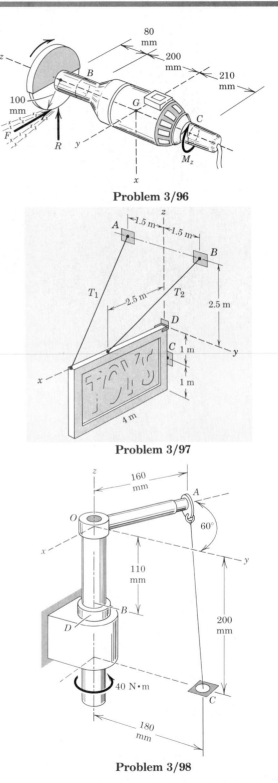

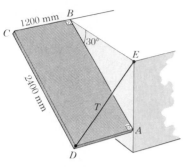

**Problem 3/99**

▶**3/99** The uniform rectangular panel *ABCD* has a mass of 40 kg and is hinged at its corners *A* and *B* to the fixed vertical surface. A wire from *E* to *D* keeps edges *BC* and *AD* horizontal. Hinge *A* can support thrust along the hinge axis *AB*, whereas hinge *B* supports force normal to the hinge axis only. Find the tension *T* in the wire and the magnitude *B* of the force supported by hinge *B*.

*Ans.* $T = 277$ N, $B = 169.9$ N

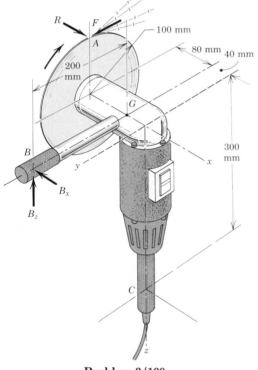

**Problem 3/100**

▶**3/100** The electric sander has a mass of 3 kg with mass center at *G* and is held in a slightly tilted position (*z*-axis vertical) so that the sanding disk makes contact at its top *A* with the surface being sanded. The sander is gripped by its handles at *B* and *C*. If the normal force *R* against the disk is maintained at 20 N and is due entirely to the force component $B_x$ (i.e., $C_x = 0$), and if the friction force *F* acting on the disk is 60 percent of *R*, determine the components of the couple *M* which must be applied to the handle at *C* to hold the sander in position. Assume that half of the weight is supported at *C*.

*Ans.* $M_x = -1.857$ N·m, $M_y = 1.411$ N·m
$M_z = -2.56$ N·m

## 3/5  PROBLEM FORMULATION AND REVIEW

In Chapter 3 we have applied our knowledge of the properties of forces, moments, and couples studied in Chapter 2 to solve problems in the equilibrium of rigid bodies. Each such body in complete equilibrium is characterized by the requirement that the vector resultant of all forces acting on it is zero ($\Sigma \mathbf{F} = \mathbf{0}$) and the vector resultant of all moments on the body about a point (or axis) is also zero ($\Sigma \mathbf{M} = \mathbf{0}$). We are guided in all of our solutions by these two requirements, which are easily comprehended physically.

As is often the case, it is not the theory but its application which presents the difficulty. The crucial steps in applying our principles of equilibrium should be quite familiar by now. They are:

1. Make an unequivocal decision as to which body in equilibrium is to be analyzed.

2. Isolate the body in question from all contacting bodies by drawing its *free-body diagram* on which *all* forces and couples that act on the isolated body from external sources are represented.

3. Observe the principle of action and reaction (Newton's third law) when assigning the sense of each force.

4. Label all forces and couples, known and unknown.

5. Choose and label reference axes, always using a right-handed set for three-dimensional analysis.

6. Check the adequacy of the constraints (supports) and match the number of unknowns with the number of available independent equations of equilibrium.

When solving an equilibrium problem, we should first check to see that the body is statically determinate. If there are more supports than are necessary to hold the body in place, the body is statically indeterminate, and the equations of equilibrium by themselves will not enable us to solve for all of the external reactions. In applying the equations of equilibrium, we choose scalar algebra, vector algebra, or graphical analysis according to both preference and experience, with vector algebra being particularly useful for many three-dimensional problems.

One of our most useful procedures is to simplify the algebra of a solution by the choice of a moment axis that eliminates as many unknowns as possible or by the choice of a direction for a force summation that avoids reference to certain unknowns. A few moments of thought to take advantage of these simplifications can save appreciable time and effort.

The principles and methods which are covered in Chapters 2 and 3 constitute the most basic part of statics. They lay the foundation for what follows not only in statics but in dynamics as well.

## REVIEW PROBLEMS

**Problem 3/101**

**3/101** A 50-kg acrobat pedals her unicycle across the taut but slightly elastic cable. If the deflection at the center of the 18-m span is 75 mm, determine the tension in the cable. Neglect the effects of the weights of the cable and unicycle.

*Ans. T* = 29.4 kN

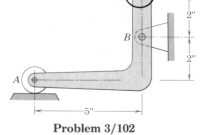

**Problem 3/102**

**3/102** Calculate the magnitude of the force supported by the pin at *B* for the bell crank loaded and supported as shown.

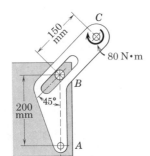

**Problem 3/103**

**3/103** The light bracket *ABC* is freely hinged at *A* and is constrained by the fixed pin in the smooth slot at *B*. Calculate the magnitude *R* of the force supported by the pin at *A* under the action of the 80-N·m applied couple.          *Ans. R* = 566 N

**Problem 3/104**

**3/104** The uniform beam weighs 50 pounds per foot of its length. Determine the reactions at the built-in support at *A*.

**3/105** The uniform bar with end rollers weighs 60 lb and is supported by the horizontal and vertical surfaces and by the wire $AC$. Calculate the tension $T$ in the wire and the reactions against the rollers at $A$ and at $B$.                 *Ans.* $T = 60.2$ lb, $A = 15$ lb

**3/106** Magnetic tape under a tension of 10 N at $D$ passes around the guide pulleys and through the erasing head at $C$ at constant speed. As a result of a small amount of friction in the bearings of the pulleys, the tape at $E$ is under a tension of 11 N. Determine the tension $T$ in the supporting spring at $B$. The plate lies in a horizontal plane and is mounted on a precision needle bearing at $A$.

**3/107** The gear reducer shown is acted on by the two couples and its weight of 200 N. Determine the vertical force supported by each of the reducer mountings at $A$ and $B$.

   *Ans.* $F_A = 486$ N down, $F_B = 686$ N up

**3/108** A freeway sign measuring 12 ft by 6 ft is supported by the single mast as shown. The sign, supporting framework, and mast together weigh 600 lb, with center of gravity 10 ft away from the vertical centerline of the mast. When the sign is subjected to the direct blast of a 75 mi/hr wind, an average pressure difference of 17.5 lb/ft² is developed between the front and back sides of the sign. Determine the magnitudes of the force and moment reactions at the base of the mast.

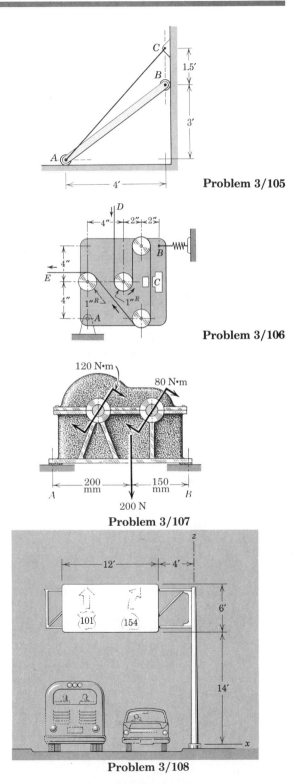

Problem 3/105

Problem 3/106

Problem 3/107

Problem 3/108

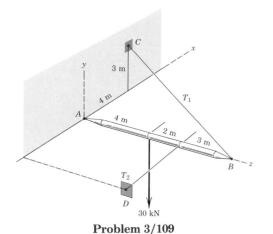

**Problem 3/109**

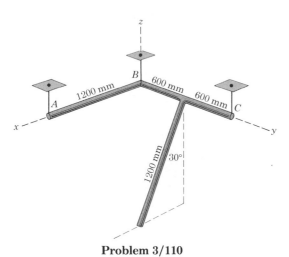

**Problem 3/110**

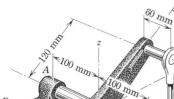

**Problem 3/111**

**3/109** If the weight of the boom is negligible compared with the applied 30-kN load, determine the cable tensions $T_1$ and $T_2$ and the force acting at the ball joint at $A$.

*Ans.* $T_1 = 45.8$ kN, $T_2 = 26.7$ kN, $A = 44.2$ kN

**3/110** Each of the three uniform 1200-mm bars has a mass of 20 kg. The bars are welded together into the configuration shown and suspended by three vertical wires. Bars $AB$ and $BC$ lie in the horizontal $x$-$y$ plane, and the third bar lies in a plane parallel to the $x$-$z$ plane. Compute the tension in each wire.

**3/111** A vertical force $P$ on the foot pedal of the bell crank is required to produce a tension $T$ of 400 N in the vertical control rod. Determine the corresponding bearing reactions at $A$ and $B$.

*Ans.* $A = 183.9$ N up, $B = 424$ N up

**3/112** The light vertical mast supports the 4-kN force and is constrained by the two cables $BC$ and $BD$ and by a ball-and-socket connection at $A$. Calculate the tensions in the cables and the magnitude of the reaction at $A$.

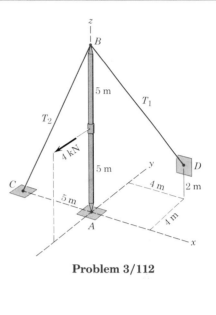

**Problem 3/112**

**3/113** The power unit of the post-hole digger supplies a torque of 4000 lb-in. to the auger. The arm $B$ is free to slide in the supporting sleeve $C$ but is not free to rotate about the horizontal axis of $C$. If the unit is free to swivel about the vertical axis of the mount $D$, determine the force exerted against the right rear wheel by the block $A$ (or $A'$), which prevents the unbraked truck from rolling. (*Hint:* View the system from above.)                          *Ans.* $A' = 41.7$ lb

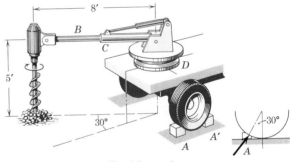

**Problem 3/113**

**3/114** Each of the plates is supported by the constraints shown and is subjected to a known force $F$. Indicate for each case which of the following categories applies:

   I. Constraints inadequate to ensure complete fixity.

   II. Constraints adequate but not excessive to hold plate in equilibrium. The three independent equations of equilibrium are sufficient to determine all unknown forces.

   III. Complete fixity with excessive constraints to ensure an equilibrium position.

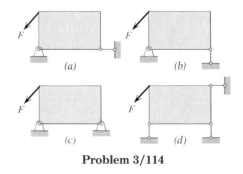

**Problem 3/114**

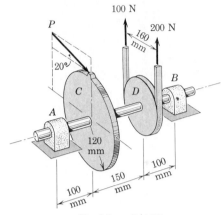

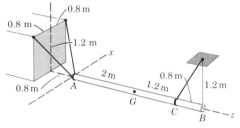

**Problem 3/115**

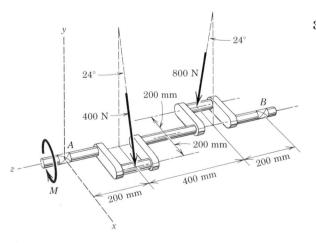

**Problem 3/116**

**Problem 3/117**

**3/115** Gear $C$ drives the V-belt pulley $D$ at a constant speed. For the belt tensions shown calculate the gear-tooth force $P$ and the magnitudes of the total forces supported by the bearings at $A$ and $B$.

$Ans.$ $P = 70.9$ N, $A = 83.3$ N, $B = 208$ N

**3/116** Explain why the 50-kg uniform circular rod cannot be in static equilibrium when in the indicated position.

**3/117** For the position shown, the crankshaft of a small two-cylinder engine is subjected to the 400-N and 800-N forces exerted by the connecting rods. If the crankshaft is in equilibrium, determine the bearing forces at $A$ and $B$ and the couple $M$ exerted on the crankshaft.

$Ans.$ $A_y = 457$ N, $A_x = -40.7$ N
$B_x = 203$ N, $B_y = 639$ N, $M = 73.1$ N·m

**3/118** Three identical steel balls, each of mass $m$, are placed in the cylindrical ring which rests on a horizontal surface and whose height is slightly greater than the radius of the balls. The diameter of the ring is such that the balls are virtually touching one another. A fourth identical ball is then placed on top of the three balls. Determine the force $P$ exerted by the ring on each of the three lower balls.

**Problem 3/118**

**3/119** The uniform lid of the rectangular container weighs 90 lb and is held by the wire $CE$ so that the lid makes an angle of $60°$ with the horizontal $x$-$y$ plane. Hinge $B$ supports thrust in the direction of the hinge axis $AB$. Hinge $A$ supports force normal to $AB$ only. Calculate the tension $T$ in the wire.

*Ans.* $T = 56.8$ lb

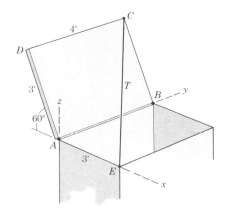

**Problem 3/119**

**3/120** The drum and shaft are welded together and have a mass of 50 kg with mass center at $G$. The shaft is subjected to a torque (couple) of 120 N·m, and the drum is prevented from rotating by the cord wrapped securely around it and attached to point $C$. Calculate the magnitudes of the forces supported by bearings $A$ and $B$.

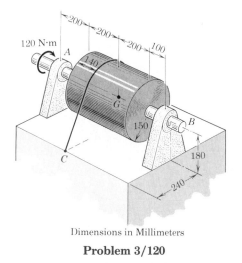

Dimensions in Millimeters

**Problem 3/120**

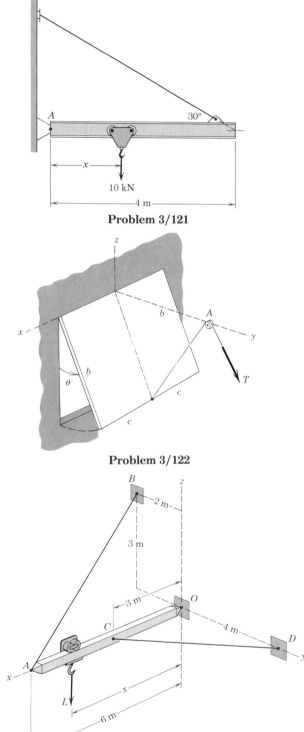

**Problem 3/121**

**Problem 3/122**

**Problem 3/123**

**\*3/121** The jib crane is designed for a maximum capacity of 10 kN, and its uniform I-beam has a mass of 200 kg. (*a*) Plot the magnitude $R$ of the force on the pin at $A$ as a function of $x$ through its operating range of $x = 0.2$ m to $x = 3.8$ m. (*b*) Determine the minimum value of $R$ and the corresponding value of $x$. (*c*) What is the least value of $R$ for which the pin at $A$ can be safely designed? (Use $g = 10$ m/s².)

*Ans.* (*b*) $R_{min} = 10.39$ kN at $x = 0.8$ m
(*c*) 18.25 kN

**\*3/122** The uniform rectangular trap door of mass $m$ is freely hinged about its upper horizontal edge and is controlled by the cable which passes over the fixed pulley at $A$. Plot the tension $T$ in the cable necessary to support the door in an open position in terms of $\theta$ from 0° to 90°. Express $T$ as a dimensionless ratio $T' = T/mg$ and prove that the limiting value of $T'$ when $\theta = 90°$ is $\frac{1}{2}$.

**\*3/123** The horizontal boom is supported by the cables $AB$ and $CD$ and by a ball-and-socket joint at $O$. To determine the influence on the reaction at $O$ of the position of the vertical load $L$ along the boom, we may neglect the weight of the boom. If $R$ represents the magnitude of the total force at $O$, determine by calculus the minimum ratio $R/L$ and the corresponding value of $x$. Then write a computer program for $R/L$ and plot the results for $0 < x < 6$ m as a check on your calculations.

*Ans.* $(R/L)_{min} = 0.951$ at $x = 0.574$ m

**3/124** The vertical position of the two 20-kg cylinders is controlled by the torque $M$ applied to the central shaft. The cords attached to the ends of the arms pass through smooth holds in a fixed surface at $A$ and $B$. Plot $M$ as a function of $\theta$ from $\theta = 0°$ to $\theta = 180°$. Determine the maximum value of $M$ and the corresponding value of $\theta$. Take $g = 10$ m/s$^2$ for your calculations.

**3/125** The 26-ft steel I-beam weighs 1820 lb and is being hoisted into position by the application of a force $P$ on the dolly to which the lower end $B$ is secured. The cable $AC$ is attached to the center of the beam and is horizontal when $B$ is directly below $C$. Determine the force $P$ required to maintain equilibrium as a function of $x$ from $x = 0$ to its value where $A'$ is directly below $C$. Find the maximum value of $P$ and the corresponding value of $x$.

*Ans.* $P_{max} = 1112$ lb at $x = 4.04$ ft

**3/126** The trap door is a uniform square panel of mass $m$ hinged about its lower edge in the horizontal $x$-$y$ plane. The control cable passes over a small pulley fixed at $A$ but free to swivel as the direction of the cable changes. Express the tension $T$ required for equilibrium at any angle $\theta$ in terms of the nondimensional variable $T' = T/mg$. Plot $T'$ versus $\theta$ from $\theta = 0°$ to $\theta = 90°$ and show that $T'$ approaches the limit of $\frac{1}{2}$ as $\theta$ approaches 90°.

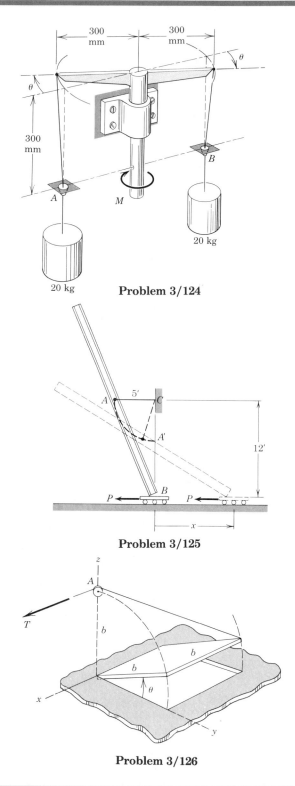

**Problem 3/124**

**Problem 3/125**

**Problem 3/126**

There are countless forms of structures that support forces, and their design requires the calculation of the forces supported by each of their major structural elements. The framework of this bridge is an example of the critical design of large structures.

# STRUCTURES

# 4

## 4/1 INTRODUCTION

In Chapter 3 we focused our attention on the equilibrium of a single rigid body or on a system of connected members which, when taken as a whole, could be treated as a single rigid body. In these problems we first drew a free-body diagram of this single body showing all forces external to the isolated body and then we applied the force and moment equations of equilibrium. In Chapter 4 our attention is directed toward the determination of the forces internal to a structure, that is, forces of action and reaction between the connected members. An engineering structure is any connected system of members built to support or transfer forces and to safely withstand the loads applied to it. In the force analysis of structures it is necessary to dismember the structure and to analyze separate free-body diagrams of individual members or combinations of members in order to determine the forces internal to the structure. This analysis calls for very careful observance of Newton's third law, which states that each action is accompanied by an equal and opposite reaction.

In Chapter 4 we shall analyze the internal forces acting in several types of structures, namely, trusses, frames, and machines. In this treatment we shall consider only *statically determinate* structures, that is, structures which do not have more supporting constraints than are necessary to maintain an equilibrium configuration. Thus, as we have already seen, the equations of equilibrium are adequate to determine all unknown reactions.

The student who has mastered the basic procedure developed in Chapter 3 of defining unambiguously the body under consideration by constructing a correct free-body diagram will have little difficulty with the analysis of statically determinate structures. The analysis of trusses, frames and machines, and beams under concentrated loads constitutes a straightforward application of the material developed in the previous two chapters.

### 4/2  *PLANE TRUSSES*

A framework composed of members joined at their ends to form a rigid structure is known as a truss. Bridges, roof supports, derricks, and other such structures are common examples of trusses. Structural members used are I-beams, channels, angles, bars, and special shapes which are fastened together at their ends by welding, riveted connections, or large bolts or pins. When the members of the truss lie essentially in a single plane, the truss is known as a *plane truss.* Plane trusses, such as those used for bridges, are commonly designed in pairs with one truss panel placed on each side of the bridge and connected to the other by cross beams that support the roadway and transfer the applied loads to the truss members. Several examples of commonly used trusses that can be analyzed as plane trusses are shown in Fig. 4/1.

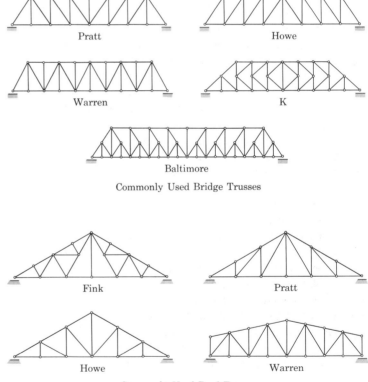

Commonly Used Bridge Trusses

Commonly Used Roof Trusses
**Figure 4/1**

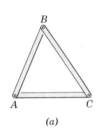

*(a)*

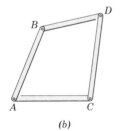

*(b)*

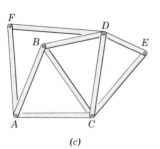

*(c)*

**Figure 4/2**

The basic element of a plane truss is the triangle. Three bars joined by pins at their ends, Fig. 4/2a, constitute a rigid frame. On the other hand, four or more bars pin-jointed to form a polygon of as many sides constitute a nonrigid frame. We can make the nonrigid

frame in Fig. 4/2*b* stable or rigid by adding a diagonal bar joining
*A* and *D* or *B* and *C* and thereby forming two triangles. The structure
may be extended by adding additional units of two end-connected
bars, such as *DE* and *CE* or *AF* and *DF*, Fig. 4/2*c*, which are pinned
to two fixed joints, and in this way the entire structure will remain
rigid. The term rigid is used in the sense of noncollapsible and also
in the sense that deformation of the members due to induced internal
strains is negligible.

Structures that are built from a basic triangle in the manner
described are known as *simple trusses*. When more members are
present than are needed to prevent collapse, the truss is statically
indeterminate. A statically indeterminate truss cannot be analyzed
by the equations of equilibrium alone. Additional members or
supports that are not necessary for maintaining the equilibrium
position are called *redundant*.

The design of a truss involves the determination of the forces
in the various members and the selection of appropriate sizes and
structural shapes to withstand the forces. Several assumptions are
made in the force analysis of simple trusses. First, we assume all
members to be *two-force members*. A two-force member is one in
equilibrium under the action of two forces only, as defined in general
terms with Fig. 3/4 in Art. 3/3. For trusses each member is normally
a straight link joining the two points of application of force. The two
forces are applied at the ends of the member and are necessarily
equal, opposite, and *collinear* for equilibrium. We see that the member may be in tension or compression, as shown in Fig. 4/3. When
we represent the equilibrium of a portion of a two-force member,
the tension *T* or compression *C* acting on the cut section is the same
for all sections. We assume here that the weight of the member is
small compared with the force it supports. If it is not, or if the small
effect of the weight is to be accounted for, the weight *W* of the
member may be replaced by two forces, each *W*/2 if the member is
uniform, with one force acting at each end of the member. These
forces, in effect, are treated as loads externally applied to the pin
connections. Accounting for the weight of a member in this way
gives the correct result for the average tension or compression along
the member but will not account for the effect of bending of the
member.

When welded or riveted connections are used to join structural
members, the assumption of a pin-jointed connection is usually
satisfactory if the centerlines of the members are concurrent at the
joint as in Fig. 4/4.

We also assume in the analysis of simple trusses that all external forces are applied at the pin connections. This condition is
satisfied in most trusses. In bridge trusses the deck is usually laid
on cross beams that are supported at the joints.

Provision for expansion and contraction due to temperature

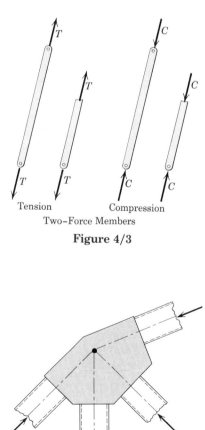

Tension                 Compression
Two–Force Members

**Figure 4/3**

**Figure 4/4**

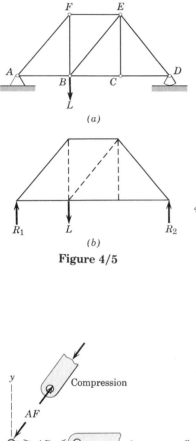

(a)

(b)

**Figure 4/5**

changes and for deformations resulting from applied loads is usually made at one of the supports for large trusses. A roller, rocker, or some kind of slip joint is provided. Trusses and frames in which no such provision is made are statically indeterminate, as explained in Art. 3/3.

Two methods for the force analysis of simple trusses will be given, and reference will be made to the simple truss shown in Fig. 4/5*a* for each of the two methods. The free-body diagram of the truss as a whole is shown in Fig. 4/5*b*. The external reactions are usually determined by computation from the equilibrium equations applied to the truss as a whole before the force analysis of the remainder of the truss is begun.

### 4/3 METHOD OF JOINTS

This method for finding the forces in the members of a simple truss consists of satisfying the conditions of equilibrium for the forces acting on the connecting pin of each joint. The method therefore deals with the equilibrium of concurrent forces, and only two independent equilibrium equations are involved. We begin the analysis with any joint where at least one known load exists and where not more than two unknown forces are present. Solution may be started with the pin at the left end, and its free-body diagram is shown in Fig. 4/6. With the joints indicated by letters, we may designate the force in each member by the two letters defining the ends of the member. The proper directions of the forces should be evident for this simple case by inspection. The free-body diagrams of portions of members *AF* and *AB* are also shown to clearly indicate the mechanism of the action and reaction. The member *AB* actually makes contact on the left side of the pin, although the force *AB* is drawn from the right side and is shown acting away from the pin. Thus, if we consistently draw the force arrows on the *same* side of the pin as the member, then tension (such as *AB*) will always be indicated by an arrow *away* from the pin, and compression (such as *AF*) will always be indicated by an arrow *toward* the pin. The magnitude of *AF* is obtained from the equation $\Sigma F_y = 0$ and *AB* is then found from $\Sigma F_x = 0$.

Joint *F* may be analyzed next, since it now contains only two unknowns, *EF* and *BF*. Joints *B*, *C*, *E*, and *D* are subsequently analyzed in that order. The free-body diagram of each joint and its corresponding force polygon which represents graphically the two equilibrium conditions $\Sigma F_x = 0$ and $\Sigma F_y = 0$ are shown in Fig. 4/7. The numbers indicate the order in which the joints are analyzed. We note that, when joint *D* is finally reached, the computed reaction $R_2$ must be in equilibrium with the forces in members *CD* and *ED*, determined previously from the two neighboring joints. This requirement will provide a check on the correctness of our work. We should also note that isolation of joint *C* quickly discloses the fact

**Figure 4/6**

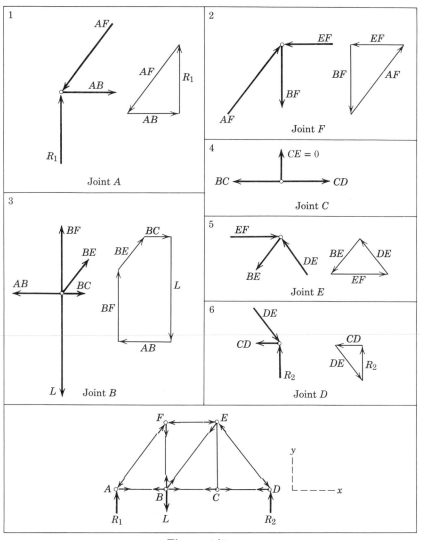

**Figure 4/7**

that the force in $CE$ is zero when the equation $\Sigma F_y = 0$ is applied. The force in this member would not be zero, of course, if an external vertical load were applied at $C$.

It is often convenient to indicate the tension $T$ and compression $C$ of the various members directly on the original truss diagram by drawing arrows away from the pins for tension and toward the pins for compression. This designation is illustrated at the bottom of Fig. 4/7.

In some instances it is not possible to initially assign the correct direction of one or both of the unknown forces acting on a given pin.

In this event we may make an arbitrary assignment. A negative value from the computation indicates that the assumed direction is incorrect.

If a simple truss has more external supports than are necessary to ensure a stable equilibrium configuration, the truss as a whole is statically indeterminate, and the extra supports constitute *external* redundancy. If the truss has more internal members than are necessary to prevent collapse, then the extra members constitute *internal* redundancy and the truss is statically indeterminate. For a truss that is statically determinate externally, there is a definite relation between the number of its members and the number of its joints necessary for internal stability without redundancy. Since we can specify the equilibrium of each joint by two scalar force equations, there are in all $2j$ such equations for a simple truss with $j$ joints. For the entire truss composed of $m$ two-force members and having the maximum of three unknown support reactions, there are in all $m + 3$ unknowns. Thus, for a simple plane truss composed of triangular elements, the equation $m + 3 = 2j$ will be satisfied if the truss is statically determinate internally.

This relation is a necessary condition for stability but it is not a sufficient condition, since one or more of the $m$ members can be arranged in such a way as not to contribute to a stable configuration of the entire truss. If $m + 3 > 2j$, there are more members than there are independent equations, and the truss is statically indeterminate internally with redundant members present. If $m + 3 < 2j$, there is a deficiency of internal members, and the truss is unstable and will collapse under load.

*Special conditions.*   We draw attention to several special conditions which occur frequently in the analysis of simple trusses. When two collinear members are under compression, as indicated in Fig. 4/8a, it is necessary to add a third member to maintain alignment of the two members and prevent buckling. We see very quickly from a force summation in the $y$-direction that the force $F_3$ in the third member must be zero and from the $x$-direction that

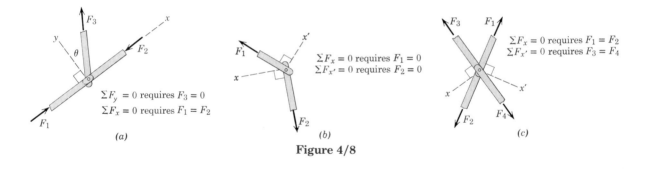

Figure 4/8

$F_1 = F_2$. This conclusion holds regardless of the angle $\theta$ and, of course, holds if the collinear members are in tension. If an external force with a component in the $y$-direction were applied to the joint, then $F_3$ would no longer be zero.

When two noncollinear members are joined as shown in Fig. 4/8*b*, then in the absence of an externally applied load at this joint, the forces in both members must be zero, as we see from the two force summations.

When two pairs of collinear members are joined as shown in Fig. 4/8*c*, the forces in each pair must be equal and opposite. This conclusion follows from the force summations indicated in the figure.

Truss panels are frequently cross-braced as shown in Fig. 4/9*a*. Such a panel is statically indeterminate if each brace is capable of supporting either tension or compression. However, when the braces

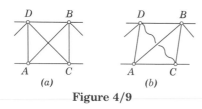

**Figure 4/9**

are flexible members incapable of supporting compression, as are cables, then only the tension member acts and the other member is disregarded. It is usually evident from the asymmetry of the loading how the panel will deflect. If the deflection is as indicated in Fig. 4/9*b*, then member *AB* should be retained and *CD* disregarded. When this choice cannot be made by inspection, we may make an arbitrary selection of the member to be retained. If the assumed tension turns out to be positive upon calculation, then the choice was correct. If the assumed tension force turns out to be negative, then the opposite member must be retained and the calculation redone.

The simultaneous solution of the equations for two unknown forces at a joint may be avoided by a careful choice of reference axes. Thus, for the joint indicated schematically in Fig. 4/10 where $L$ is known and $F_1$ and $F_2$ are unknown, a force summation in the $x$-direction eliminates reference to $F_1$ and a force summation in the $x'$-direction eliminates reference to $F_2$. When the angles involved are not easily found, then a simultaneous solution of the equations using one set of reference directions for both unknowns may be preferable.

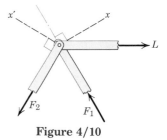

**Figure 4/10**

## Sample Problem 4/1

Compute the force in each member of the loaded cantilever truss by the method of joints.

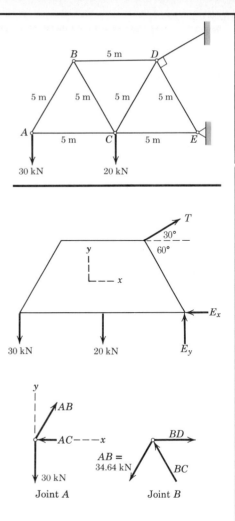

**Solution.** If it were not desired to calculate the external reactions at $D$ and $E$, the analysis for a cantilever truss could begin with the joint at the loaded end. However, this truss will be analyzed completely, so the first step will be to compute the external forces at $D$ and $E$ from the free-body diagram of the truss as a whole. The equations of equilibrium give

$[\Sigma M_E = 0]$ $\qquad\qquad$ $5T - 20(5) - 30(10) = 0$ $\qquad$ $T = 80.0$ kN

$[\Sigma F_x = 0]$ $\qquad\qquad\qquad$ $80.0 \cos 30° - E_x = 0$ $\qquad$ $E_x = 69.3$ kN

$[\Sigma F_y = 0]$ $\qquad\qquad$ $80.0 \sin 30° + E_y - 20 - 30 = 0$ $\qquad$ $E_y = 10.0$ kN

Next we draw free-body diagrams showing the forces acting on each of the connecting pins. The correctness of the assigned directions of the forces is verified when each joint is considered in sequence. There should be no question about the correct direction of the forces on joint $A$. Equilibrium requires

$[\Sigma F_y = 0]$ $\qquad$ $0.866AB - 30 = 0$ $\qquad$ $AB = 34.64$ kN $T$ $\qquad$ *Ans.*

$[\Sigma F_x = 0]$ $\qquad$ $AC - 0.5(34.64) = 0$ $\qquad$ $AC = 17.32$ kN $C$ $\qquad$ *Ans.*

① where $T$ stands for tension and $C$ stands for compression.

Joint $B$ must be analyzed next, since there are more than two unknown forces on joint $C$. The force $BC$ must provide an upward component, in which case $BD$ must balance the force to the left. Again the forces are obtained from

$[\Sigma F_y = 0]$ $\quad$ $0.866BC - 0.866(34.64) = 0$ $\quad$ $BC = 34.64$ kN $C$ $\quad$ *Ans.*

$[\Sigma F_x = 0]$ $\qquad$ $BD - 2(0.5)(34.64) = 0$ $\qquad$ $BD = 34.64$ kN $T$ $\quad$ *Ans.*

Joint $C$ now contains only two unknowns, and these are found in the same way as before:

$[\Sigma F_y = 0]$ $\qquad$ $0.866CD - 0.866(34.64) - 20 = 0$

$\qquad\qquad\qquad$ $CD = 57.74$ kN $T$ $\qquad\qquad\qquad\qquad$ *Ans.*

$[\Sigma F_x = 0]$ $\qquad$ $CE - 17.32 - 0.5(34.64) - 0.5(57.74) = 0$

$\qquad\qquad\qquad$ $CE = 63.51$ kN $C$ $\qquad\qquad\qquad\qquad$ *Ans.*

Finally, from joint $E$ there results

$[\Sigma F_y = 0]$ $\qquad$ $0.866DE = 10.00$ $\qquad$ $DE = 11.55$ kN $C$ $\qquad$ *Ans.*

and the equation $\Sigma F_x = 0$ checks.

① It should be stressed that the tension/compression designation refers to the *member*, not the joint. Note that we draw the force arrow on the same side of the joint as the member which exerts the force. In this way tension (arrow away from the joint) is distinguished from compression (arrow toward the joint).

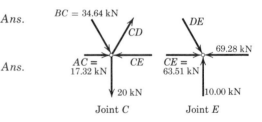

## PROBLEMS

(Solve the following problems by the method of joints. Neglect the weights of the members compared with the forces they support unless otherwise indicated. Note that 1 kip = 1000 lb.)

### *Introductory problems*

**4/1** Determine the force in each member of the loaded truss.          *Ans. AB* = 250 lb *T, AC* = 600 lb *T*
          *BC* = 650 lb *C*

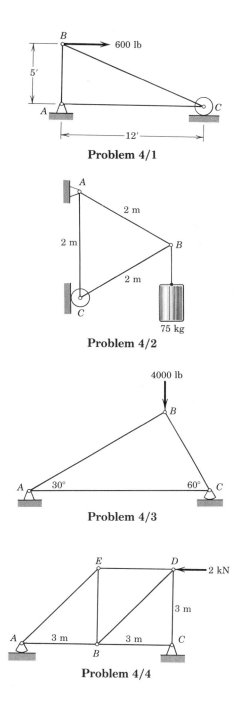

**Problem 4/1**

**4/2** Determine the force in each member of the simple equilateral truss.

**Problem 4/2**

**4/3** Determine the force in each member of the loaded truss. Explain why knowledge of the lengths of the members is unnecessary.
          *Ans. AB* = 2000 lb *C, AC* = 1732 lb *T*
          *BC* = 3460 lb *C*

**Problem 4/3**

**4/4** Calculate the force in each member of the loaded truss.

**Problem 4/4**

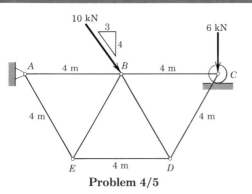

**Problem 4/5**

**4/5** Determine the force in each member of the loaded truss. Does the 6-kN load affect your results?

$Ans.\ AB = 3.69$ kN $T,\ BC = 2.31$ kN $C$
$BD = BE = 4.62$ kN $C$
$DC = DE = AE = 4.62$ kN $T$

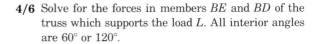

**4/6** Solve for the forces in members $BE$ and $BD$ of the truss which supports the load $L$. All interior angles are 60° or 120°.

**Problem 4/6**

*Representative problems*

**4/7** Calculate the forces in members $CG$ and $CF$ for the truss shown.

$Ans.\ CG = 2.24$ kN $T,\ CF = 1$ kN $C$

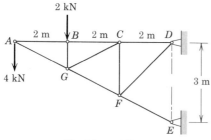

**Problem 4/7**

**4/8** If the 2-kN force acting on the truss of Prob. 4/7 were removed, identify by inspection those members in which the forces are zero. On the other hand, if the 2-kN force were aplied at $G$ instead of $B$, would there by any zero-force members?

**4/9** Calculate the force in each member of the loaded truss.

*Ans.* $AB = BC = CD = 2.11$ kN $C$
$CE = BE = 2.11$ kN $T$, $AE = 2.04$ kN $T$
$DE = 1.057$ kN $T$

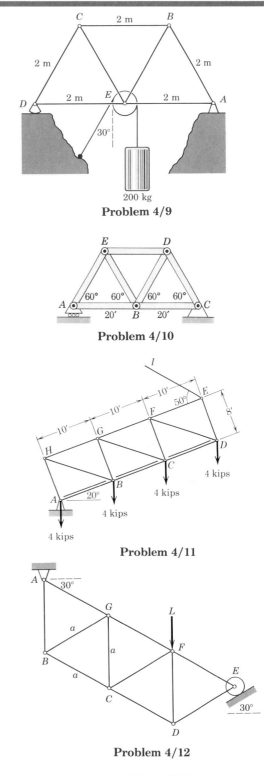

**Problem 4/9**

**4/10** Each member of the truss is a uniform 20-ft bar weighing 400 lb. Calculate the average tension or compression in each member due to the weights of the members.

**Problem 4/10**

**4/11** A drawbridge is being raised by a cable *EI*. The four joint loadings shown result from the weight of the roadway. Determine the forces in members *EF*, *DE*, *DF*, *CD*, and *FG*.

*Ans.* $EF = 5.15$ kips $C$
$DE = 6.14$ kips $T$
$DF = 3.82$ kips $C$
$CD = 1.61$ kips $T$
$FG = 8.13$ kips $C$

**Problem 4/11**

**4/12** The truss is composed of equilateral triangles of sides *a* and is loaded and supported as shown. Determine the forces in members *EF*, *DE*, and *DF*.

**Problem 4/12**

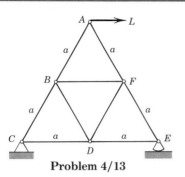

**Problem 4/13**

**4/13** The equiangular truss is loaded and supported as shown. Determine the forces in all members in terms of the horizontal load $L$.

Ans. $AB = BC = L\ T,\ AF = EF = L\ C$
$DE = CD = L/2\ T,\ BF = DF = BD = 0$

**Problem 4/14**

**4/14** Calculate the forces in members $CF$, $CG$, and $EF$ of the loaded truss.

**Problem 4/15**

**4/15** Determine the forces in members $BI$, $CI$, and $HI$ for the loaded truss. All angles are 30°, 60°, or 90°.

Ans. $BI = 2.50$ kN $T$, $CI = 2.12$ kN $T$
$HI = 2.69$ kN $T$

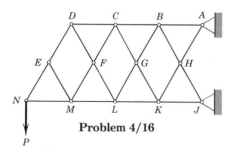

**Problem 4/16**

**4/16** Determine the forces in members $FL$, $CB$, and $HK$ of the loaded truss. Explain why the two fixed supports do not cause statical indeterminacy. All panel angles are 60° or 120°.

**4/17** A small ferris wheel is constructed of two identical trusses, one of which is shown. Member *AO* is temporarily removed for replacement. If the weight of the chairs and structural members results in a 100-lb load at each joint of the truss shown, determine the force in each member of the structure. With member *AO* replaced, could you repeat the analysis?

*Ans.* $AB = AH = 131$ lb $T$, $BO = HO = 58.6$ lb $C$
$BC = GH = 207$ lb $T$, $CO = GO = 200$ lb $C$
$CD = FG = 315$ lb $T$, $DO = FO = 341$ lb $C$
$DE = EF = 392$ lb $T$, $EO = 400$ lb $C$

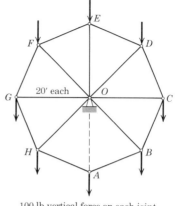

100-lb vertical force on each joint

**Problem 4/17**

**4/18** A snow load transfers the forces shown to the upper joints of a Pratt roof truss. Neglect any horizontal reactions at the supports and solve for the forces in all members.

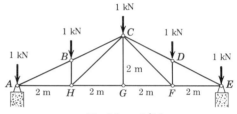

**Problem 4/18**

**4/19** The loading of Prob. 4/18 is shown applied to a Howe roof truss. Neglect any horizontal reactions at the supports and solve for the forces in all members. Compare to the results of Prob. 4/18.

*Ans.* $AB = DE = 3.35$ kN $C$
$BC = CD = 2.24$ kN $C$
$AH = EF = 3.00$ kN $T$
$BH = DF = 0$
$GH = FG = 3.00$ kN $T$
$BG = DG = 1.12$ kN $C$

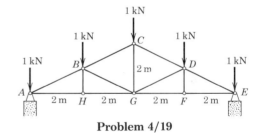

**Problem 4/19**

**4/20** Calculate the forces in all members of the loaded truss supported by the horizontal link *CG* and the pin at *A*. All panels are equilateral triangles.

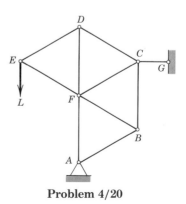

**Problem 4/20**

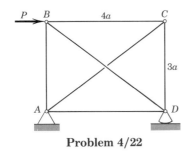

16 m

7 panels at 12 m

16 m | 24 m | 16 m

**Problem 4/21**

**4/21** The movable gantry is used to erect and prepare a 500-Mg rocket for firing. The primary structure of the gantry is approximated by the symmetrical plane truss shown, which is statically indeterminate. As the gantry is positioning a 60-Mg section of the rocket suspended from $A$, strain gage measurements indicate a compressive force of 50 kN in member $AB$ and a tensile force of 120 kN in member $CD$ due to the 60-Mg load. Calculate the corresponding forces in members $BF$ and $EF$.

*Ans.* $BF = 188.4$ kN $C$, $EF = 120$ kN $T$

**4/22** The rectangular frame is composed of four perimeter members and two cables $AC$ and $BD$ which are incapable of supporting compression. Determine the forces in all members.

$P$  $B$  $4a$  $C$

$3a$

$A$  $D$

**Problem 4/22**

**4/23** Calculate the forces in members $AB$, $BH$, and $BG$. Members $BF$ and $CG$ are cables which can support tension only.

*Ans.* $AB = 2.69$ kN $C$
$BH = 5.39$ kN $C$
$BG = 0$

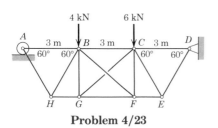

4 kN    6 kN

$A$  3 m  $B$  3 m  $C$  3 m  $D$
60° 60°           60° 60°

$H$  $G$     $F$  $E$

**Problem 4/23**

**4/24** Verify the fact that each of the trusses contains one or more elements of redundancy and propose two separate changes, either one of which would remove the redundancy and produce complete statical determinacy. All members can support compression as well as tension.

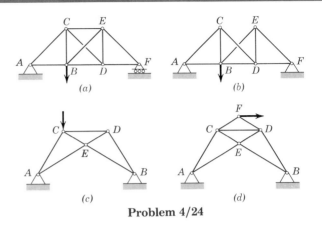

(a)                          (b)

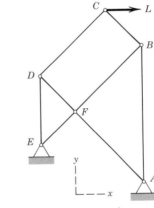

(c)                          (d)

**Problem 4/24**

**4/25** Show that the loaded truss is not statically indeterminate despite its two fixed supports at $A$ and $E$. Determine the $x$- and $y$-components of the support reactions at $A$ and $E$. All panel angles are $45°$ or $90°$.

Ans. $A_x = E_x = L/2$, $E_y = A_y = 3L/2$

**Problem 4/25**

**4/26** Verify the fact that each of the loaded trusses shown is unstable internally (nonrigid) and indicate at least two alternative ways to ensure internal stability (rigidity) for each truss by the addition of one or more members without introducing redundancy.

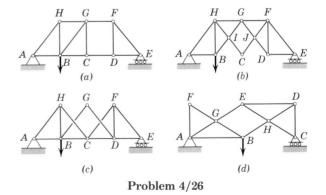

(a)                          (b)

(c)                          (d)

**Problem 4/26**

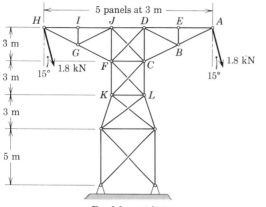

**Problem 4/27**

▶**4/27** The tower for a transmission line is modeled by the truss shown. The crossed members in the center sections of the truss may be assumed capable of supporting tension only. For the loads of 1.8 kN applied in the vertical plane, compute the forces induced in members $AB$, $DB$, and $CD$.

*Ans.* $AB = 3.89$ kN $C$, $DB = 0$, $CD = 0.93$ kN $C$

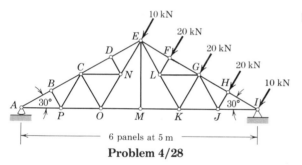

**Problem 4/28**

▶**4/28** Find the forces in members $EF$, $KL$, and $GL$ for the Fink truss shown.
*Ans.* $EF = 75.1$ kN $C$
$KL = 40$ kN $T$
$GL = 20$ kN $T$

## 4/4 *METHOD OF SECTIONS*

In the previous article on the analysis of plane trusses by the method of joints, we took advantage of only two of the three equilibrium equations, since the procedures involve concurrent forces at each joint. We may take advantage of the third or moment equation of equilibrium by selecting an entire section of the truss for the free body in equilibrium under the action of a nonconcurrent system of forces. This *method of sections* has the basic advantage that the force in almost any desired member may be found directly from an analysis of a section which has cut that member. Thus, it is not necessary to proceed with the calculation from joint to joint until the member in question has been reached. In choosing a section of the truss, we note that, in general, not more than three members whose forces are unknown may be cut, since there are only three available equilibrium relations which are independent.

The method of sections will now be illustrated for the truss in Fig. 4/5, which was used in the explanation of the previous method. The truss is shown again in Fig. 4/11*a* for ready reference. The external reactions are first computed as before, considering the truss as a whole. Now let us determine the force in the member *BE*, for example. An imaginary section, indicated by the dotted line, is passed through the truss, cutting it into two parts, Fig. 4/11*b*. This section has cut three members whose forces are initially unknown. In order for the portion of the truss on each side of the section to remain in equilibrium, it is necessary to apply to each cut member the force that was exerted on it by the member cut away. These forces, either tensile or compressive, will always be in the directions of the respective members for simple trusses composed of two-force members. The left-hand section is in equilibrium under the action of the applied load *L*, the end reaction $R_1$, and the three forces exerted on the cut members by the right-hand section which has been removed. We may usually draw the forces with their proper senses by a visual approximation of the equilibrium requirements. Thus, in balancing the moments about point *B* for the left-hand section, the force *EF* is clearly to the left, which makes it compressive, since it acts toward the cut section of member *EF*. The load *L* is greater than the reaction $R_1$, so that the force *BE* must be up and to the right to supply the needed upward component for vertical equilibrium. Force *BE* is therefore tensile, since it acts away from the cut section. With the approximate magnitudes of $R_1$ and *L* in mind we see that the balance of moments about point *E* requires that *BC* be to the right. A casual glance at the truss should lead to the same conclusion when it is realized that the lower horizontal member will stretch under the tension caused by bending. The equation of moments about joint *B* eliminates three forces from the relation, and *EF* may be determined directly. The force *BE* is calculated from the equilibrium equation

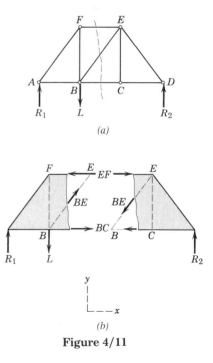

(a)

(b)

**Figure 4/11**

for the *y*-direction. Finally, we determine *BC* by balancing moments about point *E*. In this way each of the three unknowns has been determined independently of the other two.

The right-hand section of the truss, Fig. 4/11*b*, is in equilibrium under the action of $R_2$ and the same three forces in the cut members applied in the directions opposite to those for the left section. The proper sense for the horizontal forces may easily be seen from the balance of moments about points *B* and *E*.

It is essential to understand that in the method of sections an entire portion of the truss is considered a single body in equilibrium. Thus, the forces in members internal to the section are not involved in the analysis of the section as a whole. To clarify the free body and the forces acting externally on it, the section is preferably passed through the members and not the joints. We may use either section of a truss for the calculations, but the one involving the smaller number of forces will usually yield the simpler solution.

In some cases the methods of sections and joints can be combined for an efficient solution. For example, suppose we wish to find the force in a central member of a large truss. Furthermore, suppose that it is not possible to pass a section through this member without passing through at least four unknown members. It may be possible to determine the forces in nearby members by the method of sections and then progress to the unknown member by the method of joints. Such a combination of the two methods may be more expedient than exclusive use of either method.

The moment equations are used to great advantage in the method of sections. One should choose a moment center, either on or off the section, through which as many unknown forces as possible pass. It is not always possible to assign the proper sense of an unknown force when the free-body diagram of a section is initially drawn. With an arbitrary assignment made, a positive answer will verify the assumed sense and a negative result will indicate that the force is in the sense opposite to that assumed.

An alternative notation preferred by some is to assign all unknown forces arbitrarily as positive in the tension direction (away from the section) and let the algebraic sign of the answer distinguish between tension and compression. Thus, a plus sign would signify tension and a minus sign compression. On the other hand, the advantage of assigning forces in their correct sense on the free-body diagram of a section wherever possible is that doing so emphasizes the physical action of the forces more directly, and this practice is preferred in this treatment.

## Sample Problem 4/2

Calculate the forces induced in members $KL$, $CL$, and $CB$ by the 20-ton load on the cantilever truss.

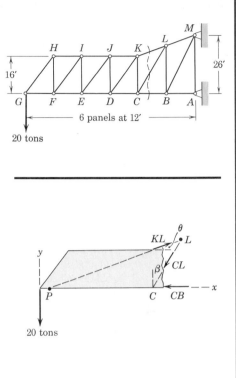

20 tons

**Solution.** Although the vertical components of the reactions at $A$ and $M$ are statically indeterminate with the two fixed supports, all members other than $AM$ are statically determinate. We may pass a section directly through members $KL$, $CL$, and $CB$ and analyze the portion of the truss to the left of this section as a statically determinate ① rigid body.

The free-body diagram of the portion of the truss to the left of the section is shown. A moment sum about $L$ quickly verifies the assignment of $CB$ as compression, and a moment sum about $C$ quickly disclosed that $KL$ is in tension. The direction of $CL$ is not quite so obvious until we observe that $KL$ and $CB$ intersect at a point $P$ to the right of $G$. A moment sum about $P$ eliminates reference to $KL$ and $CB$ and shows that $CL$ must be compressive to balance the moment of the 20-ton force about $P$. With these considerations in mind the solution becomes straightforward, as we now see how to solve for each of the three unknowns independently of the other two.

② Summing moments about $L$ requires the moment arm $BL = 16 + (26 - 16)/2 = 21$ ft. Thus,

$$[\Sigma M_L = 0] \quad 20(5)(12) - CB(21) = 0 \quad CB = 57.1 \text{ tons } C \qquad Ans.$$

Next we take moments about $C$, which requires a calculation of $\cos \theta$. From the given dimensions we see $\theta = \tan^{-1}(5/12)$ so that $\cos \theta = 12/13$. Therefore,

$$[\Sigma M_C = 0] \quad 20(4)(12) - \tfrac{12}{13}KL(16) = 0 \quad KL = 65.0 \text{ tons } T \qquad Ans.$$

Finally, we may find $CL$ by a moment sum about $P$, whose distance from $C$ is given by $PC/16 = 24/(26 - 16)$ or $PC = 38.4$ ft. We also need $\beta$ which is given by $\beta = \tan^{-1}(\overline{CB}/\overline{BL}) = \tan^{-1}(12/21) = 29.7°$ and $\cos \beta = 0.868$. We now have

③ $$[\Sigma M_P = 0] \quad 20(48 - 38.4) - CL(0.868)(38.4) = 0$$

$$CL = 5.76 \text{ tons } C \qquad Ans.$$

① We note that analysis by the method of joints would necessitate working with eight joints in order to calculate the three forces in question. Thus, the method of sections offers a considerable advantage in this case.

② We could have started with moments about $C$ or $P$ just as well.

③ We could also have determined $CL$ by a force summation in either the $x$- or $y$-direction.

## Sample Problem 4/3

Calculate the force in member $DJ$ of the Howe roof truss illustrated. Neglect any horizontal components of force at the supports.

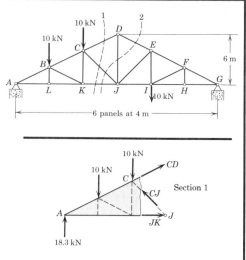

**Solution.** It is not possible to pass a section through $DJ$ without cutting four members whose forces are unknown. Although three of these cut by section 2 are concurrent at $J$ and therefore the moment equation about $J$ could be used to obtain $DE$, the force in $DJ$ cannot be obtained from the remaining two equilibrium principles. It is necessary to consider first the adjacent section 1 before analyzing section 2.

The free-body diagram for section 1 is drawn and includes the reaction of 18.3 kN at $A$, which is previously calculated from the equilibrium of the truss as a whole. In assigning the proper directions for the forces acting on the three cut members, we see that a balance of moments about $A$ eliminates the effects of $CD$ and $JK$ and clearly requires that $CJ$ be up and to the left. A balance of moments about $C$ eliminates the effect of the three forces concurrent at $C$ and indicates that $JK$ must be to the right to supply sufficient counterclockwise moment. Again it should be fairly obvious that the lower chord is under tension because of the bending tendency of the truss. Although it should also be apparent that the top chord is under compression, for purposes of illustration the force in $CD$ will be arbitrarily assigned as tension.

By the analysis of section 1, $CJ$ is obtained from

$[\Sigma M_A = 0]$     $0.707CJ(12) - 10(4) - 10(8) = 0$     $CJ = 14.1$ kN $C$

In this equation the moment of $CJ$ is calculated by considering its horizontal and vertical components acting at point $J$. Equilibrium of moments about $J$ requires

$[\Sigma M_J = 0]$     $0.894CD(6) + 18.3(12) - 10(4) - 10(8) = 0$

$CD = -18.6$ kN

The moment of $CD$ about $J$ is calculated here by considering its two components as acting through $D$. The minus sign indicates that $CD$ was assigned in the wrong direction.

Hence,                     $CD = 18.6$ kN $C$

From the free-body diagram of section 2, which now includes the known value of $CJ$, a balance of moments about $G$ is seen to eliminate $DE$ and $JK$. Thus,

$[\Sigma M_G = 0]$     $12DJ + 10(16) + 10(20) - 18.3(24) - 14.1(0.707)(12) = 0$

$DJ = 16.6$ kN $T$                     *Ans.*

Again the moment of $CJ$ is determined from its components considered to be acting at $J$. The answer for $DJ$ is positive, so that the assumed tensile direction is correct.

An alternative approach to the entire problem is to utilize section 1 to determine $CD$ and then use the method of joints applied at $D$ to determine $DJ$.

① There is no harm in assigning one or more of the forces in the wrong direction as long as the calculations are consistent with the assumption. A negative answer will show the need for reversing the direction of the force.

② If desired, the direction of $CD$ may be changed on the free-body diagram and the algebraic sign of $CD$ reversed in the calculations, or else the work may be left as it stands with a note stating the proper direction.

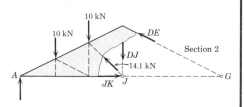

③ Observe that a section through members $CD$, $DJ$, and $DE$ could be taken that would cut only three unknown members. However, since the forces in these three members are all concurrent at $D$, a moment equation about $D$ would yield no information about them. The remaining two force equations would not be sufficient to solve for the three unknowns.

## PROBLEMS

(Use the method of sections in solving the following problems. Neglect the weight of the members compared with the forces they support.)

### *Introductory problems*

**4/29** Compute the forces in members *BC*, *CF*, and *EF*.

Ans. $BC = 2.13$ kN $C$, $CF = 0.300$ kN $T$
$EF = 1.875$ kN $T$

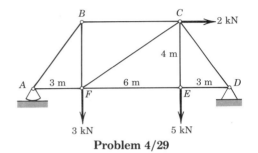

**Problem 4/29**

**4/30** Determine the forces in members *CG* and *GH*.

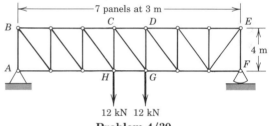

**Problem 4/30**

**4/31** Determine the forces in members *GH* and *CG* for the truss loaded and supported as shown. Does the statical indeterminacy of the supports affect your calculation?

Ans. $CG = 70.7$ kN $T$, $GH = 100$ kN $T$, No

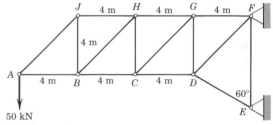

**Problem 4/31**

**4/32** Determine the forces in members *BC*, *BE*, and *BF*. The triangles are equilateral.

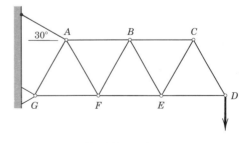

**Problem 4/32**

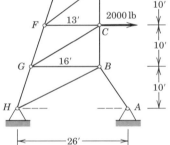

**Problem 4/33**

**4/33** Calculate the forces in members $BC$, $BE$, and $EF$. Solve for each force from an equilibrium equation which contains that force as the only unknown.

*Ans.* $BC = 21$ kN $T$, $BE = 8.41$ kN $T$
$EF = 29.5$ kN $C$

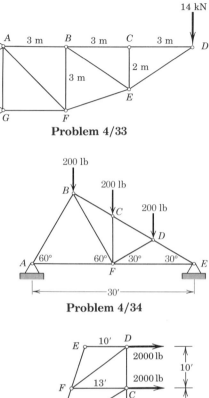

**Problem 4/34**

**4/34** The asymmetric roof truss is of the type used when a near normal angle of incidence of sunlight onto the south-facing surface $AB$ is desirable for solar energy purposes. Determine the forces in members $CD$, $CF$, and $EF$. Ignore any horizontal reactions at the supports.

## *Representative problems*

**4/35** The truss of Prob. 4/14 is repeated here. Solve for the force in member $CG$ from an equilibrium equation which contains that force as the only unknown.

*Ans.* $CG = 4170$ lb $T$

**Problem 4/35**

**4/36** Determine the force in member $DG$ of the loaded truss.

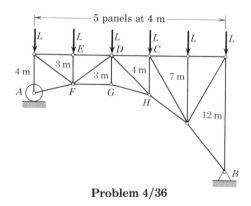

**Problem 4/36**

**4/37** Determine the forces in members *BC* and *CG* of the truss of Prob. 4/12 repeated here.

*Ans.* $BC = L/3$ T, $CG = L/3$ T

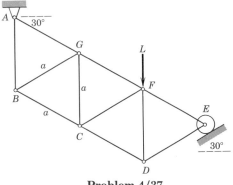

**Problem 4/37**

**4/38** Determine the forces in members *DE* and *DL*.

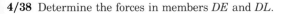

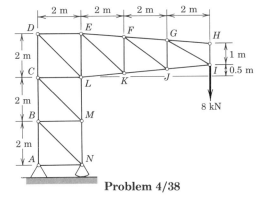

**Problem 4/38**

**4/39** Determine the forces in members *BC* and *CI*.

*Ans.* $BC = 62.5$ kN C, $CI = 34.5$ kN C

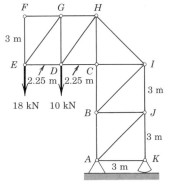

**Problem 4/39**

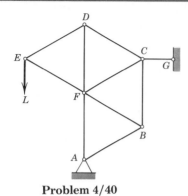

**Problem 4/40**

**4/40** The truss of Prob. 4/20 is repeated here. Calculate the forces in members $AF$, $BF$, and $BC$. After finding the force in link $CG$, solve for each of the three forces from an equilibrium equation which contains that force as the only unknown.

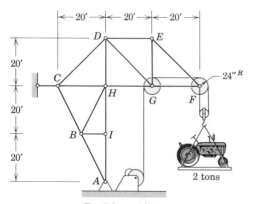

**Problem 4/41**

**4/41** Determine the forces in members $BC$, $BE$, and $EF$ of the loaded crane truss.    *Ans.* $BC = 23.1$ lb $T$
$BE = 257$ lb $T$
$EF = 1054$ lb $C$

**4/42** A crane is modeled by the truss shown. Compute the forces in members $DE$, $DG$, and $HG$ under the load of the 2-ton tractor.

**Problem 4/42**

**4/43** The signboard truss is designed to support a horizontal wind load of 800 lb. If the resultant of this load passes through point $C$, calculate the forces in members $BG$ and $BF$.

*Ans.* $BG = 400\sqrt{5}$ lb $C$, $BF = 800$ lb $T$

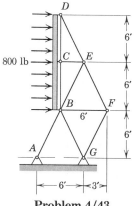

**Problem 4/43**

**4/44** Determine the force in member $BF$.

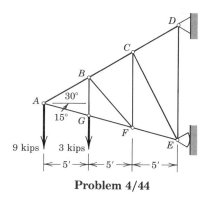

**Problem 4/44**

**4/45** The truss shown is composed of 45° right triangles. The crossed members in the center two panels are slender tie rods incapable of supporting compression. Retain the two rods which are under tension and compute the magnitudes of their tensions. Also find the force in member $MN$.

*Ans.* $FN = GM = 84.8$ kN $T$
$MN = 20$ kN $T$

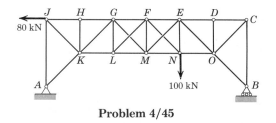

**Problem 4/45**

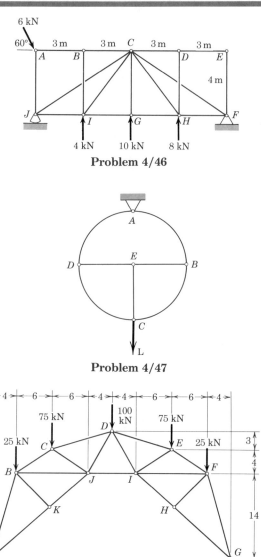

**Problem 4/46**

**Problem 4/47**

Dimensions in Meters

**Problem 4/48**

**Problem 4/49**

**4/46** The members $CJ$ and $CF$ of the loaded truss cross but are not connected to members $BI$ and $DG$. Compute the forces in members $BC$, $CJ$, $CI$, and $HI$.

**4/47** Determine the force in member $DE$ of the truss. Note that the curved members act as two-force members.
Ans. $DE = LC$

**4/48** Determine the forces in members $DE$, $EI$, $FI$, and $HI$ of the arched roof truss.

**4/49** Find the force in member $JQ$ for the Baltimore truss where all angles are 30°, 60°, 90°, or 120°.
Ans. $JQ = 57.7$ kN C

▶**4/50** Determine the force in member *DK* of the loaded overhead sign truss.     *Ans. DK* = 1 kip *T*

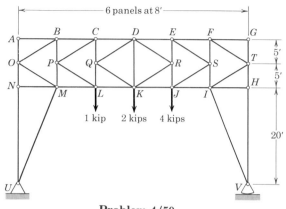

**Problem 4/50**

▶**4/51** In the traveling bridge crane shown all crossed members are slender tie rods incapable of supporting compression. Determine the forces in members *DF* and *EF* and find the horizontal reaction on the truss at *A*. Show that if *CF* = 0, *DE* = 0 also.

*Ans. DF* = 768 kN *C*, *EF* = 364 kN *C*

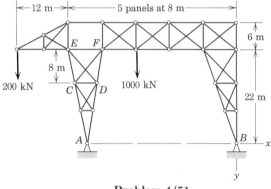

**Problem 4/51**

**4/52** Determine the force in member *DG* of the compound truss. The joints all lie on radial lines subtending angles of 15° as indicated, and the curved members act as two-force members. Distance $\overline{OC}$ = $\overline{OA}$ = $\overline{OB}$ = *R*.     *Ans. DG* = 0.569*L C*

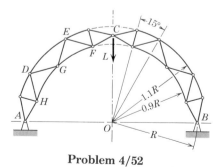

**Problem 4/52**

### 4/5 SPACE TRUSSES

A space truss is the three-dimensional counterpart of the plane truss described in the three previous articles. The idealized space truss consists of rigid links connected at their ends by ball-and-socket joints. We saw that a triangle of pin-connected bars forms the basic noncollapsible unit for the plane truss. A space truss, on the other hand, requires six bars joined at their ends to form the edges of a tetrahedron for the basic noncollapsible unit. In Fig. 4/12*a* the two bars *AD* and *BD* joined at *D* require a third support *CD* to keep the triangle *ADB* from rotating about *AB*. In Fig. 4/12*b* the supporting base is replaced by three more bars *AB*, *BC*, and *AC* to form a tetrahedron not dependent on the foundation for its own rigidity. We may form a new rigid unit to extend the structure with three additional concurrent bars whose ends are attached to three fixed joints on the existing structure. Thus, in Fig. 4/12*c* the bars *AF*, *BF*, and *CF* are attached to the foundation and therefore fix point *F* in space. Likewise point *H* is fixed in space by the bars *AH*, *DH*, and *CH*. The three additional bars *CG*, *FG*, and *HG* are attached to the three fixed points *C*, *F*, and *H* and therefore fix *G* in space. Point *E* is similarly established. We see now that the structure is entirely rigid. The two applied loads shown will result in forces in all of the members.

Ideally there must be point support, such as that given by a ball-and-socket joint, for the connections of a space truss so that there will be no bending in the members. Again, as in riveted and welded connections for plane trusses, if the centerlines of joined members intersect at a point, we may justify the assumption of two-force members under simple tension and compression.

For a space truss which is supported externally in such a way that it is statically determinate as an entire unit, a relationship exists between the number of its joints and the number of its members necessary for internal stability without redundancy. Since the equilibrium of each joint is specified by three scalar force equations, there are in all $3j$ such equations for a simple space truss with $j$ joints. For the entire truss composed of $m$ members there are $m$ unknowns plus six unknown support reactions in the general case of a statically determinate space structure. Thus, for a simple space truss composed of tetrahedral elements, the equation $m + 6 = 3j$ will be satisfied if the truss is statically determinate internally. Again, as in the case of the plane truss, this relation is a necessary condition for stability, but it is not a sufficient condition, since one or more of the $m$ members can be arranged in such a way as not to contribute to a stable configuration of the entire truss. If $m + 6 > 3j$, there are more members than there are independent equations, and the truss is statically indeterminate internally with redundant members present. If $m + 6 < 3j$, there is a deficiency of internal

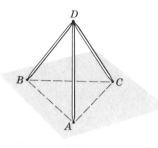

(a)

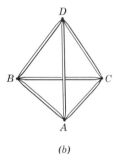

(b)

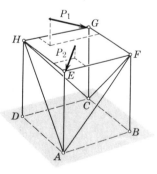

(c)

**Figure 4/12**

members, and the truss is unstable and subject to collapse under load. The foregoing relationship between the number of joints and the number of members for a space truss is very helpful in the preliminary design for such a truss, since the configuration is not nearly so obvious as in the case of a plane truss, where the geometry for statical determinacy is generally quite apparent.

The method of joints developed in Art. 4/3 for plane trusses may be extended directly to space trusses by satisfying the complete vector equation

$$\Sigma \mathbf{F} = \mathbf{0} \qquad\qquad (4/1)$$

for each joint. We normally begin the analysis at a joint where at least one known force acts and not more than three unknown forces are present. Adjacent joints on which not more than three unknown forces act may be analyzed in turn.

The above step-by-step joint technique tends to minimize the number of simultaneous equations which must be solved when the forces in all members of the space truss are to be determined. For this reason, such a judicious approach is recommended. As an alternative procedure, however, we may systematically write $3j$ joint equations by applying Eq. 4/1 to all joints of the space frame. The number of unknowns would be $m + 6$ if there are six external support reactions. If the number of equations ($3j$) equals the number of unknowns ($m + 6$), then the entire system of equations may be solved simultaneously for the unknowns. Because of the large number of coupled equations, a computer solution is usually warranted. With this latter approach, it is not necessary to begin at a joint where at least one known and no more than three unknown forces act.

The method of sections developed in the previous article may also be applied to space trusses. The two vector equations

$$\Sigma \mathbf{F} = \mathbf{0} \qquad \text{and} \qquad \Sigma \mathbf{M} = \mathbf{0}$$

must be satisfied for any section of the truss, where the zero moment sum will hold for all moment axes. Since the two vector equations are equivalent to six scalar equations, we conclude that a section should in general not be passed through more than six members whose forces are unknown. The method of sections for space trusses is not widely used, however, because a moment axis can seldom be found which eliminates all but one unknown, as in the case of plane trusses.

Vector notation for expressing the terms in the force and moment equations for space trusses is of considerable advantage and is used in the sample problem that follows.

202

## Sample Problem 4/4

The space truss consists of the rigid tetrahedron $ABCD$ anchored by a ball-and-socket connection at $A$ and prevented from any rotation about the $x$-, $y$-, or $z$-axes by the respective links 1, 2, and 3. The load $L$ is applied to joint $E$, which is rigidly fixed to the tetrahedron by the three additional links. Solve for the forces in the members at joint $E$ and indicate the procedure for the solution of the forces in the remaining members of the truss.

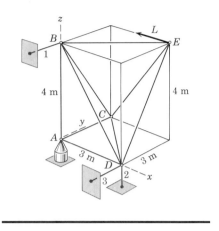

**Solution.** We note first that the truss is supported with six properly placed constraints, which are the three at $A$ and the links 1, 2, and 3. Also, with $m = 9$ members and $j = 5$ joints the condition $m + 6 = 3j$ for a sufficiency of members to provide a noncollapsible structure is satisfied.

The external reactions at $A$, $B$, and $D$ can be calculated easily as a first step, although their values will be determined from the solution of all forces on each of the joints in succession.

① We must start with a joint on which at least one known force and not more than three unknown forces act, which in this case is joint $E$. The free-body diagram of joint $E$ is shown with all force vectors arbi-② trarily assumed in their positive tension directions (away from the joint). The vector expressions for the three unknown forces are

$$\mathbf{F}_{EB} = \frac{F_{EB}}{\sqrt{2}}(-\mathbf{i} - \mathbf{j}), \quad \mathbf{F}_{EC} = \frac{F_{EC}}{5}(-3\mathbf{i} - 4\mathbf{k}), \quad \mathbf{F}_{ED} = \frac{F_{ED}}{5}(-3\mathbf{j} - 4\mathbf{k})$$

Equilibrium of joint $E$ requires

$$[\Sigma\mathbf{F} = 0] \qquad \mathbf{L} + \mathbf{F}_{EB} + \mathbf{F}_{EC} + \mathbf{F}_{ED} = 0 \qquad \text{or}$$

$$-L\mathbf{i} + \frac{F_{EB}}{\sqrt{2}}(-\mathbf{i} - \mathbf{j}) + \frac{F_{EC}}{5}(-3\mathbf{i} - 4\mathbf{k}) + \frac{F_{ED}}{5}(-3\mathbf{j} - 4\mathbf{k}) = 0$$

Rearranging terms gives

$$\left(-L - \frac{F_{EB}}{\sqrt{2}} - \frac{3F_{EC}}{5}\right)\mathbf{i} + \left(-\frac{F_{EB}}{\sqrt{2}} - \frac{3F_{ED}}{5}\right)\mathbf{j} + \left(-\frac{4F_{EC}}{5} - \frac{4F_{ED}}{5}\right)\mathbf{k} = 0$$

Equating the coefficients of the $\mathbf{i}$-, $\mathbf{j}$-, and $\mathbf{k}$-unit vectors to zero gives the three equations

$$\frac{F_{EB}}{\sqrt{2}} + \frac{3F_{EC}}{5} = -L \qquad \frac{F_{EB}}{\sqrt{2}} + \frac{3F_{ED}}{5} = 0 \qquad F_{EC} + F_{ED} = 0$$

Solving the equations gives us

$$F_{EB} = -L/\sqrt{2} \qquad F_{EC} = -5L/6 \qquad F_{ED} = 5L/6 \qquad Ans.$$

Thus, we conclude that $F_{EB}$ and $F_{EC}$ are compressive forces and $F_{ED}$ is tension.

Unless we have computed the external reactions first, we must next analyze joint $C$ with the known value of $F_{EC}$ and the three unknowns $F_{CB}$, $F_{CA}$, and $F_{CD}$. The procedure is identical with that used for joint $E$. Joints $B$, $D$, and $A$ are then analyzed in the same way and in that order, which limits the unknowns to three for each joint. The external reactions computed from these analyses must, of course, agree with the values which can be determined initially from an analysis of the truss as a whole.

① *Suggestion*: Draw a free-body diagram of the truss as a whole and verify that the external forces acting on the truss are $\mathbf{A}_x = L\mathbf{i}$, $\mathbf{A}_y = L\mathbf{j}$, $\mathbf{A}_z = (4L/3)\mathbf{k}$, $\mathbf{B}_y = 0$, $\mathbf{D}_y = -L\mathbf{j}$, $\mathbf{D}_z = -(4L/3)\mathbf{k}$.

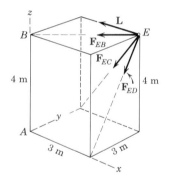

② With this assumption, a negative numerical value for a force would indicate compression.

# PROBLEMS

(In the following problems use plus for tension and minus for compression.)

**4/53** Determine the forces in members $AB$, $AC$, and $AD$.

$$Ans. \; AB = AD = -2.64 \text{ kN}$$
$$AC = -2.46 \text{ kN}$$

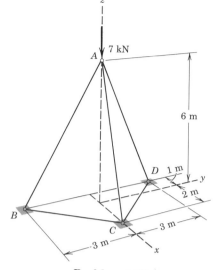

**Problem 4/53**

**4/54** The base of an automobile jackstand forms an equilateral triangle of side length 10 in. and is centered under the collar $A$. Model the structure as one with ball and sockets at all joints and determine the forces in members $BC$, $BD$, and $CD$. Neglect any horizontal reaction components under the feet $B$, $C$, and $D$.

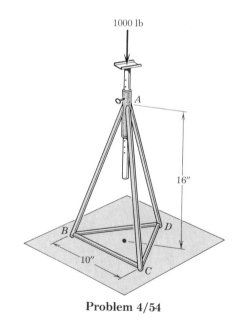

**Problem 4/54**

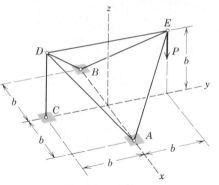

**Problem 4/55**

**4/55** The space truss in the form of a tetrahedron is supported by ball-and-socket connections at its base points $A$ and $B$ and is prevented from rotating about $AB$ by the vertical tie bar $CD$. After noting the vertical components of the reactions under the symmetrical truss at $A$ and $B$, draw a free-body diagram of the triangular configuration of links $BDE$ and determine the $x$-component of the force exerted by the foundation on the truss at $B$.

*Ans.* $B_x = P$

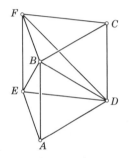

**Problem 4/56**

**4/56** The prismatic space truss has a horizontal base $ADE$ and a parallel top face $BCF$ in the shape of equal equilateral triangles which are connected by three equal vertical legs and braced by three diagonal members as shown. Show that this truss represents a stable configuration.

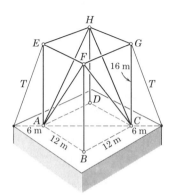

**Problem 4/57**

**4/57** The rectangular space truss 16 m in height is erected on a horizontal square base 12 m on a side. Guy wires are attached to the structure at $E$ and $G$ as shown and are tightened until the tension $T$ in each wire is 9 kN. Calculate the force $F$ in each of the diagonal members. *Ans.* $F = -3.72$ kN

**4/58** The tetrahedral space truss has a horizontal base *ABC* in the form of an isosceles triangle and legs *AD*, *BD*, and *CD* which support the mass *m* from point *D*. Each vertex of the base is suspended by a vertical wire from overhead supports. Calculate the force induced in members *AC* and *AB*.

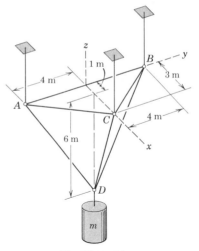

**Problem 4/58**

**4/59** Determine the force in member *BD* of the regular pyramid with square base.     *Ans.* $DB = -2.00L$

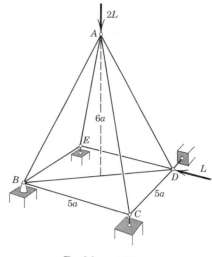

**Problem 4/59**

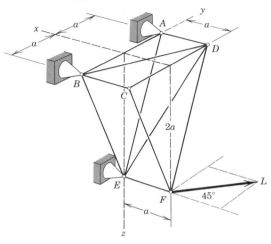

**Problem 4/60**

**4/60** The space truss shown is secured to the fixed supports at $A$, $B$, and $E$ and is loaded by the force $L$ which has equal $x$- and $y$-components but no vertical $z$-component. Show that there are a sufficient number of members to provide internal stability and that their placement is adequate for this purpose. Next determine the forces in members $CD$, $BC$, and $CE$.

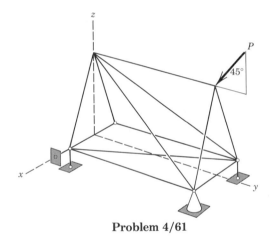

**Problem 4/61**

**4/61** For the space truss shown check the sufficiency of the supports and also the number and arrangement of the members to ensure statical determinacy, both external and internal. By inspection determine the forces in members $DC$, $CB$, and $CF$. Calculate the force in member $AF$ and the $x$-component of the reaction on the truss at $D$.

$$Ans.\ F_{AF} = \frac{\sqrt{13}}{3\sqrt{2}} P,\ D_x = \frac{P}{3\sqrt{2}}$$

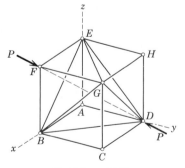

**Problem 4/62**

▶**4/62** A space truss is constructed in the form of a cube with six diagonal members shown. Verify that the truss is internally stable. If the truss is subjected to the compressive forces $P$ applied at $F$ and $D$ along the diagonal $FD$, determine the forces in members $FE$ and $EG$.

$$Ans.\ F_{FE} = -P/\sqrt{3},\ F_{EG} = P/\sqrt{6}$$

**4/63** The lengthy boom of an overhead construction crane, a portion of which is shown, is an example of a periodic structure—one which is composed of repeated and identical structural units. Use the method of sections to find the forces in members $FJ$ and $GJ$.

*Ans.* $FJ = 0$, $GJ = -70.8$ kN

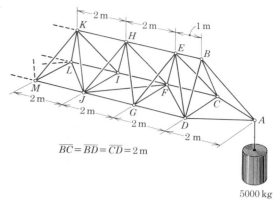

$\overline{BC} = \overline{BD} = \overline{CD} = 2\,\text{m}$

5000 kg

**Problem 4/63**

**4/64** A space truss consists of two pyramids on identical square bases in the horizontal $x$-$y$ plane with common side $DG$. The truss is loaded at the vertex $A$ by the downward force $L$ and is supported by the vertical reactions shown at its corners. All members except the two base diagonals are of the same length $b$. Take advantage of the two vertical planes of symmetry and determine the forces in $AB$ and $DA$. (Note that link $AB$ prevents the two pyramids from hinging about $DG$.)

*Ans.* $AB = -\dfrac{\sqrt{2}L}{4}$

$DA = -\dfrac{\sqrt{2}L}{8}$

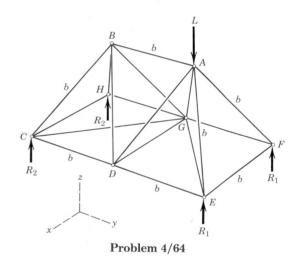

**Problem 4/64**

## 4/6 FRAMES AND MACHINES

A structure is called a *frame* or *machine* if at least one of its individual members is a *multiforce member*. A multiforce member is defined as one with three or more forces acting on it or one with two or more forces and one or more couples acting on it. Frames are structures which are designed to support applied loads and are usually fixed in position. Machines are structures which contain moving parts and are designed to transmit forces or couples from input values to output values.

Because frames and machines contain multiforce members, the forces in these members in general will *not* be in the directions of the members. Therefore, we cannot analyze these structures by the methods developed in Arts. 4/3, 4/4, and 4/5 for simple trusses composed of two-force members where the forces are in the directions of the members.

In the previous chapter the equilibrium of multiforce bodies was discussed and illustrated, but we concentrated on the equilibrium of a *single* rigid body. In this present article attention is focused on the equilibrium of *interconnected* rigid bodies which contain multiforce members. Although most such bodies may be analyzed as two-dimensional systems, there are numerous examples of frames and machines which are three-dimensional.

The forces acting on each member of a connected system are found by isolating the member with a free-body diagram and applying the established equations of equilibrium. The *principle of action and reaction* must be carefully observed when we represent the forces of interaction on the separate free-body diagrams. If the structure contains more members or supports than are necessary to prevent collapse, then, as in the case of trusses, the problem is statically indeterminate, and the principles of equilibrium, although necessary, are not sufficient for solution. Although many frames and machines are statically indeterminate, we will consider in this article only those which are statically determinate.

If the frame or machine constitutes a rigid unit by itself when removed from its supports, as in the A-frame in Fig. 4/13*a*, the

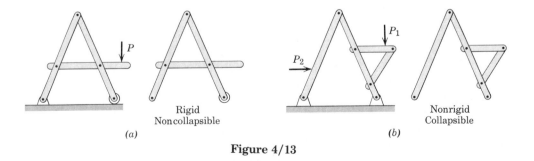

Rigid
Noncollapsible

*(a)*

Nonrigid
Collapsible

*(b)*

**Figure 4/13**

analysis is best begun by establishing all the forces external to the structure considered as a single rigid body. We then dismember the structure and consider the equilibrium of each part separately. The equilibrium equations for the several parts will be related through the terms involving the forces of interaction. If the structure is not a rigid unit by itself but depends on its external supports for rigidity, as illustrated in Fig. 4/13*b*, then the calculation of the external support reactions cannot be completed until the structure is dismembered and the individual parts are analyzed.

In most cases we find that the analysis of frames and machines is facilitated by representing the forces in terms of their rectangular components. This is particularly so when the dimensions of the parts are given in mutually perpendicular directions. The advantage of this representation is that the calculation of moment arms is accordingly simplified. In some three-dimensional problems, particularly when moments are evaluated about axes that are not parallel to the coordinate axes, we find the use of vector notation is an advantage.

It is not always possible to assign every force or its components in the proper sense when drawing the free-body diagrams, and it becomes necessary for us to make an arbitrary assignment. In any event it is *absolutely necessary* that a force be *consistently* represented on the diagrams for interacting bodies which involve the force in question. Thus, for two bodies connected by the pin $A$, Fig. 4/14*a*, the force components must be consistently represented in *opposite* directions on the separate free-body diagrams. For a ball-and-socket connection between members of a space frame, we must apply the action-and-reaction principle to all three components as shown in Fig. 4/14*b*. The assigned directions may prove to be wrong when the algebraic signs of the components are determined upon calculation. If $A_x$, for instance, should turn out to be negative, it is actually acting in the direction opposite to that originally represented. Accordingly it would be necessary for us to reverse the direction of the force on *both* members and to reverse the sign of its force terms in the equations. Or we may leave the representation as originally made, and the proper sense of the force will be understood from the negative sign. If we choose to use vector notation in labeling the forces, then we must be careful to use a plus sign for an action and a minus sign for the corresponding reaction, as shown in Fig. 4/15.

Finally, situations occasionally arise where it is necessary to solve two or more equations simultaneously in order to separate the unknowns. In most instances, however, we may avoid simultaneous solutions by careful choice of the member or group of members for the free-body diagram and by a careful choice of moment axes which will eliminate undesired terms from the equations. The method of solution described in the foregoing paragraphs is illustrated in the following sample problems.

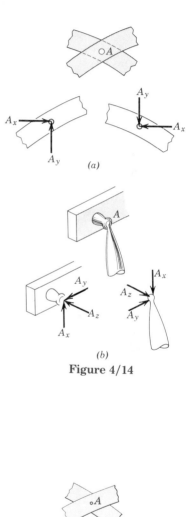

**Figure 4/14**

**Figure 4/15**

## Sample Problem  4/5

The frame supports the 400-kg load in the manner shown. Neglect the weights of the members compared with the forces induced by the load and compute the horizontal and vertical components of all forces acting on each of the members.

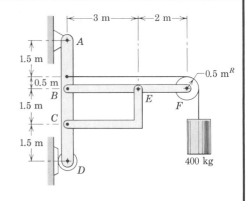

---

**Solution.**  We observe first that the three supporting members that constitute the frame form a rigid assembly that can be analyzed as a ① single unit. We also observe that the arrangement of the external supports makes the frame statically determinate.

From the free-body diagram of the entire frame we determine the external reactions, Thus,

$[\Sigma M_A = 0]$     $5.5(0.4)(9.81) - 5D = 0$     $D = 4.32$ kN

$[\Sigma F_x = 0]$          $A_x - 4.32 = 0$     $A_x = 4.32$ kN

$[\Sigma F_y = 0]$          $A_y - 3.92 = 0$     $A_y = 3.92$ kN

Next we dismember the frame and draw a separate free-body diagram of each member. The diagrams are arranged in their approximate relative positions to aid in keeping track of the common forces of interaction. The external reactions just obtained are entered onto the diagram for $AD$. Other known forces are the 3.92-kN forces exerted by the shaft of the pulley on the member $BF$, as obtained from the free-body diagram of the pulley. The cable tension of 3.92 kN is also shown acting on $AD$ at its attachment point.

Next, the components of all unknown forces are shown on the ② diagrams. Here we observe that $CE$ is a two-force member. The force components on $CE$ have equal and opposite reactions, which are shown on $BF$ at $E$ and on $AD$ at $C$. We may not recognize the actual sense of the components at $B$ at first glance, so they may be arbitrarily but consistently assigned.

The solution may proceed by use of a moment equation about $B$ or $E$ for member $BF$, followed by the two force equations. Thus,

$[\Sigma M_B = 0]$          $3.92(5) - \tfrac{1}{2}E_x(3) = 0$     $E_x = 13.08$ kN          *Ans.*

$[\Sigma F_y = 0]$     $B_y + 3.92 - 13.08/2 = 0$     $B_y = 2.62$ kN          *Ans.*

$[\Sigma F_x = 0]$     $B_x + 3.92 - 13.08 = 0$     $B_x = 9.15$ kN          *Ans.*

Positive numerical values of the unknowns mean that we assumed their directions correctly on the free-body diagrams. The value of $C_x = E_x = 13.08$ kN obtained by inspection of the free-body diagram of $CE$ is now entered onto the diagram for $AD$, along with the values of $B_x$ and $B_y$ just determined. The equations of equilibrium may now be applied to member $AD$ as a check, since all the forces acting on it have already been computed. The equations give

$[\Sigma M_C = 0]$   $4.32(3.5) + 4.32(1.5) - 3.92(2) - 9.15(1.5) = 0$

$[\Sigma F_x = 0]$          $4.32 - 13.08 + 9.15 + 3.92 + 4.32 = 0$

$[\Sigma F_y = 0]$               $-13.08/2 + 2.62 + 3.92 = 0$

① We see that the frame corresponds to the category illustrated in Fig. 4/13$a$.

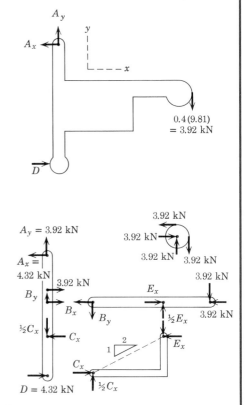

② Without this observation, the problem solution would be much longer, because the three equilibrium equations for member $BF$ would contain four unknowns: $B_x$, $B_y$, $E_x$, and $E_y$. Note that the direction of the line joining the two points of force application, and not the shape of the member, determines the direction of the forces acting on a two-force member.

# Sample Problem 4/6

Neglect the weight of the frame and compute the forces acting on all of its members.

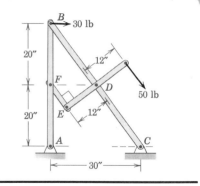

**Solution.** We note first that the frame is not a rigid unit when removed from its supports since *BDEF* is a movable quadrilateral and ① not a rigid triangle. Consequently the external reactions cannot be completely determined until the individual members are analyzed. However, we can determine the vertical components of the reactions at *A* and ② *C* from the free-body diagram of the frame as a whole. Thus,

$$[\Sigma M_C = 0] \quad 50(12) + 30(40) - 30A_y = 0 \quad A_y = 60 \text{ lb} \quad \textit{Ans.}$$

$$[\Sigma F_y = 0] \quad C_y - 50(4/5) - 60 = 0 \quad C_y = 100 \text{ lb} \quad \textit{Ans.}$$

Next we dismember the frame and draw the free-body diagram of each part. Since *EF* is a two-force member, the direction of the force at *E* on *ED* and at *F* on *AB* is known. We assume that the 30-lb force is ③ applied to the pin as a part of member *BC*. There should be no difficulty in assigning the correct directions for forces *E, F, D,* and *B_x*. The direction of *B_y*, however, may not be assigned by inspection and therefore is arbitrarily shown as downward on *AB* and upward on BC.

**Member ED.** The two unknowns are easily obtained by

$$[\Sigma M_D = 0] \quad 50(12) - 12E = 0 \quad E = 50 \text{ lb} \quad \textit{Ans.}$$

$$[\Sigma F = 0] \quad D - 50 - 50 = 0 \quad D = 100 \text{ lb} \quad \textit{Ans.}$$

**Member EF.** Clearly *F* is equal and opposite to *E* with the magnitude of 50 lb.

**Member AB.** Since *F* is now known, we solve for $B_x$, $A_x$, and $B_y$ from

$$[\Sigma M_A = 0] \quad 50(3/5)(20) - B_x(40) = 0 \quad B_x = 15 \text{ lb} \quad \textit{Ans.}$$

$$[\Sigma F_x = 0] \quad A_x + 15 - 50(3/5) = 0 \quad A_x = 15 \text{ lb} \quad \textit{Ans.}$$

$$[\Sigma F_y = 0] \quad 50(4/5) - 60 - B_y = 0 \quad B_y = -20 \text{ lb} \quad \textit{Ans.}$$

The minus sign shows that we assigned $B_y$ in the wrong direction.

**Member BC.** The results for $B_x$, $B_y$, and *D* are now transferred to ④ *BC*, and the remaining unknown $C_x$ is found from

$$[\Sigma F_x = 0] \quad 30 + 100(3/5) - 15 - C_x = 0 \quad C_x = 75 \text{ lb} \quad \textit{Ans.}$$

We may apply the remaining two equilibrium equations as a check. Thus,

$$[\Sigma F_y = 0] \quad 100 + (-20) - 100(4/5) = 0$$

$$[\Sigma M_C = 0] \quad (30 - 15)(40) + (-20)(30) = 0$$

① We see that this frame corresponds to the category illustrated in Fig. 4/13*b*.

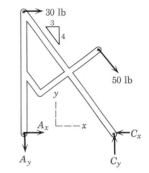

② The directions of $A_x$ and $C_x$ are not obvious initially and can be assigned arbitrarily to be corrected later if necessary.

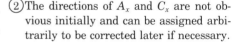
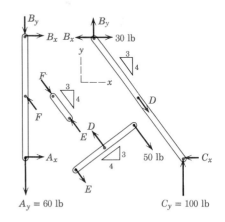

③ Alternatively the 30-lb force could be applied to the pin considered a part of *BA*, with a resulting change in the reaction $B_x$.

④ Alternatively we could have returned to the free-body diagram of the frame as a whole and found $C_x$.

## Sample Problem 4/7

The machine shown is an overload protection device which releases the load when it exceeds a predetermined value $T$. A soft metal shear pin $S$ is inserted in a hole in the lower half and is acted on by the upper half. When the total force on the pin exceeds its strength, it will break. The two halves then rotate about $A$ under the action of the tensions in $BD$ and $CD$, as shown in the second sketch, and rollers $E$ and $F$ release the eye bolt. Determine the maximum allowable tension $T$ if the pin $S$ will shear when the total force on it is 800 N. Also compute the corresponding force on the hinge pin $A$.

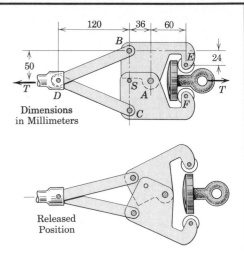

Dimensions in Millimeters

Released Position

---

**Solution.** Because of symmetry we analyze only one of the two hinged members. The upper part is chosen, and its free-body diagram along with that for the connection at $D$ is drawn. Because of symmetry the forces at $S$ and $A$ have no $x$-components. The two-force members $BD$ and $CD$ exert forces of equal magnitude $B = C$ on the connection at $D$. Equilibrium of the connection gives

① 

$[\Sigma F_x = 0] \quad B \cos \theta + C \cos \theta - T = 0 \quad 2B \cos \theta = T$

$$B = T/(2 \cos \theta)$$

From the free-body diagram of the upper part we express the equilibrium of moments about point $A$. Substituting $S = 800$ N and the expression for $B$ gives

② $[\Sigma M_A = 0]$

$$\frac{T}{2 \cos \theta}(\cos \theta)(50) + \frac{T}{2 \cos \theta}(\sin \theta)(36) - 36(800) - \frac{T}{2}(26) = 0$$

Substituting $\sin \theta/\cos \theta = \tan \theta = 5/12$ and solving for $T$ give

$$T\left(25 + \frac{5(36)}{2(12)} - 13\right) = 28\ 800$$

$$T = 1477 \text{ N} \quad \text{or} \quad T = 1.477 \text{ kN} \qquad Ans.$$

Finally, equilibrium in the $y$-direction gives us

$[\Sigma F_y = 0] \quad S - B \sin \theta - A = 0$

$$800 - \frac{1477}{2(12/13)}\frac{5}{13} - A = 0 \quad A = 492 \text{ N} \qquad Ans.$$

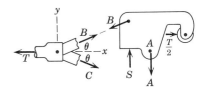

① It is always useful to recognize symmetry. Here it tells us that the forces acting on the two parts behave as mirror images of each other with respect to the $x$-axis. Thus, we cannot have an action on one member in the plus $x$-direction and its reaction on the other member in the negative $x$-direction. Consequently the forces at $S$ and $A$ have no $x$-components.

② Be careful not to forget the moment of the $y$-component of $B$. Note that our units here are newton-millimeters.

# PROBLEMS

(Unless otherwise instructed, neglect the mass of the various members in the problems that follow.)

### *Introductory problems*

**4/65** Determine the magnitude of the pin reaction at $B$ by (a) ignoring the fact that $BD$ is a two-force member and (b) recognizing that $BD$ is a two-force member. *Ans.* $B$ = 1250 N

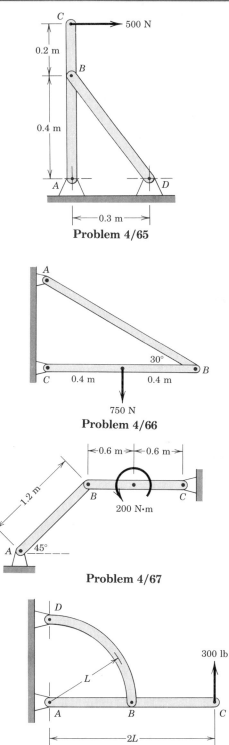

Problem 4/65

**4/66** Determine the magnitudes of the pin reactions at $A$ and $C$ for the simple frame loaded with the 750-N force.

Problem 4/66

**4/67** Determine the magnitude of the pin reaction at $C$. *Ans.* $C$ = 236 N

Problem 4/67

**4/68** Determine the magnitude of the pin reaction at $A$ for the frame loaded by the 300-lb force.

**4/69** Replace the 300-lb force in Prob. 4/68 by a 300-lb-ft clockwise couple applied at point $C$ and determine the magnitude of the pin reaction at $A$. For what value of $L$ is this pin reaction equal to that of Prob. 4/68? *Ans.* $A$ = 424/$L$, $L$ = 0.632 ft

Problem 4/68

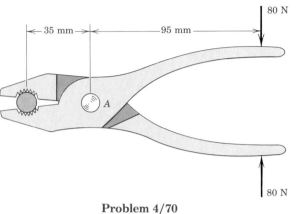

**Problem 4/70**

**4/70** For an 80-N squeeze on the handle of the pliers, determine the force $F$ applied to the round rod by each jaw. In addition, calculate the force supported by the pin at $A$.

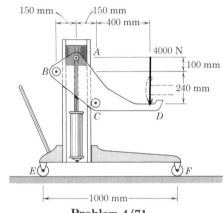

**Problem 4/71**

**4/71** The automobile bumper jack is subjected to a 4000-N downward load. Begin with a free-body diagram of $BCD$ and determine the force supported by roller $C$. Note that roller $B$ does not contact the vertical column.                         *Ans.* $C = 6470$ N

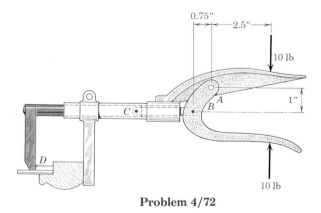

**Problem 4/72**

**4/72** The device shown in the figure is used to drive brads into picture-framing material. For a gripping force of 10 lb on the handles, determine the force $F$ exerted on the brad.

**4/73** Two uniform bars are supported in a vertical plane as shown. If the mass of $AC$ is 60 kg and that of $BD$ is 30 kg, with each center of mass at midlength, compute the force supported by the pin at $A$.

*Ans.* $A = 607$ N

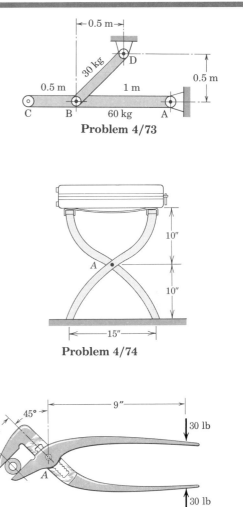

**Problem 4/73**

**4/74** A collapsible suitcase stand is constructed of two end units identical to the one shown, two connecting members (the ends of which are shown) on which the suitcase is placed, and two straps which limit the opening angle of the stand. If a 100-lb suitcase is symmetrically placed on the stand, determine the tension $T$ in each strap. Assume that the stand rests on a very smooth floor so that friction forces may be neglected.

**Problem 4/74**

### Representative problems

**4/75** Compute the force supported by the pin at $A$ for the slip-joint pliers under a grip of 30 lb.

*Ans.* $A = 157.6$ lb

**Problem 4/75**

**4/76** The "jaws-of-life" device is utilized by rescuers to pry apart wreckage, thus helping to free accident victims. If a pressure of 500 lb/in.$^2$ is developed behind the piston $P$ of area 20 in.$^2$, determine the vertical force $R$ which is exerted by the jaw tips on the wreckage for the position shown. Note that link $AB$ and its counterpart are both horizontal in the figure for this position.

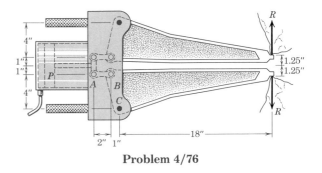

**Problem 4/76**

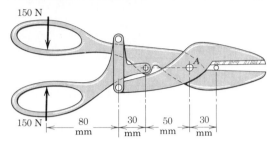

**Problem 4/77**

**4/77** Compound-lever snips, shown in the figure, are often used in place of regular tinners' snips when large cutting forces are required. For the gripping force of 150 N, what is the cutting force $P$ at a distance of 30 mm along the blade from the pin at $A$?

*Ans.* $P = 1467$ N

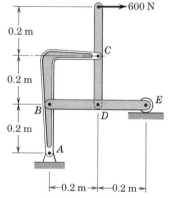

**Problem 4/78**

**4/78** Determine the force supported by the pin at $C$ for the loaded frame.

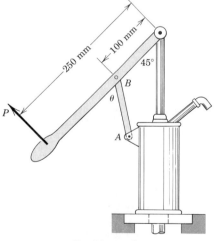

**Problem 4/79**

**4/79** For the 45° position of the pump handle with force $P$ perpendicular to the handle, determine graphically the angle $\theta$ between the handle and the compensating link $AB$ which enables the force transmitted to the plunger to be along its vertical axis.

*Ans.* $\theta = 59.0°$

**4/80** A small bolt cutter operated by hand for cutting small bolts and rods is shown in the sketch. For a hand grip $P = 150$ N, determine the force $Q$ developed by each jaw on the rod to be cut.

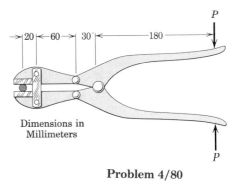

Dimensions in Millimeters

**Problem 4/80**

**4/81** In the spring clamp shown, an internal spring is coiled around the pin at $A$ and the spring ends bear against the inner surfaces of the handle halves in order to provide the desired clamping force. In the position shown, a force of magnitude $P = 6$ lb is required to release the clamp. Determine the compressive force at $B$ if $P = 0$.     *Ans.* $B = 16.50$ lb

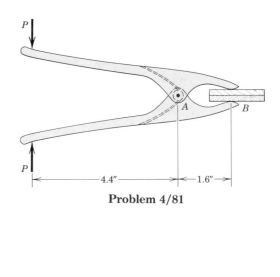

**Problem 4/81**

**4/82** The clamp is adjusted so that it exerts a pair of 200-N compressive forces on the boards between its swivel grips. Determine the force in the threaded shaft $BC$ and the magnitude of the pin reaction at $D$.

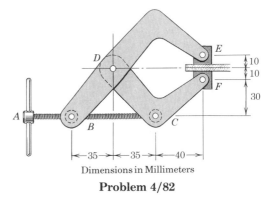

Dimensions in Millimeters

**Problem 4/82**

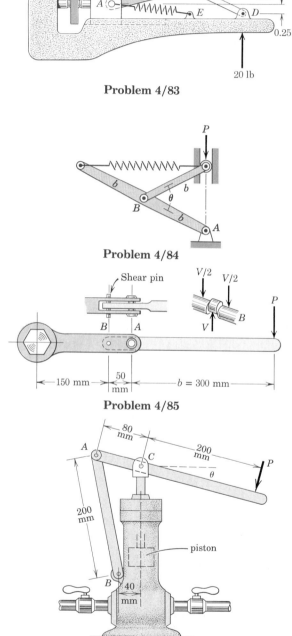

**Problem 4/83**

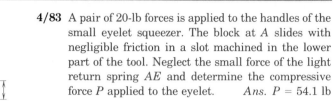

**4/83** A pair of 20-lb forces is applied to the handles of the small eyelet squeezer. The block at $A$ slides with negligible friction in a slot machined in the lower part of the tool. Neglect the small force of the light return spring $AE$ and determine the compressive force $P$ applied to the eyelet.    *Ans.* $P = 54.1$ lb

**Problem 4/84**

**4/84** The force $P$ holds the spring-loaded frame in the equilibrium position shown. For this position determine the expressions for the tension $T$ in the spring and the force supported by the pin at $B$.

**Problem 4/85**

**4/85** A limiting torque wrench is designed with a shear pin $B$ which breaks when the force on it exceeds its strength and, hence, limits the torque which can be aplied by the wrench. If the limiting strength of the shear pin is 900 N in double shear (that is, $V/2 = 450$ N), calculate the limiting torque $M$. What would be the effect on $M$ of increasing $b$, all other conditions remaining unchanged?    *Ans.* $M = 75$ N·m

**Problem 4/86**

**4/86** The figure shows a high-pressure hand pump used for boosting oil pressure in a hydraulic line. When the handle is in equilibrium at $\theta = 15°$ under the action of a force $P = 120$ N, determine the oil pressure $p$ which acts on the 46-mm-diameter piston. (Pressure on the top of the piston is atmospheric.)

**4/87** For the paper punch shown find the punching force $Q$ corresponding to a hand grip $P$.

*Ans.* $Q = P(b/a)$

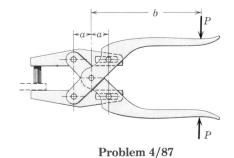

**Problem 4/87**

**4/88** Pin $B$ connects the two members of the 45° right-angled frame. Determine expressions for the forces supported by the pins at $A$ and $C$ due to the couple $M$ applied to the pin at $B$ if (*a*) pin $B$ is welded to $AB$ but free to turn in $BC$ and (*b*) if pin $B$ is welded to $CB$ but is free to turn in $AB$.

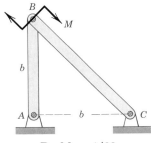

**Problem 4/88**

**4/89** The special box wrench with head $B$ swiveled at $C$ to the handle $A$ will accommodate a range of sizes of hexagonal bolt heads. For the nominal size shown where the center $O$ of the bolt and the pin $C$ are in line with the handle, compute the magnitude of the force supported by the pin at $C$ if $P = 160$ N. Assume the surface of the bolt head to be smooth.

*Ans.* $C = 1367$ N

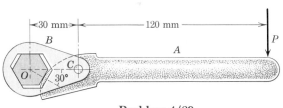

**Problem 4/89**

**4/90** Determine the vertical clamping force at $E$ in terms of the force $P$ applied to the handle of the toggle clamp.

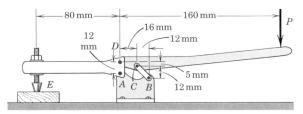

**Problem 4/90**

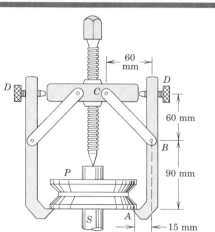

**Problem 4/91**

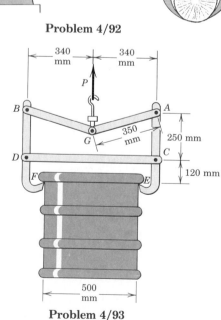

**Problem 4/92**

**Problem 4/93**

**4/91** The figure shows a wheel puller which is removing a *V*-belt pulley *P* from its tight-fitting shaft *S* by tightening the central screw. If the pulley starts to slide off the shaft when the compression in the screw has reached 1.2 kN, calculate the magnitude of the force supported by each jaw at *A*. The adjusting screws *D* support horizontal force and keep the side arms parallel with the central screw.

*Ans. A* = 0.626 kN

**4/92** In the special position shown for the log hoist, booms *AF* and *EG* are at right angles to one another and *AF* is perpendicular to *AB*. If the hoist is handling a log weighing 4800 lb, compute the force supported by the pins at *A* and *D* in this one position due to the weight of the log.

**4/93** A lifting device for transporting 135-kg steel drums in shown. Calculate the magnitude of the force exerted on the drum at *E* and *F*.

*Ans. E* = 5.19 kN

**4/94** The toggle pliers are used for a variety of clamping purposes. For the handle position given by $\alpha = 10°$ and for a handle grip of $P = 150$ N, calculate the clamping force $C$ produced.

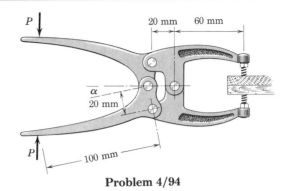

**Problem 4/94**

**4/95** The car hoist allows the car to be driven onto the platform, after which the rear wheels are raised. If the loading from both rear wheels is 1500 lb, determine the force in the hydraulic cylinder $AB$. Neglect the weight of the platform itself. Member $BCD$ is a right-angle bell crank pinned to the ramp at $C$.

*Ans.* $AB = 3970$ lb

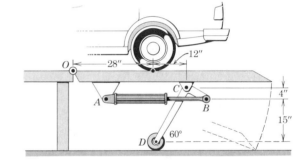

**Problem 4/95**

**4/96** The upper jaw $D$ of the toggle press slides with negligible frictional resistance along the fixed vertical column. Calculate the compressive force $R$ exerted on the cylinder $E$ and the force supported by the pin at $A$ if a force $F = 200$ N is applied to the handle at an angle $\theta = 75°$.

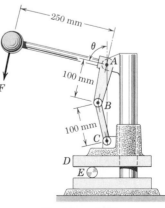

**Problem 4/96**

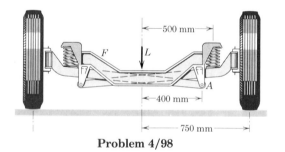

**Problem 4/97**

**4/97** The angle of elevation $\theta$ of the upper arm of a "cherry picker" is controlled by the two hydraulic cylinders attached to the upper end of the lower boom of the rig. Each piston rod of the cylinder is connected to the chain which engages the sprocket $A$, as shown in the enlarged view. The sprocket is welded to the end of the upper arm. Determine the magnitude $R$ of the total force supported by the hinge pin $B$ and the oil pressure $p$ in the upper cylinder to support the arm in the position $\theta = 30°$. A constant pressure of 80 kPa is maintained in the lower cylinder with the lower arm in the position $\beta = 60°$. The net area of the pistons subjected to hydraulic pressure is $7.85 \ (10^{-3}) \ \text{m}^2$. The mass center of the 120-kg upper boom is at midlength, and the combined mass of the hinged bucket and man is 110 kg.

*Ans.* $R = 68.2$ kN, $p = 8360$ kPa

**Problem 4/98**

**4/98** A double-axle suspension for use on small trucks is shown in the figure. The mass of the central frame $F$ is 40 kg, and the mass of each wheel and attached link is 35 kg with center of mass 680 mm from the vertical centerline. For a load $L = 12$ kN transmitted to the frame $F$, compute the total shear force supported by the pin at $A$.

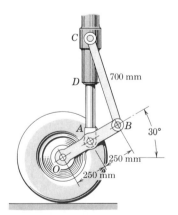

**Problem 4/99**

**4/99** The aircraft landing gear consists of a spring- and hydraulically-loaded piston and cylinder $D$ and the two pivoted links $OB$ and $CB$. If the gear is moving along the runway at a constant speed with the wheel supporting a stabilized constant load of 24 kN, calculate the total force that the pin at $A$ supports.

*Ans.* $A = 44.7$ kN

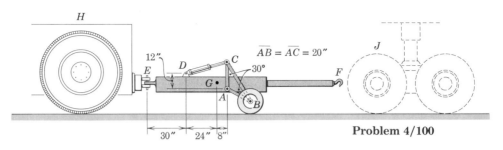

**Problem 4/100**

**4/100** An adjustable tow bar connecting the tractor unit *H*
with the landing gear *J* of a large aircraft is shown
in the figure. Adjusting the height of the hook *F* at
the end of the tow bar is accomplished by the hydraulic cylinder *CD* activated by a small hand pump (not
shown). For the nominal position shown of the triangular linkage *ABC*, calculate the force *P* supplied
by the cylinder to the pin *C* to position the tow bar.
The rig has a total weight of 100 lb and is supported
by the tractor hitch at *E*.

**4/101** A common folding chair is shown in the figure where
pin *E* is latched in position. If half the weight of a
70-kg person is supported by the side of the chair
frame shown, determine the *x*- and *y*-components of
all forces acting on member *BCE*. Assume that the
chair is resting on a smooth floor so that any horizontal
friction forces at *A* and *B* may be neglected.

    *Ans.* $B_y = 247$ N, $C_x = 85.8$ N, $C_y = -446$ N
        $E_x = -85.8$ N, $E_y = 199.1$ N

Dimensions in Millimeters
**Problem 4/101**

**4/102** Determine the shearing force *Q* applied to the bar
if a 400-N force is applied to the handle for $\theta = 30°$.
For a given applied force what value of $\theta$ gives the
greatest shear?

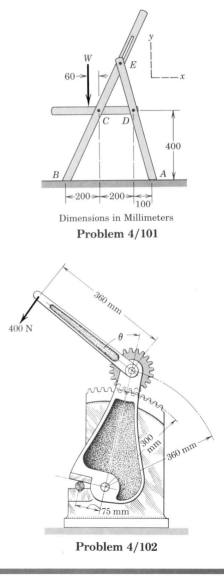

**Problem 4/102**

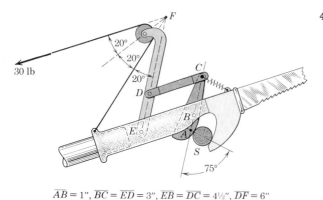

$\overline{AB} = 1''$, $\overline{BC} = \overline{ED} = 3''$, $\overline{EB} = \overline{DC} = 4\frac{1}{2}''$, $\overline{DF} = 6''$

**Problem 4/103**

**4/103** The pruning mechanism of a pole saw is shown as it cuts a branch $S$. For the particular position drawn, the actuating cord is parallel to the pole and carries a tension of 30 lb. Determine the shearing force $P$ applied to the branch by the cutter and the total force supported by the pin at $E$. The force exerted by the light return spring at $C$ is small and may be neglected.          *Ans.* $P = 338$ lb, $E = 75.1$ lb

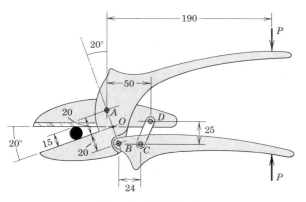

Dimensions in Millimeters
**Problem 4/104**

**4/104** For the pruning shears shown determine the force $Q$ applied to the circular branch of 15-mm diameter for a gripping force $P = 200$ N. *Suggestion:* First draw a free-body diagram of the isolated branch.

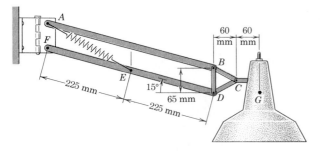

**Problem 4/105**

**4/105** The designers of lamp mechanisms, such as that shown in the figure, usually rely on joint friction to aid in maintaining static equilibrium. For the present problem, assume that sufficient friction exists at point $C$ to prevent rotation there, but ignore friction at all other joints. If the mass of the lamp fixture is 0.6 kg with mass center at $G$, determine the spring force $F_s$ necessary for equilibrium in the position shown.          *Ans.* $F_s = 45.2$ N

**4/106** The backhoe is controlled by the three hydraulic cylinders, and in the particular position shown, the hoe can apply a horizontal force $P = 10$ kN. Neglect the masses of the members and compute the magnitude of the forces supported by the pins at $A$ and $E$.

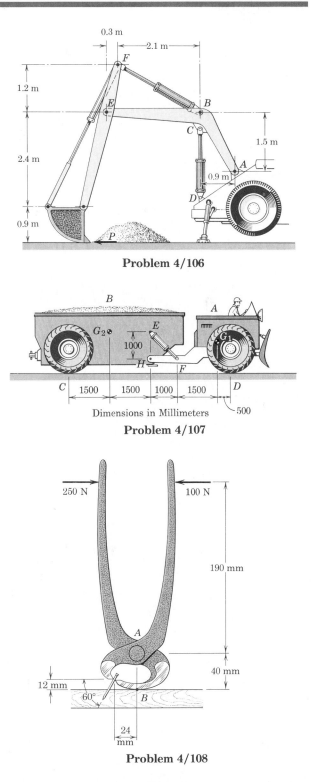

**Problem 4/106**

**4/107** The power unit $A$ of the tractor-scraper has a mass of 4 Mg with mass center at $G_1$. The fully loaded trailer-scraper unit $B$ has a mass of 24 Mg with mass center at $G_2$. Positioning of the scraper is controlled by two hydraulic cylinders $EF$, one on each side of the machine. Calculate the compression $F$ in each of the cylinders and the magnitude of the force supported by each of the pins at $H$, one on each side. Assume that the wheels are free to turn so that there are no horizontal components of force under the wheels.

*Ans.* $F = 131.8$ kN, $H = 113.9$ kN

**Problem 4/107**

**4/108** The two sharp jaws of the nail puller exert equal pulling forces parallel to the nail, and the normal forces which the jaws exert on the nail are also equal, causing no tendency to bend the nail for the configuration shown. Determine the total extracting force $T$ on the nail, the normal force $N$ which each jaw exerts on the nail, and the magnitude of the pin reaction at $A$. The surface at $B$ is rough enough to prevent slippage.

**Problem 4/108**

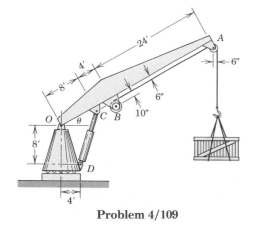

**Problem 4/109**

▶**4/109** The shipboard crane is supporting a load of 4 tons in the position shown where $\theta = 30°$. The hoisting drum $B$ is operated by a high-torque electric motor. Calculate the added compression $P$ in the hydraulic cylinder and the magnitude $R$ of the additional force supported by the pin at $O$, both due to the effect of the 4-ton load.  *Ans.* $P = 43{,}100$ lb, $R = 35{,}400$ lb

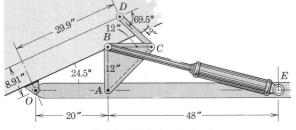

▶**4/110** The hoisting mechanism for the dump truck is shown in the enlarged view. Determine the compression $P$ in the hydraulic cylinder $BE$ and the magnitude of the force supported by the pin at $A$ for the particular position shown, where $BA$ is perpendicular to $OAE$ and link $DC$ is perpendicular to $AC$. The dump and its load together weigh 20,000 lb with center of mass at $G$. All dimensions for the indicated geometry are given on the figure.

*Ans.* $P = 26{,}930$ lb, $A = 14{,}600$ lb

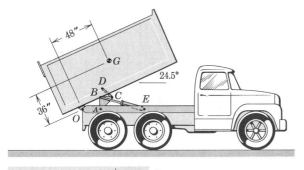

Detail of Hoisting Mechanism
**Problem 4/110**

**4/111** Determine the force acting on member *ABC* at connection *A* for the loaded space frame shown. Each connection may be treated as a ball-and-socket joint.

*Ans.* *A* = 4.25 kN

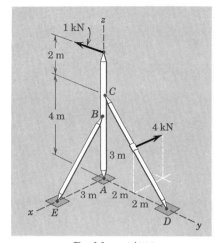

**Problem 4/111**

**4/112** The frame shown rests on a smooth horizontal surface, and the connections at *D*, *E*, *F*, *G*, and *H* act as ball-and-socket joints. Determine the total force acting on the connection at *D* under the action of the 500-lb load applied at *F*.

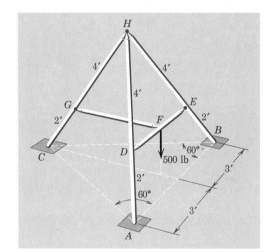

**Problem 4/112**

## *4/7 PROBLEM FORMULATION AND REVIEW*

In Chapter 4 we have applied the principles of equilibrium to two classes of problems: (*a*) simple trusses and (*b*) frames and machines. Basically, no new theory was needed since we merely drew the necessary free-body diagrams and applied our familiar equations of equilibrium. The structures dealt with in Chapter 4, however, have given us the opportunity to further develop our appreciation for a systematic approach to mechanics problems.

The most essential features of the analysis of these two classes of structures are reviewed in the statements which follow.

### *(a) Simple trusses*

1. Simple trusses are composed of two-force members joined at their ends and capable of supporting tension or compression. Each internal force, therefore, is always in the direction of its member.

2. Simple trusses are built around the basic rigid (noncollapsible) unit of the triangle for plane trusses and the tetrahedron for space trusses. Additional units of a truss are formed by adding new members, two for plane trusses and three for space trusses, attached to existing joints and joined at their ends to form a new joint.

3. The joints for simple trusses are assumed to be pin connections for plane trusses and ball-and-socket connections for space trusses. Thus, the joints can transmit force but not moment.

4. External loads are applied only at the joints.

5. Trusses are statically determinate externally when the external constraints are not in excess of those required to maintain an equilibrium position.

6. Trusses are statically determinate internally when constructed in the manner described in item (2), where internal members are not in excess of those required to prevent collapse.

7. The *method of joints* utilizes the force equations of equilibrium for each joint. Analysis normally begins at a joint where at least one force is known and not more than two forces are unknown for plane trusses or not more than three forces are unknown for space trusses.

8. The *method of sections* utilizes a free body of an entire section of a truss containing two or more joints and, in general, will involve the equilibrium of a nonconcurrent system of forces. The moment equation of equilibrium is of special value when the method of sections is used. In general the forces acting on a section which cuts more than three

unknown members of a plane truss cannot be solved for completely because there are only three independent equations of equilibrium.

9. The vector representing a force acting on a joint or a section is drawn on the same side of the joint or section as the member which transmits the force. With this convention tension is indicated when the force arrow is away from the joint or section, and compression is indicated when the arrow points toward the joint or section.

10. When the two diagonal members which brace a quadrilateral panel are flexible members incapable of supporting compression, only the one in tension is retained, and the panel remains statically determinate.

11. When two joined members under load are collinear and a third member with a different direction is joined with their connection, the force in the third member must be zero unless an external force is applied at the joint with a component in the direction of the third member.

### (b) Frames and machines

1. Frames and machines are structures which contain one or more multiforce members. A multiforce member is one which has acting on it three or more forces or two or more forces and one or more couples.

2. Frames are structures designed to support loads generally under static conditions. Machines are structures which transform input forces and moments to output forces and moments and generally involve one or more moving parts. Some structures may be classified in either category.

3. Only frames and machines which are statically determinate externally and internally are considered in this treatment.

4. If a frame or machine as a whole is a rigid (noncollapsible) unit when its external supports are removed, then the analysis is begun by computing the external reactions on the entire unit. If a frame or machine as a whole is a nonrigid (collapsible) unit when its external supports are removed, then the analysis of the external reactions cannot be completed until the structure is dismembered.

5. Forces acting in the internal connections of frames and machines are calculated by dismembering the structure and constructing a separate free-body diagram of each part. The principle of action and reaction must be *strictly* observed; otherwise, error will result.

6. The force and moment equations of equilibrium are applied to the members as needed to compute the desired unknowns.

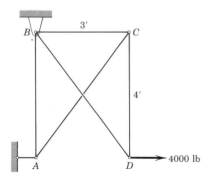

**Problem 4/113**

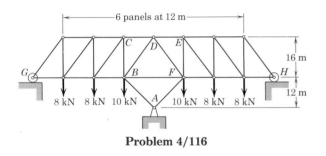

**Problem 4/114**

**Problem 4/115**

**Problem 4/116**

## REVIEW PROBLEMS

**4/113** Members $AC$ and $BD$ of the simple truss cross without touching. Verify that the truss is internally determinate and stable and calculate the force in member $AC$.                    *Ans. $AC = 6670$ lb $T$*

**4/114** Calculate the forces in members $BH$, $HI$, and $BC$ for the truss loaded by the 40- and 60-kN forces.

**4/115** Determine the force in member $AC$ in terms of the mass $m$ supported by the truss. All interior acute angles are either 30° or 60°.    *Ans. $AC = mg/3$ $C$*

**4/116** If it is known that the center pin $A$ supports one-half of the vertical loading shown, determine the force in member $BF$.

**4/117** Calculate the force in member *BG* using a free-body diagram of the rigid member *ABC*.

*Ans.  BG* = 1800 lb *C*

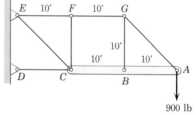

**Problem 4/117**

**4/118** Calculate the force supported by the pin at *A* under the action of the 500-N force and 300-N·m couple.

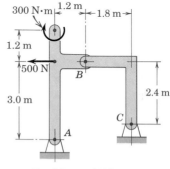

**Problem 4/118**

**4/119** Determine the forces in members *AB*, *BI*, and *CI* of the simple truss. Note that all curved members are two-force members.

*Ans.  AB* = 2.26*L T*
*BI* = *L T*
*CI* = 0.458*L T*

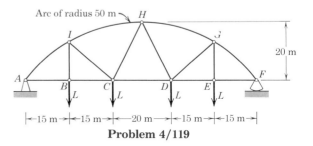

**Problem 4/119**

**4/120** The structure of Prob. 4/119 is modified in that the four curved members are replaced by the two members *AIH* and *HGF*. Instrumentation indicates the tension in members *CH* and *DH* to be 0.5*L* each. Determine the forces in members *AB*, *BI*, and *CI*. Is the problem solvable without the information about *CH*?

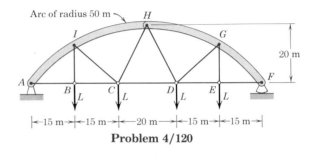

**Problem 4/120**

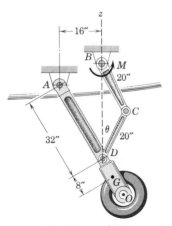

**Problem 4/121**

**4/121** The nose-wheel assembly is raised by the application of a torque $M$ to link $BC$ through the shaft at $B$. If the arm and wheel $AO$ have a combined weight of 100 lb with center of gravity at $G$, find the value of $M$ necessary to lift the wheel when $D$ is directly under $B$ at which position angle $\theta$ is 30°.

> *Ans.* $M = 1250$ lb-in. CCW

**4/122** A pneumatic cylinder pivoted at $F$ operates the lever $AB$ of the quick-acting toggle clamp which holds the work in position while it is machined. For an air pressure of 400 kPa above atmospheric pressure against the 50-mm-diameter piston, determine the clamping force at $G$ for the position $\alpha = 10°$. For this position the piston rod is perpendicular to $AB$.

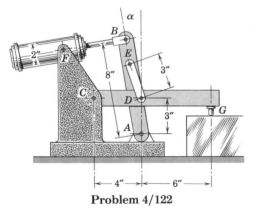

**Problem 4/122**

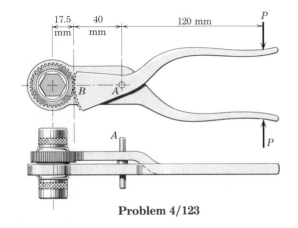

**Problem 4/123**

**4/123** An antitorque wrench is designed for use by a crewman of a spacecraft where he has no stable platform aginst which to push as he tightens a bolt. The pin $A$ fits into an adjacent hole in the structure which contains the bolt to be turned. Successive oscillations of the gear and handle unit turn the socket in one direction through the action of a ratchet mechanism. The reaction against the pin $A$ provides the "antitorque" characteristic of the tool. For a gripping force $P = 150$ N determine the torque $M$ transmitted to the bolt and the external reaction $R$ against the pin $A$ normal to the line $AB$. (One side of the tool is used for tightening and the opposite side for loosening a bolt.)

> *Ans.* $M = 7.88$ N·m, $R = 137.0$ N

**4/124** The man and his bicycle together weigh 900 N. For the position of the pedals shown, determine the force $F$ normal to the road which must be exerted on the pedal at $A$ to maintain a constant bicycle speed up the incline. Neglect the effects of the man's left foot and ignore all friction except that at the interface of the rear tire and road surface, where there is no slippage.

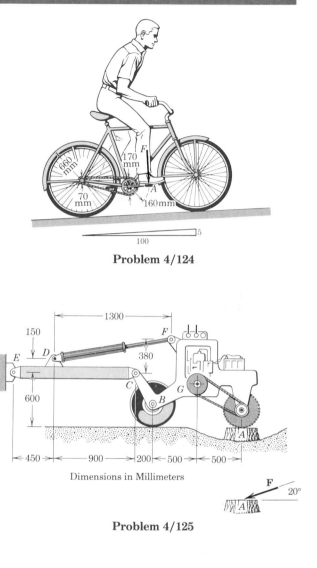

**Problem 4/124**

**4/125** The elements of a stump grinder with a total mass (exclusive of the hydraulic cylinder $DF$ and arm $CE$) of 300 kg with mass center at $G$ are shown in the figure. The mechanism for articulation about a vertical axis is omitted, and the wheels at $B$ are free to turn. For the nominal position shown, link $CE$ is horizontal and the teeth of the cutting wheel are even with the ground. If the magnitude of the force $F$ exerted by the cutter on the stump is 400 N, determine the force $P$ in the hydraulic cylinder and the magnitude of the force supported by the pin at $C$. The machine is to be treated as a two-dimensional problem. *Ans.* $P = 3170$ N, $C = 2750$ N

Dimensions in Millimeters

**Problem 4/125**

**4/126** Determine the punching force $P$ in terms of the gripping force $F$ for the rivet squeezer shown.

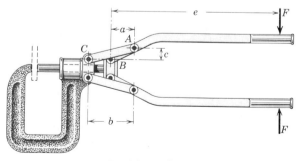

**Problem 4/126**

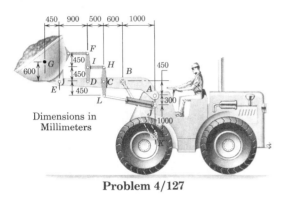

450  900  500 600  1000

F
G 600
E
450
450
450
I  H  B
D  C
L
A
300
J
1000
K

Dimensions in
Millimeters

**Problem 4/127**

**4/127** The loader has a capacity of 4 m³ and is handling shale having a density of 2.6 Mg/m³. For the particular position shown, where the arm *EB* is horizontal, find the compressive force in the hydraulic piston rod *JL* and the total shear force supported by the pin at *A*. The machine is symmetrical about a central vertical plane in the fore-and-aft direction and has two sets of the linkages shown.

*Ans.* $L = 52.0$ kN, $A = 247$ kN

4 m
0.5 m
2 m
2 m
2 m
P
θ
1 m
E
A
B
2 m
C
D
2 Mg

**Problem 4/128**

**4/128** The force *P* on the winch handle is required to support the 2-Mg load. Find the force $F_A$ supported by the pin at *A* for the handle position $\theta = 45°$.

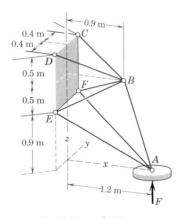

0.9 m
0.4 m
0.4 m
C
D
0.5 m
F
B
0.5 m
E
0.9 m
z
y
x
A
1.2 m
F

**Problem 4/129**

▶**4/129** Each of the landing struts for a planet exploration spacecraft is a space truss symmetrical about the vertical *x-z* plane as shown. For a landing force $F = 2.2$ kN, calculate the corresponding force in member *BE*. The assumption of static equilibrium for the truss is permissible if the mass of the truss is very small. Assume equal loads in the symmetrically placed members.    *Ans.* $F_{BE} = 1.620$ kN

### *Computer-oriented problems*

**4/130**  For a given force $P$ on the handle of the toggle clamp the clamping force $C$ increases to very large values as the angle $\theta$ decreases. For $P = 30$ lb determine the relationship between $C$ and $\theta$ and plot it as a function of $\theta$ from $\theta = 2°$ to $\theta = 30°$. Assume that the shaft slides freely in its guide.

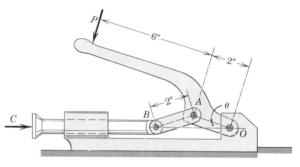

**Problem 4/130**

**4/131**  The truck with bed-mounted crane is used in the delivery of cubes of bricks. For the given position of the carrier $C$ along the boom, determine and plot the force $F$ in the hydraulic cylinder $AB$ as a function of the elevation angle $\theta$ for $0 \le \theta \le 75°$. Find the maximum value of $F$ and the corresponding angle $\theta$. Neglect the weight of the boom compared with the 12-kN load of bricks.

*Ans.* $F_{\text{max}} = 103.3$ kN at $\theta = 39.8°$

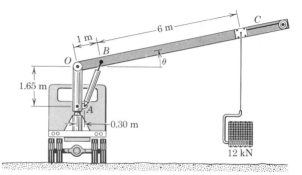

**Problem 4/131**

**4/132**  The structural members support the 3-kN load which may be applied at any angle $\theta$ from essentially $-90°$ to $+90°$. The pin at $A$ must be adequate to support the maximum force transmitted to it. Plot the force $F_A$ at $A$ as a function of $\theta$ and determine its maximum value and the corresponding angle $\theta$.

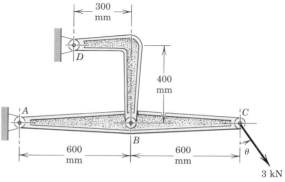

**Problem 4/132**

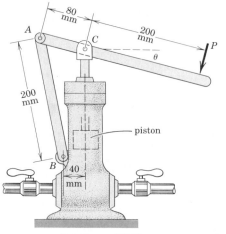

**Problem 4/133**

*4/133 For the hand pump described in Prob. 4/86 and shown again here, plot the oil pressure $p$ in kilopascals as a function of $\theta$ from $\theta = -20°$ to $\theta = +15°$ and find the minimum pressure and its corresponding value of $\theta$. The force $P$ has a constant magnitude of 120 N.    *Ans.* $P_{min} = 248$ kPa at $\theta = -15.17°$

*4/134 The type of marine crane shown is utilized for both dockside and offshore operations. Determine and plot the force in member $BC$ as a function of the boom angle $\theta$ for $0 \le \theta \le 80°$. Neglect the radius of all pulleys and the weight of the boom.

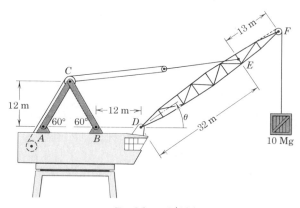

**Problem 4/134**

*4/135 The "jaws-of-life" device of Prob. 4/76 is redrawn here with its jaws open. The pressure behind the piston $P$ of area 20 in.$^2$ is maintained at 500 lb/in.$^2$ Calculate and plot the force $R$ as a function of $\theta$ for $0 \le \theta \le 45°$, where $R$ is the vertical force acting on the wreckage as shown. Determine the maximum value of $R$ and the corresponding value of the jaw angle. See the figure of Prob. 4/76 for dimensions and the geometry associated with the condition $\theta = 0$. Note that link $AB$ and its counterpart are both horizontal in the figure for $\theta = 0$ but do not remain horizontal as the jaws open.
    *Ans.* $R_{max} = 1314$ lb at $\theta = 45°$

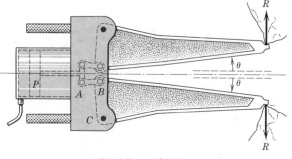

**Problem 4/135**

When forces are continuously distributed over a region of a structure, the cumulative effect of this distribution must be calculated. The cables of this suspension bridge support the weight of the roadway distributed along its length.

# DISTRIBUTED FORCES

# 5

## 5/1 INTRODUCTION

In the previous chapters we have treated all forces as concentrated along their lines of action and at their points of application. This treatment has provided a reasonable model for the forces with which we have dealt. Actually, "concentrated" forces do not exist in the exact sense, since every external force applied mechanically to a body is distributed over a finite contact area however small. The force exerted by the pavement on an automobile tire, for instance, is applied to the tire over its entire area of contact, Fig. 5/1a, which may be appreciable if the tire is soft. When the dimension $b$ of the contact area is negligible compared with the other pertinent dimensions, such as the distance between wheels, then replacement of the actual distributed forces of contact by their resultant $R$ considered as a concentrated force raises no question when we are analyzing the forces acting on the car as a whole. Even the force of contact between a hardened steel ball and its race in a loaded ball bearing, Fig. 5/1b, will be applied over a finite contact area, the dimensions of which, of course, are extremely small. The forces applied to a two-force member of a truss, Fig. 5/1c, are applied over an actual area of contact of the pin against the hole and internally across the cut section in a manner similar to that indicated. In these and other similar examples we have no hesitation in treating the forces as concentrated when analyzing their external effects on bodies as a whole.

If, on the other hand, we are interested in finding the distribution of *internal* forces in the material of the body in the neighborhood of the contact location where the internal stresses and strains may be appreciable, then we no longer treat the load as concentrated but would be obliged to take the actual distribution into account. This type of problem requires a knowledge of the properties of the material and belongs in more advanced treatments of the mechanics of materials and the theories of elasticity and plasticity.

When forces are applied over a region whose dimensions are not negligible compared with other pertinent dimensions, then we must account for the actual manner in which the force is distributed

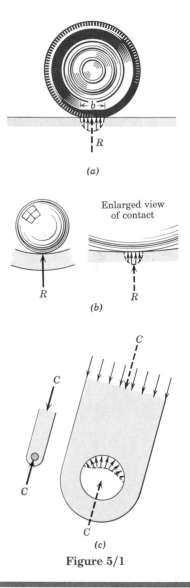

(a)

Enlarged view of contact

(b)

(c)

Figure 5/1

by summing up the effects of the distributed force over the entire region. We carry out this process by using the procedures of mathematical integration. For this purpose we need to know the intensity of the force at any location. There are three categories into which such problems fall.

***Line distribution.***    When a force is distributed along a line, as in the continuous vertical load supported by a suspended cable, Fig. 5/2*a*, the intensity *w* of the loading is expressed as force per unit length of line, newtons per meter (N/m), or pounds per foot (lb/ft).

***Area distribution.***    When a force is distributed over an area, as with the hydraulic pressure of water against the inner face of a section of dam, Fig. 5/2*b*, the intensity is expressed as force per unit area. This intensity is known as *pressure* for the action of fluid forces and *stress* for the internal distribution of forces in solids. The basic unit for pressure or stress in SI is the newton per square meter (N/m$^2$), which is also called the *pascal* (Pa). This unit, however, is too small for most applications (6895 Pa = 1 lb/in.$^2$), and the kilopascal (kPa), which equals $10^3$ Pa, is more commonly used for fluid pressure and the megapascal, which equals $10^6$ Pa, is used for stress. In the U.S. customary system of units both fluid pressure and mechanical stress are commonly expressed in pounds per square inch (lb/in.$^2$).

***Volume distribution.***    A force which is distributed over the volume of a body is known as a *body force*. The most common body force is the force of gravitational attraction which acts on all elements of mass of a body. The determination of the forces on the supports of the heavy cantilevered structure in Fig. 5/2*c*, for example, would require accounting for the distribution of gravitational force throughout the structure. The intensity of gravitational force is the *specific weight* $\rho g$, where $\rho$ is the density (mass per unit volume) and *g* is the acceleration due to gravity. The units for $\rho g$ are (kg/m$^3$)(m/s$^2$) = N/m$^3$ in SI and lb/ft$^3$ or lb/in.$^3$ in the U.S. customary system.

The body force due to the earth's gravitational attraction (weight) is by far the most commonly encountered distributed force. Section A of the chapter deals with the determination of the point in a body through which the resultant gravitational force acts and the associated geometric properties of lines, areas, and volumes. Section B of the chapter treats the important problems of distributed forces which act on and in beams and flexible cables and distributed forces which fluids exert on exposed surfaces.

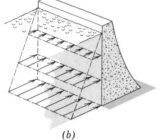

(a)

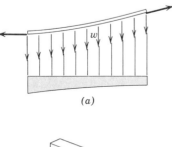

(b)

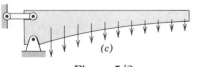

(c)

**Figure 5/2**

# SECTION A.  CENTERS OF MASS AND CENTROIDS

## 5/2  *CENTER OF MASS*

Consider a three-dimensional body of any size and shape, having a mass $m$. If we suspend the body, as shown in Fig. 5/3, from any point such as $A$, the body will be in equilibrium under the action of the tension in the cord and the resultant $W$ of the gravitational forces acting on all particles of the body. This resultant is clearly collinear with the cord, and it will be assumed that we mark its position by drilling a hypothetical hole of negligible size along its line of action. We repeat the experiment by suspending the body from other points such as $B$ and $C$, and in each instance we mark the line of action of the resultant force. For all practical purposes these lines of action will be concurrent at a single point $G$, which is known as the *center of gravity* of the body. An exact analysis, however, would take into account the fact that the directions of the gravity forces for the various particles of the body differ slightly because they converge toward the center of attraction of the earth. Also, since the particles are at different distances from the earth, the intensity of the earth's force field is not exactly constant over the body. These considerations lead to the conclusion that the lines of action of the gravity force resultants in the experiments just described will not quite be concurrent, and therefore no unique center of gravity exists in the exact sense. This condition is of no practical importance as long as we deal with bodies whose dimensions are small compared with those of the earth. We therefore assume a uniform and parallel field of force due to the earth's gravitational attraction, and this condition results in the concept of a unique center of gravity.

To determine mathematically the location of the center of gravity of any body, Fig. 5/4a, we apply the *principle of moments* (see Art. 2/6) to the parallel system of gravitational forces to locate its resultant. The moment of the resultant gravitational force $W$ about

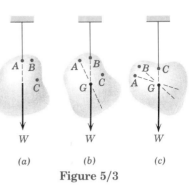

(a)        (b)        (c)

**Figure 5/3**

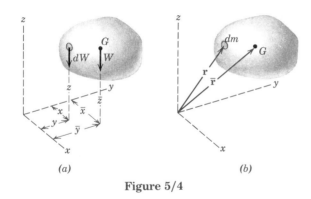

(a)                              (b)

**Figure 5/4**

any axis equals the sum of the moments about the same axis of the gravitational forces $dW$ acting on all particles treated as infinitesimal elements of the body. The resultant of the gravitational forces acting on all elements is the weight of the body and is given by the sum $W = \int dW$. If we apply the moment principle about the $y$-axis, for example, the moment about this axis of the elemental weight is $x\,dW$, and the sum of these moments for all elements of the body is $\int x\,dW$. This sum of moments must equal $W\bar{x}$, the moment of the sum. Therefore, $\bar{x}W = \int x\,dW$. With similar expressions for the other two components, we may express the coordinates of the center of gravity $G$ as

$$\bar{x} = \frac{\int x\,dW}{W} \qquad \bar{y} = \frac{\int y\,dW}{W} \qquad \bar{z} = \frac{\int z\,dW}{W} \qquad \textbf{(5/1}a\textbf{)}$$

To visualize the physical moments of the gravity forces to obtain the third equation, we may reorient the body and attached axes so that the $z$-axis is horizontal. It is essential to recognize that the numerator of each of these expressions represents the *sum of the moments*, whereas the product of $W$ and the corresponding coordinate of $G$ represent the *moment of the sum*. This moment principle finds repeated use throughout mechanics.

With the substitution of $W = mg$ and $dW = g\,dm$, the expressions for the coordinates of the center of gravity become

$$\bar{x} = \frac{\int x\,dm}{m} \qquad \bar{y} = \frac{\int y\,dm}{m} \qquad \bar{z} = \frac{\int z\,dm}{m} \qquad \textbf{(5/1}b\textbf{)}$$

Equations 5/1$b$ may be expressed in vector form with the aid of Fig. 5/4$b$, where the elemental mass and the position of $G$ are located by their respective position vectors $\mathbf{r} = x\mathbf{i} + y\mathbf{j} + z\mathbf{k}$ and $\bar{\mathbf{r}} = \bar{x}\mathbf{i} + \bar{y}\mathbf{j} + \bar{z}\mathbf{k}$. Thus, Eqs. 5/1$b$ are the components of the single vector equation

$$\bar{\mathbf{r}} = \frac{\int \mathbf{r}\,dm}{m} \qquad \textbf{(5/2)}$$

The density $\rho$ of a body is its mass per unit volume. Thus, the mass of a differential element of volume $dV$ becomes $dm = \rho\,dV$. In the event that $\rho$ is not constant throughout the body but can be expressed as a function of the coordinates of the body, it will be necessary to account for this variation in the calculation of both the numerators and denominators of Eqs. 5/1b. We may then write these expressions as

$$\bar{x} = \frac{\int x\rho \, dV}{\int \rho \, dV} \qquad \bar{y} = \frac{\int y\rho \, dV}{\int \rho \, dV} \qquad \bar{z} = \frac{\int z\rho \, dV}{\int \rho \, dV} \qquad (5/3)$$

Equations 5/1b, 5/2, and 5/3 contain no reference to gravitational effects since $g$ no longer appears. Therefore, they define a unique point in the body which is a function solely of the distribution of mass. This point is known as the *center of mass*, and clearly it coincides with the center of gravity as long as the gravity field is treated as uniform and parallel. It is meaningless for us to speak of the center of gravity of a body that is removed from the earth's gravitational field, since no gravitational forces would act on the body. The body would, however, still possess its unique center of mass. For the most part we will make reference henceforth to the center of mass rather than to the center of gravity. Also, the center of mass has a special significance in calculating the dynamic response of a body to unbalanced forces. This class of problems is discussed at length in *Vol. 2 Dynamics*.

In most problems the calculation of the position of the center of mass may be simplified by an intelligent choice of reference axes. In general the axes should be placed so as to simplify the equations of the boundaries as much as possible. Thus, polar coordinates will be useful for bodies having circular boundaries. Another important clue may be taken from considerations of symmetry. Whenever there exists a line or plane of symmetry in a homogeneous body, a coordinate axis or plane should be chosen to coincide with this line or plane. The center of mass will always lie on such a line or plane, since the moments due to symmetrically located elements will always cancel, and the body can be considered composed of pairs of these elements. Thus, the center of mass $G$ of the homogeneous right-circular cone of Fig. 5/5a will lie somewhere on its central axis, which is a line of symmetry. The center of mass of the half right-circular cone lies on its plane of symmetry, Fig. 5/5b. The center of mass of the half ring in Fig. 5/5c lies in both of its planes of symmetry and therefore is situated on line $AB$. The location of $G$ is always facilitated by the observation of symmetry when it exists.

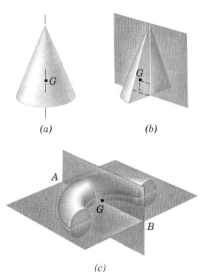

(a)                    (b)

(c)

**Figure 5/5**

## 5/3 CENTROIDS OF LINES, AREAS, AND VOLUMES

When the density $\rho$ of a body is uniform throughout, it will be a constant factor in both the numerators and denominators of Eqs. 5/3 and will therefore cancel. The expressions that remain define a purely geometrical property of the body, since any reference to its mass properties has disappeared. The term *centroid* is used when the calculation concerns a geometrical shape only. When speaking of an actual physical body, we use the term *center of mass*. If the density is uniform throughout the body, the positions of the centroid and center of mass are identical, whereas if the density varies, these two points will, in general, not coincide.

The calculations of centroids fall within three distinct categories, depending on whether the shape of the body involved can be modeled as a line, an area, or a volume.

*(a) Lines.* For a slender rod or wire of length $L$, cross-sectional area $A$, and density $\rho$, Fig. 5/6, the body approximates a line segment, and $dm = \rho A \, dL$. If $\rho$ and $A$ are constant over the length of the rod, the coordinates of the center of mass also become the coordinates of the centroid $C$ of the line segment, which, from Eqs. 5/1$b$, may be written

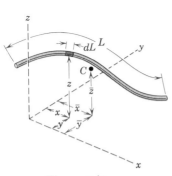

Figure 5/6

$$\bar{x} = \frac{\int x \, dL}{L} \qquad \bar{y} = \frac{\int y \, dL}{L} \qquad \bar{z} = \frac{\int z \, dL}{L} \qquad \textbf{(5/4)}$$

It should be noted that, in general, the centroid $C$ will not lie on the line. If the rod lies in a single plane, such as the $x$-$y$ plane, only two coordinates will require calculation.

*(b) Areas.* When a body of density $\rho$ has a small but constant thickness $t$, it can be modeled as a surface area $A$, Fig. 5/7. The mass of an element becomes $dm = \rho t \, dA$. Again, if $\rho$ and $t$ are constant over the entire area, the coordinates of the center of mass of the body also become the coordinates of the centroid $C$ of the surface area, and from Eqs. 5/1$b$ may be written

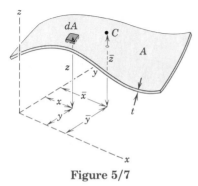

Figure 5/7

$$\bar{x} = \frac{\int x \, dA}{A} \qquad \bar{y} = \frac{\int y \, dA}{A} \qquad \bar{z} = \frac{\int z \, dA}{A} \qquad \textbf{(5/5)}$$

The numerators in Eqs. 5/5 are known as the *first moments of area.** If the surface is curved, as illustrated in Fig. 5/7 with the shell segment, all three coordinates will be involved. Here again the centroid $C$ for the curved surface will in general not lie on the

---

* Second moments of areas (moments of first moments) appear later in our discussion of second moments of area, also called area moments of inertia, in Appendix A.

surface. If the area is a flat surface in, say, the $x$-$y$ plane, only the coordinates of $C$ in that plane will be unknown.

*(c) Volumes.* For a general body of volume $V$ and density $\rho$, the element has a mass $dm = \rho\,dV$. The density $\rho$ cancels if it is constant over the entire volume, and the coordinates of the center of mass also become the coordinates of the centroid $C$ of the body. From Eqs. 5/3 or 5/1$b$ they become

$$\bar{x} = \frac{\int x\,dV}{V} \qquad \bar{y} = \frac{\int y\,dV}{V} \qquad \bar{z} = \frac{\int z\,dV}{V} \qquad \textbf{(5/6)}$$

*(d) Choice of element for integration.* As is often the case, the principal difficulty in a theory lies not in its concepts but in the procedures for applying it. With mass centers and centroids the concept of the moment principle is simple enough; the difficulties reside primarily with the choice of the differential element and with setting up the integrals. In particular there are five guidelines to be specially observed.

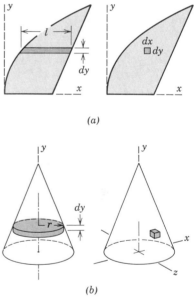

(a)

*(1) Order of element* Whenever possible, a first-order differential element should be selected in preference to a higher-order element so that only one integration will be required to cover the entire figure. Thus, in Fig. 5/8$a$ a first-order horizontal strip of area $dA = l\,dy$ will require only one integration with respect to $y$ to cover the entire figure. The second-order element $dx\,dy$ will require two integrations, first with respect to $x$ and second with respect to $y$, to cover the figure. As a further example, for the solid cone in Fig. 5/8$b$ we choose a first-order element in the form of a circular slice of volume $dV = \pi r^2\,dy$, which requires only one integration, in preference to choosing a third-order element $dV = dx\,dy\,dz$, which would require three awkward integrations.

(b)

**Figure 5/8**

*(2) Continuity* Whenever possible, we choose an element which can be integrated in one continuous operation to cover the figure. Thus, the horizontal strip in Fig. 5/8$a$ would be preferable to the vertical strip in Fig. 5/9, which, if used, would require two separate integrals because of the discontinuity in the expression for the height of the strip at $x = x_1$.

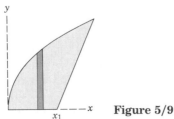

**Figure 5/9**

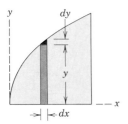

**Figure 5/10**

*(3) Discarding higher-order terms* Higher-order terms may always be dropped compared with lower-order terms (see Art. 1/7). Thus, the vertical strip of area under the curve in Fig. 5/10 is given by the first-order term $dA = y \, dx$, and the second-order triangular area $\frac{1}{2}dx \, dy$ is discarded. In the limit, of course, there is no error.

*(4) Choice of coordinates* As a general rule, we choose the coordinate system which best matches the boundaries of the figure. Thus, the boundaries of the area in Fig. 5/11a are most easily described in rectangular coordinates, whereas the boundaries of the circular sector of Fig. 5/11b are best suited to polar coordinates.

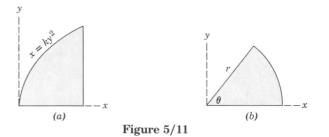

**Figure 5/11**

*(5) Centroidal coordinate of element* When a first- or second-order differential element is adopted, it is essential to use the *coordinate of the centroid of the element* for the moment arm in setting up the moment of the differential element. Thus, for the horizontal strip of area in Fig. 5/12a the moment of $dA$ about the $y$-axis is $x_c \, dA$, where $x_c$ is the $x$-coordinate of the centroid $C$ of the element. Note that $x_c$ is *not* the $x$ which describes the boundaries of the area. In the $y$-direction for this element the moment arm $y_c$ of the centroid of the element is the same, in the limit, as the $y$-coordinates of the two boundaries.

As a second example, consider the solid half-cone of Fig. 5/12b with the semicircular slice of differential thickness as the element

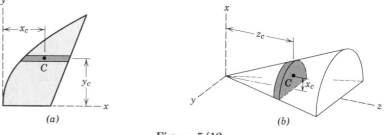

**Figure 5/12**

of volume. The moment arm for the element in the $x$-direction is the distance $x_c$ to the centroid of the face of the element and not the $x$-distance to the boundary of the element. On the other hand, in the $z$-direction the moment arm $z_c$ of the centroid of the element is the same as the $z$-coordinate of the element.

With these examples in mind, we rewrite Eqs. 5/5 and 5/6 in the form

$$\bar{x} = \frac{\int x_c \, dA}{A}$$

$$\bar{y} = \frac{\int y_c \, dA}{A} \qquad\qquad (5/5a)$$

$$\bar{z} = \frac{\int z_c \, dA}{A}$$

and

$$\bar{x} = \frac{\int x_c \, dV}{V}$$

$$\bar{y} = \frac{\int y_c \, dV}{V} \qquad\qquad (5/6a)$$

$$\bar{z} = \frac{\int z_c \, dV}{V}$$

It is *essential* to recognize that the subscript $c$ serves as a reminder that the moment arms appearing in the numerators of the integral expressions for moments are *always* the coordinates of the *centroids* of the particular elements chosen.

At this point it would be well for the student to make certain that he or she understands clearly the principle of moments, which was introduced in Art. 2/4. It is essential that the physical meaning of this principle be recognized as it is applied to the system of parallel weight forces depicted in Fig. 5/4a. With the equivalence clearly in mind between the moment of the resultant weight $W$ and the sum (integral) of the moments of the elemental weights $dW$, a mistake in setting up the necessary mathematics will be far less likely to occur. Recognition of the principle of moments will provide assurance that the correct expression will be used for the moment arm $x_c$, $y_c$, or $z_c$ of the centroid of the particular differential element chosen. Also, with the physical picture of the principle of moments in mind, Eqs. 5/4, 5/5, and 5/6, which are geometric relationships, will be recognized as descriptive also of homogeneous physical bodies, where the density $\rho$ has canceled. If the density of the body in question is not constant but varies throughout the body as some function of the coordinates, then it will not cancel from the numerator and denominator of the mass-center expressions. In this event, we must use Eqs. 5/3 as explained earlier.

Sample Problems 5/1 through 5/5 which follow are carefully chosen to illustrate the application of Eqs. 5/4, 5/5, and 5/6 for calculating the location of the centroid for line segments (slender rods), areas (thin flat plates), and volumes (homogeneous solids). The five considerations in item ($d$) are illustrated in detail in these sample problems. They are basic to setting up definite integrals not only for the calculation of centroids but for other problems where integration is required. For this reason, the sample problems should be thoroughly mastered.

Section C/10 of Appendix C contains a table of integrals which includes those needed for the problems in this and subsequent chapters. A summary of the centroidal coordinates for some of the commonly used shapes is given in Tables D/3 and D/4, Appendix D.

## Sample Problem 5/1

**Centroid of a circular arc.** Locate the centroid of a circular arc as shown in the figure.

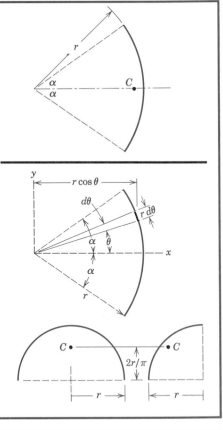

**Solution.** Choosing the axis of symmetry as the $x$-axis makes $\bar{y} = 0$. A differential element of arc has the length $dL = r\,d\theta$ expressed in polar coordinates, and the $x$-coordinate of the element is $r\cos\theta$.

(1) Applying the first of Eqs. 5/4 and substituting $L = 2\alpha r$ give

$$[L\bar{x} = \int x\,dL] \qquad (2\alpha r)\bar{x} = \int_{-\alpha}^{\alpha} (r\cos\theta)r\,d\theta$$

$$2\alpha r\bar{x} = 2r^2 \sin\alpha$$

$$\bar{x} = \frac{r\sin\alpha}{\alpha} \qquad\qquad Ans.$$

For a semicircular arc $2\alpha = \pi$, which gives $\bar{x} = 2r/\pi$. By symmetry we see immediately that this result also applies to the quarter-circular arc when the measurement is made as shown.

① It should be perfectly evident that polar coordinates are preferable to rectangular coordinates to express the length of a circular arc.

## Sample Problem 5/2

**Centroid of a triangular area.** Determine the distance $\bar{h}$ from the base of a triangle of altitude $h$ to the centroid of its area.

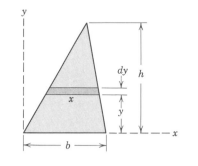

**Solution.** The $x$-axis is taken to coincide with the base. A differential
(1) strip of area $dA = x\,dy$ is chosen. By similar triangles $x/(h - y) = b/h$. Applying the second of Eqs. 5/5a gives

$$[A\bar{y} = \int y_c\,dA] \qquad \frac{bh}{2}\bar{y} = \int_0^h y\,\frac{b(h - y)}{h}\,dy = \frac{bh^2}{6}$$

and

$$\bar{y} = \frac{h}{3} \qquad\qquad Ans.$$

This same result holds with respect to either of the other two sides of the triangle considered a new base with corresponding new altitude. Thus, the centroid lies at the intersection of the medians, since the distance of this point from any side is one-third the altitude of the triangle with that side considered the base.

① We save one integration here by using the first-order element of area. Recognize that $dA$ must be expressed in terms of the integration variable $y$; hence, $x = f(y)$ is required.

## Sample Problem 5/3

**Centroid of the area of a circular sector.** Locate the centroid of the area of a circular sector with respect to its vertex.

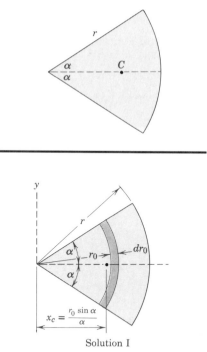

*Solution I.* The x-axis is chosen as the axis of symmetry, and $\bar{y}$ is therefore automatically zero. We may cover the area by moving an element in the form of a partial circular ring, as shown in the figure, from the center to the outer periphery. The radius of the ring is $r_0$ and its thickness is $dr_0$, so that its area is $dA = 2r_0\alpha \, dr_0$.

The x-coordinate to the centroid of the element from Sample Problem 5/1 is $x_c = r_0 \sin \alpha/\alpha$, where $r_0$ replaces $r$ in the formula. Thus, the first of Eqs. 5/5a gives

$$[A\bar{x} = \int x_c \, dA] \qquad \frac{2\alpha}{2\pi}(\pi r^2)\bar{x} = \int_0^r \left(\frac{r_0 \sin \alpha}{\alpha}\right)(2r_0\alpha \, dr_0)$$

$$r^2\alpha\bar{x} = \tfrac{2}{3}r^3 \sin \alpha$$

$$\bar{x} = \frac{2}{3}\frac{r \sin \alpha}{\alpha} \qquad\qquad Ans.$$

*Solution II.* The area may also be covered by swinging a triangle of differential area about the vertex and through the total angle of the sector. This triangle, shown in the illustration, has an area $dA = (r/2)(r\,d\theta)$, where higher-order terms are neglected. From Sample Problem 5/2 the centroid of the triangular element of area is two-thirds of its altitude from its vertex, so that the x-coordinate to the centroid of the element is $x_c = \tfrac{2}{3}r \cos \theta$. Applying the first of Eqs. 5/5a gives

$$[A\bar{x} = \int x_c \, dA] \qquad (r^2\alpha)\bar{x} = \int_{-\alpha}^{\alpha} (\tfrac{2}{3}r \cos \theta)(\tfrac{1}{2}r^2 \, d\theta)$$

$$r^2\alpha\bar{x} = \tfrac{2}{3}r^3 \sin \alpha$$

and as before

$$\bar{x} = \frac{2}{3}\frac{r \sin \alpha}{\alpha} \qquad\qquad Ans.$$

For a semicircular area $2\alpha = \pi$, which gives $\bar{x} = 4r/3\pi$. By symmetry we see immediately that this result also applies to the quarter-circular area where the measurement is made as shown.

It should be noted that, if we had chosen a second-order element $r_0 \, dr_0 \, d\theta$, one integration with respect to $\theta$ would yield the ring with which Solution I began. On the other hand, integration with respect to $r_0$ initially would give the triangular element with which Solution II began.

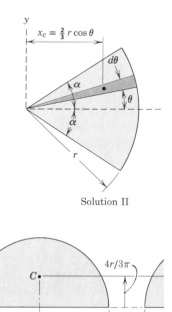

Solution I

① Note carefully that we must distinguish between the variable $r_0$ and the constant $r$.

Solution II

② Be careful not to use $r_0$ as the centroidal coordinate for the element.

## Sample Problem 5/4

Locate the centroid of the area under the curve $x = ky^3$ from $x = 0$ to $x = a$.

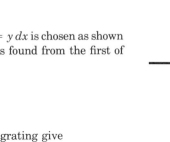

**Solution I.** A vertical element of area $dA = y\,dx$ is chosen as shown in the figure. The $x$-coordinate of the centroid is found from the first of Eqs. 5/5a. Thus,

①  $[A\bar{x} = \int x_c\,dA]$  $\qquad \bar{x}\int_0^a y\,dx = \int_0^a xy\,dx$

Substituting $y = (x/k)^{1/3}$ and $k = a/b^3$ and integrating give

$$\frac{3ab}{4}\bar{x} = \frac{3a^2b}{7} \qquad \bar{x} = \tfrac{4}{7}a \qquad\qquad Ans.$$

In solving for $\bar{y}$ from the second of Eqs. 5/5a, the coordinate to the centroid of the rectangular element is $y_c = y/2$, where $y$ is the height of the strip governed by the equation of the curve $x = ky^3$. Thus, the moment principle becomes

$[A\bar{y} = \int y_c\,dA]$  $\qquad \frac{3ab}{4}\bar{y} = \int_0^a \left(\frac{y}{2}\right) y\,dx$

Substituting $y = b(x/a)^{1/3}$ and integrating give

$$\frac{3ab}{4}\bar{y} = \frac{3ab^2}{10} \qquad \bar{y} = \tfrac{2}{5}b \qquad\qquad Ans.$$

**Solution II.** The horizontal element of area shown in the lower figure may be employed in place of the vertical element. The $x$-coordinate to the centroid of the rectangular element is seen to be $x_c = x + \frac{1}{2}(a - x) = (a + x)/2$ which is simply the average of the coordinates $a$ and $x$ of the ends of the strip. Hence,

$[A\bar{x} = \int x_c\,dA]$  $\qquad \bar{x}\int_0^b (a - x)\,dy = \int_0^b \left(\frac{a + x}{2}\right)(a - x)\,dy$

The value of $\bar{y}$ is found from

$[A\bar{y} = \int y_c\,dA]$  $\qquad \bar{y}\int_0^b (a - x)\,dy = \int_0^b y(a - x)\,dy$

where $y_c = y$ for the horizontal strip. The evaluation of these integrals will check the previous results for $\bar{x}$ and $\bar{y}$.

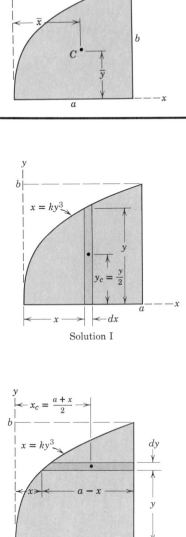

Solution I

Solution II

① Note that $x_c = x$ for the vertical element.

## Sample Problem 5/5

**Hemispherical volume.** Locate the centroid of the volume of a hemisphere of radius $r$ with respect to its base.

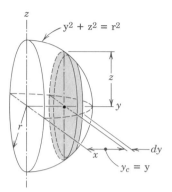

    *Solution I.* With the axes chosen as shown in the figure, $\bar{x} = \bar{z} = 0$ by symmetry. The most convenient element is a circular slice of thickness $dy$ parallel to the $x$-$z$ plane. Since the hemisphere intersects the $y$-$z$ plane in the circle $y^2 + z^2 = r^2$, the radius of the circular slice is $z = +\sqrt{r^2 - y^2}$. The volume of the elemental slice becomes

$$dV = \pi(r^2 - y^2)\, dy$$

The second of Eqs. 5/6a requires

$$[V\bar{y} = \int y_c\, dV] \qquad \bar{y} \int_0^r \pi(r^2 - y^2)\, dy = \int_0^r y\pi(r^2 - y^2)\, dy$$

where $y_c = y$. Integrating gives

$$\tfrac{2}{3}\pi r^3 \bar{y} = \tfrac{1}{4}\pi r^4 \qquad \bar{y} = \tfrac{3}{8}r \qquad\qquad Ans.$$

Solution I

    *Solution II.* Alternatively we may use for our differential element a cylindrical shell of length $y$, radius $z$, and thickness $dz$, as shown in the lower figure. By expanding the radius of the shell from zero to $r$, we cover the entire volume. By symmetry the centroid of the elemental shell lies at its center, so that $y_c = y/2$. The volume of the element is $dV = (2\pi z\, dz)(y)$. Expressing $y$ in terms of $z$ from the equation of the circle gives $y = +\sqrt{r^2 - z^2}$. Using the value of $\tfrac{2}{3}\pi r^3$ computed in Solution I for the volume of the hemisphere and substituting in the second of Eqs. 5/6a give us

$$[V\bar{y} = \int y_c\, dV] \qquad (\tfrac{2}{3}\pi r^3)\bar{y} = \int_0^r \frac{\sqrt{r^2 - z^2}}{2}(2\pi z\sqrt{r^2 - z^2})\, dz$$

$$= \int_0^r \pi(r^2 z - z^3)\, dz = \frac{\pi r^4}{4}$$

$$\bar{y} = \tfrac{3}{8}r \qquad\qquad Ans.$$

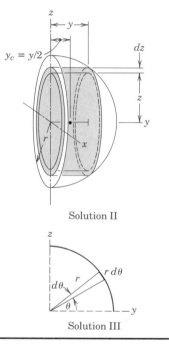

Solution II

Solution III

    Solutions I and II are of comparable use since each involves an element of simple shape and requires integration with respect to one variable only.

    *Solution III.* As an alternative, we could use the angle $\theta$ as our variable with limits of 0 and $\pi/2$. The radius of either element would become $r \sin \theta$, whereas the thickness of the slice in Solution I would be $dy = (r\, d\theta) \sin \theta$ and that of the shell in Solution II would be $dz = (r\, d\theta) \cos \theta$. The length of the shell would be $y = r \cos \theta$.

① Can you identify the higher-order element of volume which is omitted from the expression for $dV$?

# PROBLEMS

### *Introductory problems*

**5/1** Place your pencil on the position of your best visual estimate of the centroid of the triangular area. Check the horizontal position of your estimate by referring to the results of Sample Problem 5/2.

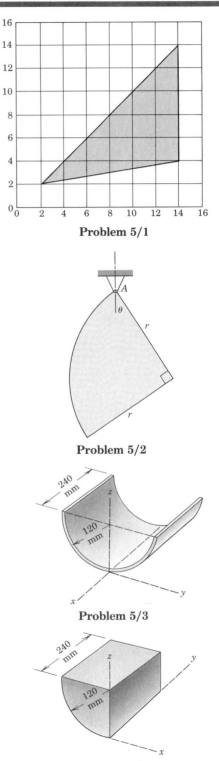

**Problem 5/1**

**5/2** The uniform quarter-circular plate is supported by a small bearing at $A$ and is free to swing in the vertical plane. Calculate the angle made by the upper edge of the plate with the vertical for the equilibrium position.

**Problem 5/2**

**5/3** Specify the $x$- and $z$-coordinates of the center of mass of the semicylindrical shell.

$$Ans. \ \bar{x} = 120 \text{ mm}, \ \bar{z} = 43.6 \text{ mm}$$

**Problem 5/3**

**5/4** Specify the $x$-, $y$-, and $z$-coordinates of the mass center of the quadrant of the homogeneous solid cylinder.

**Problem 5/4**

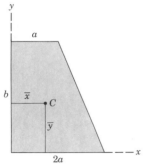

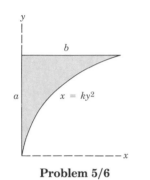

**Problem 5/5**

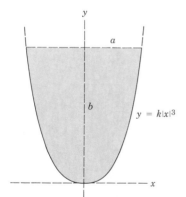

**Problem 5/6**

**Problem 5/7**

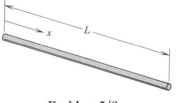

**Problem 5/8**

**5/5** Determine the $x$- and $y$-coordinates of the centroid $C$ of the trapezoidal area.

*Ans.* $\bar{x} = 7a/9, \bar{y} = 4b/9$

**5/6** Determine the coordinates of the centroid of the shaded area.

**5/7** Determine the coordinates of the centroid of the shaded area.      *Ans.* $\bar{x} = 0, \bar{y} = 4b/7$

**5/8** The slender rod is made of an experimental material whose density varies with position according to $\rho = \rho_0 e^{-ax}$. Determine the location of the center of mass of the rod. Evaluate your answer for $a = 1/L$.

**5/9** Determine the coordinates of the centroid of the shaded area.      *Ans.* $\bar{x} = 1.44, \bar{y} = 0.361k$

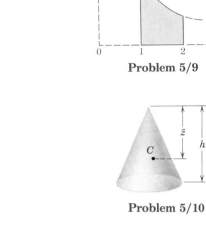

**Problem 5/9**

**5/10** Find the distance $\bar{z}$ from the vertex of the right-circular cone to the centroid of its volume.

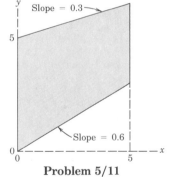

**Problem 5/10**

### *Representative problems*

**5/11** By direct integration, determine the coordinates of the centroid of the trapezoidal area.
       *Ans.* $\bar{x} = 2.35, \bar{y} = 3.56$

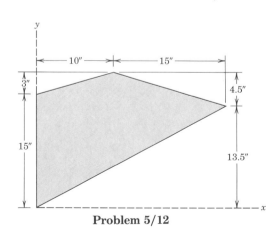

**Problem 5/11**

**5/12** By direct integration determine the $x$-coordinate of the centroid of the shaded area.

**Problem 5/12**

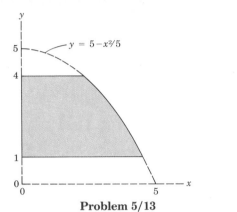

Problem 5/13

**5/13** Locate the centroid of the shaded area shown.
*Ans.* $\bar{x} = 1.797, \bar{y} = 2.34$

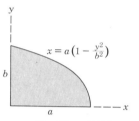

Problem 5/14

**5/14** Locate the centroid of the shaded area.

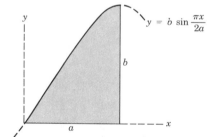

Problem 5/15

**5/15** Determine the coordinates of the centroid of the shaded area. *Ans.* $\bar{x} = 2a/\pi, \bar{y} = \pi b/8$

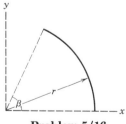

Problem 5/16

**5/16** Determine the coordinate of the centroid of the circular arc using the coordinates shown. Reconcile your results with those of Sample Problem 5/1.

**5/17** Determine the coordinates of the centroid of the area of the circular sector using the coordinates shown. Reconcile your results with those of Sample Problem 5/3.

$$\text{Ans. } \bar{x} = \frac{2r}{3\beta} \sin \beta, \ \bar{y} = \frac{2r}{3\beta} (1 - \cos \beta)$$

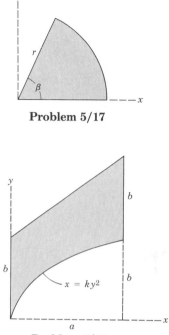

**Problem 5/17**

**5/18** Determine the $y$-coordinate of the centroid of the shaded area.

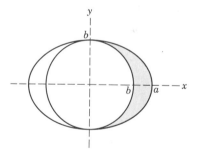

**Problem 5/18**

**5/19** Locate the centroid of the shaded area between the ellipse and the circle.

$$\text{Ans. } \bar{x} = \frac{4}{3\pi} (a + b)$$

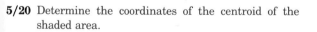

**Problem 5/19**

**5/20** Determine the coordinates of the centroid of the shaded area.

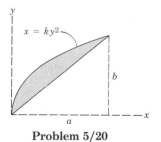

**Problem 5/20**

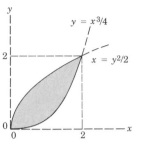

$$y = x^{3/4}$$
$$x = y^2/2$$

**Problem 5/21**

**5/21** Locate the centroid of the shaded area between the two curves. *Ans.* $\bar{x} = \frac{24}{25}$, $\bar{y} = \frac{6}{7}$

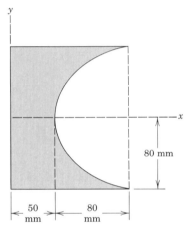

80 mm

50 mm    80 mm

**Problem 5/22**

**5/22** The figure represents a flat piece of sheet metal symmetrical about the $x$-axis and having a curved boundary in the form of a parabola. Calculate the $x$-coordinate of the centroid of the figure.

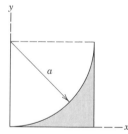

$a$

**Problem 5/23**

**5/23** Locate the centroid of the area shown in the figure by direct integration. (*Caution:* Observe carefully the proper sign of the radical involved.)

$$Ans. \ \bar{x} = \frac{2a}{3(4 - \pi)}, \ \bar{y} = \frac{10 - 3\pi}{3(4 - \pi)} a$$

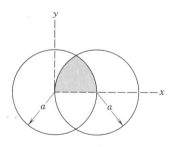

$a$    $a$

**Problem 5/24**

**5/24** Determine the $y$-coordinate of the centroid of the shaded area shown. (Observe the caution cited with Prob. 5/23.)

**5/25** Calculate the distance $\bar{h}$ measured from the base to the centroid of the volume of the frustum of the right-circular cone. *Ans.* $\bar{h} = \frac{11}{56}h$

**5/26** Use the results of Sample Problem 5/2 and determine by inspection the distance $\bar{h}$ from the centroid of the lateral area of any cone or pyramid of altitude $h$ to the base of the figure.

**5/27** The thickness of the triangular plate varies linearly with $y$ from a value $t_0$ along its base $y = 0$ to $2t_0$ at $y = h$. Determine the $y$-coordinate of the center of mass of the plate. *Ans.* $\bar{y} = 3h/8$

**5/28** Determine the $z$-coordinate of the centroid of the volume obtained by revolving the shaded triangular area about the $z$-axis through 360°.

**5/29** Determine the $z$-coordinate of the volume generated by revolving the shaded area around the $z$-axis through 360°. *Ans.* $\bar{z} = 5a/8$

**5/30** Determine the $z$-coordinate of the centroid of the volume obtained by revolving the shaded area under the parabola about the $z$-axis through 180°.

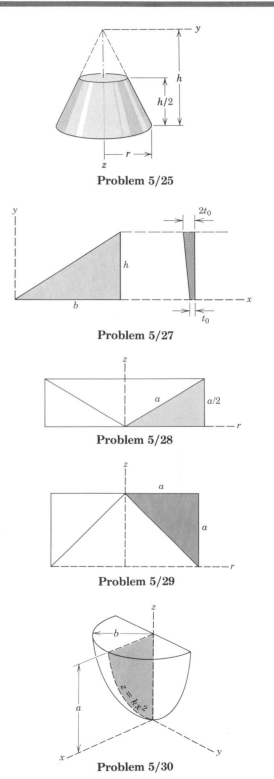

Problem 5/25

Problem 5/27

Problem 5/28

Problem 5/29

Problem 5/30

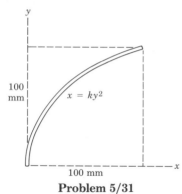

**Problem 5/31**

**5/31** The homogeneous slender rod has a uniform cross section and is bent into the shape shown. Calculate the $y$-coordinate of the mass center of the rod. (*Reminder:* A differential arc length is $dL = \sqrt{(dx)^2 + (dy)^2} = \sqrt{1 + (dx/dy)^2}\, dy$.)

*Ans.* $\bar{y} = 57.4$ mm

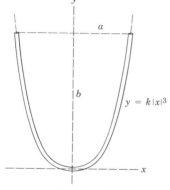

**Problem 5/32**

**5/32** Set up, but do not evaluate, the integrals for the $x$- and $y$-coordinates of the centroid of the uniform slender rod bent into the shape shown.

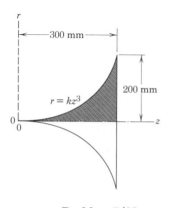

**Problem 5/33**

**5/33** Locate the mass center of the homogeneous solid body whose volume is determined by revolving the shaded area through 360° about the $z$-axis.

*Ans.* $\bar{z} = 263$ mm

**5/34** Determine the $x$-coordinate of the centroid of the volume described in Prob. 5/30.

**5/35** Determine the coordinates of the centroid of the volume obtained by revolving the shaded area about the $z$-axis through the 90° angle.

$$Ans. \ \bar{x} = \bar{y} = \left(\frac{4}{\pi} - \frac{3}{4}\right) a, \bar{z} = a/4$$

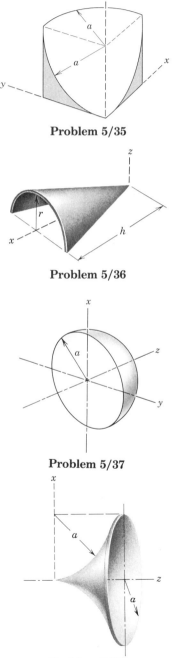

**Problem 5/35**

**5/36** Determine the position of the center of mass of the homogeneous thin conical shell shown.

**Problem 5/36**

**5/37** The density of a sphere decreases linearly with radial location from $2\rho_0$ at its center to $\rho_0$ at the surface $r = a$. If the sphere is cut on a diametral plane as shown, determine the distance $\bar{z}$ from this base plane to the center of mass of the hemisphere.

$$Ans. \ \bar{z} = 9a/25$$

**Problem 5/37**

▶**5/38** Locate the center of mass of the homogeneous bell-shaped shell of uniform but negligible thickness.

$$Ans. \ \bar{z} = \frac{a}{\pi - 2}$$

**Problem 5/38**

▶**5/39** Determine the position of the centroid of the volume within the bell-shaped shell of Prob. 5/38.

$$Ans. \ \bar{z} = \frac{a}{2(10 - 3\pi)}$$

▶**5/40** Locate the center of mass $G$ of the steel half ring. (*Hint*: Choose an element of volume in the form of a cylindrical shell whose intersection with the plane of the ends is shown.)

$$Ans. \ \bar{r} = \frac{a^2 + 4R^2}{2\pi R}$$

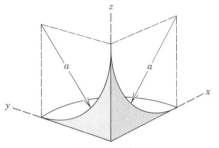

**Problem 5/40**

▶**5/41** Determine the $x$- and $y$-coordinates of the centroid of the volume generated by rotating the shaded area about the $z$-axis through 90°.

$$Ans. \ \bar{x} = \bar{y} = 0.242a$$

**Problem 5/41**

▶**5/42** Determine the $z$-coordinate of the centroid of the volume described in Prob. 5/41.

$$Ans. \ \bar{z} = 0.1308a$$

## 5/4 COMPOSITE BODIES AND FIGURES; APPROXIMATIONS

When a body or figure can be conveniently divided into several parts whose mass centers are easily determined, we may use the principle of moments where each part is treated as a finite element of the whole. Thus, for such a body, illustrated schematically in Fig. 5/13, whose parts have masses $m_1$, $m_2$, $m_3$ with the respective mass-center coordinates $\bar{x}_1$, $\bar{x}_2$, $\bar{x}_3$ in the $x$-direction, for example, the moment principle gives

$$(m_1 + m_2 + m_3)\overline{X} = m_1\bar{x}_1 + m_2\bar{x}_2 + m_3\bar{x}_3$$

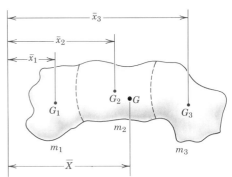

**Figure 5/13**

where $\overline{X}$ is the $x$-coordinate of the center of mass of the whole. Similar relations hold for the other two coordinate directions. We generalize, then, for a body of any number of parts and express the sums in condensed form and obtain the mass-center coordinates

$$\boxed{\overline{X} = \frac{\Sigma\, m\bar{x}}{\Sigma\, m} \qquad \overline{Y} = \frac{\Sigma\, m\bar{y}}{\Sigma\, m} \qquad \overline{Z} = \frac{\Sigma\, m\bar{z}}{\Sigma\, m}} \qquad \textbf{(5/7)}$$

Analogous relations hold for composite lines, areas, and volumes, where the $m$'s are replaced by $L$'s, $A$'s, and $V$'s, respectively. It should be pointed out that if a hole or cavity is considered one of the component parts of a composite body or figure, the corresponding mass represented by the cavity or hole is treated as a negative quantity.

Frequently in practice the boundaries of an area or volume are not expressible in terms of simple geometrical shapes or in shapes that can be represented mathematically. For such cases we find it necessary to resort to a method of approximation. As an example, consider the problem of locating the centroid $C$ of the irregular area shown in Fig. 5/14. The area may be divided into strips of width $\Delta x$ and variable height $h$. The area $A$ of each strip, such as the one shown in color, is $h\,\Delta x$ and is multiplied by the coordinates $x_c$ and $y_c$ to its *centroid* to obtain the moments of the element of area. The sum of the moments for all strips divided by the total area of the

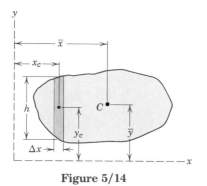

**Figure 5/14**

strips will give the corresponding centroidal coordinate. A systematic tabulation of the results will permit an orderly evaluation of the total area $\Sigma A$, the sums $\Sigma Ax_c$ and $\Sigma Ay_c$, and the centroidal coordinates

$$\bar{x} = \frac{\Sigma Ax_c}{\Sigma A} \qquad \bar{y} = \frac{\Sigma Ay_c}{\Sigma A}$$

The accuracy of the approximation will be increased by decreasing the widths of the strips used. In all cases the average height of the strip should be estimated in approximating the areas. Although it is usually of advantage to use elements of constant width, it is not necessary to do so. In fact we may use elements of any size and shape which approximate the given area to satisfactory accuracy.

In locating the centroid of an irregular volume, the problem may be reduced to one of determining the centroid of an area. Consider the volume shown in Fig. 5/15, where the magnitudes $A$

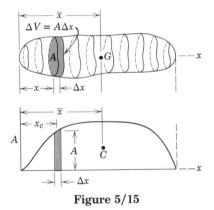

**Figure 5/15**

of the cross-sectional areas normal to the $x$-direction are plotted against $x$ as shown. A vertical strip of area under the curve is $A \, \Delta x$, which equals the corresponding element of volume $\Delta V$. Thus, the area under the plotted curve represents the volume of the body, and the $x$-coordinate of the centroid of the area under the curve is given by

$$\bar{x} = \frac{\Sigma(A \, \Delta x)x_c}{\Sigma A \, \Delta x} \qquad \text{which equals} \qquad \bar{x} = \frac{\Sigma Vx_c}{\Sigma V}$$

for the centroid of the actual volume.

## Sample Problem 5/6

Locate the centroid of the shaded area.

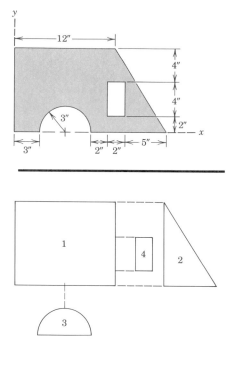

**Solution.** The composite area is divided into the four elementary shapes shown in the lower figure. The centroid locations of all these shapes may be obtained from Table D/3. Note that the areas of the "holes" (parts 3 and 4) are taken as negative in the following table:

| PART | $A$ in.$^2$ | $\bar{x}$ in. | $\bar{y}$ in. | $\bar{x}A$ in.$^3$ | $\bar{y}A$ in.$^3$ |
|---|---|---|---|---|---|
| 1 | 120 | 6 | 5 | 720 | 600 |
| 2 | 30 | 14 | 10/3 | 420 | 100 |
| 3 | −14.14 | 6 | 1.273 | −84.8 | −18 |
| 4 | −8 | 12 | 4 | −96 | −32 |
| TOTALS | 127.9 | | | 959 | 650 |

The area counterparts to Eqs. 5/7 are now applied and yield

$$\left[\bar{X} = \frac{\Sigma A\bar{x}}{\Sigma A}\right] \qquad \bar{X} = \frac{959}{127.9} = 7.50 \text{ in.} \qquad Ans.$$

$$\left[\bar{Y} = \frac{\Sigma A\bar{y}}{\Sigma A}\right] \qquad \bar{Y} = \frac{650}{127.9} = 5.08 \text{ in.} \qquad Ans.$$

## Sample Problem 5/7

Approximate the $x$-coordinate of the volume centroid of a body whose length is 1 m and whose cross-sectional area varies with $x$ as shown in the figure.

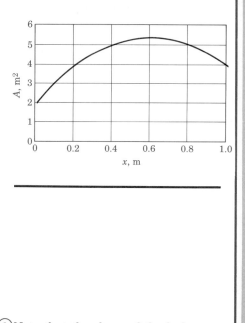

**Solution.** The body is divided into five sections. For each section, the average area, volume, and centroid location are determined and entered in the following table:

| INTERVAL | $A_{av}$ m$^2$ | Volume $V$ m$^3$ | $\bar{x}$ m | $V\bar{x}$ m$^4$ |
|---|---|---|---|---|
| 0–0.2 | 3 | 0.6 | 0.1 | 0.060 |
| 0.2–0.4 | 4.5 | 0.90 | 0.3 | 0.270 |
| 0.4–0.6 | 5.2 | 1.04 | 0.5 | 0.520 |
| 0.6–0.8 | 5.2 | 1.04 | 0.7 | 0.728 |
| 0.8–1.0 | 4.5 | 0.90 | 0.9 | 0.810 |
| TOTALS | | 4.48 | | 2.388 |

① $\left[\bar{X} = \dfrac{\Sigma V\bar{x}}{\Sigma V}\right] \qquad \bar{X} = \dfrac{2.388}{4.48} = 0.533 \text{ m} \qquad Ans.$

① Note that the shape of the body as a function of $y$ and $z$ does not affect $\bar{X}$.

## Sample Problem 5/8

Locate the center of mass of the bracket-and-shaft combination. The vertical face is made from sheet metal which has a mass of 25 kg/m². The material of the horizontal base has a mass of 40 kg/m², and the steel shaft has a density of 7.83 Mg/m³.

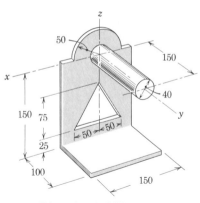

Dimensions in Millimeters

**Solution.** The composite body may be considered to be composed of the five elements shown in the lower portion of the illustration. The triangular part will be taken as a negative mass. For the reference axes indicated it is clear by symmetry that the x-coordinate of the center of mass is zero.

The mass $m$ of each part is easily calculated and should need no further explanation. For Part 1 we have from Sample Problem 5/3

$$\bar{z} = \frac{4r}{3\pi} = \frac{4(50)}{3\pi} = 21.2 \text{ mm}$$

For Part 3 from Sample Problem 5/2 we see that the centroid of the triangular mass is one-third of its altitude above its base. Measurement from the coordinate axes becomes

$$\bar{z} = -[150 - 25 - \tfrac{1}{3}(75)] = -100 \text{ mm}$$

The y- and z-coordinates to the mass centers of the remaining parts should be evident by inspection. The terms involved in applying Eqs. 5/7 are best handled in the form of a table as follows:

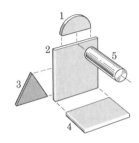

| PART | $m$ kg | $\bar{y}$ mm | $\bar{z}$ mm | $m\bar{y}$ kg·mm | $m\bar{z}$ kg·mm |
|------|--------|--------------|--------------|------------------|------------------|
| 1 | 0.098 | 0 | 21.2 | 0 | 2.08 |
| 2 | 0.562 | 0 | −75.0 | 0 | −42.19 |
| 3 | −0.094 | 0 | −100.0 | 0 | 9.38 |
| 4 | 0.600 | 50.0 | −150.0 | 30.0 | −90.00 |
| 5 | 1.476 | 75.0 | 0 | 110.7 | 0 |
| TOTALS | 2.642 | | | 140.7 | −120.73 |

Equations 5/7 are now applied and the results are

$$\left[ \bar{Y} = \frac{\Sigma m\bar{y}}{\Sigma m} \right] \qquad \bar{Y} = \frac{140.7}{2.642} = 53.3 \text{ mm} \qquad\qquad Ans.$$

$$\left[ \bar{Z} = \frac{\Sigma m\bar{z}}{\Sigma m} \right] \qquad \bar{Z} = \frac{-120.73}{2.642} = -45.7 \text{ mm} \qquad\qquad Ans.$$

# PROBLEMS

## *Introductory problems*

**5/43** Determine the coordinates of the centroid of the trapezoidal area shown.

Ans. $\overline{X}$ = 233 mm, $\overline{Y}$ = 333 mm

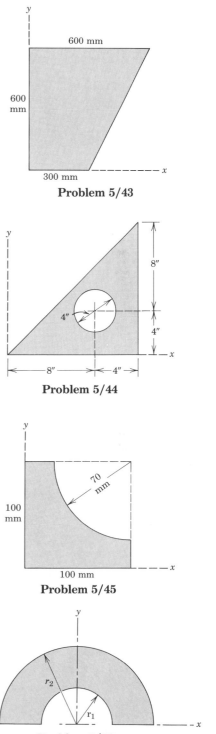

**Problem 5/43**

**5/44** Calculate the $x$- and $y$-coordinates of the centroid of the shaded area.

**Problem 5/44**

**5/45** Determine the $x$- and $y$-coordinates of the centroid of the shaded area.     Ans. $\overline{X} = \overline{Y} = 37.3$ mm

**Problem 5/45**

**5/46** Express the $y$-coordinate of the centroid of the area in terms of $r_1$ and $r_2$.

**Problem 5/46**

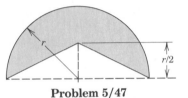

**Problem 5/47**

**5/47** Determine the distance $\overline{H}$ from the horizontal base diameter to the centroid of the shaded area in terms of $r$.  *Ans.* $\overline{H} = 0.545r$

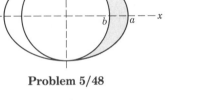

**Problem 5/48**

**5/48** Determine by the method of this article the $x$-coordinate of the centroid of the shaded area between the ellipse and the circle of Prob. 5/19, shown again here. (Refer to Table D/3 in Appendix D for the properties of an elliptical area.)

**5/49** Calculate the $y$-coordinate of the centroid of the shaded area.  *Ans.* $\overline{Y} = 9.93$ in.

**Problem 5/49**

### Representative problems

**5/50** Calculate the coordinates of the centroid of the shaded area.

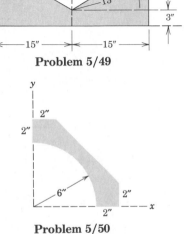

**Problem 5/50**

**5/51** Determine the distance $\overline{H}$ from the bottom of the base plate to the centroid of the built-up structural section shown.                    *Ans.* $\overline{H} = 39.3$ mm

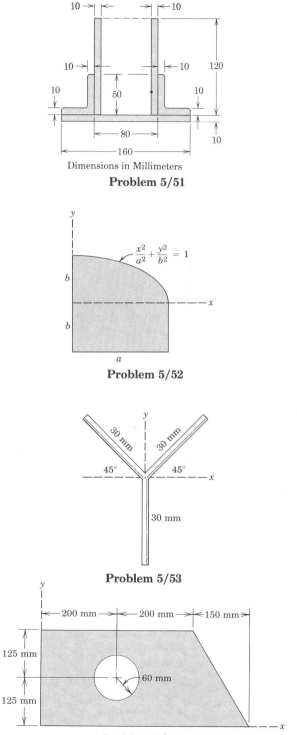

Dimensions in Millimeters

**Problem 5/51**

**5/52** Determine the coordinates of the centroid of the shaded area.

**Problem 5/52**

**5/53** The two upper lengths of the welded Y-shaped assembly of uniform slender rods have a mass per unit length of 0.3 kg/m, while the lower length has a mass of 0.5 kg/m. Locate the mass center of the assembly.          *Ans.* $\overline{Y} = -1.033$ mm, $\overline{X} = 0$

**Problem 5/53**

**5/54** Determine the coordinates of the centroid of the shaded area.

**Problem 5/54**

**Problem 5/55**

**5/55** The rigidly connected unit consists of a 2-kg circular disk, a 1.5-kg round shaft, and a 1-kg square plate. Determine the z-coordinate of the mass center of the unit.      *Ans.* $\overline{Z}$ = 70 mm

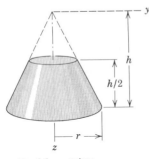

**Problem 5/56**

**5/56** A uniform rod is bent into the shape shown and pivoted about $O$. Find the value of $a$ in terms of the radius $r$ so that the straight section will remain horizontal.

**Problem 5/57**

**5/57** By the method of this article determine the height $\overline{H}$ from the base to the mass center of the frustum of the solid cone of Prob. 5/25, repeated here.      *Ans.* $\overline{H} = \frac{11}{56}h$

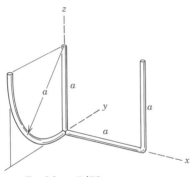

**Problem 5/58**

**5/58** Determine the coordinates of the mass center of the welded assembly of uniform slender rods made from the same bar stock.

**5/59** Determine the coordinates of the center of mass of the bracket, which is made from a plate of uniform thickness.        *Ans.* $\overline{X} = -8.3$ mm
$$\overline{Y} = -31.4 \text{ mm}$$
$$\overline{Z} = 10.3 \text{ mm}$$

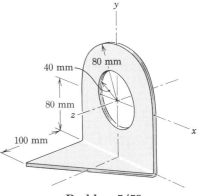

**Problem 5/59**

**5/60** An underwater instrument is modeled as shown in the figure. Determine the coordinates of the centroid of this composite volume.

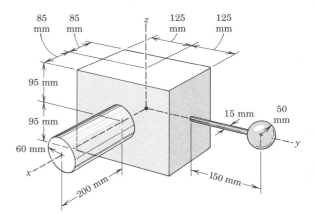

**Problem 5/60**

**5/61** The welded assembly is made of a uniform rod weighing 0.370 lb per foot of length and the semicircular plate weighing 8 lb per square foot. Calculate the coordinates of the center of gravity of the assembly.    *Ans.* $\overline{X} = 1.595$ in., $\overline{Y} = 0$, $\overline{Z} = 1.559$ in.

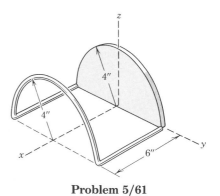

**Problem 5/61**

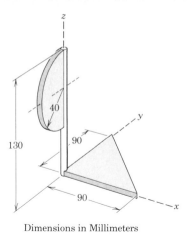

Dimensions in Millimeters
**Problem 5/62**

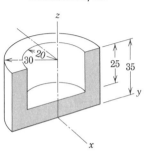

Dimensions in Millimeters
**Problem 5/63**

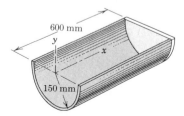

**Problem 5/64**

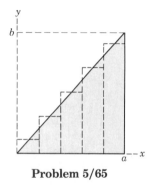

**Problem 5/65**

**5/62** The masses of the three parts of the welded assembly consisting of the triangular plate, uniform rod, and semicircular plate are, respectively, 0.210, 0.110, and 0.125 kg. Calculate the coordinates of the mass center.

**5/63** Calculate the coordinates of the mass center of the metal die casting shown.
*Ans.* $\overline{X} = -14.71$ mm, $\overline{Z} = 15.17$ mm

**5/64** Determine the position of the center of mass of the cylindrical shell with a closed semicircular end. The shell is made from sheet metal with a mass of 24 kg/m², and the end is made from metal plate with a mass of 36 kg/m².

**5/65** As an example of the accuracy involved in graphical and finite difference approximations, calculate the percentage error $e$ in determining the $x$-coordinate of the centroid of the triangular area by using the five approximating rectangles of width $a/5$ in place of the triangle.
*Ans.* $e = 1.00\%$ low

**5/66** The figure shows the underwater cross-sectional area $A$ a distance $x$ aft of the bow at the waterline of a sailboat hull. The variation of $A$ with $x$ is shown in the graph for a particular hull. Determine the distance $\overline{X}$ aft of point $P$ to the center of buoyancy of the hull (centroid of the displaced volume of water). The location of the center of buoyancy is a critical parameter in the design of the hull.

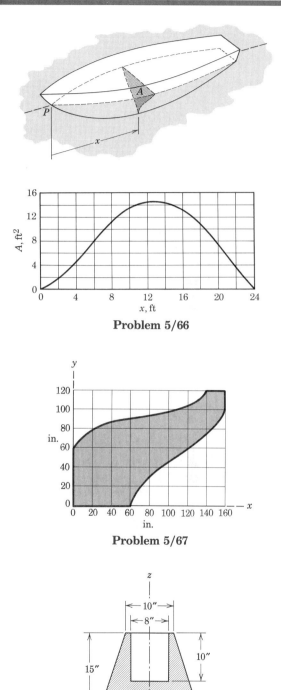

**Problem 5/66**

**5/67** A sheet-metal pattern has the shape shown. Estimate the location of the centroid of the area visually and note its coordinates. Then check your estimate by a calculation using the superimposed grid.

*Ans.* $\overline{x} = 68.4$ in., $\overline{y} = 58.0$ in.

**Problem 5/67**

**5/68** A homogeneous solid of revolution, shown in section, consists of a frustum of a right circular cone containing a cylindrical hole of 8-in. diameter. Calculate the height $\overline{Z}$ of the mass center above the base.

**Problem 5/68**

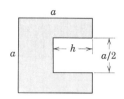

**Problem 5/69**

**5/69** A rectangular piece is removed from the square metal plate of side $a$. Determine the value of $h$ which will result in the mass center of the remaining plate being as far to the left as possible.

*Ans.* $h = 0.586a$

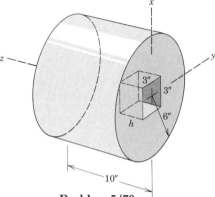

**Problem 5/70**

**5/70** Determine the depth $h$ of the square hole in the solid circular cylinder for which the $z$-coordinate of the mass center will have the maximum possible value.

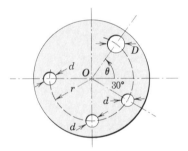

**Problem 5/71**

▶**5/71** The circular disk rotates about an axis through its center $O$ and has three holes of diameter $d$ positioned as shown. A fourth hole is to be drilled in the disk at the same radius $r$ so that the disk will be in balance (mass center at $O$). Determine the required diameter $D$ of the new hole and its angular position.

*Ans.* $D = 1.227d$, $\theta = 84.9°$

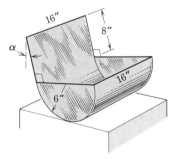

**Problem 5/72**

▶**5/72** A cylindrical container with an extended rectangular back and semicircular ends is all fabricated from the same sheet-metal stock. Calculate the angle $\alpha$ made by the back with the vertical when the container rests in an equilibrium position on a horizontal surface.

*Ans.* $\alpha = 39.6°$

## 5/5  THEOREMS OF PAPPUS*

A very simple method exists for calculating the surface area generated by revolving a plane curve about a nonintersecting axis in the plane of the curve. In Fig. 5/16 the line segment of length $L$ in the $x$-$y$ plane generates a surface when revolved about the $x$-axis. An element of this surface is the ring generated by $dL$. The area of this ring is its circumference times its slant height or $dA = 2\pi y\,dL$ and the total area is then

$$A = 2\pi \int y\,dL$$

But since $\bar{y}L = \int y\,dL$, the area becomes

$$\boxed{A = 2\pi\bar{y}L} \tag{5/8}$$

where $\bar{y}$ is the $y$-coordinate of the centroid $C$ for the line of length $L$. Thus, the generated area is the same as the lateral area of a right-circular cylinder of length $L$ and radius $\bar{y}$.

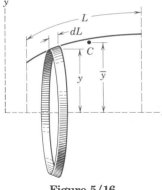

**Figure 5/16**

In the case of a volume generated by revolving an area about a nonintersecting line in its plane, an equally simple relation exists for finding the volume. An element of the volume generated by revolving the area $A$ about the $x$-axis, Fig. 5/17, is the elemental ring of cross section $dA$ and radius $y$. The volume of the element is its circumference times $dA$ or $dV = 2\pi y\,dA$, and the total volume is

$$V = 2\pi \int y\,dA$$

---

* Attributed to Pappus of Alexandria, a Greek geometer who lived in the third century A.D. The theorems often bear the name of Guldinus (Paul Guldin, 1577–1643), who claimed original authorship, although the works of Pappus were apparently known to him.

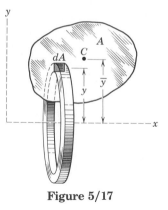

**Figure 5/17**

But since $\bar{y}A = \int y\,dA$, the volume becomes

$$V = 2\pi\bar{y}A \qquad (5/9)$$

where $\bar{y}$ is the $y$-coordinate of the centroid $C$ of the revolved area $A$. Thus, the generated volume is obtained by multiplying the generating area by the circumference of the circular path described by its centroid.

The two theorems of Pappus, expressed by Eqs. 5/8 and 5/9, are useful not only in determining areas and volumes of revolution, but they are also employed to find the centroids of plane curves and plane areas when the corresponding areas and volumes due to revolution of these figures about a nonintersecting axis are known. Dividing the area or volume by $2\pi$ times the corresponding line segment length or plane area will give the distance from the centroid to the axis.

In the event that a line or an area is revolved through an angle $\theta$ less than $2\pi$, the generated surface or volume may be found by replacing $2\pi$ by $\theta$ in Eqs. 5/8 and 5/9. Thus, the more general relations are

$$A = \theta\bar{y}L \qquad (5/8a)$$

and

$$V = \theta\bar{y}A \qquad (5/9a)$$

where $\theta$ is expressed in radians.

# PROBLEMS

*Introductory problems*

**5/73** Using the methods of this article, determine the volume and total surface area of a right circular cylinder of radius $r$ and height $h$.
$$Ans.\ V = \pi r^2 h,\ A = 2\pi r(r + h)$$

**5/74** From the known surface area $A = 4\pi r^2$ of a sphere of radius $r$, determine the radial distance $\bar{r}$ to the centroid of the semicircular arc used to generate the surface.

**5/75** Determine the volume $V$ and lateral area $A$ of a right-circular cone of base radius $r$ and altitude $h$.
$$Ans.\ V = \tfrac{1}{3}\pi r^2 h,\ A = \pi r\sqrt{h^2 + r^2}$$

**5/76** Determine the volume $V$ generated by revolving the elliptical area through 180° about the $z$-axis.

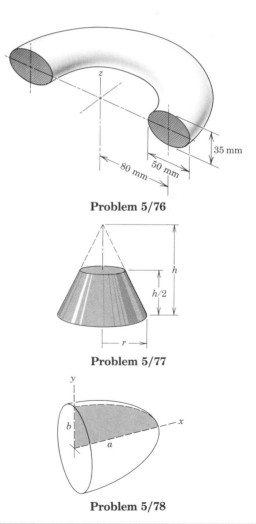

35 mm

80 mm  50 mm

**Problem 5/76**

**5/77** Determine the lateral surface area $A$ of the frustum of the right-circular cone.
$$Ans.\ A = \tfrac{3}{4}\pi r\sqrt{r^2 + h^2}$$

$h$

$h/2$

$r$

**Problem 5/77**

**5/78** Obtain the volume $V$ of the semiellipsoid of revolution obtained by revolving the area of the elliptical quadrant about the $x$-axis. Refer to the centroidal information given in Table D/3 in Appendix D.

$y$

$b$

$a$

$x$

**Problem 5/78**

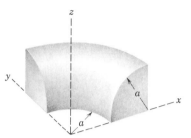

**Problem 5/79**

### Representative problems

**5/79** Determine the volume $V$ generated by revolving the quarter-circular area about the $z$-axis through an angle of 90°.

$$Ans. \quad V = \frac{\pi a^3}{12}(3\pi - 2)$$

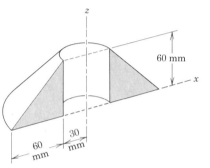

**Problem 5/80**

**5/80** Calculate the volume $V$ of the solid generated by revolving the 60-mm right-triangular area about the $z$-axis through 180°.

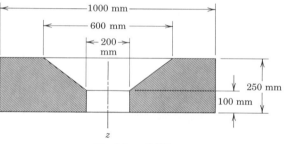

**Problem 5/81**

**5/81** Calculate the volume formed by completely revolving the cross-sectional area shown about the $z$-axis of symmetry. $Ans. \quad V = 0.1728 \text{ m}^3$

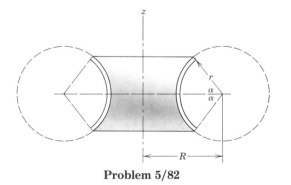

**Problem 5/82**

**5/82** A thin shell, shown in section, has the form generated by revolving the arc about the $z$-axis through 360°. Determine the surface area $A$ of one of the two sides of the shell.

**5/83** A steel die, shown in section, has the form of a solid generated by revolving the shaded area around the z-axis. Calculate the mass $m$ of the die.

*Ans.* $m = 84.5$ kg

**5/84** The two circular arcs $AB$ and $BC$ are revolved about the vertical axis to obtain the surface of revolution shown. Compute the area $A$ of this surface.

**5/85** A reducing fitting for a high-pressure oil system is shown in section. Calculate the lateral surface area of the fitting (i.e., the area generated by revolving $AB$ completely about the central axis). Also calculate the volume of the fitting.

*Ans.* $A = 438$ mm², $V = 933$ mm³

**5/86** Determine the surface area of one side of the bell-shaped shell of Prob. 5/38, shown again here, using the theorem of Pappus.

**5/87** Determine the volume within the bell-shaped shell shown with Prob. 5/86. The results cited in Prob. 5/23 may be used for this problem.

*Ans.* $V = \dfrac{\pi a^3}{6} (10 - 3\pi)$

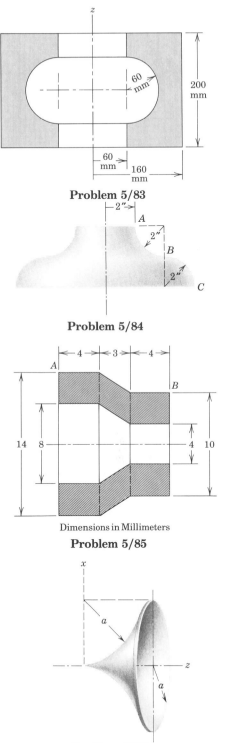

Problem 5/83

Problem 5/84

Dimensions in Millimeters

Problem 5/85

Problem 5/86

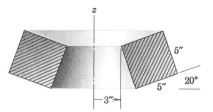

**Problem 5/88**

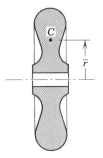

**Problem 5/89**

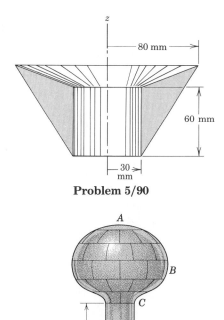

**Problem 5/90**

**Problem 5/91**

**5/88** Compute the volume $V$ and total surface area $A$ of the complete ring whose cross section is shown.

**5/89** A hand-operated control wheel made of aluminum has the proportions shown in the cross-sectional view. The area of the total section shown is 15 200 mm$^2$, and the wheel has a mass of 10.0 kg. Calculate the distance $\bar{r}$ to the centroid of the half-section. The aluminum has a density of 2.69 Mg/m$^3$.

*Ans.* $\bar{r} = 77.8$ mm

**5/90** The cross section of an aluminum casting in the form of a solid of revolution about the $z$-axis is shown. Calculate the mass $m$ of the casting.

**5/91** The water storage tank is a shell of revolution and is to be sprayed with two coats of paint which has a coverage of 500 ft$^2$ per gallon. The engineer (who remembers mechanics) consults a scale drawing of the tank and determines that the curved line $ABC$ has a length of 34 ft and that its centroid is 8.2 ft from the centerline of the tank. How many gallons of paint will be used for the tank including the vertical cylindrical column?

*Ans.* 8.82 gal

**5/92** Find the volume $V$ of the solid generated by revolving the shaded area about the $z$-axis through 90°.

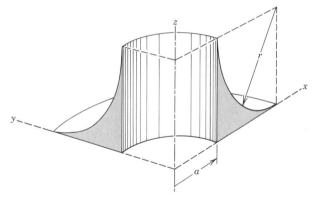

**Problem 5/92**

**5/93** Calculate the mass $m$ of concrete required to construct the arched dam shown. Concrete has a density of 2.40 Mg/m³.      *Ans.* $m = 1.126(10^6)$ Mg

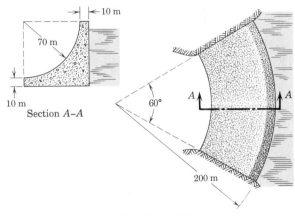

**Problem 5/93**

**5/94** The shaded area is bounded by one half-cycle of a sine wave and the axis of the sine wave. Determine the volume generated by completely revolving the area about the $x$-axis.

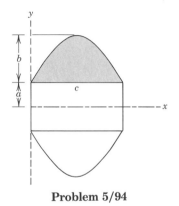

**Problem 5/94**

## SECTION B.  SPECIAL TOPICS

### 5/6  BEAMS—EXTERNAL EFFECTS

Structural members that offer resistance to bending due to applied loads are known as beams. Most beams are long prismatical bars, and the loads are usually applied normal to the axes of the bars. Beams are undoubtedly the most important of all structural members, and the basic theory underlying their design must be thoroughly understood. The analysis of the load-carrying capacities of beams consists, first, in establishing the equilibrium requirements of the beam as a whole and any portion of it considered separately. Second, the relations between the resulting forces and the accompanying internal resistance of the beam to support these forces are established. The first part of this analysis requires the application of the principles of statics, while the second part of the problem involves the strength characteristics of the material and is usually treated under the heading of the mechanics of solids or the mechanics of materials. This article is concerned with the *external* loading and reactions acting on a beam. In Art. 5/7 we calculate the distribution along the beam of the *internal* force and moment.

**Types of beams.**   Beams supported in such a way that their external support reactions can be calculated by the methods of statics alone are called *statically determinate* beams. A beam that has more supports than are necessary to provide equilibrium is said to be *statically indeterminate*, and it is necessary to consider the load-deformation properties of the beam in addition to the equations of statical equilibrium to determine the support reactions. In Fig. 5/18 are shown examples of both types of beams. In this article we will analyze only statically determinate beams.

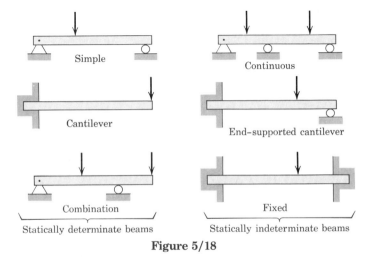

Simple

Continuous

Cantilever

End–supported cantilever

Combination

Fixed

Statically determinate beams          Statically indeterminate beams

**Figure 5/18**

Beams may also be identified by the type of external loading they support. The beams in Fig. 5/18 are supporting concentrated loads, whereas the beam in Fig. 5/19 is supporting a distributed load. The intensity $w$ of a distributed load may be expressed as force per unit length of beam. The intensity may be constant or variable, continuous or discontinuous. The intensity of the loading in Fig. 5/19 is constant from $C$ to $D$ and variable from $A$ to $C$ and from $D$ to $B$. The intensity is discontinuous at $D$, where it changes magnitude abruptly. Although the intensity itself is not discontinuous at $C$, the rate of change of intensity $dw/dx$ is discontinuous.

***Distributed loads.*** Loading intensities which are constant or which vary linearly are easily handled. Figure 5/20 illustrates the three most common cases and the resultants of the distributed loads in each case.

In cases $a$ and $b$ of Fig. 5/20 we see that the resultant load $R$ is represented by the area formed by the intensity $w$ (force per unit length of beam) and the length $L$ over which the force is distributed. The resultant passes through the centroid of this area.

In the $c$-part of Fig. 5/20, the trapezoidal area is broken into a rectangular and a triangular area, and the corresponding resultants $R_1$ and $R_2$ of these subareas are determined separately. Note that a single resultant could be determined by using the composite technique for finding centroids which was discussed in Art. 5/4; usually, the determination of a single resultant is unnecessary.

For a more general load distribution, Fig. 5/21, we must start with a differential increment of force $dR = w\,dx$. The total load $R$ is then the sum of the differential forces, or

$$R = \int w\,dx$$

As before, the resultant $R$ is located at the centroid of the area under consideration. The $x$-coordinate of this centroid is found by the principle of moments $R\bar{x} = \int xw\,dx$, or

$$\bar{x} = \frac{\int xw\,dx}{R}$$

For the distribution of Fig. 5/21, the vertical coordinate of the centroid need not be found.

Once the distributed loads have been reduced to their equivalent concentrated loads, the external reactions acting on the beam may be found by a straightforward static analysis as developed in Chapter 3.

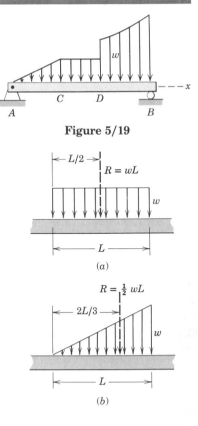

**Figure 5/19**

**Figure 5/20**

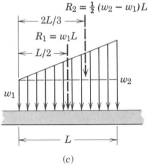

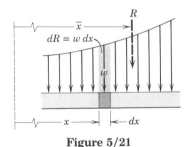

**Figure 5/21**

## Sample Problem 5/9

Determine the equivalent concentrated load(s) and external reactions for the simply supported beam which is subjected to the distributed load shown.

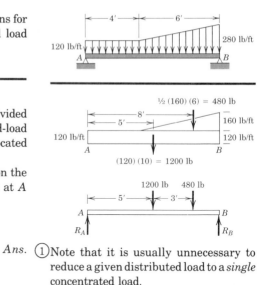

① **Solution.** The area associated with the load distribution is divided into the rectangular and triangular areas shown. The concentrated-load values are determined by computing the areas, and these loads are located at the centroids of the respective areas.

Once the concentrated loads are determined, they are placed on the free-body diagram of the beam along with the external reactions at $A$ and $B$. Using principles of equilibrium, we have

$[\Sigma M_A = 0]$ $\quad 1200(5) + 480(8) - R_B(10) = 0$

$$R_B = 984 \text{ lb} \qquad \textit{Ans.}$$

$[\Sigma M_B = 0]$ $\quad R_A(10) - 1200(5) - 480(2) = 0$

$$R_A = 696 \text{ lb} \qquad \textit{Ans.}$$

① Note that it is usually unnecessary to reduce a given distributed load to a *single* concentrated load.

## Sample Problem 5/10

Determine the reaction at the support $A$ of the loaded cantilever beam.

① **Solution.** The constants in the load distribution are found to be $w_0 = 1000$ N/m and $k = 2$ N/m⁴. The load $R$ is then

$$R = \int w \, dx = \int_0^8 (1000 + 2x^3) \, dx = \left(1000x + \frac{x^4}{2}\right)\Big|_0^8 = 10\ 048 \text{ N}$$

② The $x$-coordinate of the centroid of the area is found by

$$\bar{x} = \frac{\int xw \, dx}{R} = \frac{1}{10\ 048} \int_0^8 x(1000 + 2x^3) \, dx$$

$$= \frac{1}{10\ 048} \left(500x^2 + \tfrac{2}{5}x^5\right)\Big|_0^8 = 4.49 \text{ N}$$

From the free-body diagram of the beam, we have

$[\Sigma M_A = 0]$ $\quad M_A - (10\ 048)(4.49) = 0$

$$M_A = 45\ 100 \text{ N·m} \qquad \textit{Ans.}$$

$[\Sigma F_y = 0]$ $\quad A_y = 10\ 048 \text{ N} \qquad \textit{Ans.}$

Note that $A_x = 0$ by inspection.

① Use caution with the units of the constants $w_0$ and $k$.

② The student should recognize that the calculation of $R$ and its location $\bar{x}$ is simply an application of centroids as treated in Art. 5/3.

# PROBLEMS

## *Introductory problems*

**5/95** Determine the reactions at $A$ and $B$ for the beam subjected to the uniform load distribution.

Ans. $R_A = 1.35$ kN, $R_B = 0.45$ kN

**5/96** Determine the reactions at the supports $A$ and $B$ for the beam loaded as shown.

**5/97** Calculate the supporting force $R_A$ and moment $M_A$ at $A$ for the loaded cantilever beam.

Ans. $R_A = 2.4$ kN, $M_A = 14.4$ kN·m CCW

**5/98** Calculate the reactions at $A$ and $B$ for the beam subjected to the triangular load distribution.

**5/99** Determine the reactions at the built-in end of the beam subjected to the triangular load distribution.

Ans. $R_A = \dfrac{w_0 l}{2}$, $M_A = \dfrac{w_0 l^2}{6}$ CCW

**5/100** Calculate the reactions at $A$ and $B$ for the beam loaded as shown.

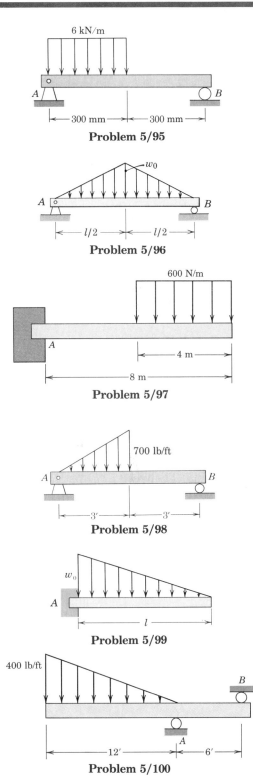

6 kN/m

$A$        $B$

300 mm    300 mm

**Problem 5/95**

$w_0$

$A$        $B$

$l/2$    $l/2$

**Problem 5/96**

600 N/m

$A$

4 m

8 m

**Problem 5/97**

700 lb/ft

$A$        $B$

3′    3′

**Problem 5/98**

$w_0$

$A$

$l$

**Problem 5/99**

400 lb/ft

$B$

12′    6′

$A$

**Problem 5/100**

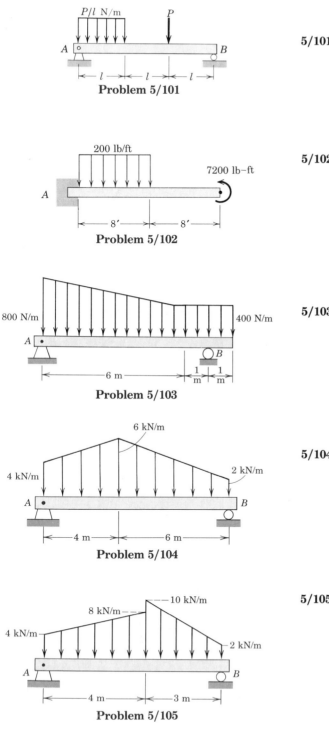

**Problem 5/101**

**Problem 5/102**

**Problem 5/103**

**Problem 5/104**

**Problem 5/105**

**5/101** Compute the reactions at $A$ and $B$ for the beam subjected to a combination of distributed and point loads. *Ans.* $R_A = 7P/6$, $R_B = 5P/6$

**5/102** Find the reaction at $A$ due to the uniform loading and the applied couple.

**5/103** Determine the reactions at the supports of the beam which is loaded as shown. *Ans.* $R_A = 2230$ N, $R_B = 2170$ N

**5/104** Calculate the support reactions at $A$ and $B$ for the beam subjected to the two linearly varying load distributions.

**5/105** Calculate the supporting reactions at $A$ and $B$ for the beam subjected to the two linearly distributed loads. *Ans.* $R_A = 21.1$ kN, $R_B = 20.9$ kN

**5/106** Determine the force and moment reactions at *A* for the beam which is subjected to the load combination shown.

**5/107** Determine the reactions at points *A* and *B* of the beam subjected to the elliptical and uniform load distributions. At which surface, upper or lower, is the reaction at *A* exerted?

$\quad\quad\quad$ *Ans.* $A = 5.15$ kN, $B = 5.37$ kN, upper

**5/108** Determine the reactions at points *A* and *B* of the inclined beam subjected to the vertical load distribution shown. The value of the load distribution at the right end of the beam is 5 kN per *horizontal* meter.

**5/109** Determine the force and moment reactions at the support *A* of the built-in beam which is subjected to the sine-wave load distribution.

$\quad\quad\quad$ *Ans.* $R_A = \dfrac{2w_0 l}{\pi}$, $M_A = \dfrac{w_0 l^2}{\pi}$ CCW

**5/110** A cantilever beam supports the variable load shown. Calculate the supporting force $R_A$ and moment $M_A$ at *A*.

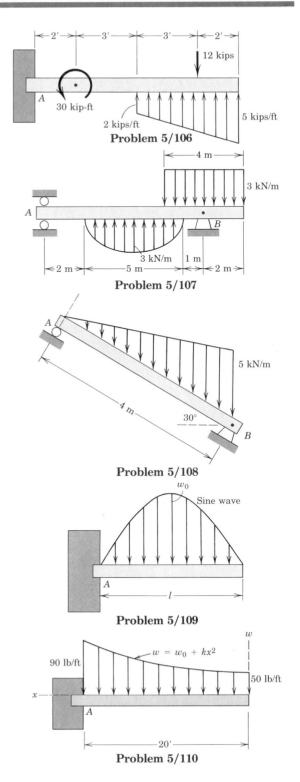

**Problem 5/106**

**Problem 5/107**

**Problem 5/108**

**Problem 5/109**

**Problem 5/110**

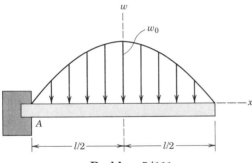

**Problem 5/111**

**5/111** The cantilever beam is subjected to a parabolic distribution of load symmetrical about the middle of the beam. Determine the supporting force $R_A$ and moment $M_A$ acting on the beam at $A$.

$$Ans. \; R_A = \frac{2w_0 l}{3}, \; M_A = \frac{w_0 l^2}{3} \; \text{CCW}$$

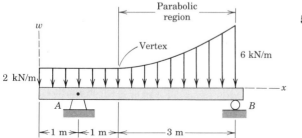

**Problem 5/112**

**5/112** A beam is subjected to the variable loading shown. Calculate the support reactions at $A$ and $B$.

Parabolic region

Vertex

**Problem 5/113**

**5/113** Determine the reactions at the supports of the beam which is acted on by the combination of uniform and parabolic loading distributions.

$$Ans. \; R_A = 7 \; \text{kN}, \; R_B = 7 \; \text{kN}$$

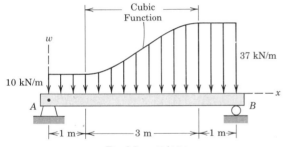

**Problem 5/114**

▶**5/114** The transition between the loads of 10 kN/m and 37 kN/m is accomplished by means of a cubic function of form $w = k_0 + k_1 x + k_2 x^2 + k_3 x^3$, the slope of which is zero at its end points $x = 1$ m and $x = 4$ m. Determine the reactions at $A$ and $B$.

$$Ans. \; R_A = 43.1 \; \text{kN}, \; R_B = 74.4 \; \text{kN}$$

## 5/7  *BEAMS—INTERNAL EFFECTS*

In the previous article we were concerned with the reduction of a distributed force to one or more equivalent concentrated forces and the subsequent determination of the external reactions acting on the beam. In this article we introduce internal beam effects and apply principles of statics to calculate the internal shear force and bending moment as functions of location along the beam.

*(a) Shear, bending, and torsion.*  In addition to supporting tension or compression a beam can resist shear, bending, and torsion. These three effects are illustrated in Fig. 5/22. The force $V$ is called the *shear force,* the couple $M$ is known as the *bending moment,* and the couple $T$ is called a *torsional moment.* These effects represent the vector components of the resultant of the forces acting on a transverse section of the beam as shown in the lower part of the figure.

We direct our attention now primarily to the shear force $V$ and bending moment $M$ caused by forces applied to the beam in a single plane. The conventions for positive values of shear $V$ and bending moment $M$ shown in Fig. 5/23 are the ones generally used. By the principle of action and reaction we note that the directions of $V$ and $M$ are reversed on the two sections. It is frequently impossible to tell without calculation whether the shear and moment at a particular section are positive or negative. For this reason it will be found advisable to represent $V$ and $M$ in their positive directions on the free-body diagrams and let the algebraic signs of the calculated values indicate the proper directions.

As an aid to the physical interpretation of the bending couple $M$, consider the beam shown in Fig. 5/24 bent by the two equal and opposite positive moments applied at the ends. The cross section of the beam is taken to be that of an H-section with a very narrow center web and heavy top and bottom flanges. For this beam we may neglect the load carried by the small web compared with that carried by the two flanges. It should be perfectly clear that the upper flange of the beam is shortened and is under compression while the lower flange is lengthened and is under tension. The resultant of the two forces, one tensile and the other compressive, acting on any

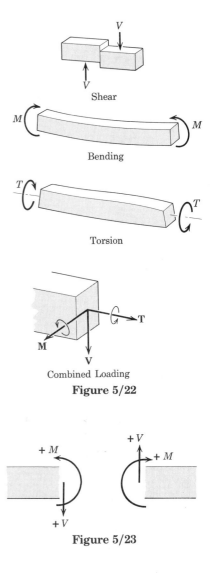

Shear

Bending

Torsion

Combined Loading

**Figure 5/22**

**Figure 5/23**

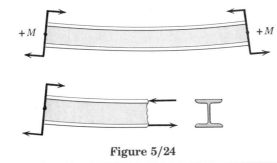

**Figure 5/24**

section is a couple and has the value of the bending moment on the section. If a beam having some other cross-sectional shape were loaded in the same way, the distribution of force over the cross section would be different, but the resultant would be the same couple.

The variation of shear force $V$ and bending moment $M$ over the length of a beam provides information necessary for the design analysis of the beam. In particular, the maximum magnitude of the bending moment is usually the primary consideration in the design or selection of a beam, and its value and position should be determined. The variations in shear and moment are best shown graphically, and the expressions for $V$ and $M$ when plotted against distance along the beam give the *shear-force* and *bending-moment diagrams* for the beam.

The first step in the determination of the shear and moment relations is to establish the values of all external reactions on the beam by applying the equations of equilibrium to a free-body diagram of the beam as a whole. Next, we isolate a portion of the beam, either to the right or to the left of an arbitrary transverse section, with a free-body diagram, and apply the equations of equilibrium to this isolated portion of the beam. These equations will yield expressions for the shear force $V$ and bending moment $M$ acting at the cut section on the part of the beam isolated. The part of the beam which involves the smaller number of forces, either to the right or to the left of the arbitrary section, usually yields the simpler solution. We should avoid using a transverse section that coincides with the location of a concentrated load or couple, as such a position represents a point of discontinuity in the variation of shear or bending moment. Finally, it is important to note that the calculations for $V$ and $M$ on each section chosen should be consistent with the positive convention illustrated in Fig. 5/23.

*(b) General loading, shear, and moment relationships.*    Certain general relationships may be established for any beam with distributed loads which will aid greatly in the determination of the shear and moment distributions along the beam. Figure 5/25 represents a portion of a loaded beam, and an element $dx$ of the beam is isolated. The loading $w$ represents the force per unit length of beam. At the location $x$ the shear $V$ and moment $M$ acting on the element are drawn in their positive directions. On the opposite side of the element where the coordinate is $x + dx$, these quantities are also shown in their positive directions but must be labeled $V + dV$ and $M + dM$, since the changes in $V$ and $M$ with $x$ are required. The applied loading $w$ may be considered constant over the length of the element, since this length is a differential quantity and the effect of any change in $w$ disappears in the limit compared with the effect of $w$ itself.

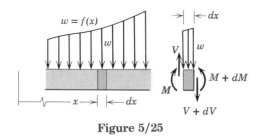

**Figure 5/25**

Equilibrium of the element requires that the sum of the vertical force be zero. Thus, we have

$$V - w\,dx - (V + dV) = 0$$

or

$$\boxed{w = -\frac{dV}{dx}} \qquad\qquad \textbf{(5/10)}$$

We see from Eq. 5/10 that the slope of the shear diagram must everywhere be equal to the negative of the value of the applied loading. Equation 5/10 holds on either side of a concentrated load but not at the concentrated load by reason of the discontinuity produced by the abrupt change in shear.

We may now express the shear force $V$ in terms of the loading $w$ by integrating Eq. 5/10. Thus,

$$\int_{V_0}^{V} dV = -\int_{x_0}^{x} w\,dx$$

or

$$V = V_0 + \text{(the negative of the area under}$$
$$\text{the loading curve from } x_0 \text{ to } x)$$

In this expression $V_0$ is the shear force at $x_0$ and $V$ is the shear force at $x$. Summing up the area under the loading curve is usually a simple way to construct the shear force diagram.

Equilibrium of the element in Fig. 5/25 also requires that the moment sum be zero. Summing moments about the left side of the element gives

$$M + w\,dx\,\frac{dx}{2} + (V + dV)\,dx - (M + dM) = 0$$

The two $M$'s cancel, and the terms $w(dx)^2/2$ and $dV\,dx$ may be dropped, since they are differentials of higher order than those that remain. This leaves merely

$$\boxed{V = \frac{dM}{dx}} \qquad\qquad \textbf{(5/11)}$$

which expresses the fact that the shear everywhere is equal to the slope of the moment curve. Equation 5/11 holds on either side of a concentrated couple but not at the concentrated couple by reason of the discontinuity caused by the abrupt change in moment.

We may now express the moment $M$ in terms of the shear $V$ by integrating Eq. 5/11. Thus,

$$\int_{M_0}^{M} dM = \int_{x_0}^{x} V \, dx$$

or

$$M = M_0 + \text{(area under the shear diagram from } x_0 \text{ to } x\text{)}$$

In this expression $M_0$ is the bending moment at $x_0$ and $M$ is the bending moment at $x$. For beams where there is no externally applied moment $M_0$ at $x_0 = 0$, the total moment at any section equals the area under the shear diagram up to that section. Summing up the area under the shear diagram is usually the simplest way to construct the moment diagram.

When $V$ passes through zero and is a continuous function of $x$ with $dV/dx \neq 0$, the bending moment $M$ will be a maximum or a minimum, since $dM/dx = 0$ at such a point. Critical values of $M$ also occur when $V$ crosses the zero axis discontinuously, which occurs for beams under concentrated loads.

We observe from Eqs. 5/10 and 5/11 that the degree of $V$ in $x$ is one higher than that of $w$. Also $M$ is of one higher degree in $x$ than is $V$. Consequently, $M$ is two degrees higher in $x$ than $w$. Thus for a beam loaded by $w = kx$, which is of the first degree in $x$, the shear $V$ is of the second degree in $x$ and the bending moment $M$ is of the third degree in $x$.

Equations 5/10 and 5/11 may be combined to yield

$$\boxed{\frac{d^2M}{dx^2} = -w} \tag{5/12}$$

Thus, if $w$ is a known function of $x$, the moment $M$ may be obtained by two integrations, provided that the limits of integration are properly evaluated each time. This method is usable only if $w$ is a continuous function of $x$.*

When bending in a beam occurs in more than a single plane, a separate analysis in each plane may be carried out. The results may then be combined vectorially.

---

* When $w$ is a discontinuous function of $x$, it is possible to introduce a special set of expressions called *singularity functions* which permit writing analytical expressions for shear $V$ and moment $M$ over a range of discontinuities. These functions are not discussed in this book.

## Sample Problem 5/11

Determine the shear and moment distributions produced in the simple beam by the 4-kN concentrated load.

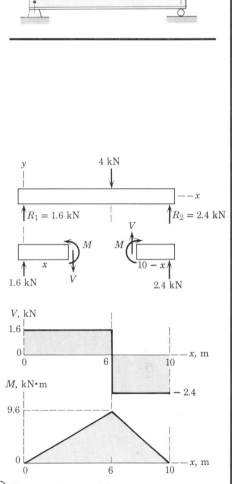

**Solution.** From the free-body diagram of the entire beam we find the support reactions, which are

$$R_1 = 1.6 \text{ kN} \qquad R_2 = 2.4 \text{ kN}$$

A section of the beam of length $x$ is next isolated with its free-body diagram on which we show the shear $V$ and the bending moment $M$ in their positive directions. Equilibrium gives

$$[\Sigma F_y = 0] \qquad 1.6 - V = 0 \qquad V = 1.6 \text{ kN}$$

$$[\Sigma M_{R_1} = 0] \qquad M - 1.6x = 0 \qquad M = 1.6x$$

① These values of $V$ and $M$ apply to all sections of the beam to the left of the 4-kN load.

A section of the beam to the right of the 4-kN load is next isolated with its free-body diagram on which $V$ and $M$ are shown in their positive directions. Equilibrium requires

$$[\Sigma F_y = 0] \qquad V + 2.4 = 0 \qquad V = -2.4 \text{ kN}$$

$$[\Sigma M_{R_2} = 0] \qquad -(2.4)(10 - x) + M = 0 \qquad M = 2.4(10 - x)$$

These results apply only to sections of the beam to the right of the 4-kN load.

The values of $V$ and $M$ are plotted as shown. The maximum bending moment occurs where the shear changes direction. As we move in the positive $x$-direction starting with $x = 0$, we see that the moment $M$ is merely the accumulated area under the shear diagram.

① We must be careful not to take our section at a concentrated load (such as $x = 6$ m) since the shear and moment relations involve discontinuities at such positions.

## Sample Problem 5/12

The cantilever beam is subjected to the load intensity (force per unit length) which varies as $w = w_0 \sin (\pi x/l)$. Determine the shear force $V$ and bending moment $M$ as functions of the ratio $x/l$.

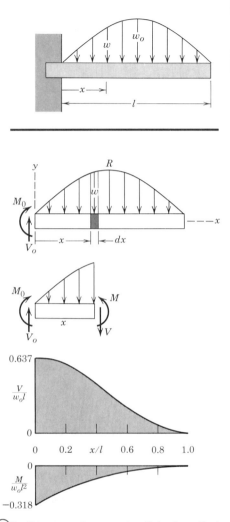

*Solution.* The free-body diagram of the entire beam is drawn first so that the shear force $V_0$ and bending moment $M_0$ which act at the supported end at $x = 0$ can be computed. By convention $V_0$ and $M_0$ are shown in their positive mathematical sense. A summation of vertical forces for equilibrium gives

$$[\Sigma F_y = 0] \qquad V_0 - \int_0^l w \, dx = 0 \qquad V_0 = \int_0^l w_0 \sin \frac{\pi x}{l} \, dx = \frac{2w_0 l}{\pi}$$

① A summation of moments about the left end at $x = 0$ for equilibrium gives

$$[\Sigma M = 0] \qquad -M_0 - \int_0^l x(w \, dx) = 0 \qquad M_0 = -\int_0^l w_0 x \sin \frac{\pi x}{l} \, dx$$

$$M_0 = \frac{-w_0 l^2}{\pi^2} \left[ \sin \frac{\pi x}{l} - \frac{\pi x}{l} \cos \frac{\pi x}{l} \right]_0^l = -\frac{w_0 l^2}{\pi}$$

From a free-body diagram of an arbitrary section of length $x$, integration of Eq. 5/10 permits us to find the shear force internal to the beam. Thus,

② $[dV = -w \, dx]$ $\qquad \displaystyle\int_{V_0}^V dV = -\int_0^x w_0 \sin \frac{\pi x}{l} \, dx$

$$V - V_0 = \left[ \frac{w_0 l}{\pi} \cos \frac{\pi x}{l} \right]_0^x \qquad V - \frac{2w_0 l}{\pi} = \frac{w_0 l}{\pi} \left( \cos \frac{\pi x}{l} - 1 \right)$$

or in dimensionless form

$$\frac{V}{w_0 l} = \frac{1}{\pi} \left( 1 + \cos \frac{\pi x}{l} \right) \qquad Ans.$$

The bending moment is obtained by integration of Eq. 5/11, which gives

$[dM = V \, dx]$ $\qquad \displaystyle\int_{M_0}^M dM = \int_0^x \frac{w_0 l}{\pi} \left( 1 + \cos \frac{\pi x}{l} \right) dx$

$$M - M_0 = \frac{w_0 l}{\pi} \left[ x + \frac{l}{\pi} \sin \frac{\pi x}{l} \right]_0^x$$

$$M = -\frac{w_0 l^2}{\pi} + \frac{w_0 l}{\pi} \left[ x + \frac{l}{\pi} \sin \frac{\pi x}{l} - 0 \right]$$

or in dimensionless form

$$\frac{M}{w_0 l^2} = \frac{1}{\pi} \left( \frac{x}{l} - 1 + \frac{1}{\pi} \sin \frac{\pi x}{l} \right) \qquad Ans.$$

The variations of $V/w_0 l$ and $M/w_0 l^2$ with $x/l$ are shown in the bottom figures. The negative values of $M/w_0 l^2$ indicate that physically the bending moment is in the direction opposite to that shown.

① In this case of symmetry it is clear that the resultant $R = V_0 = 2w_0 l/\pi$ of the load distribution acts at midspan, so that the moment requirement is simply $M_0 = -Rl/2 = -w_0 l^2/\pi$. The minus sign tells us that physically the bending moment at $x = 0$ is opposite to that represented on the free-body diagram.

② The free-body diagram serves to remind us that the integration limits for $V$ as well as for $x$ must be accounted for. We see that the expression for $V$ is positive, so that the shear force is as represented on the free-body diagram.

## Sample Problem 5/13

Draw the shear-force and bending-moment diagrams for the loaded beam and determine the maximum moment $M$ and its location $x$ from the left end.

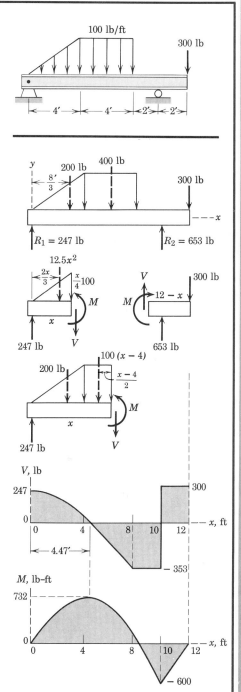

**Solution.** The support reactions are most easily obtained by considering the resultants of the distributed loads as shown on the free-body diagram of the beam as a whole. The first interval of the beam is analyzed from the free-body diagram of the section for $0 < x < 4$ ft. A vertical summation of forces and a moment summation about the cut section yield

$[\Sigma F_y = 0]$ $\qquad V = 247 - 12.5x^2$

$[\Sigma M = 0]$ $\quad M + (12.5x^2)\dfrac{x}{3} - 247x = 0 \quad M = 247x - 4.167x^3$

These values of $V$ and $M$ hold for $0 < x < 4$ ft and are plotted for that interval in the shear and moment diagrams shown.

From the free-body diagram of the section for which $4 < x < 8$ ft, equilibrium in the vertical direction and a moment sum about the cut section give

$[\Sigma F_y = 0]$ $\quad V + 100(x - 4) + 200 - 247 = 0 \quad V = 447 - 100x$

$[\Sigma M = 0]$ $\quad M + 100(x - 4)\dfrac{x - 4}{2} + 200[x - \tfrac{2}{3}(4)] - 247x = 0$

$\qquad M = -266.7 + 447x - 50x^2$

These values of $V$ and $M$ are plotted on the shear and moment diagrams for the interval $4 < x < 8$ ft.

The analysis of the reminder of the beam is continued from the free-body diagram of the portion of the beam to the right of a section in the next interval. It should be noted that $V$ and $M$ are represented in their positive directions. A vertical force summation and a moment summation about the section yield

$\qquad V = -353$ lb $\qquad$ and $\qquad M = 2930 - 353x$

These values of $V$ and $M$ are plotted on the shear and moment diagrams for the interval $8 < x < 10$ ft.

The last interval may be analyzed by inspection. The shear is constant at $+300$ lb, and the moment follows a straight-line relation beginning with zero at the right end of the beam.

The maximum moment occurs at $x = 4.47$ ft, where the shear curve crosses the zero axis, and the magnitude of $M$ is obtained for this value of $x$ by substitution into the expression for $M$ for the second interval. The maximum moment is

$\qquad\qquad M = 732$ lb-ft $\qquad\qquad\qquad$ *Ans.*

As before, note that the moment $M$ at any section equals the area under the shear diagram up to that section. For instance, for $x < 4$ ft,

$[\Delta M = \displaystyle\int V\,dx]$ $\quad M - 0 = \displaystyle\int_0^x (247 - 12.5x^2)\,dx$

and, as above, $\qquad M = 247x - 4.167x^3$

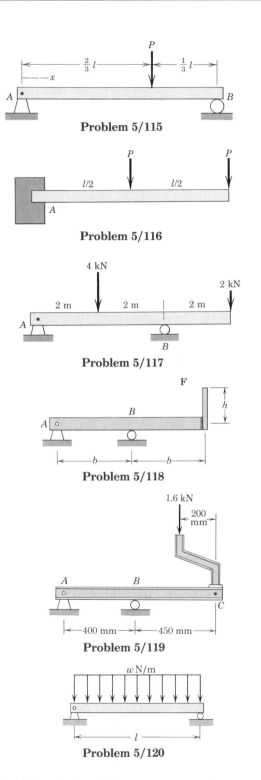

**Problem 5/115**

**Problem 5/116**

**Problem 5/117**

**Problem 5/118**

**Problem 5/119**

**Problem 5/120**

## PROBLEMS

### Introductory problems

**5/115** Determine the shear-force and bending-moment distributions produced in the beam by the concentrated load. What are the values of the shear and moment when $x = l/2$?          *Ans.* $V = P/3$, $M = Pl/6$

**5/116** Draw the shear and moment diagrams for the loaded cantilever beam.

**5/117** Draw the shear and moment diagrams for the loaded beam and determine the distance $d$ to the right of $A$ where the moment is zero.          *Ans.* $d = 2.67$ m

**5/118** Draw the shear and moment diagrams for the beam loaded by the force $F$ applied to the strut welded to the beam as shown.

**5/119** The angle strut is welded to the end $C$ of the I-beam and supports the 1.6-kN vertical force. Determine the bending moment at $B$ and the distance $x$ to the left of $C$ at which the bending moment is zero. Also construct the moment diagram for the beam.
          *Ans.* $M_B = -0.40$ kN·m, $x = 0.2$ m

**5/120** Draw the shear and moment diagrams for the uniformly loaded beam and find the maximum bending moment $M$.

**5/121** Draw the shear and moment diagrams for the loaded beam and determine the maximum value $M_{max}$ of the moment. *Ans.* $M_{max} = 3000$ lb-ft

300 lb/ft

A

B

|←4'→|←4'→|←4'→|

**Problem 5/121**

### Representative problems

**5/122** Construct the bending-moment diagram for the cantilevered shaft $AB$ of the rigid unit shown.

750 N

150 mm

150 mm

y

C

A

B

z

75 mm

100 mm

x

500 N

**Problem 5/122**

**5/123** The I-beam supports the 1000-lb force and the 2000-lb-ft couple, applied to the 2-ft strut welded to the end of the beam. Calculate the shear $V$ and moment $M$ at the section midway between $A$ and $B$. *Ans.* $V = 1467$ lb, $M = -200$ lb-ft

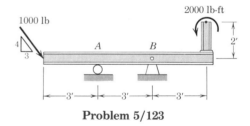

1000 lb

2000 lb-ft

A

B

2'

4
3

|←3'→|←3'→|←3'→|

**Problem 5/123**

**5/124** Construct the shear and moment diagrams for the beam loaded by the 2-kN force and the 1.6-kN·m couple.

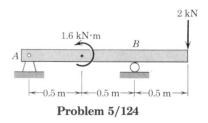

2 kN

1.6 kN·m

B

A

|←0.5 m→|←0.5 m→|←0.5 m→|

**Problem 5/124**

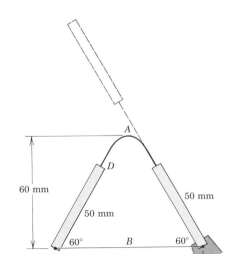

**Problem 5/125**

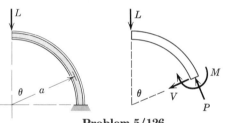

**Problem 5/126**

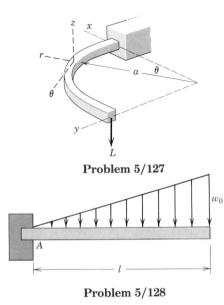

**Problem 5/127**

**Problem 5/128**

**5/125** A spacecraft sensing element consists of two flat bars rigidly connected to the thin spring $A$ which is bent in the stored position of the sensor but returns to its unbent deployed (dotted) position when the restraining wire $B$ is released. If the tension in the wire is 20 N in the restrained position, determine the corresponding bending moment $M$, compression $C$, and shear $V$ in the spring where it attaches to the bar at $D$. Also find the bending moment $M_A$ in the spring at $A$. Consider the spacecraft to be remote from the earth.

*Ans.* $M = 0.866$ N·m, $C = 10$ N,
$V = 17.32$ N, $M_A = 1.2$ N·m

**5/126** A curved cantilever beam has the form of a quarter-circular arc. Determine the expressions for the shear $V$ and the bending moment $M$ as functions of $\theta$.

**5/127** Write expressions for the torsional moment $T$ and bending moment $M$ in the curved quarter-circular beam under the end load $L$. Use a notation consistent with the right-handed $r$-$\theta$-$z$ coordinate system where positive moment vectors are taken in the direction of the positive coordinates.

*Ans.* $M = -La \cos \theta$, $T = -La(1 - \sin \theta)$

**5/128** Draw the shear and moment diagrams for the linearly loaded cantilever beam and specify the bending moment $M_A$ at the support $A$.

**5/129** Determine the shear and moment diagrams for the beam of Prob. 5/97 repeated here. Specify the shear $V$ and moment $M$ at the middle section of the beam.

Ans. $V = 4.8 - 0.6x$ kN
$M = -0.3(8 - x)^2$ kN·m
$V = 2.4$ kN, $M = -4.8$ kN·m

**5/130** Draw the shear and moment diagrams for the beam of Prob. 5/100 repeated here and specify the shear $V$ and moment $M$ at a section 6 ft to the left of the support at $A$.

**5/131** Draw the shear and moment diagrams for the beam of Prob. 5/110 repeated here and specify the shear $V$ and moment $M$ at the midlength of the beam.

Ans. $V = 533$ lb, $M = -2580$ lb-ft

**5/132** Derive expressions for the shear $V$ and moment $M$ in terms of $x$ for the cantilever beam of Prob. 5/111 shown again here.

**5/133** Determine the maximum bending moment $M$ and the corresponding value of $x$ in the crane beam and indicate the section where this moment acts.

Ans. $M_A = \dfrac{L}{4l}(l - a)^2, \ x = \dfrac{a + l}{2}$

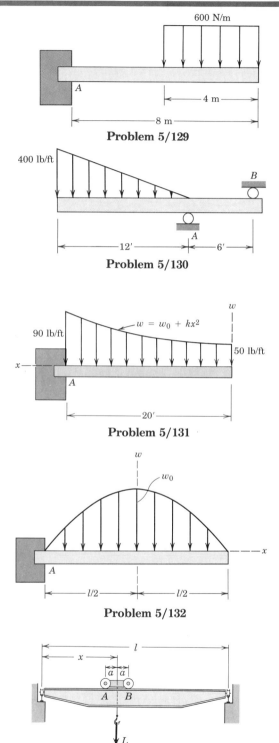

600 N/m

$A$

4 m

8 m

**Problem 5/129**

400 lb/ft

$B$

12'

$A$

6'

**Problem 5/130**

$w$

$w = w_0 + kx^2$

90 lb/ft

50 lb/ft

$x$

$A$

20'

**Problem 5/131**

$w$

$w_0$

$A$

$l/2$

$l/2$

$x$

**Problem 5/132**

$l$

$x$

$a$ $a$

$A$ $B$

$L$

**Problem 5/133**

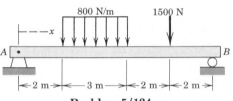

800 N/m    1500 N

$A$

$B$

|← 2 m →|← 3 m →|← 2 m →|← 2 m →|

**Problem 5/134**

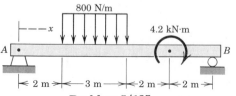

800 N/m

4.2 kN·m

$A$

$B$

|← 2 m →|← 3 m →|← 2 m →|← 2 m →|

**Problem 5/135**

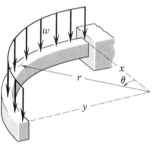

**Problem 5/136**

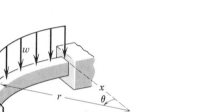

**Problem 5/137**

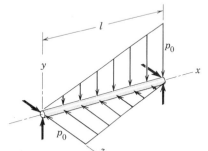

**Problem 5/138**

**5/134** Plot the shear and moment diagrams for the beam loaded with both the distributed and point loads. What are the values of the shear and moment at $x = 6$ m? Determine the maximum bending moment $M_{max}$.

**5/135** Repeat Prob. 5/134, where the 1500-N load has been replaced by the 4.2 kN·m couple.

*Ans.* $V = -1400$ N, $M = 0$, $M_{max} = 2800$ N·m

▶**5/136** The curved cantilever beam in the form of a quarter-circular arc supports a load of $w$ N/m applied along the curve of the beam on its upper surface. Determine the magnitudes of the torsional moment $T$ and bending moment $M$ in the beam as functions of $\theta$.

*Ans.* $T = wr^2 \left( \dfrac{\pi}{2} - \theta - \cos \theta \right)$

$M = wr^2 (1 - \sin \theta)$

▶**5/137** The end-supported shaft is subjected to the linearly varying loads in mutually perpendicular planes. Determine the expression for the resultant bending moment $M$ in the shaft.

*Ans.* $M = \dfrac{p_0}{6l} x(l - x) \sqrt{5l^2 - 2lx + 2x^2}$

▶**5/138** The uniform quarter-circular member of mass $m$ lies in the vertical plane and is hinged at $A$ and supported against the vertical wall by its small roller at $B$. For any section $S$, write expressions for the shear force $V$, compression $C$, and bending moment $M$ due to the weight of the member.

*Ans.* $V = \dfrac{2mg}{\pi} (\theta \sin \theta - \cos \theta)$

$C = \dfrac{2mg}{\pi} (\theta \cos \theta + \sin \theta)$

$M = \dfrac{2mgr}{\pi} \theta \cos \theta$

# 5/8 *FLEXIBLE CABLES*

One important type of structural member is the flexible cable which is used in suspension bridges, transmission lines, messenger cables for supporting heavy trolley or telephone lines, and many other applications. In the design of these structures it is necessary to know the relations involving the tension, span, sag, and length of the cables. We determine these quantities by examining the cable as a body in equilibrium. In the analysis of flexible cables we assume that any resistance offered to bending is negligible. This assumption means that the force in the cable is always in the direction of the cable.

Flexible cables may support a series of distinct concentrated loads, as shown in Fig. 5/26*a*, or they may support loads that are continuously distributed over the length of the cable, as indicated by the variable-intensity loading *w* in Fig. 5/26*b*. In some instances the weight of the cable is negligible compared with the loads it supports, and in other cases the weight of the cable may be an appreciable load or the sole load, in which case it cannot be neglected. Regardless of which of these conditions is present, the equilibrium requirements of the cable may be formulated in the same manner.

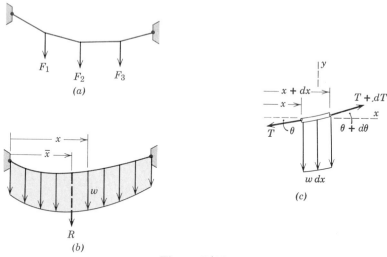

**Figure 5/26**

*(a) General relationships.* If the intensity of the variable and continuous load applied to the cable of Fig. 5/26*b* is expressed as *w* units of force per unit of horizontal length *x*, then the resultant *R* of the vertical loading is

$$R = \int dR = \int w \, dx$$

where the integration is taken over the desired interval. We find the position of $R$ from the moment principle, so that

$$R\bar{x} = \int x \, dR \qquad \bar{x} = \frac{\int x \, dR}{R}$$

The elemental load $dR = w \, dx$ is represented by an elemental strip of vertical length $w$ and width $dx$ of the shaded area of the loading diagram, and $R$ is represented by the total area. It follows from the foregoing expressions that $R$ passes through the *centroid* of the shaded area.

The equilibrium condition of the cable will be satisfied if each infinitesimal element of the cable is in equilibrium. The free-body diagram of a differential element is shown in Fig. 5/26c. At the general position $x$ the tension in the cable is $T$, and the cable makes an angle $\theta$ with the horizontal $x$-direction. At the section $x + dx$ the tension is $T + dT$, and the angle is $\theta + d\theta$. Note that the changes in both $T$ and $\theta$ are taken to be positive with a positive change in $x$. The vertical load $w \, dx$ completes the free-body diagram. The equilibrium of vertical and horizontal forces requires, respectively, that

$$(T + dT) \sin (\theta + d\theta) = T \sin \theta + w \, dx$$

$$(T + dT) \cos (\theta + d\theta) = T \cos \theta$$

The trigonometric expansion for the sine and cosine of the sum of two angles and the substitutions $\sin d\theta = d\theta$ and $\cos d\theta = 1$, which hold in the limit as $d\theta$ approaches zero, yield

$$(T + dT)(\sin \theta + \cos \theta \, d\theta) = T \sin \theta + w \, dx$$

$$(T + dT)(\cos \theta - \sin \theta \, d\theta) = T \cos \theta$$

Dropping the second-order terms and simplifying give us

$$T \cos \theta \, d\theta + dT \sin \theta = w \, dx$$

$$-T \sin \theta \, d\theta + dT \cos \theta = 0$$

which we may write as

$$d(T \sin \theta) = w \, dx \qquad \text{and} \qquad d(T \cos \theta) = 0$$

The second relation expresses the fact that the horizontal component of $T$ remains unchanged, which is clear from the free-body diagram. If we introduce the symbol $T_0 = T \cos \theta$ for this constant horizontal force, we may then substitute $T = T_0/\cos \theta$ into the first of the two equations just obtained and get $d(T_0 \tan \theta) = w \, dx$. But $\tan \theta = dy/dx$, so that the equilibrium equation may be written in the form

$$\boxed{\frac{d^2y}{dx^2} = \frac{w}{T_0}} \qquad\qquad \textbf{(5/13)}$$

Equation 5/13 is the *differential equation* for the flexible cable. The solution to the equation is that functional relation $y = f(x)$ which satisfies the equation and also satisfies the conditions at the fixed ends of the cable, called *boundary conditions*. This relationship defines the shape of the cable, and we will use it to solve two important and limiting cases of cable loading.

*(b) Parabolic cable.*   When the intensity of vertical loading $w$ is constant, the description closely approximates a suspension bridge where the uniform weight of the roadway may be expressed by the constant $w$. The mass of the cable itself is not distributed uniformly with the horizontal but is relatively small and its weight is neglected. For this limiting case we will prove that the cable hangs in a parabolic arc. We start with a cable which is suspended, first, from two points $A$ and $B$ that are not on the same horizontal line, Fig. 5/27$a$. Our origin of coordinates is taken at the lowest point of the cable where

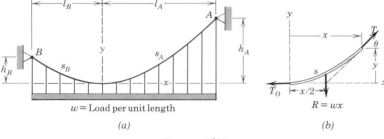

$w = $ Load per unit length

(a)                                        (b)

**Figure 5/27**

the horizontal tension is $T_0$. Integration of Eq. 5/13 once with respect to $x$ gives

$$\frac{dy}{dx} = \frac{wx}{T_0} + C$$

where $C$ is a constant of integration. For the coordinate axes chosen, $dy/dx = 0$ when $x = 0$, so that $C = 0$. Hence,

$$\frac{dy}{dx} = \frac{wx}{T_0}$$

which defines the slope of the curve as a function of $x$. One further integration yields

$$\int_0^y dy = \int_0^x \frac{wx}{T_0}\,dx \qquad \text{or} \qquad \boxed{y = \frac{wx^2}{2T_0}} \qquad\qquad \textbf{(5/14)}$$

Alternatively, you should be able to obtain the identical results with the indefinite integral together with the evaluation of the constant of integration. Equation 5/14 gives the shape of the cable, which we see is a vertical parabola. The constant horizontal component of cable tension becomes the cable tension at the origin.

Inserting the corresponding values $x = l_A$ and $y = h_A$ in Eq. 5/14 gives

$$T_0 = \frac{wl_A^2}{2h_A} \qquad \text{so that} \qquad y = h_A(x/l_A)^2$$

The tension $T$ is found from a free-body diagram of a finite portion of the cable, shown in Fig. 5/27$b$, which requires that

$$T = \sqrt{T_0^2 + w^2x^2}$$

Elimination of $T_0$ gives

$$T = w\sqrt{x^2 + (l_A^2/2h_A)^2} \tag{5/15}$$

The maximum tension occurs where $x = l_A$ and is

$$T_{\max} = wl_A\sqrt{1 + (l_A/2h_A)^2} \tag{5/15a}$$

We obtain the length $s_A$ of the cable from the origin to point $A$ by integrating the expression for a differential length $ds = \sqrt{(dx)^2 + (dy)^2}$. Thus,

$$\int_0^{s_A} ds = \int_0^{l_A} \sqrt{1 + (dy/dx)^2}\, dx = \int_0^{l_A} \sqrt{1 + (wx/T_0)^2}\, dx$$

Although we can integrate this expression in closed form, for computational purposes it is more convenient to express the radical as a convergent series and then integrate it term by term. For this purpose we use the binomial expansion

$$(1 + x)^n = 1 + nx + \frac{n(n-1)}{2!}x^2 + \frac{n(n-1)(n-2)}{3!}x^3 + \cdots$$

which converges for $x^2 < 1$. Replacing $x$ in the series by $(wx/T_0)^2$ and setting $n = \frac{1}{2}$ give the expression

$$s_A = \int_0^{l_A} \left(1 + \frac{w^2x^2}{2T_0^2} - \frac{w^4x^4}{8T_0^4} + \cdots\right) dx$$

$$= l_A \left(1 + \frac{2}{3}\left(\frac{h_A}{l_A}\right)^2 - \frac{2}{5}\left(\frac{h_A}{l_A}\right)^4 + \cdots\right) \tag{5/16}$$

This series is convergent for values of $h_A/l_A < \frac{1}{2}$, which applies to most cases in practice.

The relationships which apply to the section of the cable from the origin to point $B$ may be obtained merely by replacing $h_A$, $l_A$, and $s_A$ by $h_B$, $l_B$, and $s_B$, respectively.

For a suspension bridge where the supporting towers are on the same horizontal line, Fig. 5/28, the total span is $L = 2l_A$, the sag is $h = h_A$, and the total length of the cable is $S = 2s_A$. With these substitutions, the maximum tension and the total length become

$$T_{max} = \frac{wL}{2}\sqrt{1 + (L/4h)^2} \tag{5/15b}$$

$$S = L\left[1 + \frac{8}{3}\left(\frac{h}{L}\right)^2 - \frac{32}{5}\left(\frac{h}{L}\right)^4 + \cdots\right] \tag{5/16a}$$

This series converges for all values of $h/L < \frac{1}{4}$. In most cases $h$ is much smaller than $L/4$, so that the three terms of Eq. 5/16a give a sufficiently accurate approximation.

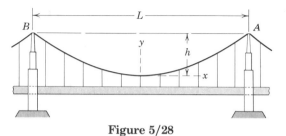

**Figure 5/28**

*(c) Catenary cable.*    Consider now a uniform cable, Fig. 5/29a, suspended from two points $A$ and $B$ and hanging under the action of its own weight only. The free-body diagram of a finite portion of the cable of length $s$ measured from the origin is shown in the $b$-part of the figure. This free-body diagram differs from that in Fig. 5/27b in that the total vertical force supported is equal to the weight of the section of cable of length $s$ in place of the load distributed uniformly with respect to the horizontal. If the cable has a weight $\mu$ per unit of its length, the resultant $R$ of the load is $R = \mu s$, and the incremental vertical load $w\,dx$ of Fig. 5/26c is replaced by $\mu\,ds$. With this replacement the differential relation, Eq. 5/13, for the cable becomes

$$\boxed{\frac{d^2y}{dx^2} = \frac{\mu}{T_0}\frac{ds}{dx}} \tag{5/17}$$

Since $s = f(x,y)$, it is necessary to change this equation to one containing only the two variables.

We may substitute the identity $(ds)^2 = (dx)^2 + (dy)^2$ to obtain

$$\frac{d^2y}{dx^2} = \frac{\mu}{T_0}\sqrt{1 + \left(\frac{dy}{dx}\right)^2} \tag{5/18}$$

Equation 5/18 is the differential equation of the curve (catenary) assumed by the cable. Solution of this equation is facilitated by the substitution $p = dy/dx$, which gives

$$\frac{dp}{\sqrt{1 + p^2}} = \frac{\mu}{T_0} \, dx$$

Integrating this equation gives us

$$\ln (p + \sqrt{1 + p^2}) = \frac{\mu}{T_0} x + C$$

The constant $C$ is zero since $dy/dx = p = 0$ when $x = 0$. Substituting $p = dy/dx$, changing to exponential form, and clearing the equation of the radical give

$$\frac{dy}{dx} = \frac{e^{\mu x/T_0} - e^{-\mu x/T_0}}{2} = \sinh \frac{\mu x}{T_0}$$

where the hyperbolic function* is introduced for convenience. The slope may be integrated to obtain

$$y = \frac{T_0}{\mu} \cosh \frac{\mu x}{T_0} + K$$

The integration constant $K$ is evaluated from the boundary condition $x = 0$ when $y = 0$. This substitution requires that $K = -T_0/\mu$, and hence,

$$y = \frac{T_0}{\mu} \left( \cosh \frac{\mu x}{T_0} - 1 \right) \tag{5/19}$$

Equation 5/19 is the equation of the curve (catenary) assumed by the cable hanging under the action of its weight only.

From the free-body diagram in Fig. 5/29*b* we see that $dy/dx = \tan \theta = \mu s/T_0$. Thus, from the previous expression for the slope,

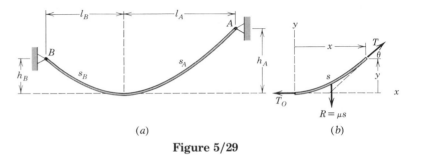

(a)                    (b)

**Figure 5/29**

* See Arts. C/8 and C/10, Appendix C, for the definition and integral of hyperbolic functions.

$$s = \frac{T_0}{\mu} \sinh \frac{\mu x}{T_0} \tag{5/20}$$

We obtain the tension $T$ in the cable from the equilibrium triangle of the forces in Fig. 5/29b. Thus,

$$T^2 = \mu^2 s^2 + T_0{}^2$$

which, upon combination with Eq. 5/20, becomes

$$T^2 = T_0{}^2 \left(1 + \sinh^2 \frac{\mu x}{T_0}\right) = T_0{}^2 \cosh^2 \frac{\mu x}{T_0}$$

or

$$T = T_0 \cosh \frac{\mu x}{T_0} \tag{5/21}$$

We may also express the tension in terms of $y$ with the aid of Eq. 5/19, which, when substituted into Eq. 5/21, gives

$$T = T_0 + \mu y \tag{5/22}$$

Equation 5/22 shows us that the increment in cable tension from that at the lowest position depends only on $\mu y$.

Most problems dealing with the catenary involve solutions of Eqs. 5/19 through 5/22, which may be handled by a graphical approximation or solved by computer. The procedure for a graphical or computer solution is illustrated in Sample Problem 5/15 following this article.

The solution of catenary problems where the sag-to-span ratio is small may be approximated by the relations developed for the parabolic cable. A small sag-to-span ratio means a tight cable, and the uniform distribution of weight along the cable is not much different from the same load intensity distributed uniformly along the horizontal.

Many problems dealing with both the catenary and parabolic cable involve suspension points that are not on the same level. In such cases we may apply the relations just developed to the part of the cable on each side of the lowest point.

## Sample Problem 5/14

The light cable supports a mass of 12 kg per meter of horizontal length and is suspended between the two points on the same level 300 m apart. If the sag is 60 m, find the tension at midlength, the maximum tension, and the total length of the cable.

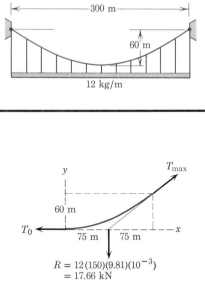

**Solution.** With a uniform horizontal distribution of load, the solution of part (b) of Art. 5/8 applies, and we have a parabolic shape for the cable. For $h = 60$ m, $L = 300$ m, and $w = 12(9.81)(10^{-3})$ kN/m the relation following Eq. 5/14 with $l_A = L/2$ gives for the midlength tension

$$\left[ T_0 = \frac{wL^2}{8h} \right] \qquad T_0 = \frac{0.1177(300)^2}{8(60)} = 22.07 \text{ kN} \qquad Ans.$$

The maximum tension occurs at the supports and is given by Eq. 5/15b. Thus,

① $$\left[ T_{\max} = \frac{wL}{2} \sqrt{1 + \left( \frac{L}{4h} \right)^2} \right]$$

$$T_{\max} = \frac{12(9.81)(10^{-3})(300)}{2} \sqrt{1 + \left( \frac{300}{4(60)} \right)^2} = 28.27 \text{ kN} \qquad Ans.$$

The sag-to-span ratio is $60/300 = 1/5 < 1/4$. Therefore, the series expression developed in Eq. 5/16a is convergent, and we may write for the total length

$$S = 300 \left[ 1 + \frac{8}{3} \left( \frac{1}{5} \right)^2 - \frac{32}{5} \left( \frac{1}{5} \right)^4 + \cdots \right]$$

$$= 300[1 + 0.1067 - 0.01024 + \cdots]$$

$$= 329 \text{ m} \qquad Ans.$$

① *Suggestion:* Check the value of $T_{\max}$ directly from the free-body diagram of the right-hand half of the cable, from which a force polygon may be drawn.

# Sample Problem  5/15

Replace the cable of Sample Problem 5/14, which is loaded uniformly along the horizontal, by a cable which has a mass of 12 kg per meter of its own length and supports its own weight only. The cable is suspended between two points on the same level 300 m apart and has a sag of 60 m. Find the tension at midlength, the maximum tension, and the total length of the cable.

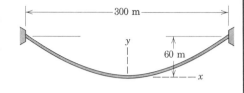

**Solution.**   With a load distributed uniformly along the length of the cable, the solution of part (c) of Art. 5/8 applies, and we have a catenary shape of the cable. Equations 5/20 and 5/21 for the cable length and tension both involve the minimum tension $T_0$ at midlength, which must be found from Eq. 5/19. Thus, for $x = 150$ m, $y = 60$ m, and $\mu = 12(9.81)(10^{-3}) = 0.1177$ kN/m, we have

$$60 = \frac{T_0}{0.1177}\left[\cosh\frac{(0.1177)(150)}{T_0} - 1\right]$$

or

$$\frac{7.063}{T_0} = \cosh\frac{17.66}{T_0} - 1$$

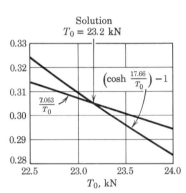

Solution
$T_0 = 23.2$ kN

This equation can be solved graphically. We compute the expression on each side of the equals sign and plot it as a function of various values of $T_0$. The intersection of the two curves establishes the equality and determines the correct value of $T_0$. This plot is shown in the figure accompanying this problem and yields the solution

$$T_0 = 23.2 \text{ kN}$$

Alternatively, we may write the equation as

$$f(T_0) = \cosh\frac{17.66}{T_0} - \frac{7.063}{T_0} - 1 = 0$$

and set up a computer program to calculate the value(s) of $T_0$ which renders $f(T_0) = 0$. See Art. C/11 of Appendix C for an explanation of one applicable numerical method.

The maximum tension occurs for maximum $y$ and from Eq. 5/22 is

$$T_{\max} = 23.2 + (0.1177)(60) = 30.2 \text{ kN} \qquad Ans.$$

① From Eq. 5/20 the total length of the cable becomes

$$2s = 2\,\frac{23.2}{0.1177}\sinh\frac{(0.1177)(150)}{23.2} = 330 \text{ m} \qquad Ans.$$

① Note that the solution of Sample Problem 5/14 for the parabolic cable gives a very close approximation to the values for the catenary even though we have a fairly large sag. The approximation is even better for smaller sag-to-span ratios.

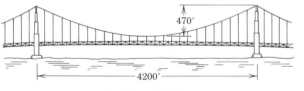

**Problem 5/139**

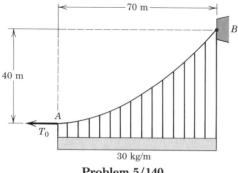

**Problem 5/140**

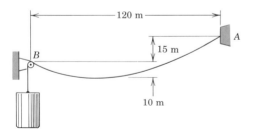

**Problem 5/141**

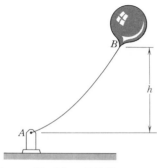

**Problem 5/142**

## PROBLEMS

(The problems marked with an asterisk (*) involve transcendental equations which may be solved with a computer or by graphical methods.)

### *Introductory problems*

**5/139**  The Golden Gate Bridge in San Francisco has a main span of 4200 ft, a sag of 470 ft, and a total static loading of 21,300 lb per lineal foot of horizontal measurement. The weight of both of the main cables is included in this figure and is assumed to be uniformly distributed along the horizontal. The angle made by the cable with the horizontal at the top of the tower is the same on each side of each tower. Calculate the midspan tension $T_0$ in each of the main cables and the compressive force $C$ exerted by each cable on the top of each tower.
$$Ans. \ T_0 = 50.0(10^6) \ \text{lb}, \ C = 44.7(10^6) \ \text{lb}$$

**5/140**  Calculate the tension $T_0$ in the cable at $A$ necessary to support the load distributed uniformly with respect to the horizontal. Also find the angle $\theta$ made by the cable with the horizontal at the attachment point $B$.

**5/141**  A cable weighing 40 newtons per meter of length is suspended from point $A$ and passes over the small pulley at $B$. Determine the mass $m$ of the attached cylinder which will produce a sag of 10 m. With the small sag-to-span ratio, approximation as a parabolic cable may be used.
$$Ans. \ m = 480 \ \text{kg}$$

**5/142**  An advertising balloon is moored to a post with a cable which has a mass of 0.12 kg/m. In a wind the cable tensions at $A$ and $B$ are 110 N and 230 N, respectively. Determine the height $h$ of the balloon.

**5/143** A horizontal 350-mm-diameter water pipe is supported over a ravine by the cable shown. The pipe and the water within it have a combined mass of 1400 kg per meter of its length. Calculate the compression $C$ exerted by the cable on each support. The angles made by the cable with the horizontal are the same on both sides of each support.

*Ans.* $C = 549$ kN

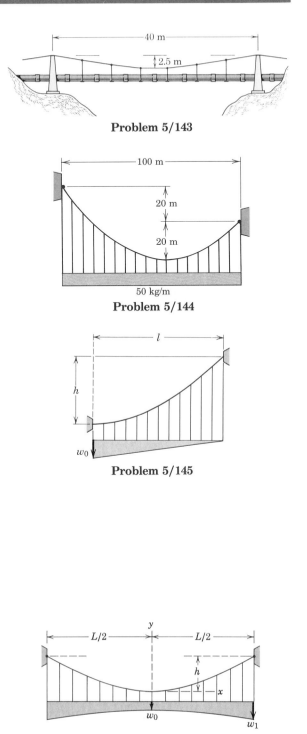

**Problem 5/143**

### Representative problems

**5/144** A cable supports a load of 50 kg/m uniformly distributed with respect to the horizontal and is suspended from the two fixed points located as shown. Determine the maximum and minimum tensions $T$ and $T_0$ in the cable.

**Problem 5/144**

**5/145** A cable of negligible mass is suspended from the fixed points shown and has a zero slope at its lower end. If the cable supports a unit load $w$ which decreases uniformly with $x$ from $w_0$ to zero as indicated, determine the equation of the curve assumed by the cable.

*Ans.* $y = \dfrac{3hx^2}{2l^2}\left(1 - \dfrac{x}{3l}\right)$

**Problem 5/145**

**5/146** Expand Eq. 5/19 in a power series for cosh $(\mu x/T_0)$ and show that the equation for the parabola, Eq. 5/14, is obtained by taking only the first two terms in the series. (See Art. C/8, Appendix C, for the series expansion of the hyperbolic function.)

**5/147** The light cable is suspended from two points a distance $L$ apart and on the same horizontal line. If the load per unit of horizontal distance supported by the cable varies from $w_0$ at the center to $w_1$ at the ends in accordance with the relation $w = a + bx^2$, derive the equation for the sag $h$ of the cable in terms of the midspan tension $T_0$.

*Ans.* $h = \dfrac{L^2}{48T_0}(5w_0 + w_1)$

**Problem 5/147**

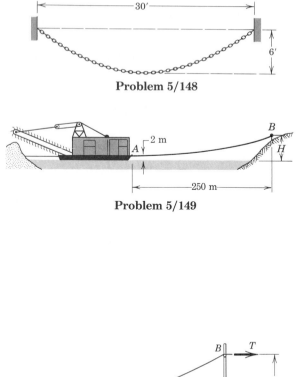

**Problem 5/148**

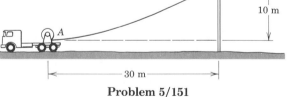

**Problem 5/149**

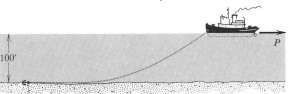

**Problem 5/151**

**Problem 5/152**

**Problem 5/153**

*5/148 Find the total length $L$ of chain which will have a sag of 6 ft when suspended from two points on the same horizontal line 30 ft apart.

5/149 A floating dredge is anchored in position with a single stern cable which has a horizontal direction at the attachment $A$ and extends a horizontal distance of 250 m to an anchorage $B$ on shore. A tension of 300 kN is required in the cable at $A$. If the cable has a mass of 22 kg per meter of its length, compute the required height $H$ of the anchorage above water level and find the length of cable between $A$ and $B$.
*Ans.* $H = 24.5$ m, $s = 251$ m

5/150 Work Prob. 5/149 using the relations for a parabolic cable as an approximation and compare the results with those cited for Prob. 5/149.

*5/151 Calculate the tension $T$ required to steadily pull the cable over a roller support on the utility pole. Neglect the effects of friction at the support. The cable, which is horizontal at $A$, has a mass of 3 kg/m. Also determine the length of cable from $A$ to $B$.
*Ans.* $T = 1665$ N, $s = 32.1$ m

5/152 In setting its anchor in 100 ft of water, a small power boat reverses its propeller, which gives a reverse thrust $P = 800$ lb. A total of 400 ft of anchor chain from anchor to bow has been released. The chain weighs 1.63 lb/ft, and the upward force due to water buoyancy is 0.21 lb/ft. Calculate the length $l$ of chain in contact with the bottom.

*5/153 A rope 40 m in length is suspended between two points which are separated by a horizontal distance of 10 m. Compute the distance $h$ to the lowest part of the loop.
*Ans.* $h = 18.53$ m

**\*5/154** Numerous small flotation devices are attached to the cable, and the difference between buoyancy and weight results in a net upward force of 30 newtons per meter of cable length. Determine the force $T$ which must be applied to cause the cable configuration shown.

**Problem 5/154**

**\*5/155** A flexible cable is secured to point $A$ and passes over a small pulley at $B$, which is 600 ft higher than $A$. If it requires a tension $T = 12,000$ lb at $B$ to make $\alpha = 0$ at $A$, determine the weight $\mu$ of the cable per foot of its length.       *Ans.* $\mu = 6.35$ lb/ft

**Problem 5/155**

**5/156** The blimp is moored to the ground winch in a gentle wind with 100 m of 12-mm cable which has a mass of 0.51 kg/m. A torque of 400 N·m on the drum is required to start winding in the cable. At this condition the cable makes an angle of 30° with the vertical as it approaches the winch. Calculate the height $H$ of the blimp. The diameter of the drum is 0.5 m.

**Problem 5/156**

**\*5/157** In preparing to spray-clean a wall, a person arranges a hose as shown in the figure. The hose is horizontal at $A$ and has a mass of 0.75 kg/m when empty and 1.25 kg/m when full of water. Determine the necessary tension $T$ and angle $\theta$ for both the empty and full hose.
      *Ans.* $T = 63.0$ N (empty), $T = 105.0$ N (full)
          $\theta = 40.0°$ in both cases

**Problem 5/157**

**\*5/158** A cable weighing 10 lb/ft is attached to point $A$ and passes over the small pulley at $B$ on the same horizontal line with $A$. Determine the sag $h$ and length $S$ of the cable between $A$ and $B$ if a tension of 2500 lb is applied to the cable over the pulley.

**Problem 5/158**

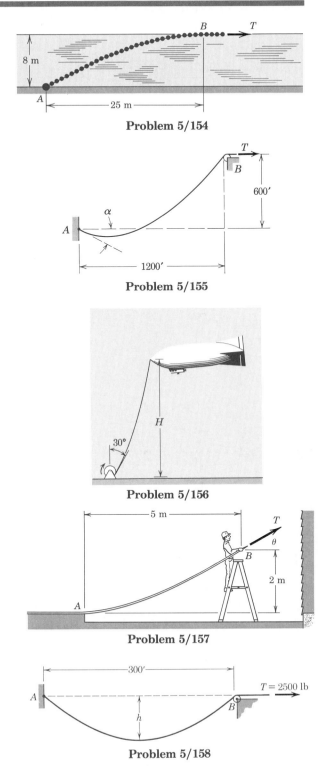

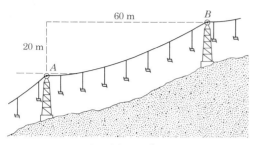

**Problem 5/159**

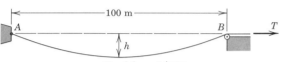

**Problem 5/160**

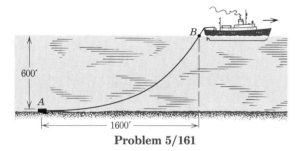

**Problem 5/161**

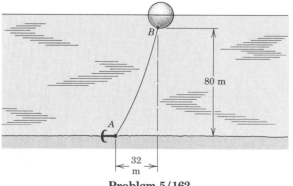

**Problem 5/162**

**\*5/159**  The moving cable for a ski lift has a mass of 10 kg/m and carries equally spaced chairs and passengers, whose added mass is 20 kg/m when averaged over the length of the cable. The cable leads horizontally from the supporting guide wheel at $A$. Calculate the tension in the cable at $A$ and $B$ and the length $s$ of the cable between $A$ and $B$.

*Ans.* $T_A = 27.4$ kN, $T_B = 33.3$ kN, $s = 64.2$ m

**\*5/160**  A cable which weighs 50 newtons per meter of length is secured at point $A$ and passes over the small pulley at $B$ on the same level under a tension $T$. Determine the minimum value of $T$ to support the cable and the corresponding deflection $h$. (*Hint:* Treat $(T - \mu h)$ for point $B$ as one variable.)

**\*5/161**  A cable ship tows a plow $A$ during a survey of the ocean floor for later burial of a telephone cable. The ship maintains a constant low speed with the plow at a depth of 600 ft and with a sufficient length of cable so that it leads horizontally from the plow, which is 1600 ft astern of the ship. The tow cable has an effective weight of 3.10 lb/ft when the buoyancy of the water is accounted for. Also, the forces on the cable due to movement through the water are neglected at the low speed. Compute the horizontal force $T_0$ applied to the plow and the maximum tension in the cable. Also find the length of the tow cable from point $A$ to point $B$.

*Ans.* $T_0 = 6900$ lb, $T_{max} = 8760$ lb, $s = 1740$ ft

**\*5/162**  A spherical buoy used to mark the course for a sailboat race is shown in the figure. There is a water current from left to right which causes a horizontal drag on the buoy; the effect of the current on the cable can be neglected. The length of the cable between points $A$ and $B$ is 87 m, and the effective cable mass is 2 kg/m when the buoyancy of the cable is accounted for. Determine the tensions at both $A$ and $B$.

## 5/9  *FLUID STATICS*

In the work so far, we have directed attention to the action of forces on and between solid bodies. In this article we shall consider the equilibrium of bodies subjected to forces due to fluid pressures. A fluid is any continuous substance which, when at rest, is unable to support shear force. A shear force is one tangent to the surface on which it acts and is developed when differential velocities exist between adjacent layers of fluids. Thus, a fluid at rest can exert only normal forces on a bounding surface. Fluids may be either gaseous or liquid. The statics of fluids is generally referred to as *hydrostatics* when the fluid is a liquid and as *aerostatics* when the fluid is a gas.

*(a) Fluid pressure.*  The pressure at any given point in a fluid is the same in all directions (Pascal's law). We may prove this fact by considering the equilibrium of an infinitesimal triangular prism of fluid as shown in Fig. 5/30. The fluid pressures normal to the faces of the element are taken to be $p_1, p_2, p_3$, and $p_4$ as shown. With force equal to pressure times area, the equilibrium of forces in the x- and y-directions gives

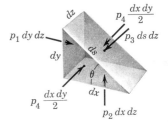

$$p_2 \, dx \, dz = p_3 \, ds \, dz \cos \theta \qquad p_1 \, dy \, dz = p_3 \, ds \, dz \sin \theta$$

Since $ds \sin \theta = dy$ and $ds \cos \theta = dx$, these equations require that

$$p_1 = p_2 = p_3 = p$$

**Figure 5/30**

By rotating the element through 90°, we see that $p_4$ is also equal to the other pressures. Hence, the pressure at any point in a fluid is the same in all directions. In this analysis it is unnecessary to account for the weight of the fluid element since, when the weight per unit volume (density $\rho$ times $g$) is multiplied by the volume of the element, a differential quantity of third order results which disappears in the limit compared with the second-order pressure-force terms.

In all fluids at rest the pressure is a function of the vertical dimension. To determine this function, we consider the forces acting on a differential element of a vertical column of fluid of cross-sectional area $dA$, as shown in Fig. 5/31. The positive direction of vertical measurement $h$ is taken downward. The pressure on the upper face is $p$, and that on the lower face is $p$ plus the change in $p$, or $p + dp$. The weight of the element equals $\rho g$ multiplied by its volume. The normal forces on the lateral surface, which are horizontal and do not affect the balance of forces in the vertical direction, are not shown. Equilibrium of the fluid element in the $h$-direction requires

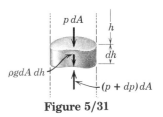

**Figure 5/31**

$$p \, dA + \rho g \, dA \, dh - (p + dp) \, dA = 0$$

$$dp = \rho g \, dh \qquad\qquad (5/23)$$

This differential relation shows us that the pressure in a fluid increases with depth or decreases with increased elevation. Equation

5/23 holds for both liquids and gases and agrees with our common observations of air and water pressures.

Fluids that are essentially incompressible are called liquids, and for most practical purposes we may consider their density $\rho$ constant for every part of the liquid.* With $\rho$ a constant, Eq. 5/23 may be integrated as it stands, and the result is

$$p = p_0 + \rho g h \qquad \qquad \textbf{(5/24)}$$

The pressure $p_0$ is the pressure on the surface of the liquid where $h = 0$. If $p_0$ is due to atmospheric pressure and the measuring instrument records only the increment above atmospheric pressure,[†] the measurement gives what is known as "gage pressure" and is $p = \rho g h$.

The common unit for pressure in SI is the kilopascal (kPa), which is the same as a kilonewton per square meter ($10^3$ N/m²). In computing pressure, if we use Mg/m³ for $\rho$, m/s² for $g$, and m for $h$, then the product $\rho g h$ gives us pressure in kPa directly. For example, the pressure at a depth of 10 m in fresh water is

$$p = \rho g h = \left(1.0 \, \frac{\text{Mg}}{\text{m}^3}\right)\left(9.81 \, \frac{\text{m}}{\text{s}^2}\right)(10 \text{ m}) = 98.1 \left(10^3 \, \frac{\text{kg·m}}{\text{s}^2} \, \frac{1}{\text{m}^2}\right)$$

$$= 98.1 \text{ kN/m}^2 = 98.1 \text{ kPa}$$

In the U.S. customary system, fluid pressure is generally expressed in pounds per square inch (lb/in.²) or occasionally in pounds per square foot (lb/ft²). Thus, at a depth of 10 ft in fresh water the pressure is

$$p = \rho g h = \left(62.4 \, \frac{\text{lb}}{\text{ft}^3}\right)\left(\frac{1}{1728} \, \frac{\text{ft}^3}{\text{in.}^3}\right)(120 \text{ in.}) = 4.33 \text{ lb/in.}^2$$

***(b) Hydrostatic pressure on submerged rectangular surfaces.*** A surface submerged in a liquid, such as a gate valve in a dam, or the wall of a tank, is subjected to fluid pressure acting normal to its surface and distributed over its area. In problems where fluid forces are appreciable, we must determine the resultant force due to the distribution of pressure on the surface and the position at which this resultant acts. For systems that are open to the earth's atmosphere, the atmospheric pressure $p_0$ acts over all surfaces and, hence, yields a zero resultant. In such cases, then, we need to consider only the gage pressure $p = \rho g h$ which is the increment above atmospheric pressure.

Consider the special but common case of the action of hydrostatic pressure on the surface of a rectangular plate submerged in a liquid.

---

* See Table D/1, Appendix D, for table of densities.
† Atmospheric pressure at sea level may be taken to be 101.3 kPa or 14.7 lb/in.²

Figure 5/32a shows such a plate 1-2-3-4 with its top edge horizontal and with the plane of the plate making an arbitrary angle $\theta$ with the vertical plane. The horizontal surface of the liquid is represented by the $x$-$y'$ plane. The fluid pressure (gage) acting normal to the plate at point 2 is represented by the arrow 6-2 and equals $\rho g$ times the vertical distance from the liquid surface to point 2. This same pressure acts at all points along the edge 2-3. At point 1 on the lower edge, the fluid pressure equals $\rho g$ times the depth of point 1, and this pressure is the same at all points along edge 1-4. The variation of pressure $p$ over the area of the plate is governed by the linear depth relationship, and we see, therefore, that it is represented by the arrow $p$, shown in Fig. 5/32b, which varies linearly from the value 6-2 to the value 5-1. The resultant force produced by this pressure distribution is represented by $R$, which acts at some point $P$ known as the *center of pressure*.

We see clearly that the conditions which prevail at the vertical section 1-2-6-5 in Fig. 5/32a are identical to those at section

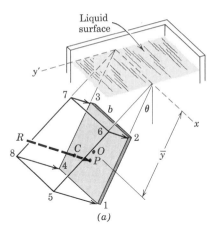

(a)

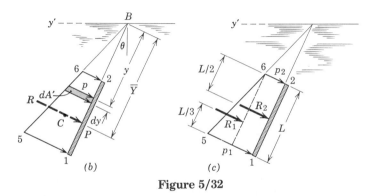

(b)          (c)

**Figure 5/32**

4-3-7-8 and at every other vertical section normal to the plate. Thus, we may analyze the problem from the two-dimensional view of a vertical section as shown in Fig. 5/32*b* for section 1-2-6-5. For this section the pressure distribution is trapezoidal. If *b* is the horizontal width of the plate measured normal to the plane of the figure (dimension 2-3 in Fig. 5/32*a*), an element of plate area over which the pressure $p = \rho g h$ acts is $dA = b\,dy$, and an increment of the resultant force is $dR = p\,dA = bp\,dy$. But $p\,dy$ is merely the shaded increment of trapezoidal area $dA'$, so that $dR = b\,dA'$. Therefore, we may express the resultant force acting on the entire plate as the trapezoidal area 1-2-6-5 times the width *b* of the plate,

$$R = b \int dA' = bA'$$

Care must be taken not to confuse the physical area *A* of the plate with the geometrical area $A'$ defined by the trapezoidal distribution of pressure.

The trapezoidal area representing the pressure distribution is easily expressed by using its average altitude. Therefore, the resultant force *R* may be written in terms of the average pressure $p_{av} = \frac{1}{2}(p_1 + p_2)$ times the plate area *A*. The average pressure is also the pressure that exists at the average depth, measured to the centroid *O* of the plate. Therefore, an alternative expression for *R* is

$$R = p_{av}A = \rho g \overline{h} A$$

where $\overline{h} = \overline{y} \cos \theta$.

We obtain the line of action of the resultant force *R* from the principle of moments. Using the *x*-axis (point *B* in Fig. 5/32*b*) as the moment axis yields $R\overline{Y} = \int y(pb\,dy)$. Substituting $p\,dy = dA'$ and $R = bA'$ and canceling *b* give

$$\overline{Y} = \frac{\int y\,dA'}{\int dA'}$$

which is simply the expression for the centroidal coordinate of the trapezoidal area $A'$. In the two-dimensional view, therefore, the resultant *R* passes through the centroid *C* of the trapezoidal area defined by the pressure distribution in the vertical section. Clearly $\overline{Y}$ also locates the centroid *C* of the truncated prism 1-2-3-4-5-6-7-8 in Fig. 5/32*a* through which the resultant actually passes.

For a trapezoidal distribution of pressure, we may simplify the calculation by dividing the trapezoid into a rectangle and a triangle, Fig. 5/32*c*, and the force represented by each part is considered separately. The force represented by the rectangular portion acts at the center *O* of the plate and is $R_2 = p_2 A$, where *A* is the area 1-2-3-4 of the plate. The force $R_1$ represented by the triangular increment of pressure distribution is $\frac{1}{2}(p_1 - p_2)A$ and acts through the centroid of the triangular portion as shown.

*(c) Hydrostatic pressure on cylindrical surfaces.* For a submerged curved surface the resultant $R$ caused by distributed pressure involves more calculation than for a flat surface. As an example, consider the submerged cylindrical surface shown in Fig. 5/33$a$ where the elements of the curved surface are parallel to the horizontal surface $x$-$y'$ of the liquid. Vertical sections perpendicular to the surface all disclose the same curve $AB$ and the same pressure distribution. Hence, the two-dimensional representation in Fig. 5/33$b$ may be used. To find $R$ by a direct integration, it would be necessary to integrate the $x$- and $y$-components of $dR$ along the curve $AB$, because $dR$ continuously changes direction. Thus,

$$R_x = b \int (p \, dL)_x = b \int p \, dy \quad \text{and} \quad R_y = b \int (p \, dL)_y = b \int p \, dx$$

A moment equation would now be required if it were desired to establish the position of $R$.

A second method for finding $R$ is usually much simpler. The equilibrium of the block of liquid $ABC$ directly above the surface, shown in Fig. 5/33$c$, is considered. The resultant $R$ then appears as the equal and opposite reaction of the surface on the block of liquid. The resultants of the pressures along $AC$ and $CB$ are $P_y$ and $P_x$, respectively, and are easily obtained. The weight $W$ of the liquid block is calculated from the area $ABC$ of its section multiplied by the constant dimension $b$ and by $\rho g$. The weight $W$ passes through the centroid of area $ABC$. The equilibrant $R$ is then determined completely from the equilibrium equations which we apply to the free-body diagram of the fluid block.

*(d) Hydrostatic pressure on flat surfaces of any shape.* Figure 5/34$a$ shows a flat plate of any shape submerged in a liquid. The horizontal surface of the liquid is the plane $x$-$y'$, and the plane of the plate makes an angle $\theta$ with the vertical. The force acting on a differential strip of area $dA$ parallel to the surface of the liquid is $dR = p \, dA = \rho g h \, dA$. The pressure $p$ has the same magnitude throughout the length of the strip, since there is no change of depth along the strip. We obtain the total force acting on the exposed area $A$ by integration, which gives

$$R = \int dR = \int p \, dA = \rho g \int h \, dA$$

Substituting the centroidal relation $\bar{h} A = \int h \, dA$ gives us

$$\boxed{R = \rho g \bar{h} A} \qquad \qquad \textbf{(5/25)}$$

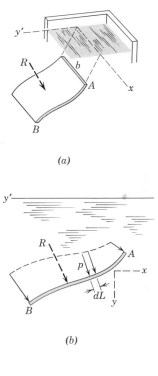

(a)

(b)

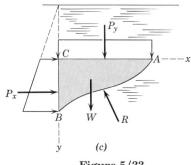

(c)

**Figure 5/33**

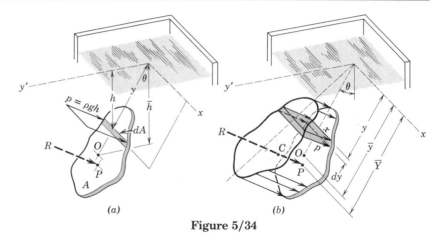

**Figure 5/34**

The quantity $\rho g\overline{h}$ is the pressure that exists at the depth of the centroid $O$ of the area and is the average pressure over the area.

We may also represent the resultant $R$ geometrically by the volume of the figure shown in Fig. 5/34*b*. Here the fluid pressure $p$ is represented as a dimension normal to the plate regarded as a base. We see that the resulting volume is a truncated right cylinder. The force $dR$ acting on the differential area $dA = x\,dy$ is represented by the elemental volume $p\,dA$ shown by the shaded slice, and the total force is represented by the total volume of the cylinder. We see from Eq. 5/25 that the average altitude of the truncated cylinder is the average pressure $\rho g\overline{h}$ which exists at a depth corresponding to the centroid $O$ of the area exposed to pressure. For problems where the centroid $O$ or the volume $V$ is not readily apparent, a direct integration may be performed to obtain $R$. Thus,

$$R = \int dR = \int p\,dA = \int \rho g h x\,dy$$

where the depth $h$ and the length $x$ of the horizontal strip of differential area must be expressed in terms of $y$ to carry out the integration.

The second requirement of the analysis of fluid pressure is the determination of any needed moments of the pressure forces. Using the principle of moments with the $x$-axis of Fig. 5/34*b* as the moment axis gives

$$R\overline{Y} = \int y\,dR \qquad \text{or} \qquad \overline{Y} = \frac{\displaystyle\int y(px\,dy)}{\displaystyle\int px\,dy} \tag{5/26}$$

This second relation satisfies the definition of the coordinate $\overline{Y}$ to the centroid of the volume $V$ of the pressure-area truncated cylinder, and it is concluded, therefore, that the resultant $R$ passes through the centroid $C$ of the volume described by the plate area as base and the linearly varying pressure as the perpendicular coordinate. The point $P$ at which $R$ is applied to the plate is the center of pressure. We note carefully that the center of pressure $P$ and the centroid $O$ of the plate area are *not* the same.

*(e) Buoyancy.* The principle of buoyancy, the discovery of which is credited to Archimedes, is easily explained in the following manner for any fluid, gaseous or liquid, in equilibrium. Consider a portion of the fluid defined by an imaginary closed surface, as illustrated by the irregular dotted boundary in Fig. 5/35*a*. If the body of the fluid could be sucked out from within the closed cavity and replaced simultaneously by the forces that it exerted on the boundary of the cavity, Fig. 5/35*b*, there would be no disturbance of the equilibrium of the surrounding fluid. Furthermore, a free-body diagram of the fluid portion before removal, Fig. 5/35*c*, shows us that the resultant of the pressure forces distributed over its surface must be equal and opposite to its weight $mg$ and must pass

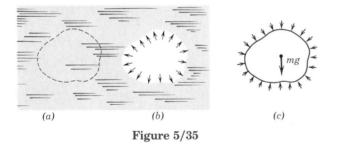

(a)　　　　　　　　(b)　　　　　　　　(c)

**Figure 5/35**

through the center of mass of the fluid element. If we replace the fluid element by a body of the same dimensions, the surface forces acting on the body held in this position will be identical with those acting on the fluid element. Thus, we see that the resultant force exerted on the surface of an object immersed in a fluid is equal and opposite to the weight of fluid displaced and passes through the center of mass of the displaced fluid. This *resultant force* is called the force of *buoyancy*

$$F = \rho g V \qquad\qquad (5/27)$$

where $\rho$ is the density of the fluid, $g$ is the acceleration due to gravity, and $V$ is the volume of the fluid displaced. In the case of a liquid whose density is constant, the center of mass of the displaced liquid coincides with the centroid of the displaced volume.

From the foregoing discussion we see that when the density of an object is less than the density of the fluid in which it is fully immersed, there will be an imbalance of force in the vertical direction, and the object will rise. When the immersing fluid is a liquid, the object continues to rise until it comes to the surface of the liquid and then comes to rest in an equilibrium position, assuming that the density of the new fluid above the surface is less than the density of the object. In the case of the surface boundary between a liquid and a gas, such as water and air, the effect of the gas pressure on that portion of the floating object above the liquid is balanced by the added pressure in the liquid due to the action of the gas on its surface.

One of the most important problems involving buoyancy is the determination of the stability of a floating object. This analysis may be illustrated by considering a ship's hull shown in cross section in an upright position in Fig. 5/36*a*. Point *B* is the centroid of the displaced volume and is known as the *center of buoyancy*. The resultant of the forces exerted on the hull by the water pressure is the buoyancy force *F*. Force *F* passes through *B* and is equal and opposite to the weight *W* of the ship. If the ship is caused to list

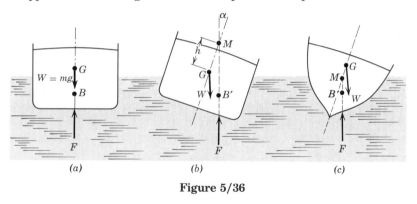

**Figure 5/36**

through an angle $\alpha$, Fig. 5/36*b*, the shape of the displaced volume changes, and the center of buoyancy shifts to some new position such as *B'*. The point of intersection of the vertical line through *B'* with the centerline of the ship is called the *metacenter M*, and the distance *h* of *M* above the center of mass *G* is known as the *metacentric height*. For most hull shapes we find that the metacentric height remains practically constant for angles of list up to about 20°. When *M* is above *G*, as in Fig. 5/36*b*, there is clearly a righting moment which tends to bring the ship back to its upright position. The magnitude of this moment for any particular angle of list is a measure of the stability of the ship. If *M* is below *G*, as for the hull of Fig. 5/36*c*, the moment accompanying the list is in the direction to increase the list. This is clearly a condition of instability and must be avoided in the design of any ship.

## Sample Problem 5/16

A rectangular plate, shown in vertical section $AB$, is 4 m high and 6 m wide (normal to the plane of the paper) and blocks the end of a fresh-water channel 3 m deep. The plate is hinged about a horizontal axis along its upper edge through $A$ and is restrained from opening by the fixed ridge $B$ that bears horizontally against the lower edge of the plate. Find the force $B$ exerted on the plate by the ridge.

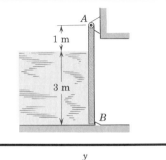

**Solution.** The free-body diagram of the plate is shown in section and includes the vertical and horizontal components of the force at $A$, the unspecified weight $W = mg$ of the plate, the unknown horizontal force $B$, and the resultant $R$ of the triangular distribution of pressure against the vertical face.

The density of fresh water is $\rho = 1.000$ Mg/m$^3$ so that the average pressure is

① $[p_{av} = \rho g \bar{h}]$     $p_{av} = 1.000(9.81)(\tfrac{3}{2}) = 14.72$ kPa

The resultant $R$ of the pressure forces against the plate becomes

$[R = p_{av}A]$     $R = (14.72)(3)(6) = 265$ kN

This force acts through the centroid of the triangular distribution of pressure, which is 1 m above the bottom of the plate. A zero moment summation about $A$ establishes the unknown force $B$. Thus,

$[\Sigma M_A = 0]$     $3(265) - 4B = 0$     $B = 198.7$ kN     *Ans.*

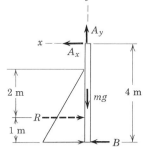

①Note that the units of pressure $\rho g h$ are

$$\left(10^3\ \frac{\text{kg}}{\text{m}^3}\right)\left(\frac{\text{m}}{\text{s}^2}\right)(\text{m}) = \left(10^3\ \frac{\text{kg}\cdot\text{m}}{\text{s}^2}\right)\left(\frac{1}{\text{m}^2}\right)$$

$$= \text{kN/m}^2 = \text{kPa}.$$

## Sample Problem 5/17

The air space in the closed fresh-water tank is maintained at a pressure of 0.80 lb/in.$^2$ (above atmospheric). Determine the resultant force $R$ exerted by the air and water on the end of the tank.

**Solution.** The pressure distribution on the end surface is shown, where $p_0 = 0.80$ lb/in.$^2$ The specific weight of fresh water is $\mu = \rho g = 62.4/1728 = 0.0361$ lb/in.$^3$ so that the increment of pressure $\Delta p$ due to the water is

$$\Delta p = \mu\ \Delta h = 0.0361(30) = 1.083\ \text{lb/in.}^2$$

① The resultant forces $R_1$ and $R_2$ due to the rectangular and triangular distributions of pressure, respectively, are

$$R_1 = p_0 A_1 = 0.80(38)(25) = 760\ \text{lb}$$

$$R_2 = \Delta p_{av} A_2 = \frac{1.083}{2}(30)(25) = 406\ \text{lb}$$

The resultant is then $R = R_1 + R_2 = 760 + 406 = 1166$ lb.     *Ans.*

We locate $R$ by applying the moment principle about $A$ noting that $R_1$ acts through the center of the 38-in. depth and that $R_2$ acts through the centroid of the triangular pressure distribution 20 in. below the surface of the water and $20 + 8 = 28$ in. below $A$. Thus,

$[Rh = \Sigma M_A]$     $1166h = 760(19) + 406(28)$     $h = 22.1$ in.     *Ans.*

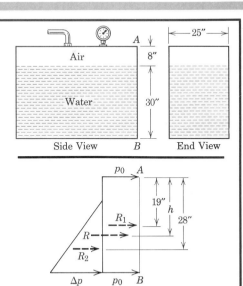

①Dividing the pressure distribution into these two parts is decidedly the simplest way in which to make the calculation.

## Sample Problem 5/18

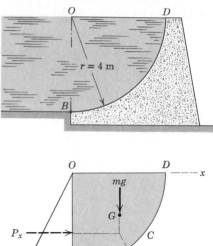

Determine completely the resultant force $R$ exerted on the cylindrical dam surface by the water. The density of fresh water is 1.000 Mg/m³, and the dam has a length $b$, normal to the paper, of 30 m.

---

**Solution.** The circular block of water $BDO$ is isolated and its free-body diagram is drawn. The force $P_x$ is

①
$$P_x = \rho g \bar{h} A = \frac{\rho g r}{2} br = \frac{(1.000)(9.81)(4)}{2}(30)(4) = 2350 \text{ kN}$$

The weight $W$ of the water passes through the mass center $G$ of the quarter-circular section and is

$$mg = \rho g V = (1.000)(9.81)\frac{\pi(4)^2}{4}(30) = 3700 \text{ kN}$$

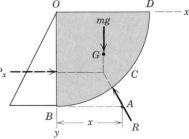

Equilibrium of the section of water requires

$[\Sigma F_x = 0]$       $R_x = P_x = 2350 \text{ kN}$

$[\Sigma F_y = 0]$       $R_y = mg = 3700 \text{ kN}$

The resultant force $R$ exerted by the fluid on the dam is equal and opposite to that shown acting on the fluid and is

$[R = \sqrt{R_x{}^2 + R_y{}^2}]$    $R = \sqrt{(2350)^2 + (3700)^2} = 4380 \text{ kN}$    *Ans.*

① See note ① in Sample Problem 5/16 if there is any question about the units for $\rho g \bar{h}$.

The $x$-coordinate of the point $A$ through which $R$ passes may be found from the principle of moments. Using $B$ as a moment center gives

$$P_x\frac{r}{3} + mg\frac{4r}{3\pi} - R_y x = 0, \quad x = \frac{2350\left(\frac{4}{3}\right) + 3700\left(\frac{16}{3\pi}\right)}{3700} = 2.55 \text{ m} \quad Ans.$$

②   **Alternative solution.** The force acting on the dam surface may be obtained by a direct integration of the components

$$dR_x = p \, dA \cos\theta \quad\text{and}\quad dR_y = p \, dA \sin\theta$$

where $p = \rho g h = \rho g r \sin\theta$ and $dA = b(r \, d\theta)$. Thus,

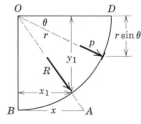

$$R_x = \int_0^{\pi/2} \rho g r^2 b \sin\theta \cos\theta \, d\theta = -\rho g r^2 b\left[\frac{\cos 2\theta}{4}\right]_0^{\pi/2} = \tfrac{1}{2}\rho g r^2 b$$

$$R_y = \int_0^{\pi/2} \rho g r^2 b \sin^2\theta \, d\theta = \rho g r^2 b\left[\frac{\theta}{2} - \frac{\sin 2\theta}{4}\right]_0^{\pi/2} = \tfrac{1}{4}\pi\rho g r^2 b$$

② This approach by integration is feasible here mainly because of the simple geometry of the circular arc.

Thus, $R = \sqrt{R_x{}^2 + R_y{}^2} = \tfrac{1}{2}\rho g r^2 b\sqrt{1 + \pi^2/4}$. Substituting the numerical values gives

$$R = \tfrac{1}{2}(1.000)(9.81)(4^2)(30)\sqrt{1 + \pi^2/4} = 4380 \text{ kN} \quad Ans.$$

Since $dR$ always passes through point $O$, we see that $R$ also passes through $O$ and, therefore, the moments of $R_x$ and $R_y$ about $O$ must cancel. So we write $R_x y_1 = R_y x_1$, which gives us

$$x_1/y_1 = R_x/R_y = (\tfrac{1}{2}\rho g r^2 b)/(\tfrac{1}{4}\pi\rho g r^2 b) = 2/\pi$$

By similar triangles we see that

$$x/r = x_1/y_1 = 2/\pi \quad\text{and}\quad x = 2r/\pi = 2(4)/\pi = 2.55 \text{ m} \quad Ans.$$

## Sample Problem 5/19

Determine the resultant force $R$ exerted on the semicircular end of the water tank shown in the figure if the tank is filled to capacity. Express the result in terms of the radius $r$ and the water density $\rho$.

**Solution I.** We will obtain $R$ first by a direct integration. With a horizontal strip of area $dA = 2x\ dy$ acted on by the pressure $p = \rho gy$, the increment of the resultant force is $dR = p\ dA$ so that

$$R = \int p\ dA = \int \rho gy(2x\ dy) = 2\rho g \int_0^r y\sqrt{r^2 - y^2}\ dy$$

Integrating gives $\qquad\qquad R = \tfrac{2}{3}\rho gr^3 \qquad\qquad$ *Ans.*

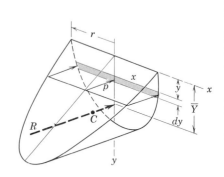

The location of $R$ is determined by using the principle of moments. Taking moments about the $x$-axis gives

$$\left[R\overline{Y} = \int y\ dR\right] \qquad \tfrac{2}{3}\rho gr^3\overline{Y} = 2\rho g \int_0^r y^2\sqrt{r^2 - y^2}\ dy$$

Integrating gives $\qquad \tfrac{2}{3}\rho gr^3\overline{Y} = \dfrac{\rho gr^4}{4}\dfrac{\pi}{2} \qquad$ and $\qquad \overline{Y} = \dfrac{3\pi}{16}r \qquad$ *Ans.*

**Solution II.** We may use Eq. 5/25 directly to find $R$, where the average pressure is $\rho g\overline{h}$ and $\overline{h}$ is the coordinate to the centroid of the area over which the pressure acts. For a semicircular area $\overline{h} = 4r/(3\pi)$.

$$[R = \rho g\overline{h}A] \qquad\qquad R = \rho g\dfrac{4r}{3\pi}\dfrac{\pi r^2}{2} = \tfrac{2}{3}\rho gr^3 \qquad\qquad \textit{Ans.}$$

which is the volume of the pressure-area figure.

The resultant $R$ acts through the centroid $C$ of the volume defined by
① the pressure-area figure. Calculation of the centroidal distance $\overline{Y}$ involves the same integral obtained in *Solution I*.

① Be very careful not to make the mistake of assuming that $R$ passes through the centroid of the area over which the pressure acts.

## Sample Problem 5/20

A buoy in the form of a uniform 8-m pole 0.2 m in diameter has a mass of 200 kg and is secured at its lower end to the bottom of a fresh-water lake with 5 m of cable. If the depth of the water is 10 m, calculate the angle $\theta$ made by the pole with the horizontal.

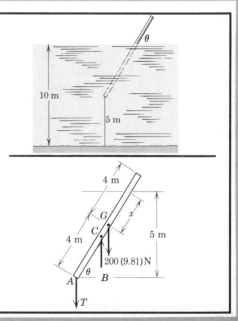

**Solution.** The free-body diagram of the buoy shows its weight acting through $G$, the vertical tension $T$ in the anchor cable, and the buoyancy force $B$ which passes through centroid $C$ of the submerged portion of the buoy. Let $x$ be the distance from $G$ to the waterline. The density of fresh water is $\rho = 10^3$ kg/m³, so that the buoyancy force is

$$[B = \rho gV] \qquad\qquad B = 10^3(9.81)\pi(0.1)^2(4 + x)\ \text{N}$$

Moment equilibrium, $\Sigma M_A = 0$, about $A$ gives

$$200(9.81)(4\cos\theta) - [10^3(9.81)\pi(0.1)^2(4 + x)]\dfrac{4 + x}{2}\cos\theta = 0$$

Thus, $\quad x = 3.14$ m $\quad$ and $\quad \theta = \sin^{-1}\left(\dfrac{5}{4 + 3.14}\right) = 44.5° \qquad$ *Ans.*

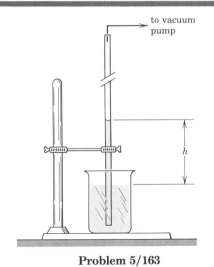

**Problem 5/163**

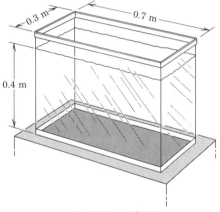

**Problem 5/164**

**Problem 5/165**

## PROBLEMS

### *Introductory problems*

**5/163** Determine the maximum height $h$ to which a vacuum pump can cause the fresh water to rise. Assume standard atmospheric pressure of $1.0133(10^5)$ Pa. Repeat your calculations for mercury.

*Ans.* $h = 10.33$ m (water)
$h = 0.761$ m (mercury)

**5/164** Specify the magnitude and location of the resultant force which acts on each side and the bottom of the aquarium due to the fresh water inside it.

**5/165** The submersible diving chamber has a total mass of 6.7 Mg including personnel, equipment, and ballast. When the chamber is lowered to a depth of 1.2 km in the ocean, the cable tension is 8 kN. Compute the total volume $V$ displaced by the chamber.

*Ans.* $V = 5.71$ m$^3$

**5/166** A deep-submersible diving chamber in the form of a spherical shell 1500 mm in diameter is ballasted with lead so that its weight slightly exceeds its buoyancy. Atmospheric pressure is maintained within the sphere during an ocean dive to a depth of 3 km. The thickness of the shell is 25 mm. For this depth calculate the compressive stress $\sigma$ which acts on a diametral section of the shell, as indicated in the right-hand view.

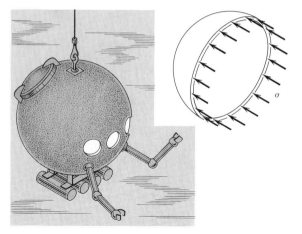

**Problem 5/166**

**5/167** The forms for a small concrete retaining wall are shown in section. There is a brace *BC* for every 1.5 m of wall length. Assuming that the joints at *A*, *B*, and *C* act as hinged connections, compute the compression in each brace *BC*. Wet concrete may be treated as a liquid with a density of 2400 kg/m³.

*Ans.* $C = 95.5$ kN

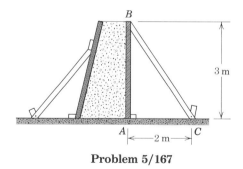

**Problem 5/167**

**5/168** Fresh water in a channel is contained by the uniform 2.5-m plate freely hinged at *A*. If the gate is designed to open when the depth of the water reaches 0.8 m as shown in the figure, what must the weight $w$ (in newtons per meter of horizontal length into the paper) of the gate be?

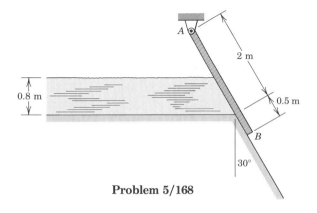

**Problem 5/168**

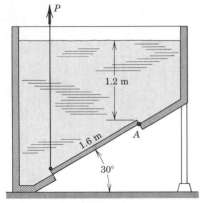

**Problem 5/169**

**5/169**  The cross section of a fresh-water tank with a slanted bottom is shown. A rectangular door 1.6 m by 0.8 m (normal to the plane of the figure) in the bottom of the tank is hinged at $A$ and is opened against the pressure of the water by the cable under a tension $P$ as shown. Calculate $P$.        *Ans.* $P = 12.57$ kN

### Representative problems

**5/170**  The homogeneous block of density $\rho$ is floating between two liquids of densities $\rho_1 < \rho$ and $\rho_2 > \rho$. Determine an expression for the distance $b$ that the block protrudes into the top liquid.

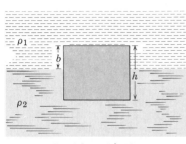

**Problem 5/170**

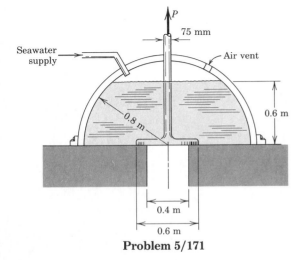

**Problem 5/171**

**5/171**  When the sea-water level inside the hemispherical chamber reaches the 0.6-m level shown in the figure, the plunger is lifted, allowing a surge of sea water to enter the vertical pipe. For this fluid level $(a)$ determine the average pressure $\sigma$ supported by the seal area of the valve before force is applied to lift the plunger and $(b)$ determine the force $P$ (in addition to the force needed to support its weight) required to lift the plunger. Assume atmospheric pressure in all airspaces and in the seal area when contact ceases under the action of $P$.
        *Ans.* $\sigma = 10.74$ kPa, $P = 1.687$ kN

**5/172** The cloth, shrouds, basket, fuel, burner, and sand
ballast for a hot-air balloon weigh 1050 lb, and its
four passengers together with their gear weigh
420 lb. When filled, the balloon has a volume of
180,000 ft³. If the atmospheric temperature and
specific weight at launch are 60°F and 0.0763 lb/ft³,
respectively, determine the average temperature
$T_h$ which must be reached by the hot air inside in
order to produce a tension of 300 lb in the single
mooring line prior to release. (Recall that for conditions of constant pressure, the specific weight varies inversely as the absolute temperature in degrees Rankine and that °R equals 460 + °F.)

**Problem 5/172**

**5/173** A channel-marker buoy consists of an 8-ft hollow
steel cylinder 12 in. in diameter weighing 180 lb and
anchored to the bottom with a cable as shown. If
$h = 2$ ft at high tide, calculate the tension $T$ in the
cable. Also find the value of $h$ when the cable goes
slack as the tide drops. The weight density of sea
water is 64 lb/ft³. Assume the buoy is weighted at
its base so that it remains vertical.

*Ans.* $T = 121.6$ lb, $h = 4.42$ ft

**5/174** If the center of mass of the buoy of Prob. 5/173 is in
the geometric center of the cylinder, calculate the
angle $\theta$ which would be made by the buoy axis with
the vertical when the water surface is 5 ft above the
lower end of the cylinder. Neglect the diameter
compared with the length when locating the center
of buoyancy.

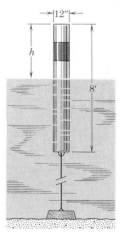

**Problem 5/173**

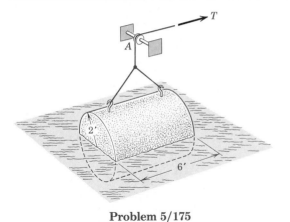

**Problem 5/175**

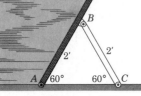

**Problem 5/176**

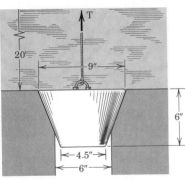

**Problem 5/177**

**Problem 5/178**

**5/175** The solid concrete cylinder 6 ft long and 4 ft in diameter is supported in a half-submerged position in fresh water by a cable which passes over a fixed pulley at $A$. Compute the tension $T$ in the cable. The cylinder is water-proofed by a plastic coating. (Consult Table D/1, Appendix D, as needed.)

*Ans.* $T = 8960$ lb

**5/176** The triangular and rectangular sections are being considered for a small fresh-water concrete dam. From the standpoint of resistance to overturning about $C$, which section will require less concrete, and how much less per foot of dam length? Concrete weighs 150 lb/ft$^3$.

**5/177** A fresh-water channel 10 ft wide (normal to the plane of the paper) is blocked at its end by a rectangular barrier, shown in section $ABD$. Supporting struts $BC$ are spaced every 2 ft along the 10-ft width. Determine the compression $C$ in each strut. Neglect the weights of the members. *Ans.* $C = 666$ lb

**5/178** The cast-iron plug seals the drain pipe of an open fresh-water tank which is filled to a depth of 20 ft. Determine the tension $T$ required to remove the plug from its tapered hole. Atmospheric pressure exists in the drainpipe and in the seal area as the plug is being removed. Neglect mechanical friction between the plug and its supporting surface.

**5/179** The hydraulic cylinder operates the toggle which closes the vertical gate against the pressure of fresh water on the opposite side. The gate is rectangular with a horizontal width of 2 m perpendicular to the paper. For a depth $h = 3$ m of water, calculate the required oil pressure $p$ which acts on the 150-mm-diameter piston of the hydraulic cylinder.

*Ans.* $p = 7.49$ MPa

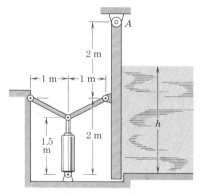

**Problem 5/179**

**5/180** The barge crane of rectangular proportions has a 12-by 30-ft cross section over its entire length of 80 ft. If the maximum permissible submergence and list in sea water are represented by the position shown, determine the corresponding maximum safe load $w$ that the barge can handle at the 20-ft extended position of the boom. Also find the total displacement $W$ in long tons of the unloaded barge (1 long ton equals 2240 lb). The distribution of machinery and ballast places the center of gravity $G$ of the barge, minus the load $w$, at the center of the hull.

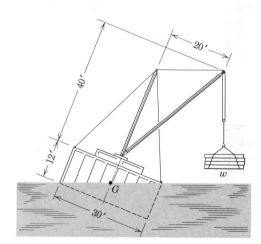

**Problem 5/180**

**5/181** The quonset hut is subjected to a horizontal wind, and the pressure $p$ against the circular roof is approximated by $p_0 \cos \theta$. The pressure is positive on the windward side of the hut and is negative on the leeward side. Determine the total horizontal shear $Q$ on the foundation per unit length of roof measured normal to the paper. *Ans.* $Q = \frac{1}{2}\pi r p_0$

**Problem 5/181**

**5/182** The hull of a floating oil-drilling platform consists of two rectangular pontoons and six cylindrical columns that support the working platform. When ballasted, the entire structure has a displacement of 26,000 tons (expressed in long tons of 2240 lb). Calculate the total draft $h$ of the structure when moored in the ocean. The specific weight of salt water is 64 lb/ft$^3$. Neglect the vertical components of the mooring forces.

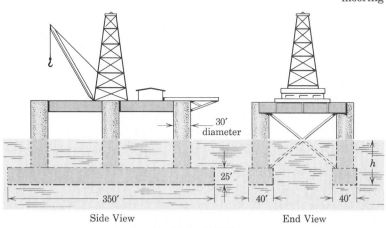

30' diameter

25'

350'

40' 40'

$h$

Side View                           End View

**Problem 5/182**

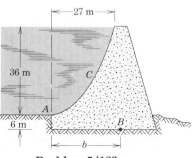

27 m

36 m

$C$

$A$

6 m

$b$

**Problem 5/183**

**5/183** The fresh-water side of a concrete dam has the shape of a vertical parabola with vertex at $A$. Determine the position $b$ of the base point $B$ through which acts the resultant force of the water against the dam face $C$.                      *Ans.* $b = 28.1$ m

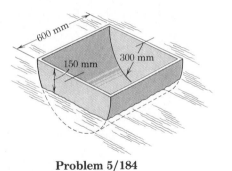

600 mm

150 mm

300 mm

**Problem 5/184**

**5/184** The semicylindrical steel shell with closed ends has a mass of 26.6 kg. Determine the mass $m$ of the lead ballast which must be placed in the shell so that it floats in fresh water at its half-radius depth of 150 mm.

**5/185** A dam consists of the flat-plate barriers $A$ and $B$ whose masses are small. Supporting struts $C$ and $D$ are placed every 3 meters of dam section. A mud sample drawn up to the surface has a density of 1.6 Mg/m$^3$. Determine the compression in $C$ and $D$. All joints may be assumed to be hinged.

*Ans.* $D = 88.3$ kN, $C = 474$ kN

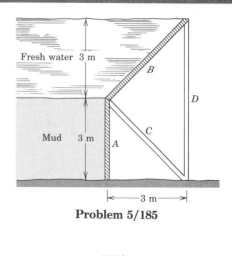

**Problem 5/185**

**5/186** The rectangular gate shown in section is 10 ft long (perpendicular to the paper) and is hinged about its upper edge $B$. The gate divides a channel leading to a fresh-water lake on the left and a salt-water tidal basin on the right. Calculate the torque $M$ on the shaft of the gate at $B$ required to prevent the gate from opening when the salt-water level drops to $h = 3$ ft.

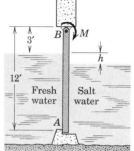

**Problem 5/186**

**5/187** A structure for observation of sea life beneath the ice in polar waters consists of the cylindrical viewing chamber connected to the surface by the cylindrical shaft open at the top for ingress and egress. Ballast is carried in the rack below the chamber. To ensure a stable condition for the structure, it is necessary that its legs bear on the ice with a force that is at least 15 percent of the total buoyancy force of the submerged structure. If the structure less ballast has a mass of 5.7 Mg, calculate the required mass $m$ of lead ballast. The density of lead is 11.37 Mg/m$^3$.

*Ans.* $m = 4.24$ Mg

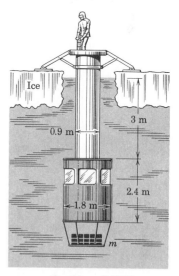

**Problem 5/187**

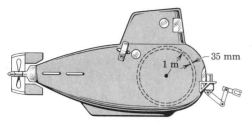

**Problem 5/188**

**5/188** The deep-submersible research vessel has a passenger compartment in the form of a spherical steel shell with a mean radius of 1.000 m and a thickness of 35 mm. Calculate the mass of lead ballast which the vessel must carry so that the combined weight of the steel shell and lead ballast exactly cancels the combined buoyancy of these two parts alone. (Consult Table D/1, Appendix D, as needed.)

**Problem 5/189**

**5/189** The cover plate for an access opening in a fresh-water tank is bolted in place and the tank is filled to the level shown. Compute the increase in tension in each bolt at $A$ and $B$ due to filling the tank. (Assume no change in the gasket pressure between the cover plate and the tank so that the force due to water pressure is resisted entirely by the increase in the bolt tensions.)

*Ans.* $T_A = 80$ N, $T_B = 96.6$ N

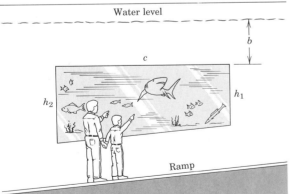

**Problem 5/190**

**5/190** The trapezoidal viewing window in a sea-life aquarium has the dimensions shown. With the aid of appropriate diagrams and coordinates describe two methods by which the resultant force $R$ on the glass due to water pressure and the vertical location of $R$ could be found if numerical values were supplied.

**5/191** The 3-m plank shown in section has a density of 800 kg/m$^3$ and is hinged about a horizontal axis through its upper edge $O$. Calculate the angle $\theta$ assumed by the plank with the horizontal for the level of fresh water shown.     *Ans.* $\theta = 48.2°$

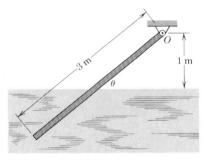

**Problem 5/191**

**5/192** Determine the total force $R$ exerted on the triangular window by the fresh water in the tank. The water level is even with the top of the window. Also determine the distance $H$ from $R$ to the water level.

**Problem 5/192**

**5/193** The sphere is used as a valve to close the hole in the fresh-water tank. As the depth $h$ decreases, the tension $T$ required to open the valve decreases because the downward force on the sphere decreases with less pressure. Determine the depth $h$ for which $T$ equals the weight of the sphere.
     *Ans.* $h = 9.33$ in.

**5/194** The accurate determination of the vertical position of the center of mass $G$ of a ship is difficult to achieve by calculation. It is more easily obtained by a simple inclining experiment on the loaded ship. With reference to the figure, a known external mass $m_0$ is placed a distance $d$ from the centerline, and the angle of list $\theta$ is measured by means of the deflection of a plumb bob. The displacement of the ship and the location of the metacenter $M$ are known. Calculate the metacentric height $\overline{GM}$ for a 12 000-t ship inclined by a 27-t mass placed 7.8 m from the centerline if a 6-m plumb line is deflected a distance $a = 0.2$ m. The mass $m_0$ is at a distance $b = 1.8$ m above $M$. [Note that the metric ton (t) equals 1000 kg and is the same as the megagram (Mg).]
     *Ans.* $\overline{GM} = 0.530$ m

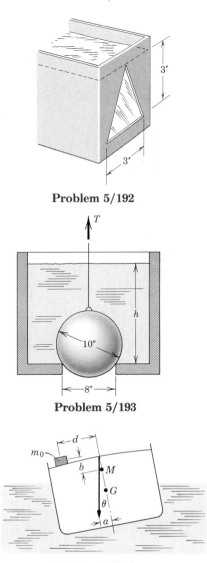

**Problem 5/193**

**Problem 5/194**

## 5/10  PROBLEM FORMULATION AND REVIEW

In Chapter 5 we have treated various common examples of forces distributed throughout volumes, over areas, and along lines. In all of these problems we are often interested in the resultant of the distributed forces and the location of the resultant.

In finding the resultant, we begin by multiplying the intensity of the force by the appropriate element of volume, area, or length in terms of which the intensity is expressed. We then sum up (integrate) the incremental forces over the region involved to obtain their resultant.

In locating the line of action of the resultant, we use the principle of moments. Here we equate the sum of the moments about a convenient axis of all of the increments of force to the moment of the resultant about the same axis. Then we solve for the unknown moment arm of the resultant.

When force is distributed throughout a mass, as in the case of gravitational attraction, the intensity is the force of attraction $\rho g$ per unit of volume, where $\rho$ is the density and $g$ is the gravitational acceleration. For bodies whose density is constant, we saw in Section A that $\rho g$ cancels when the moment principle is applied. This leaves us with a strictly geometric problem of finding the centroid of the figure, which then coincides with the mass center of the physical body whose boundary defines the figure. For flat plates and shells which are homogeneous and have constant thickness, the problem becomes one of finding the properties of an area. For slender rods and wires of uniform density and constant cross section, the problem becomes one of finding the properties of a line segment.

For problems which require the integration of differential relationships, we cite four considerations which should be observed. First, select a coordinate system which provides the simplest description of the boundaries of the region of integration. Second, eliminate higher-order differential quantities whenever lower-order differential quantities will remain. Third, choose a first-order differential element in preference to a second-order element and a second-order element in preference to a third-order element. Fourth, wherever possible choose a differential element which avoids discontinuities within the region of integration.

In Section B of Chapter 5 we made use of the foregoing observations along with our principles of equilibrium to solve for the effects of distributed forces in beams, cables, and fluids. In beams and cables we expressed the force intensity as force per unit length. For fluids we expressed the force intensity as force per unit area, or pressure. Although these three problems are physically quite different, in their formulation they embody the common elements cited.

## REVIEW PROBLEMS

**5/195** Determine the *x*- and *y*-coordinates of the centroid of the shaded area. *Ans.* $\bar{x} = 37/84, \bar{y} = 13/30$

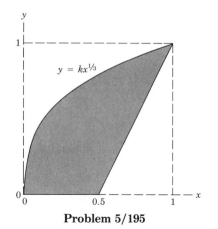

**Problem 5/195**

**5/196** The disk of uniform thickness is composed of equal sectors of the materials shown. Determine the location of the mass center of the disk.

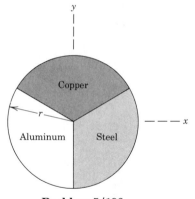

**Problem 5/196**

**5/197** The assembly consists of four rods cut from the same bar stock. The curved member is a circular arc of radius *b*. Determine the *y*- and *z*-coordinates of the mass center of the assembly.

*Ans.* $\overline{Y} = 0.461b, \overline{Z} = 0.876b$

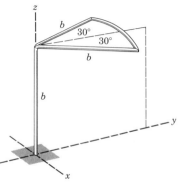

**Problem 5/197**

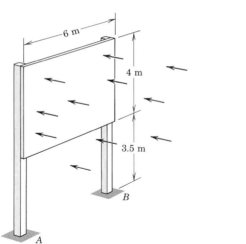

**Problem 5/198**

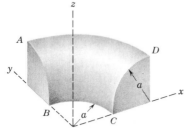

**Problem 5/199**

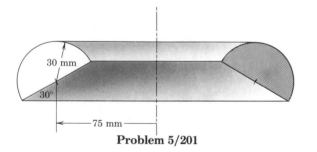

**Problem 5/200**

**Problem 5/201**

**5/198** A signboard is supported by two posts embedded in concrete at $A$ and $B$. Determine the moment $M$ which the concrete exerts on each post at $A$ and $B$ during a storm when the wind velocity is 100 km/h. The air pressure (called stagnation pressure) against the vertical surface corresponding to this wind velocity is 1.4 kPa.

**5/199** Compute the volume $V$ of the solid generated by revolving the right triangle about the $z$-axis through 180°. *Ans.* $V = 3619$ mm³

**5/200** Determine the area $A$ of the curved surface $ABCD$ of the solid of revolution shown.

**5/201** Determine the volume generated by rotating the semicircular area through 180°. *Ans.* $V = 361\ 000$ mm³

**5/202** Sketch the shear and moment diagrams for each of the four beams loaded and supported as shown.

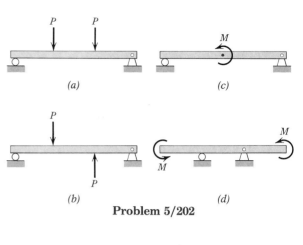

(a)

(c)

(b)

(d)

**Problem 5/202**

**5/203** Determine the $y$-coordinate of the centroid of the volume obtained by revolving the shaded area about the $x$-axis through 180°.

$$Ans. \ \bar{y} = \frac{15a}{14\pi}$$

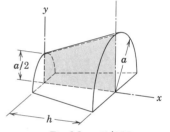

**Problem 5/203**

**5/204** Draw the shear and moment diagrams for the beam, which supports the unit load of 50 lb per foot of beam length distributed over its midsection.

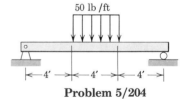

50 lb /ft

4'    4'    4'

**Problem 5/204**

**5/205** A uniform rectangular plate has three holes drilled in it—one of diameter $3d/2$ and two of diameter $d$. If a fourth hole of diameter $d$ is to be drilled so that the center of mass of the plate is at point $O$, determine the $x$- and $y$-coordinates of the center of the fourth hole.          $Ans. \ x = 2.25r, y = -0.250r$

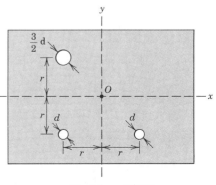

$\frac{3}{2} d$

**Problem 5/205**

80 mm
200 mm
*z*
*B*
*A*
*x*
*y*
400 mm

**Problem 5/206**

**5/206** An open-ended mailbox is formed from a single piece of steel sheet for the bottom and closed end, with a 90° bend along *AB*. The sides and curved top are formed from a single piece of fiberglass sheet bonded to the steel. Locate the mass center of the mailbox, ignoring the bonding material and neglecting the effects of the thickness of the sheet materials. The density of the steel is 7 kg/m² and that of the fiberglass is 0.5 kg/m².

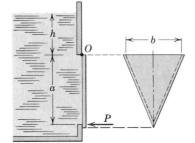

*h*
*r*

**Problem 5/207**

**5/207** A solid floating object is composed of a hemisphere and a cone of equal radius *r* made from the same homogeneous material. If the object floats with the center of the hemisphere above the water surface, find the maximum altitude *h* which the cone may have before the object will no longer float in the upright position illustrated.    *Ans.* $h = r\sqrt{3}$

*h*
*O*
*a*
*b*
*P*

**Problem 5/208**

**5/208** A flat plate seals a triangular opening in the vertical wall of a tank of liquid of density $\rho$. The plate is hinged about the upper edge *O* of the triangle. Determine the force *P* required to hold the gate in a closed position against the pressure of the liquid.

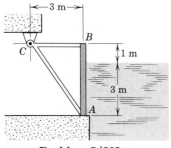

3 m
*B*
*C*
1 m
3 m
*A*

**Problem 5/209**

**5/209** The figure shows the cross section of a rectangular gate 4 m high and 6 m long (perpendicular to the paper) which blocks a fresh-water channel. The gate has a mass of 8.5 Mg and is hinged about a horizontal axis through *C*. Compute the vertical force *P* exerted by the foundation on the lower edge *A* of the gate. Neglect the mass of the frame to which the gate is attached.    *Ans.* $P = 348$ kN

**5/210** A cable is suspended from points $A$ and $B$ on the same horizontal line and supports a total load $W$ uniformly distributed along the horizontal. Determine the length $S$ of the cable. (Recall that the convergence of the series of Eq. 5/16$a$ requires that the sag-to-span ratio be less than 1/4.)

**5/211** Determine the maximum bending moment $M_{max}$ for the loaded beam and specify the distance $x$ to the right of end $A$ where $M_{max}$ exists.

*Ans.* $M_{max} = 186.4$ N·m, $x = 0.879$ m

**5/212** Locate the center of mass of the thin spherical shell that is formed by cutting out one-eighth of a complete shell.

**5/213** As part of a preliminary design study, the effects of wind loads on a 900-ft building are investigated. For the parabolic distribution of wind pressure shown in the figure, compute the force and moment reactions at the base $A$ of the building due to the wind load. The depth of the building is 200 ft.

*Ans.* $A = 1.440(10^6)$ lb, $M = 7.78(10^8)$ lb-ft

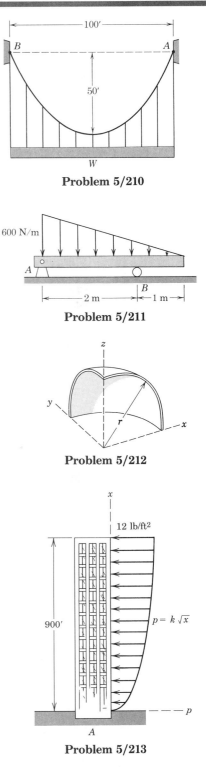

**Problem 5/210**

**Problem 5/211**

**Problem 5/212**

**Problem 5/213**

▶**5/214** Regard the tall building of Prob. 5/213 to be a uniform upright beam. Determine and plot the shear force and bending moment in the structure as a function of the height $x$ above the ground. Evaluate your expressions at $x = 450$ ft.

$$Ans. \quad V = 1.440(10^6) - \frac{160}{3} x^{3/2} \text{ lb}$$

$$M = 7.776(10^8) - 1.440(10^6)x - \frac{64}{3} x^{5/2} \text{ lb-ft}$$

$$V|_{x=450'} = 0.931(10^6) \text{ lb}$$

$$M|_{x=450'} = 2.21(10^8) \text{ lb-ft}$$

### *Computer-oriented problems*

*\*5/215** Construct the shear and moment diagrams for the loaded beam of Prob. 5/112, repeated here. Determine the maximum values of the shear and moment and their locations on the beam.

$$Ans. \quad V_{max} = 1900 \text{ lb at } x = 0$$
$$M_{max} = 9080 \text{ lb-ft at } x = 9.63 \text{ ft}$$

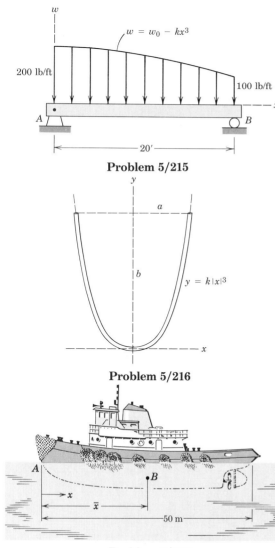

**Problem 5/215**

**Problem 5/216**

**Problem 5/217**

*\*5/216** The figure for Prob. 5/32 is repeated here. If $a = 2$ and $b = 8$, determine the $y$-coordinate of the centroid of the slender rod.

*\*5/217** The center of buoyancy $B$ of a ship's hull is the centroid of its submerged volume. The underwater cross-sectional areas $A$ of the transverse sections of the tugboat hull shown are tabulated for every five meters of waterline length. With an appropriate computer program determine to the nearest 0.5 m the distance $\bar{x}$ of $B$ aft of point $A$.

| $x$, m | $A$, m² | $x$, m | $A$, m² |
|---|---|---|---|
| 0 | 0 | 30 | 23.8 |
| 5 | 7.1 | 35 | 19.5 |
| 10 | 15.8 | 40 | 12.5 |
| 15 | 22.1 | 45 | 5.1 |
| 20 | 24.7 | 50 | 0 |
| 25 | 25.1 | | |

$$Ans. \quad \bar{x} = 24 \text{ m}$$

**5/218** A homogeneous charge of solid propellant for a rocket is in the shape of the circular cylinder formed with a concentric hole of depth $x$. For the dimensions shown, plot $\overline{X}$, the $x$-coordinate of the mass center of the propellant, as a function of the depth $x$ of the hole from $x = 0$ to $x = 600$ mm. Determine the maximum value of $\overline{X}$ and show that it is equal to the corresponding value of $x$.

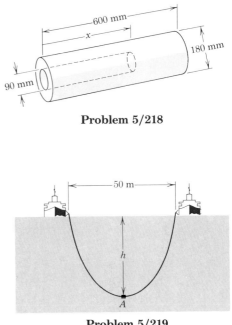

**Problem 5/218**

**5/219** An underwater detection instrument $A$ is attached to the midpoint of a 100-m cable suspended between two ships 50 m apart. Determine the depth $h$ of the instrument, which has negligible mass. Does the result depend on the mass of the cable or on the density of the water?          *Ans.* $h = 39.8$ m

**Problem 5/219**

**5/220** As a preliminary step in the construction of a tramway across a scenic river gorge, a 505-m cable with a mass of 12 kg/m is strung between points $A$ and $B$. Determine the horizontal distance $x$ to the right of point $A$ to the lowest point on the cable and compute the tensions at points $A$ and $B$.

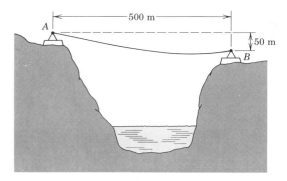

**Problem 5/220**

When contacting surfaces tend to slip one on the other, friction forces are produced and must be accounted for. The rock climber descending the face of the cliff depends on friction between the rope and his rappeling device to control his descent.

# FRICTION

# 6

## 6/1  INTRODUCTION

In the preceding chapters the forces of action and reaction between contacting surfaces have for the most part been assumed to act normal to the surfaces. This assumption characterizes the interaction between smooth surfaces and was illustrated in example 2 of Fig. 3/1. Although in many instances this ideal assumption involves only a relatively small error, there are a great many problems wherein we must consider the ability of contacting surfaces to support tangential as well as normal forces. Tangential forces generated between contacting surfaces are known as friction forces and are present to some degree with the interaction between all real surfaces. Whenever a tendency exists for one contacting surface to slide along another surface, we find that the friction forces developed are always in a direction to oppose this tendency.

In some types of machines and processes we desire to minimize the retarding effect of friction forces. Examples are bearings of all types, power screws, gears, the flow of fluids in pipes, and the propulsion of aircraft and missiles through the atmosphere. In other situations we wish to maximize the use of friction, as in brakes, clutches, belt drives, and wedges. Wheeled vehicles depend on friction for both starting and stopping, and ordinary walking depends on friction between the shoe and the ground. Friction forces are present throughout nature and exist to a considerable extent in all machines no matter how accurately constructed or carefully lubricated. A machine or process in which friction is small enough to be neglected is often referred to as *ideal*. When friction must be taken into account, the machine or process is termed *real*. In all real cases where sliding motion between parts occurs, the friction forces result in a loss of energy which is dissipated in the form of heat. In addition to the generation of heat and the accompanying loss of energy, friction between mating parts causes wear.

## SECTION A.  FRICTIONAL PHENOMENA

### 6/2  TYPES OF FRICTION

There are a number of separate types of frictional resistance encountered in mechanics, and we will mention each of these types briefly in this article prior to a more detailed account of the most common type of friction in the next article.

***(a) Dry friction.***   Dry friction is encountered when the unlubricated surfaces of two solids are in contact under a condition of sliding or tendency to slide. A friction force tangent to the surfaces of contact is developed both during the interval leading up to impending slippage and while slippage takes place. The direction of the force always opposes the motion or impending motion. This type of friction is also called *Coulomb* friction. The principles of dry or Coulomb friction were developed largely from the experiments of Coulomb in 1781 and from the work of Morin from 1831 to 1834. Although a comprehensive theory of dry friction is not yet fully developed, an analytical model sufficient to handle the vast majority of problems in dry friction is available and is described in Art. 6/3 which follows. This model forms the basis for most of this chapter.

***(b) Fluid friction.***   Fluid friction is developed when adjacent layers in a fluid (liquid or gas) are moving at different velocities. This motion gives rise to frictional forces between fluid elements, and these forces depend on the relative velocity between layers. When there is no such relative velocity, there is no fluid friction. Fluid friction depends not only on the velocity gradients within the fluid but also on the viscosity of the fluid, which is a measure of its resistance to shearing action between fluid layers. Fluid friction is treated in the study of fluid mechanics and will not be developed further in this book.

***(c) Internal friction.***   Internal friction is found in all solid materials that are subjected to cyclical loading. For highly elastic materials the recovery from deformation occurs with very little loss of energy caused by internal friction. For materials which have low limits of elasticity and which undergo appreciable plastic deformations during loading, the amount of internal friction that accompanies this deformation may be considerable. The mechanism of internal friction is associated with the action of shear deformation, and the student should consult a reference on materials science for a detailed description of this shear mechanism. Since this book deals primarily with the external effects of forces, we shall not be concerned with internal friction in the work which follows.

## 6/3  DRY FRICTION

The remainder of this chapter will be devoted to a description of the effects of dry friction acting on the exterior surfaces of rigid bodies. We shall explain the mechanism of dry friction with the aid of a very simple experiment.

*(a) Mechanism of friction.*  Consider a solid block of mass $m$ resting on a horizontal surface, as shown in Fig. 6/1$a$. The contacting surfaces possess a certain amount of roughness. The experiment will involve the application of a horizontal force $P$ which will vary continuously from zero to a value sufficient to move the block and give it an appreciable velocity. The free-body diagram of the block for any value of $P$ is shown in Fig. 6/1$b$, and the tangential friction force exerted by the plane on the block is labeled $F$. This friction force will *always* be in a direction to oppose motion or the tendency toward motion of the body on which it acts. There is also a normal force $N$ which in this case equals $mg$, and the total force $R$ exerted by the supporting surface on the block is the resultant of $N$ and $F$.

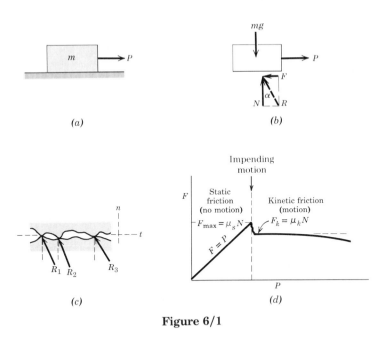

**Figure 6/1**

A magnified view of the irregularities of the mating surfaces, Fig. 6/1$c$, will aid in visualizing the mechanical action of friction. Support is necessarily intermittent and exists at the mating humps. The direction of each of the reactions on the block $R_1$, $R_2$, $R_3$, etc., will depend not only on the geometric profile of the irregularities but also on the extent of local deformation at each contact point. The total normal force $N$ is merely the sum of the $n$-components of the $R$'s, and the total frictional force $F$ is the sum of the $t$-components of the $R$'s. When the surfaces are in relative motion, we can see that the contacts are more nearly along the tops of the humps, and the $t$-components of the $R$'s will be smaller than when the surfaces are at rest relative to one another. This consideration helps to explain the well-known fact that the force $P$ necessary to maintain motion is generally less than that required to start the block when the irregularities are more nearly in mesh.

Assume now that we perform the experiment indicated and record the friction force $F$ as a function of $P$. The resulting experimental relation is indicated in Fig. 6/1*d*. When $P$ is zero, equilibrium requires that there be no friction force. As $P$ is increased, the friction force must be equal and opposite to $P$ as long as the block does not slip. During this period the block is in equilibrium, and all forces acting on the block must satisfy the equilibrium equations. Finally, we reach a value of $P$ which causes the block to slip and to move in the direction of the applied force. At this same time the friction force drops slightly and rather abruptly to a somewhat lower value. Here it remains essentially constant for an interval but then drops off still more with higher velocities.

The region up to the point of slippage or impending motion is known as the range of *static friction*, and the value of the friction force is determined by the *equations of equilibrium*. This force may have any value from zero up to and including, in the limit, the maximum value. For a given pair of mating surfaces we find that this maximum value of static friction $F_{\max}$ is proportional to the normal force $N$. Hence, we may write

$$\boxed{F_{\max} = \mu_s N} \qquad (6/1)$$

where $\mu_s$ is the proportionality constant, known as the *coefficient of static friction*. We must observe carefully that this equation describes only the *limiting* or *maximum* value of the static friction force and *not* any lesser value. Thus, the equation applies *only* to cases where it is known that motion is impending with the friction force at its peak value. As a reminder, then, for a condition of static equilibrium when motion is *not* impending, we see that the static friction force is

$$F < \mu_s N$$

After slippage occurs, a condition of *kinetic friction* accompanies the ensuing motion. Kinetic friction force is usually somewhat less than the maximum static friction force. We find also that the kinetic friction force $F_k$ is proportional to the normal force. Hence,

$$F_k = \mu_k N \qquad (6/2)$$

where $\mu_k$ is the *coefficient of kinetic friction*. It follows that $\mu_k$ is generally less than $\mu_s$. As the velocity of the block increases, the kinetic friction coefficient decreases somewhat, and when high velocities are reached, the effect of lubrication by an intervening fluid film may become appreciable. Coefficients of friction depend greatly on the exact condition of the surfaces, as well as on the velocity, and are subject to a considerable measure of uncertainty.

Because of the variability of the conditions governing the action of friction, in engineering practice it is frequently difficult to distinguish between a static and a kinetic coefficient, especially in the region of transition between impending motion and motion. Well-greased screw threads under mild loads, for example, will often exhibit comparable frictional retardation, whether they are on the verge of turning or whether they are in motion. In the literature of engineering practice we frequently find expressions for maximum static friction and for kinetic friction written simply as $F = \mu N$, and it is understood from the problem at hand whether maximum static friction or kinetic friction is described. Although we will frequently distinguish between the static and kinetic coefficients, in other cases no distinction will be made, and the friction coefficient will be written simply as $\mu$. In those cases the student must decide which of the friction conditions, maximum static friction for impending motion or kinetic friction, is involved. We emphasize again that many problems involve a static friction force which is less than the maximum value at impending motion, and therefore under these conditions the friction relation Eq. 6/1 cannot be used.

From Fig. 6/1c we observe that for rough surfaces there is a greater possibility for large angles between the reactions and the *n*-direction than for smoother surfaces. Thus, a friction coefficient reflects the roughness of a pair of mating surfaces and incorporates a geometric property of both mating contours. With this geometric model of the mechanical action of friction, we shall refer to mating surfaces as being "smooth" when the friction forces which they can support are negligibly small. It is meaningless to speak of a coefficient of friction for a single surface.

The direction of the resultant $R$ in Fig. 6/1b measured from the direction of $N$ is specified by $\tan \alpha = F/N$. When the friction force reaches its limiting static value $F_{\max}$, the angle $\alpha$ reaches a maximum value $\phi_s$. Thus,

$$\tan \phi_s = \mu_s$$

When slippage occurs, the angle $\alpha$ will have a value $\phi_k$ corresponding to the kinetic friction force. In like manner

$$\tan \phi_k = \mu_k$$

In practice the expression $\tan \phi = \mu$ is frequently seen where, again, the coefficient of friction may refer to either the static or the kinetic case, depending on the particular problem at hand. The angle $\phi_s$ is known as the *angle of static friction*, and the angle $\phi_k$ is called the *angle of kinetic friction*. This friction angle for each case clearly defines the limiting position of the total reaction $R$ between two contacting surfaces. If motion is impending, $R$ must be one element of a right-circular cone of vertex angle $2\phi_s$, as shown in Fig. 6/2. If motion is not impending, $R$ will be within the cone. This cone of vertex angle $2\phi_s$ is known as the *cone of static friction* and represents the locus of possible positions for the reaction $R$ at impending motion. If motion occurs, the angle of kinetic friction applies, and the reaction must lie on the surface of a slightly different cone of vertex angle $2\phi_k$. This cone is the *cone of kinetic friction*.

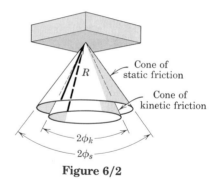

Figure 6/2

Further experiment shows that the friction force is essentially independent of the apparent or projected area of contact. The true contact area is much smaller than the projected value, since only the peaks of the contacting surface irregularities support the load. Relatively small normal loads result in high stresses at these contact points. As the normal force increases, the true contact area also increases as the material undergoes yielding, crushing, or tearing at the points of contact. A comprehensive theory of dry friction must go beyond the mechanical explanation presented here. For example, there is some evidence to support the theory that molecular attraction may be an important cause of friction under conditions where the mating surfaces are in very intimate contact. Other factors that influence dry friction are the generation of high local temperatures and adhesion at contact points, relative hardness of mating surfaces, and the presence of thin surface films of oxide, oil, dirt, or other substances.

***(b) Types of friction problems.***   From what has been said we may now recognize three distinct types of problems which are encountered in dry friction. The very first step in the solution of a friction problem is to identify which of these categories applies.

(1)   In the *first* type the condition of impending motion is known to exist. Here a body which is in equilibrium is on the verge of slipping, and the friction force equals the limiting static friction $F_{\max} = \mu_s N$. The equations of equilibrium will, of course, also hold.

(2)   In the *second* type of problem, neither the condition of impending motion nor that of motion is known to exist. To determine the actual friction conditions, we first assume static equilibrium and then solve for the friction force $F$ necessary for equilibrium. Three possible outcomes are possible:

(a)   $F < (F_{\max} = \mu_s N)$: We conclude that the friction force necessary for equilibrium can be supported, and therefore the body is in static equilibrium as assumed. It should be stressed that the *actual* friction force $F$ is *less than* the limiting value $F_{\max}$ given by Eq. 6/1 and that $F$ is determined *solely* by the equations of equilibrium.

(b)   $F = (F_{\max} = \mu_s N)$: Since the friction force $F$ is at its maximum value $F_{\max}$, motion impends as discussed in Case (1). The assumption of static equilibrium is valid.

(c)   $F > (F_{\max} = \mu_s N)$: Clearly this condition is impossible, since the surfaces cannot support more force than the maximum $\mu_s N$. Therefore, the assumption of equilibrium is invalid, the motion occurs. The friction force $F$ is equal to $\mu_k N$ from Eq. 6/2.

(3)   In the *third* type of problem, relative motion is known to exist between the contacting surfaces, and here the kinetic coefficient of friction clearly applies. For this case Eq. 6/2 will always give the kinetic friction force directly.

The foregoing discussion applies to all dry contacting surfaces and, to a limited extent, to moving surfaces which are partially lubricated. Some typical values of the coefficients of friction are given in Table D/1, Appendix D. These values are only approximate and are subject to considerable variation, depending on the exact conditions prevailing. They may be used, however, as typical examples of the magnitudes of frictional effects. When a reliable calculation involving friction is required, it is often desirable to determine the appropriate friction coefficient by experiment wherein the surface conditions of the problem are duplicated as closely as possible.

352

## Sample Problem 6/1

Determine the maximum angle $\theta$ which the adjustable incline may have with the horizontal before the block of mass $m$ begins to slip. The coefficient of static friction between the block and the inclined surface is $\mu_s$.

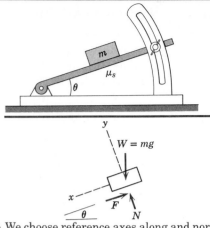

**Solution.** The free-body diagram of the block shows its weight $W = mg$, the normal force $N$, and the friction force $F$ exerted by the incline on the block. The friction force acts in the direction to oppose the slipping which would occur if no friction were present.

① Equilibrium in the x- and y-directions requires

$[\Sigma F_x = 0]$ $\qquad mg \sin\theta - F = 0 \qquad F = mg \sin\theta$

$[\Sigma F_y = 0]$ $\qquad -mg \cos\theta + N = 0 \qquad N = mg \cos\theta$

Dividing the first equation by the second gives $F/N = \tan\theta$. Since the maximum angle occurs when $F = F_{max} = \mu_s N$, for impending motion we have

$$\mu_s = \tan\theta_{max} \quad \text{or} \quad \theta_{max} = \tan^{-1}\mu_s \qquad Ans.$$

② 

① We choose reference axes along and normal to the direction of $F$ to avoid resolving both $F$ and $N$ into components.

② This problem describes a very simple way to determine a static coefficient of friction. The maximum value of $\theta$ is known as the *angle of repose*.

## Sample Problem 6/2

Determine the range of values which the mass $m_0$ may have so that the 100-kg block shown in the figure will neither start moving up the plane nor slip down the plane. The coefficient of static friction for the contact surfaces is 0.30.

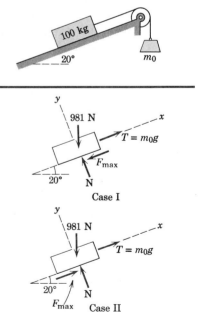

**Solution.** The maximum value of $m_0$ will be given by the requirement for motion impending up the plane. The friction force on the block therefore acts down the plane, as shown in the free-body diagram of the block for Case I in the figure. With the weight $mg = 100(9.81) = 981$ N, the equations of equilibrium give

$[\Sigma F_y = 0]$ $\qquad N - 981 \cos 20° = 0 \qquad N = 922$ N

$[F_{max} = \mu_s N]$ $\qquad F_{max} = 0.30(922) = 277$ N

$[\Sigma F_x = 0]$ $\quad m_0(9.81) - 277 - 981 \sin 20° = 0 \qquad m_0 = 62.4$ kg $\quad Ans.$

① The minimum value of $m_0$ is determined when motion is impending down the plane. The friction force on the block will act up the plane to oppose the tendency to move, as shown in the free-body diagram for Case II. Equilibrium in the x-direction requires

$[\Sigma F_x = 0]$ $\quad m_0(9.81) + 277 - 981 \sin 20° = 0 \qquad m_0 = 6.0$ kg $\quad Ans.$

Thus, $m_0$ may have any value from 6.0 to 62.4 kg, and the block will remain at rest.

In both cases equilibrium requires that the resultant of $F_{max}$ and $N$ be concurrent with the 981-N weight and the tension $T$.

① We see from the results of Sample Problem 6/1 that the block would slide down the incline without the restraint of attachment to $m_0$ since $\tan 20° > 0.30$. Thus, a value of $m_0$ will be required to maintain equilibrium.

## Sample Problem 6/3

Determine the magnitude and direction of the friction force acting on the 100-kg block shown if, first, $P = 500$ N and, second, $P = 100$ N. The coefficient of static friction is 0.20, and the coefficient of kinetic friction is 0.17. The forces are applied with the block initially at rest.

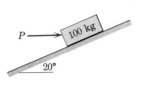

**Solution.** There is no way of telling from the statement of the problem whether the block will remain in equilibrium or whether it will begin to slip following the application of $P$. It is therefore necessary that we make an assumption, so we will take the friction force to be up the plane, as shown by the solid arrow. From the free-body diagram a balance of forces in both $x$- and $y$-directions gives

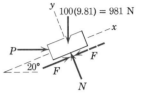

$[\Sigma F_x = 0]$      $P \cos 20° + F - 981 \sin 20° = 0$

$[\Sigma F_y = 0]$      $N - P \sin 20° - 981 \cos 20° = 0$

**Case I.** $P = 500$ N
Substitution into the first of the two equations gives

$$F = -134 \text{ N}$$

The negative sign tells us that *if* the block is in equilibrium, the friction force acting on it is in the direction opposite to that assumed and therefore is down the plane, as represented by the dotted arrow. We cannot reach a conclusion on the magnitude of $F$, however, until we verify that the surfaces are capable of supporting 134 N of friction force. This may be done by substituting $P = 500$ N into the second equation, which gives

$$N = 1093 \text{ N}$$

The maximum static friction force that the surfaces can support is then

$[F_{\max} = \mu_s N]$      $F_{\max} = 0.20(1093) = 219$ N

Since this force is greater than that required for equilibrium, we conclude that the assumption of equilibrium was correct. The answer is, then,

$$F = 134 \text{ N down the plane} \qquad Ans.$$

**Case II.** $P = 100$ N
Substitution into the two equilibrium equations gives

$$F = 231 \text{ N} \qquad N = 956 \text{ N}$$

But the maximum possible static friction force is

$[F_{\max} = \mu_s N]$      $F_{\max} = 0.20(956) = 191$ N

① It follows that 231 N of friction cannot be supported. Therefore, equilibrium cannot exist, and we obtain the correct value of the friction force by using the kinetic coefficient of friction accompanying the motion down the plane. Hence, the answer is

$[F_k = \mu_k N]$      $F = 0.17(956) = 163$ N up the plane      *Ans.*

① We should note that even though $\Sigma F_x$ is no longer equal to zero, equilibrium does exist in the $y$-direction, so that $\Sigma F_y = 0$. Therefore, the normal force $N$ is 956 N whether or not the block is in equilibrium.

354

## Sample Problem 6/4

The homogeneous rectangular block of mass $m$, width $b$, and height $H$ is placed on the horizontal surface and subjected to a horizontal force $P$ which moves the block along the surface with a constant velocity. The coefficient of kinetic friction between the block and the surface is $\mu_k$. Determine (a) the greatest value that $h$ may have so that the block will slide without tipping over and (b) the location of a point $C$ on the bottom face of the block through which the resultant of the friction and normal forces acts if $h = H/2$.

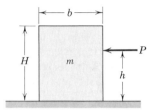

**Solution.** (a) With the block on the verge of tipping, we see that the entire reaction between the plane and the block will necessarily be at $A$. The free-body diagram of the block shows this condition. Since slipping occurs, the friction force is the limiting value $\mu_k N$, and the angle $\theta$ becomes $\theta = \tan^{-1}\mu_k$. The resultant of $F_k$ and $N$ passes through a point $B$ through which $P$ must also pass, since three coplanar forces in equilibrium are concurrent. Hence, from the geometry of the block

$$\tan\theta = \mu_k = \frac{b/2}{h} \qquad h = \frac{b}{2\mu_k} \qquad Ans.$$

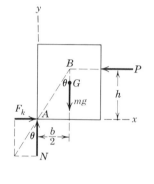

If $h$ were greater than this value, moment equilibrium about $A$ would not be satisfied, and the block would tip over.

Alternatively, we may find $h$ by combining the equilibrium requirements for the $x$- and $y$-directions with the moment-equilibrium equation about $A$. Thus,

$$[\Sigma F_y = 0] \qquad N - mg = 0 \qquad N = mg$$

$$[\Sigma F_x = 0] \qquad F_k - P = 0 \qquad P = F_k = \mu_k N = \mu_k mg$$

$$[\Sigma M_A = 0] \quad Ph - mg\frac{b}{2} = 0 \qquad h = \frac{mgb}{2P} = \frac{mgb}{2\mu_k mg} = \frac{b}{2\mu_k} \qquad Ans.$$

(b) With $h = H/2$ we see from the free-body diagram for case (b) that the resultant of $F_k$ and $N$ passes through a point $C$ which is a distance $x$ to the left of the vertical centerline through $G$. The angle $\theta$ is still $\theta = \phi = \tan^{-1}\mu_k$ as long as the block is slipping. Thus, from the geometry of the figure we have

$$\frac{x}{H/2} = \tan\theta = \mu_k \qquad so \qquad x = \mu_k H/2 \qquad Ans.$$

If we were to replace $\mu_k$ by the static coefficient $\mu_s$, then our solutions would describe the conditions under which the block is (a) on the verge of tipping and (b) on the verge of slipping, both from a rest position.

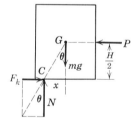

① Recall that the equilibrium equations apply to a body moving with a constant velocity (zero acceleration) just as well as to a body at rest.

② Alternatively, we could equate the moments about $G$ to zero, which would give us $F(H/2) - Nx = 0$. Thus, with $F_k = \mu_k N$ we get $x = \mu_k H/2$.

## Sample Problem 6/5

The three flat blocks are positioned on the 30° incline as shown, and a force $P$ parallel to the incline is applied to the middle block. The upper block is prevented from moving by a wire which attaches it to the fixed support. The coefficient of static friction for each of the three pairs of mating surfaces is shown. Determine the maximum value which $P$ may have before any slipping takes place.

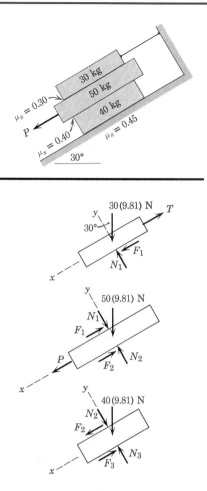

**Solution.** The free-body diagram of each block is drawn. The friction forces are assigned in the directions to oppose the relative motion which ① would occur if no friction were present. There are two possible conditions for impending motion. Either the 50-kg block slips and the 40-kg block remains in place, or the 50- and 40-kg blocks move together with slipping occurring between the 40-kg block and the incline.

The normal forces, which are in the $y$-direction, may be determined without reference to the friction forces, which are all in the $x$-direction. Thus,

$$[\Sigma F_y = 0] \quad (30\text{-kg}) \quad N_1 - 30(9.81) \cos 30° = 0 \qquad N_1 = 255 \text{ N}$$

$$(50\text{-kg}) \quad N_2 - 50(9.81) \cos 30° - 255 = 0 \quad N_2 = 680 \text{ N}$$

$$(40\text{-kg}) \quad N_3 - 40(9.81) \cos 30° - 680 = 0 \quad N_3 = 1019 \text{ N}$$

We shall assume arbitrarily that only the 50-kg block slips, so that the 40-kg block remains in place. Thus, for impending slippage at both surfaces of the 50-kg block we have

$$[F_{\max} = \mu_s N] \quad F_1 = 0.30(255) = 76.5 \text{ N} \qquad F_2 = 0.40(680) = 272 \text{ N}$$

The assumed equilibrium of forces at impending motion for the 50-kg block gives

$$[\Sigma F_x = 0] \quad P - 76.5 - 272 + 50(9.81) \sin 30° = 0 \qquad P = 103.1 \text{ N}$$

We now check on the validity of our initial assumption. For the 40-kg block with $F_2 = 272$ N the friction force $F_3$ would be given by

$$[\Sigma F_x = 0] \quad 272 + 40(9.81) \sin 30° - F_3 = 0 \qquad F_3 = 468 \text{ N}$$

But the maximum possible value of $F_3$ is $F_3 = \mu_s N_3 = 0.45(1019) = $ 459 N. Thus, 468 N cannot be supported and our initial assumption was wrong. We conclude, therefore, that slipping occurs first between the 40-kg block and the incline. With the corrected value $F_3 = 459$ N equilibrium of the 40-kg block for its impending motion requires

② $[\Sigma F_x = 0] \qquad F_2 + 40(9.81) \sin 30° - 459 = 0 \qquad F_2 = 263 \text{ N}$

Equilibrium of the 50-kg block gives, finally,

$$[\Sigma F_x = 0] \qquad P + 50(9.81) \sin 30° - 263 - 76.5 = 0$$

$$P = 93.8 \text{ N} \qquad\qquad Ans.$$

Thus, with $P = 93.8$ N, motion impends for the 50-kg and 40-kg blocks as a unit.

① In the absence of friction the middle block, under the influence of $P$, would have a greater movement than the 40-kg block, and the friction force $F_2$ will be in the direction to oppose this motion as shown.

② We see now that $F_2$ is less than $\mu_s N_2 = 272$ N.

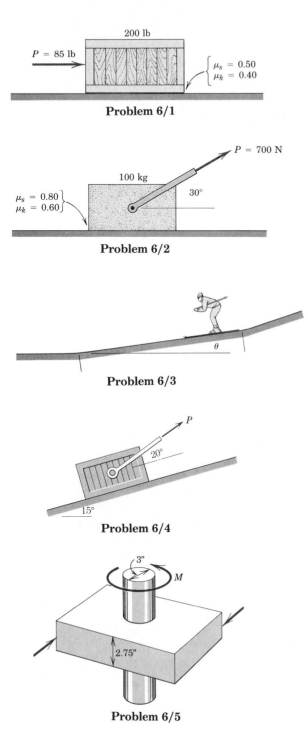

**Problem 6/1**

**Problem 6/2**

**Problem 6/3**

**Problem 6/4**

**Problem 6/5**

## PROBLEMS

### *Introductory problems*

**6/1** The 85-lb force $P$ is applied to the 200-lb crate, which is stationary before the force is applied. Determine the magnitude and direction of the friction force $F$ exerted by the horizontal surface on the crate.

*Ans. $F$ = 85 lb to the left*

**6/2** The 700-N force is applied to the 100-kg block, which is stationary before the force is applied. Determine the magnitude and direction of the friction force $F$ exerted by the horizontal surface on the block.

**6/3** The designer of a ski resort wishes to have a portion of a beginner's slope on which the skier's speed will remain fairly constant. Tests indicate the average coefficients of friction between skis and snow to be $\mu_s$ = 0.10 and $\mu_k$ = 0.08. What should be the slope angle $\theta$ of the constant-speed section?

*Ans. $\theta$ = 4.57°*

**6/4** The coefficients of static and kinetic friction between the 100-kg block and the inclined plane are 0.30 and 0.20, respectively. Determine (*a*) the friction force $F$ acting on the block when $P$ is applied with a magnitude of 200 N to the block at rest, (*b*) the force $P$ required to initiate motion up the incline from rest, and (*c*) the friction force $F$ acting on the block if $P$ = 600 N.

**6/5** The 3-in.-diameter steel shaft has a force fit in the hole in the cast iron block. If a torque $M$ = 400 lb-in. is required to start turning the shaft in the fixed block, calculate the compressive stress (pressure) $\sigma$ between the contacting surfaces. The coefficient of static friction between the parts is known to be 0.35. *Ans. $\sigma$ = 29.4 lb/in.²*

**6/6** The 1.2-kg wooden block is used for level support of the 9-kg can of paint. Determine the magnitude and direction of (*a*) the friction force exerted by the roof surface on the wooden block and (*b*) the total force exerted by the roof surface on the wooden block.

**6/7** The uniform pole of length *l* and mass *m* is leaned against the vertical wall as shown. If the coefficient of static friction between the supporting surfaces and the ends of the poles is 0.25, calculate the maximum angle $\theta$ at which the pole may be placed before it starts to slip.                    *Ans.* $\theta = 28.1°$

**6/8** The light bar is used to support the 50-kg block in its vertical guides. If the coefficient of static friction is 0.30 at the upper end of the bar and 0.40 at the lower end of the bar, find the friction force acting at each end for *x* = 75 mm. Also find the maximum value of *x* for which the bar will not slip.

**6/9** The tongs are used to handle hot steel tubes that are being heat-treated in an oil bath. For a 20° jaw opening, what is the minimum coefficient of static friction between the jaws and the tube that will enable the tongs to grip the tube without slipping?                    *Ans.* $\mu_s = 0.176$

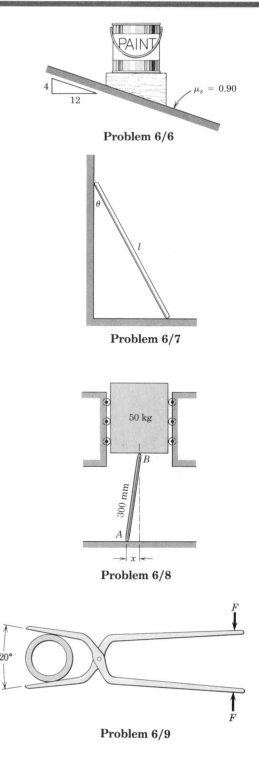

**Problem 6/6**

**Problem 6/7**

**Problem 6/8**

**Problem 6/9**

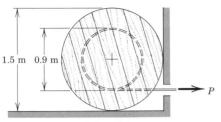

**Problem 6/10**

### Representative problems

**6/10** An 80-kg reel of cable is positioned against a vertical wall as shown. If the coefficients of static and kinetic friction for both pairs of contacting surfaces are 0.30 and 0.25, respectively, calculate the horizontal force $P$ in the cable required to keep the reel turning at a constant rate.

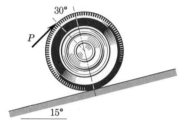

**Problem 6/11**

**6/11** The force $P$ is applied tangentially to the rim of the 25-kg wheel at the position shown to prevent rolling down the incline. The coefficient of friction between the wheel and the incline is 0.30. Determine the friction force $F$ that the incline exerts on the wheel.
*Ans. $F$ = 34.0 N*

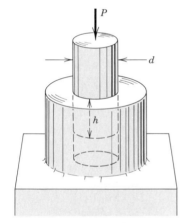

**Problem 6/12**

**6/12** If it requires a force $P$ to insert the cylindrical plug of diameter $d$ in the fixed collar to a depth $h$ against the friction of a tight fit, show that the torque $M$ required to turn the cylinder in the collar is $M = Pd/2$.

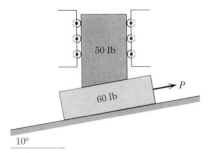

**Problem 6/13**

**6/13** Calculate the force $P$ required to initiate motion of the 60-lb block up the 10° incline. The coefficient of static friction for each pair of surfaces is 0.30.
*Ans. $P$ = 57.1 lb*

**6/14** The strut *AB* of negligible mass is hinged to the horizontal surface at *A* and to the uniform 25-kg wheel at *B*. Determine the minimum couple *M* applied to the wheel which will cause it to slip if the coefficient of static friction between the wheel and the surface is 0.40.

**6/15** If the couple *M* of Prob. 6/14 were applied in a clockwise rather than counterclockwise direction, calculate its minimum value to cause the wheel to slip.             *Ans.* $M = 36.8$ N·m

**6/16** The 180-lb man with center of gravity *G* supports the 75-lb drum as shown. Find the greatest distance *x* at which the man can position himself without slipping if the coefficient of static friction between his shoes and the ground is 0.40.

**6/17** Two men are sliding a 100-kg crate up an incline. If the lower man pushes horizontally with a force of 500 N and if the coefficient of kinetic friction is 0.40, determine the tension *T* which the upper man must exert in the rope to maintain motion of the crate.
                                                        *Ans.* $T = 465$ N

**6/18** The 300-lb crate with mass center at *G* is supported on the horizontal surfaces by a skid at *A* and a roller at *B*. If a force *P* of 60 lb is required to initiate motion of the crate, determine the coefficient of static friction at *A*.

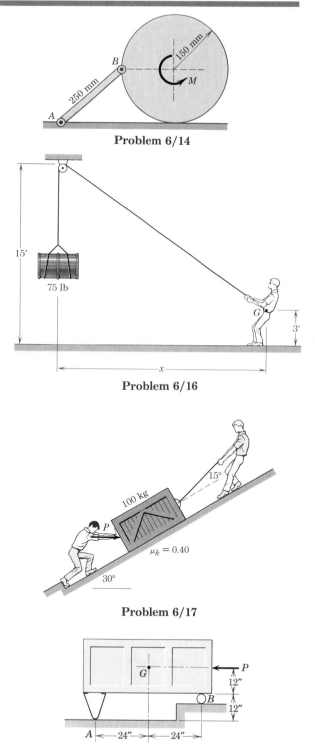

**Problem 6/14**

**Problem 6/16**

**Problem 6/17**

**Problem 6/18**

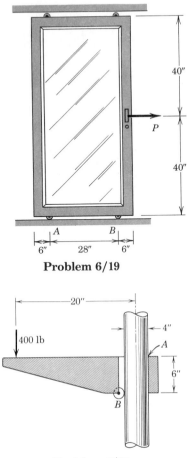

**Problem 6/19**

**6/19** The sliding door rolls on the two small lower wheels $A$ and $B$. Under normal conditions the upper wheels do not touch their horizontal guide. (*a*) Compute the force $P$ required to slide the door at a steady speed if wheel $A$ becomes "frozen" and does not turn in its bearing. (*b*) Rework the problem if wheel $B$ becomes frozen instead of wheel $A$. The coefficient of kinetic friction between a frozen wheel and the supporting surface is 0.30, and the center of mass of the 140-lb door is at its geometric center. Neglect the small diameter of the wheels.

*Ans.* (*a*) $P = 14.7$ lb, (*b*) $P = 36.8$ lb

**6/20** The figure shows the cross section of a loaded bracket which is supported on the fixed shaft by the roller at $B$ and by friction at the corner $A$. The coefficient of static friction is 0.40. Neglect the weight of the bracket and show that the bracket will remain in place. Find the friction force $F$.

**Problem 6/20**

**6/21** The section of railing of mass $m$ fits loosely in the vertical grooves in the side posts and can be removed by lifting it with a force $P$ which exceeds $mg$. When $P$ is applied a distance $x$ from the center, the section will rotate slightly counterclockwise as it starts to rise, and corners $A$ and $C$ will contact the bottom of the grooves with corners $B$ and $D$ free. If the coefficient of static friction between the corners and the bottoms of the grooves is $\mu_s$, determine the maximum value of $x$ so that the section will not bind. (If $x$ exceeds this value, the angle between each corner force and the horizontal normal to the grooves becomes less than $\phi_s = \tan^{-1} \mu_s$ and slipping cannot occur no matter how large $P$ may be.)

$$Ans.\ x = \frac{b}{2\mu_s}\left(1 - \frac{mg}{P}\right)$$

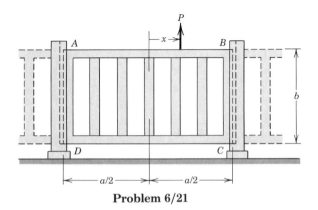

**Problem 6/21**

**6/22** Two automobiles, both of which have the mass center located as shown midway between the front and rear axles, are identical except that one is front-wheel-drive and the other is rear-wheel-drive. The cars are driven at constant speed up ramps of various inclinations. From a theoretical viewpoint, which car could climb the ramp of higher inclination angle $\theta$? Justify your answer.

**6/23** The rectangular steel yoke is used to prevent slippage between the two boards under tensile loads $P$. If the coefficients of static friction between the yoke and the board surfaces and between the boards are all 0.30, determine the maximum value of $h$ for which there is no slipping. For $P = 800$ N, determine the corresponding normal force $N$ between the two boards if impending motion occurs at all surfaces.

*Ans.* $h = 156.6$ mm, $N = 1333$ N

**6/24** The 50-kg solid cylinder is being rolled slowly up the 15° incline by the force $P$ applied parallel to the incline to a cord wrapped around the cylinder. Calculate the friction force $F$ between the cylinder and the incline. The coefficient of static friction $\mu_s$ is 0.20.

**6/25** Determine the distance $s$ to which the 90-kg painter can climb without causing the 4-m ladder to slip at its lower end $A$. The top of the 15-kg ladder has a small roller, and at the ground the coefficient of static friction is 0.25. The mass center of the painter is directly above his feet.

*Ans.* $s = 2.55$ m

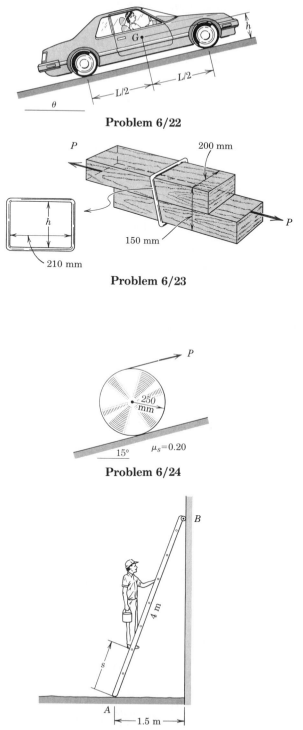

**Problem 6/22**

**Problem 6/23**

**Problem 6/24**

**Problem 6/25**

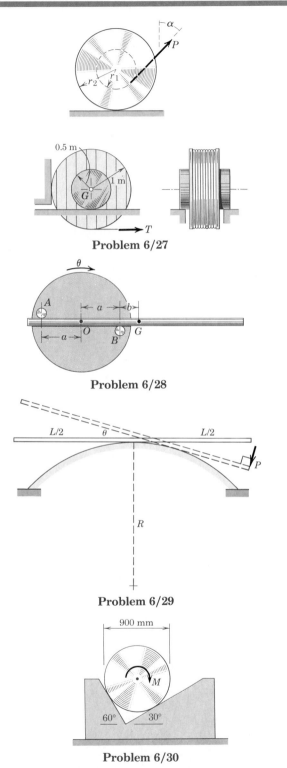

**Problem 6/27**

**Problem 6/28**

**Problem 6/29**

**Problem 6/30**

**6/26** The wheel shown will roll to the left when the angle $\alpha$ of the cord is small. When $\alpha$ is large, the wheel rolls to the right. Determine by inspection from the geometry of the free-body diagram the angle $\alpha$ for which the wheel will not roll in either direction. If the coefficient of static friction is $\mu_s$ and the mass of the wheel is $m$, determine the value of $P$ for which the wheel will slip for the critical value of $\alpha$.

**6/27** Calculate the force $T$ required to rotate the 200-kg reel of telephone cable that rests on its hubs and bears against a vertical wall. The coefficient of kinetic friction for each pair of contacting surfaces is 0.60.

*Ans.* $T = 727$ N

**6/28** The uniform rod with center of mass at $G$ is supported by the pegs $A$ and $B$, which are fixed in the wheel. If the coefficient of friction between the rods and pegs is $\mu$, determine the angle $\theta$ through which the wheel may be slowly turned about its horizontal axis through $O$, starting from the position shown, before the rod begins to slip. Neglect the diameter of the rod compared with the other dimensions.

**6/29** The uniform slender rod of mass $m$ and length $L$ is initially at rest in a centered horizontal position on the fixed circular surface of radius $R = 0.6L$. If a force $P$ normal to the bar is gradually applied to its end until the bar begins to slip at the angle $\theta = 20°$, determine the coefficient of static friction $\mu_s$.

*Ans.* $\mu_s = 0.212$

**6/30** Calculate the magnitude of the clockwise couple $M$ required to turn the 50-kg cylinder in the supporting block shown. The coefficient of kinetic friction is 0.30.

**6/31** The semicylindrical shell of mass $m$ and radius $r$ is rolled through an angle $\theta$ by the horizontal force $P$ applied to its rim. If the coefficient of friction is $\mu_s$, determine the angle $\theta$ at which the shell slips on the horizontal surface as $P$ is gradually increased. What value of $\mu_s$ will permit $\theta$ to reach 90°?

$$Ans. \ \theta = \sin^{-1}\left(\frac{\pi\mu_s}{2 - \pi\mu_s}\right), \ \mu_{90°} = 0.318$$

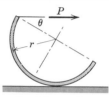

**Problem 6/31**

**6/32** The three identical rollers are stacked on a horizontal surface as shown. If the coefficient of static friction $\mu_s$ is the same for all pairs of contacting surfaces, find the minimum value of $\mu_s$ for which the rollers will not slip.

**Problem 6/32**

**6/33** The single-lever block brake prevents rotation of the flywheel under a counterclockwise torque $M$. Find the force $P$ required to prevent rotation if the coefficient of static friction is $\mu_s$. Explain what would happen if the geometry permitted $b$ to equal $\mu_s e$.

$$Ans. \ P = \frac{M}{rl}\left(\frac{b}{\mu_s} - e\right)$$

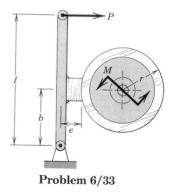

**Problem 6/33**

**6/34** Determine the force $P$ required to move the uniform 50-kg plank from its rest position shown if the coefficient of static friction at both contact locations is 0.50.

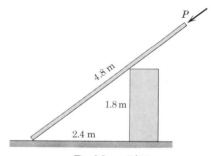

**Problem 6/34**

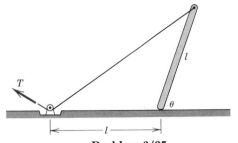

**Problem 6/35**

**6/35** The uniform slender rod is slowly lowered from the upright position ($\theta = 90°$) by means of the cord attached to its upper end and passing over the small fixed pulley. If the rod is observed to slip at its lower end when $\theta = 40°$, determine the coefficient of static friction at the horizontal surface.

*Ans.* $\mu_s = 0.761$

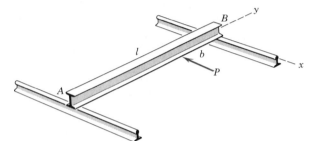

**Problem 6/36**

**6/36** The uniform I-beam of mass $m$ is supported at its ends on two fixed horizontal rails as shown. Determine the maximum horizontal force $P$ which can be applied without causing the beam to slip and find the corresponding value of the friction force at $A$. The coefficient of static friction between the beam and the rails is $\mu_s$. Also, take $b < l/2$.

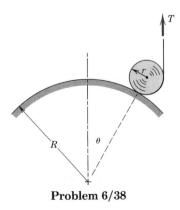

**Problem 6/37**

**6/37** What force $P$ must the two men exert on the rope to slide the uniform 20-ft plank on the overhead rack? The plank weighs 200 lb, and the coefficient of kinetic friction between the plank and each support is 0.50.

*Ans.* $P = 162.3$ lb

**6/38** Determine the maximum value of the angle $\theta$ at which the cylinder of mass $m$ and radius $r$ may be kept in equilibrium by application of a vertical force $T$ as shown. The applicable coefficient of static friction is $\mu_s$. Evaluate the normal force $N$ and the tension $T$ in the cord for the maximum value of $\theta$.

**Problem 6/38**

**6/39** A man pedals his bicycle up a 5 percent grade on a slippery road at a steady speed. The man and bicycle have a combined mass of 82 kg with mass center at $G$. If his rear wheel is on the verge of slipping, determine the coefficient of friction $\mu_s$ between the rear tire and the road. If the coefficient of friction were doubled, what would be the friction force $F$ acting on the rear wheel? (Why may we neglect friction under the front wheel?)

*Ans.* $\mu_s = 0.082$, $F = 40.2$ N

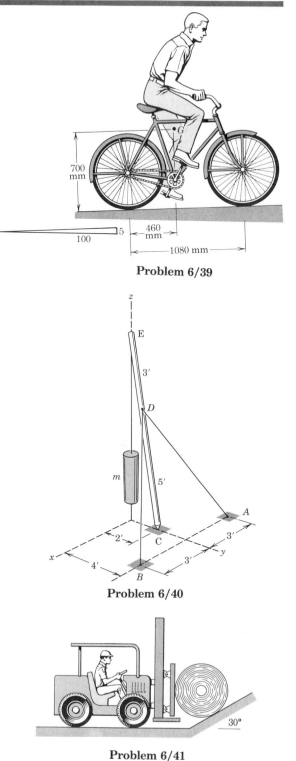

700 mm

460 mm

100   5

1080 mm

**Problem 6/39**

**6/40** The light 8-ft pole is supported at end $C$ by a rough horizontal surface in the $x$-$y$ plane and secured by guy wires from $D$ to $A$ and $B$. Determine the minimum coefficient of static friction $\mu_s$ between the pole and the supporting surface necessary to prevent the pole from slipping. Solve algebraically and then check your solution graphically.

$z$

$E$

$3'$

$D$

$m$

$5'$

$A$

$2'$

$C$

$3'$

$x$

$4'$

$B$

$3'$

$y$

**Problem 6/40**

**6/41** The industrial truck is used to move the solid 1200-kg roll of paper up the 30° incline. If the coefficients of static and kinetic friction between the roll and the vertical barrier of the truck and between the roll and the incline are both 0.40, compute the required tractive force $P$ between the tires of the truck and the horizontal surface.

*Ans.* $P = 22.1$ kN

30°

**Problem 6/41**

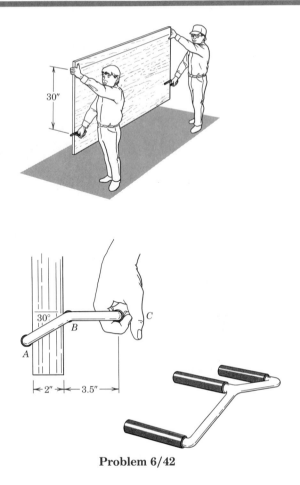

**Problem 6/42**

**6/42** Two workers are carrying a 2-in.-thick panel by means of panel carriers, one of which is shown in the detail figures. The vertical panel is steadied by equal horizontal forces applied by the left hands of the workers. Determine the minimum coefficient of static friction between the panel and the carriers for which there will be no slippage. The carrier grips at $A$, $B$, and $C$ do not rotate on the carrier frame. Note that each worker must apply both a vertical and a horizontal force to the carrier handle. Assume that each worker supports half the weight of the panel.

**6/43** A uniform bar of mass $m$ and length $l$ rests on a horizontal surface with its mass evenly distributed along its length. If the coefficient of static friction between the bar and the supporting surface is $\mu_s$, write an expression for the horizontal force $P$, applied at the end of the bar, required to move the bar and find the distance $a$ to axis $O$ about which the bar is observed to rotate. (*Hint:* The normal force under the bar is uniformly distributed over the length of the bar. The friction force acting on the bar is, therefore, uniformly distributed over the length of each of the two sections on either side of $O$.)

*Ans.* $P = 0.414\mu_s mg$, $a = 0.293l$

**Problem 6/43**

**6/44** A cement bucket of mass $m$ is at rest on the $10°$ incline when an overhead crane begins to increase the tension in the nonvertical cable. If the coefficient of static friction between the bucket and incline is 0.70 and if tipping is not a consideration, find the tension $T$ at which slipping begins and the direction $\beta$ of initial movement of the bucket.

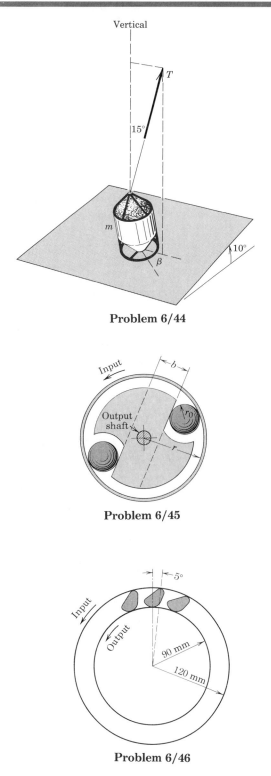

**Problem 6/44**

**▶6/45** The elements of a unidirectional mechanical clutch are shown. A torque $M$ applied to the outer ring is transmitted to the output shaft through frictional interaction between the outer ring and the balls and between the balls and the inner driven member. If the direction of rotation of the outer ring is reversed, the wedging action of the balls is absent, and no torque can be transmitted to the output shaft. For given values of $r$, $r_0$, and coefficient of friction $\mu$ which is applicable for both pairs of interacting surfaces, specify the minimum dimension $b$ of the inner member which will permit the transmission of torque without slipping.     *Ans.* $b = 2[(r - r_0) \cos 2\phi - r_0]$
where $\phi = \tan^{-1} \mu$

**Problem 6/45**

**▶6/46** The figure shows the elements of a unidirectional clutch which permits the transmission of counterclockwise torque from the outer input disk (circle) through the frictional interaction of the connecting rockers, called sprags. A reversal of the direction of motion disengages the sprags, thus preventing any torque transmission. Calculate the minimum coefficient of friction $\mu_{\min}$ which is applicable for the contacting surfaces which will enable the clutch to engage without slipping.     *Ans.* $\mu_{\min} = 0.354$

**Problem 6/46**

# SECTION B.  APPLICATIONS OF FRICTION IN MACHINES

In Section B we investigate the action of friction in various machine applications. Because these conditions are normally either those of limiting static or kinetic friction, we will use the variable $\mu$ (rather than $\mu_s$ or $\mu_k$) in our general developments. Then, depending on whether motion is impending or actually occurring, $\mu$ can be interpreted as either the static or kinetic coefficient of friction.

## 6/4  WEDGES

A wedge is one of the simplest and most useful of machines and is used as a means of producing small adjustments in the position of a body or as a means of applying large forces. Wedges are largely dependent on friction. When sliding of a wedge is impending, the resultant force on each sliding surface of the wedge will be inclined from the normal to the surface by an amount equal to the friction angle. The component of the resultant along the surface is the friction force, which is always in the direction to oppose the motion of the wedge relative to the mating surfaces.

Figure 6/3a shows a wedge that is used to position or lift a large mass $m$, where the vertical loading is $mg$. The coefficient of friction for each pair of surfaces is $\mu = \tan \phi$. The force $P$ required to start the wedge is found from the equilibrium triangles of the forces on the load and on the wedge. The free-body diagrams are shown in Fig. 6/3b, where the reactions are inclined at an angle $\phi$ from their respective normals and are in the direction to oppose the motion. The mass of the wedge is neglected. From these diagrams we may write the force equilibrium conditions by equating to zero the sum of the force vectors acting on each body. The solutions of these equations are shown in the $c$-part of the figure, where $R_2$ is found first in the upper diagram using the known value of $mg$. The force $P$ is then found from the lower triangle once the value of $R_2$ has been established.

If $P$ is removed and the wedge remains in place, equilibrium of the wedge requires that the equal reactions $R_1$ and $R_2$ must be collinear as shown in Fig. 6/4, where the wedge angle $\alpha$ is taken to be less than $\phi$. The $a$-part of the figure represents impending slippage at the upper surface, and the $c$-part of the figure represents impending slippage at the lower surface. In order for the wedge to slide out of its space, slippage must occur at *both* surfaces simultaneously; otherwise, the wedge is *self-locking*, and there is a finite range of possible intermediate angular positions of $R_1$ and $R_2$ for which the wedge will remain in place. Figure 6/4b illustrates this range and shows that simultaneous slippage is not possible if $\alpha < 2\phi$. The student is encouraged to construct additional diagrams for the case

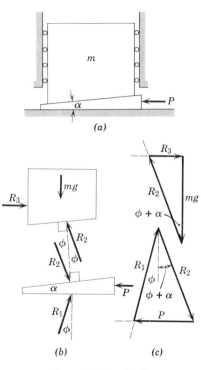

Forces to Raise Load

**Figure 6/3**

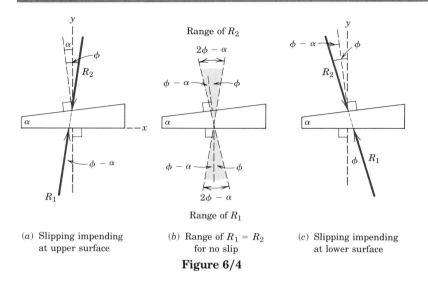

(a) Slipping impending at upper surface

(b) Range of $R_1 = R_2$ for no slip

(c) Slipping impending at lower surface

**Figure 6/4**

where $\alpha > \phi$ and verify that the wedge is self-locking as long as $\alpha < 2\phi$.

If the wedge is self-locking and is to be withdrawn, a pull $P$ on the wedge will be required. In this event, to oppose the new impending motion, the reactions $R_1$ and $R_2$ would act on the opposite sides of their normals from those when the wedge was inserted. The solution would then proceed in a manner similar to that described for the case of raising the load. The free-body diagrams and vector polygons for this condition are shown in Fig. 6/5.

Wedge problems lend themselves to graphical solutions as indicated in the three figures. The accuracy of a graphical solution is easily held within tolerances consistent with the uncertainty of friction coefficients. Algebraic solutions may also be obtained from the trigonometry of the equilibrium polygons.

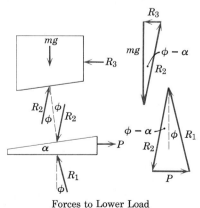

Forces to Lower Load

**Figure 6/5**

### 6/5 SCREWS

Screws are used for fastenings and for transmitting power or motion. In each case the friction developed in the threads largely determines the action of the screw. For transmitting power or motion the square thread is more efficient than the V-thread, and the analysis illustrated here is confined to the square thread.

Consider the square-threaded jack, Fig. 6/6, under the action of the axial load $W$ and a moment $M$ applied about the axis of the screw. The screw has a lead $L$ (advancement per revolution) and a mean radius $r$. The force $R$ exerted by the thread of the jack frame on a small representative portion of the thread of the screw is shown on the free-body diagram of the screw. Similar reactions exist on all segments of the screw thread where contact occurs with the thread

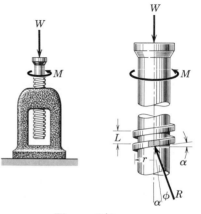

**Figure 6/6**

of the base. If $M$ is just sufficient to turn the screw, the thread of the screw will slide around and up on the fixed thread of the frame. The angle $\phi$ made by $R$ with the normal to the thread will be the angle of friction, so that $\tan \phi = \mu$. The moment of $R$ about the vertical axis of the screw is $Rr \sin(\alpha + \phi)$, and the total moment due to all reactions on the threads is $\Sigma Rr \sin(\alpha + \phi)$. Since $r \sin(\alpha + \phi)$ appears in each term, we may factor it out. The moment equilibrium equation for the screw becomes

$$M = [r \sin(\alpha + \phi)] \Sigma R$$

Equilibrium of forces in the axial direction further requires

$$W = \Sigma R \cos(\alpha + \phi) = [\cos(\alpha + \phi)] \Sigma R$$

Combining the expressions for $M$ and $W$ gives

$$M = Wr \tan(\alpha + \phi) \qquad (6/3)$$

We determine the helix angle $\alpha$ by unwrapping the thread of the screw for one complete turn where we see immediately that $\alpha = \tan^{-1}(L/2\pi r)$.

We may use the unwrapped thread of the screw as an alternative model to simulate the action of the entire screw, as shown in Fig. 6/7a. The equivalent force required to push the movable thread up the fixed incline is $P = M/r$, and the triangle of force vectors gives Eq. 6/3 immediately.

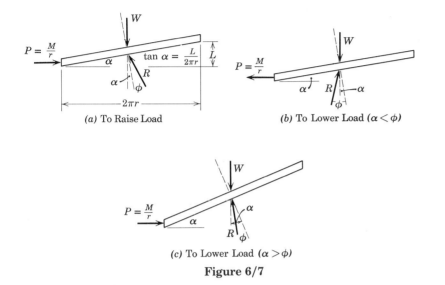

(a) To Raise Load          (b) To Lower Load ($\alpha < \phi$)

(c) To Lower Load ($\alpha > \phi$)

**Figure 6/7**

If the moment $M$ is removed, the friction force changes direction so that $\phi$ is measured to the other side of the normal to the thread. The screw will remain in place and be self-locking, provided that $\alpha < \phi$, and will be on the verge of unwinding if $\alpha = \phi$.

To lower the load by unwinding the screw, we must reverse the direction of $M$ as long as $\alpha < \phi$. This condition is illustrated in Fig. 6/7b for our simulated thread on the fixed incline, and we see that an equivalent force $P = M/r$ must be applied to the thread to pull it down the incline. Therefore, from the triangle of vectors we get the moment required to lower the screw, which is

$$M = Wr \tan (\phi - \alpha) \qquad (6/3a)$$

If $\alpha > \phi$, the screw will unwind by itself, and we see from Fig. 6/7c that the moment required to prevent unwinding would be

$$M = Wr \tan (\alpha - \phi) \qquad (6/3b)$$

## Sample Problem 6/6

The horizontal position of the 500-kg rectangular block of concrete is adjusted by the 5° wedge under the action of the force **P**. If the coefficient of static friction for both pairs of wedge surfaces is 0.30 and if the coefficient of static friction between the block and the horizontal surface is 0.60, determine the least force $P$ required to move the block.

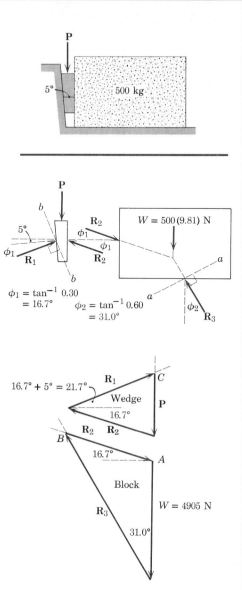

**(1)** **Solution.** The free-body diagrams of the wedge and the block are drawn with the reactions $\mathbf{R}_1$, $\mathbf{R}_2$, and $\mathbf{R}_3$ inclined with respect to their normals by the amount of the friction angles for impending motion. The friction angle for limiting static friction is given by $\phi = \tan^{-1} \mu$. Each of the two friction angles is computed and shown on the diagram.

We start our vector diagram expressing the equilibrium of the block at a convenient point $A$ and draw the only known vector, the weight **W** of the block. Next we add $\mathbf{R}_3$, whose 31.0° inclination from the vertical is now known. The vector $\mathbf{R}_2$, whose 16.7° inclination from the horizontal is also known, must close the polygon for equilibrium. Thus, point $B$ on the lower polygon is determined by the intersection of the known directions of $\mathbf{R}_3$ and $\mathbf{R}_2$ and their magnitudes become known.

For the wedge we draw $\mathbf{R}_2$, which is now known, and add $\mathbf{R}_1$, whose direction is known. The directions of $\mathbf{R}_1$ and **P** intersect at $C$, thus giving us the solution for the magnitude of **P**.

**(2)** **Algebraic solution.** The simplest choice of reference axes for calculation purposes is, for the block, in the direction $a$-$a$ normal to $\mathbf{R}_3$ and, for the wedge, in the direction $b$-$b$ normal to $\mathbf{R}_1$. The angle between $\mathbf{R}_2$ and the $a$-direction is $16.7° + 31.0° = 47.7°$. Thus, for the block

$[\Sigma F_a = 0]$     $500(9.81) \sin 31.0° - R_2 \cos 47.7° = 0$

$$R_2 = 3747 \text{ N}$$

For the wedge the angle between $\mathbf{R}_2$ and the $b$-direction is $90° - (2\phi_1 + 5°) = 51.6°$, and the angle between **P** and the $b$-direction is $\phi_1 + 5° = 21.7°$. Thus,

$[\Sigma F_b = 0]$     $3747 \cos 51.6° - P \cos 21.7° = 0$

$$P = 2505 \text{ N} \qquad Ans.$$

**Graphical solution.** The accuracy of a graphical solution is well within the uncertainty of the friction coefficients and provides a simple and direct result. By laying off the vectors to a reasonable scale following the sequence described, the magnitudes of **P** and the **R**'s are easily scaled directly from the diagrams.

**(1)** Be certain to note that the reactions are inclined from their normals in the direction to oppose the motion. Also, we note the equal and opposite reactions $\mathbf{R}_2$.

**(2)** It should be evident that we avoid simultaneous equations by eliminating reference to $\mathbf{R}_3$ for the block and $\mathbf{R}_1$ for the wedge.

## Sample Problem 6/7

The single-threaded screw of the vise has a mean diameter of 1 in. and has 5 square threads per inch. The coefficient of static friction in the threads is 0.20. A 60-lb pull applied normal to the handle at $A$ produces a clamping force of 1000 lb between the jaws of the vise. (*a*) Determine the frictional moment $M_B$ developed at $B$ due to the thrust of the screw against the body of the jaw. (*b*) Determine the force $Q$ applied normal to the handle at $A$ required to loosen the vise.

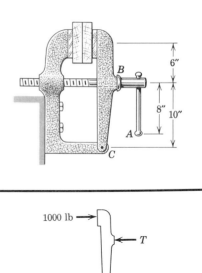

**Solution.** From the free-body diagram of the jaw we first obtain the tension $T$ in the screw.

$$[\Sigma M_C = 0] \qquad 1000(16) - 10T = 0 \qquad T = 1600 \text{ lb}$$

The helix angle $\alpha$ and the friction angle $\phi$ for the thread are given by

$$\alpha = \tan^{-1}\frac{L}{2\pi r} = \tan^{-1}\frac{1/5}{2\pi(0.5)} = 3.64°$$

$$\phi = \tan^{-1}\mu = \tan^{-1} 0.20 = 11.31°$$

where the mean radius of the thread is $r = 0.5$ in.

(*a*) **To tighten.** The isolated screw is simulated by the free-body diagram shown where all of the forces acting on the threads of the screw are represented by a single force $R$ inclined at the friction angle $\phi$ from the normal to the thread. The moment applied about the screw axis is $60(8) = 480$ lb-in. in the clockwise direction as seen from the front of the vise. The frictional moment $M_B$ due to the friction forces acting on the collar at $B$ is in the counterclockwise direction to oppose the impending motion. From Eq. 6/3 with $T$ substituted for $W$ the net moment acting on the screw is

$$M = Tr \tan(\alpha + \phi)$$

$$480 - M_B = 1600(0.5) \tan(3.64° + 11.31°)$$

$$M_B = 266 \text{ lb-in.} \qquad\qquad Ans.$$

(*b*) **To loosen.** The free-body diagram of the screw on the verge of being loosened is shown with $R$ acting at the friction angle from the normal in the direction to counteract the impending motion. Also shown is the frictional moment $M_B = 266$ lb-in. acting in the clockwise direction to oppose the motion. The angle between $R$ and the screw axis is now $\phi - \alpha$, and we use Eq. 6/3*a* with the net moment equal to the applied moment $M'$ minus $M_B$. Thus,

$$M = Tr \tan(\phi - \alpha)$$

$$M' - 266 = 1600(0.5) \tan(11.31° - 3.64°)$$

$$M' = 374 \text{ lb-in.}$$

Thus, the force on the handle required to loosen the vise is

$$Q = M'/d = 374/8 = 46.8 \text{ lb} \qquad\qquad Ans.$$

① Be careful to calculate the helix angle correctly. Its tangent is the lead $L$ (advancement per revolution) divided by the mean circumference $2\pi r$ and not by the diameter $2r$.

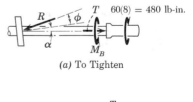

(*a*) To Tighten

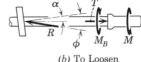

(*b*) To Loosen

② Note that $R$ swings to the opposite side of the normal as the impending motion reverses direction.

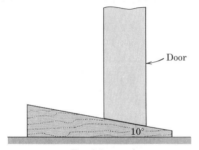

**Problem 6/47**

**Problem 6/48**

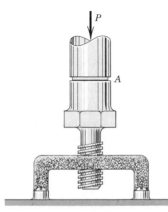

**Problem 6/49**

# PROBLEMS

### *Introductory problems*

**6/47** Because of the presence of spring-loaded door hinges (not shown), a horizontal force component (to the right) must be applied by the wedge to the door to hold it in an open position. Determine the required coefficient of friction $\mu_s$ between the wedge and the floor. Assume that the wedge is light and that the wedge-door interface is smooth.

*Ans.* $\mu_s = 0.176$

**6/48** If the coefficient of friction between the steel wedge and the moist fibers of the newly cut stump is 0.20, determine the maximum angle $\alpha$ which the wedge may have and not pop out of the wood after being driven by the sledge.

**6/49** The device shown is used for coarse adjustment of the height of an experimental apparatus without a change in its horizontal position. Because of the slipjoint at $A$, turning the screw does not rotate the cylindrical leg above $A$. The mean diameter of the thread is $\frac{3}{8}$ in. and the coefficient of friction is 0.15. For a conservative design which neglects friction at the slipjoint, what should be the minimum number $N$ of threads per inch to ensure that the screw does not turn by itself under the weight of the apparatus?

*Ans.* $N = 5.66$

**6/50** Determine the force $P$ required to force the 10° wedge under the 90-kg uniform crate which rests against the small stop at $A$. The coefficient of friction for all surfaces is 0.40.

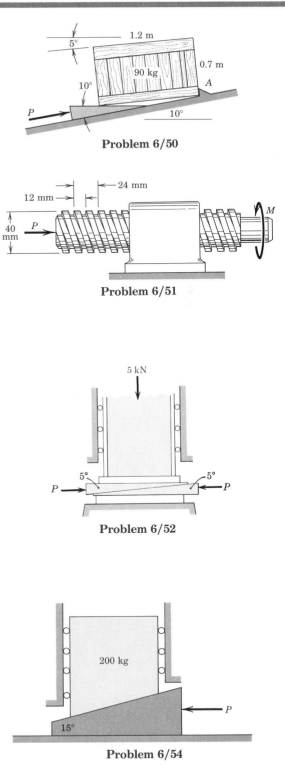

**Problem 6/50**

**6/51** The 40-mm-diameter screw has a double square thread with a pitch of 12 mm and a lead of 24 mm. The screw and its mating threads in the fixed block are graphite-lubricated and have a friction coefficient of 0.15. If a torque $M = 60$ N·m is applied to the right-hand portion of the shaft, determine (*a*) the force $P$ required to advance the shaft to the right and (*b*) the force $P$ which would allow the shaft to move to the left at a constant speed.

*Ans.* $P = 75.3$ kN, $P = 8.55$ kN

**Problem 6/51**

### *Representative problems*

**6/52** The two 5° wedges shown are used to adjust the position of the column under a vertical load of 5 kN. Determine the magnitude of the forces $P$ required to raise the column if the coefficient of friction for all surfaces is 0.40.

**Problem 6/52**

**6/53** If the loaded column of Prob. 6/52 is to be lowered, calculate the horizontal forces $P'$ required to withdraw the wedges.          *Ans.* $P' = 3.51$ kN

**6/54** Determine the magnitude and direction of the force $P$ required to (*a*) raise and (*b*) lower the 200-kg mass. The coefficient of friction is $\mu = 0.10$ for all contacting surfaces. What is the minimum value $\mu_{min}$ required for the wedge to be self-locking? The mass of the wedge is negligible.

**Problem 6/54**

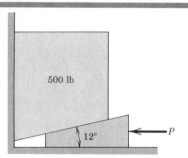

500 lb

12°

**Problem 6/55**

**6/55** Determine the horizontal force $P$ required to raise the 500-lb block. The coefficient of friction for all surfaces is 0.40.                    *Ans.* $P = 730$ lb

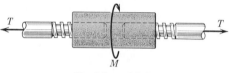

$T$                    $T$

$M$

**Problem 6/56**

**6/56** The threaded collar is used to connect two shafts, both with right-hand threads on their ends. The shafts are under a tension $T = 8$ kN. If the threads have a mean diameter of 16 mm and a lead of 4 mm, calculate the torque $M$ required to turn the collar in either direction with the shafts prevented from turning. The coefficient of friction is 0.24.

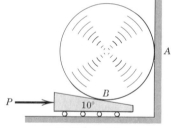

$A$

$P$ $B$

10°

**Problem 6/57**

**6/57** Calculate the horizontal force $P$ on the light 10° wedge necessary to initiate movement of the 40-kg cylinder. The coefficient of static friction for both pairs of contacting surfaces is 0.25. Also determine the friction force $F_B$ at point $B$. (*Caution:* Check carefully your assumption of where slipping occurs.)
                    *Ans.* $P = 98.6$ N, $F_B = 24.6$ N

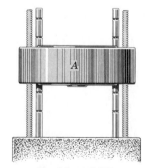

$A$

**Problem 6/58**

**6/58** The movable head of a universal testing machine has a mass of 2.2 Mg and is elevated into testing position by two 78-mm-diameter lead screws, each with a single thread and a lead of 13 mm. If the coefficient of friction in the threads is 0.25, how much torque $M$ must be supplied to each screw (*a*) to raise the head and (*b*) to lower the head? The inner loading columns are not attached to the head during positioning.

**6/59** The two 4° wedges are used to position the vertical column under a load $L$. What is the least value of the coefficient of friction $\mu_2$ for the bottom pair of surfaces for which the column may be raised by applying a single horizontal force $P$ to the upper wedge?          *Ans.* $\mu_2 = 0.378$

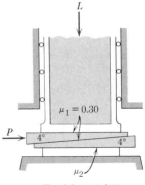

**Problem 6/59**

**6/60** Two shafts connected by a flat 5° tapered cotter, as shown by the two views in the figure, are under a constant tension $T$ of 200 lb. Find the force $P$ required to move the cotter and take up any slack in the joint. The coefficient of friction between the cotter and the sides of the slots is 0.20. Neglect any horizontal friction between the shafts.

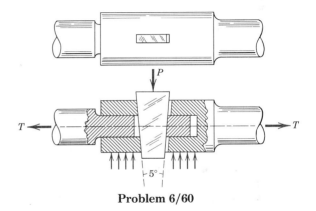

**Problem 6/60**

**6/61** The detent mechanism consists of the spring-loaded plunger with a spherical end which positions the horizontal bar by engaging the spaced notches. If the spring exerts a force of 40 N on the plunger in the position shown and a force $P = 60$ N is required to move the detent bar against the plunger, calculate the coefficient of friction between the plunger and the detent. It is known from earlier tests that the coefficient of friction between the light bar and the horizontal surface is 0.30. Assume that the plunger is well lubricated and accurately fitted so that the friction between it and its guide is negligible.         *Ans.* $\mu = 0.368$

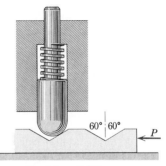

**Problem 6/61**

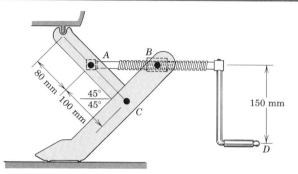

**Problem 6/62**

**6/62** The jack shown is used to lift small unit-body cars. The screw is threaded into the collar pivoted at $B$, and the shaft turns in a ball thrust bearing at $A$. The thread has a mean diameter of 10 mm and a lead (advancement per revolution) of 2 mm. The coefficient of friction for the threads is 0.20. Determine the force $P$ normal to the handle at $D$ required (*a*) to raise a mass of 500 kg from the position shown and (*b*) to lower the load from the same position. Neglect friction in the pivot and bearing at $A$.

**Problem 6/63**

**6/63** The bench hold-down clamp is being used to clamp two boards together while they are being glued. What torque $M$ must be applied to the screw in order to produce a 200-lb compression between the boards? The $\frac{1}{2}$-in.-diameter single-thread screw has 12 square threads per inch, and the coefficient of friction in the threads may be taken to be 0.20. Neglect any friction in the small ball contact at $A$ and assume that the contact force at $A$ is directed along the axis of the screw. What torque $M'$ is required to loosen the clamp? *Ans.* $M = 48.2$ lb-in., $M' = 27.4$ lb-in.

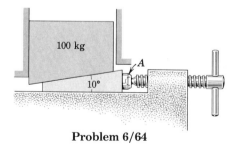

**Problem 6/64**

**6/64** The vertical position of the 100-kg block is adjusted by the screw-activated wedge. Calculate the moment $M$ which must be applied to the handle of the screw to raise the block. The single-threaded screw has square threads with a mean diameter of 30 mm and advances 10 mm for each complete turn. The coefficient of friction for the screw threads is 0.25, and the coefficient of friction for all mating surfaces of the block and wedge is 0.40. Neglect friction at the ball joint $A$.

**6/65** Calculate the moment $M'$ which must be applied to the handle of the screw of Prob. 6/64 to withdraw the wedge and lower the 100-kg load.

*Ans.* $M' = 3.02$ N·m

▶**6/66** Replace the square thread of the screw jack in Fig. 6/6 by a V-thread as indicated in the figure accompanying this problem and determine the moment $M$ on the screw required to raise the load $W$. The force $R$ acting on a representative small section of the thread is shown with its relevant projections. The vector $R_1$ is the projection of $R$ in the plane of the figure containing the axis of the screw. The analysis is begun with an axial force and a moment summation and includes substitutions for the angles $\gamma$ and $\beta$ in terms of $\theta$, $\alpha$, and the friction angle $\phi = \tan^{-1}\mu$. The helix angle of the single thread is exaggerated for clarity.

$$Ans.\ M = Wr\ \frac{\tan\alpha + \mu\sqrt{1 + \tan^2\dfrac{\theta}{2}\cos^2\alpha}}{1 - \mu\tan\alpha\sqrt{1 + \tan^2\dfrac{\theta}{2}\cos^2\alpha}}$$

$$\text{where } \alpha = \tan^{-1}\frac{L}{2\pi r}$$

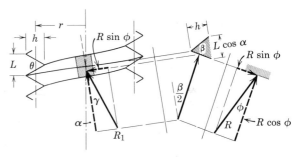

**Problem 6/66**

### 6/6  JOURNAL BEARINGS

A journal bearing is one that gives lateral support to a shaft in contrast to axial or thrust support. For dry bearings and for many partially lubricated bearings we may apply the principles of dry friction, which give us a satisfactory approximation for design purposes. A dry or partially lubricated journal bearing with contact or near contact between the shaft and the bearing is shown in Fig. 6/8, where the clearance between the shaft and bearing is greatly exaggerated to clarify the action. As the shaft begins to turn in the

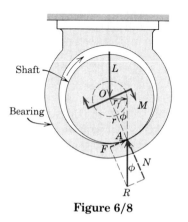

**Figure 6/8**

direction shown, it rolls up the inner surface of the bearing until slippage occurs. Here it remains in a more or less fixed position during rotation. The torque $M$ required to maintain rotation and the radial load $L$ on the shaft will cause a reaction $R$ at the contact point $A$. For vertical equilibrium $R$ must equal $L$ but will not be collinear with it. Thus, $R$ will be tangent to a small circle of radius $r_f$ called the *friction circle*. The angle between $R$ and its normal component $N$ is the friction angle $\phi$. Equating the sum of the moments about $A$ to zero gives

$$M = Lr_f = Lr \sin \phi \qquad (6/4)$$

For a small coefficient of friction the angle $\phi$ is small, and the sine and tangent may be interchanged with only small error. Since $\mu = \tan \phi$, a good approximation to the torque is

$$M = \mu Lr \qquad (6/4a)$$

This relation gives the amount of torque or moment applied to the shaft which is necessary to overcome friction for a dry or partially lubricated journal bearing.

## 6/7  THRUST BEARINGS; DISK FRICTION

Friction between circular surfaces under distributed normal pressure is encountered in pivot bearings, clutch plates, and disk brakes. To examine this application, we consider the two flat circular disks of Fig. 6/9 whose shafts are mounted in bearings (not shown) so that they can be brought into contact under the axial force $P$. The maximum torque that this clutch can transmit will be equal to the torque $M$ required to slip one disk against the other. If $p$ is the

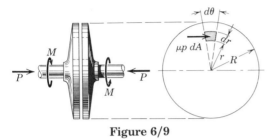

**Figure 6/9**

normal pressure at any location between the plates, the frictional force acting on an elemental area is $\mu p\, dA$, where $\mu$ is the friction coefficient and $dA$ is the area $r\, dr\, d\theta$ of the element. The moment of this elemental friction force about the shaft axis is $\mu p r\, dA$, and the total moment becomes

$$M = \int \mu p r\, dA$$

where we evaluate the integral over the area of the disk. To carry out this integration, we must know the variation of $\mu$ and $p$ with $r$.

In the following examples we shall consider that $\mu$ is constant. Furthermore, if the surfaces are new, flat, and well supported, it is reasonable to assume that the pressure $p$ is constant and uniformly distributed so that $\pi R^2 p = P$. Substituting the constant value of $p$ in the expression for $M$ gives us

$$M = \frac{\mu P}{\pi R^2} \int_0^{2\pi} \int_0^R r^2\, dr\, d\theta = \tfrac{2}{3}\mu P R \qquad (6/5)$$

We may interpret this result as equivalent to the moment due to a friction force $\mu P$ acting at a distance $\tfrac{2}{3}R$ from the shaft center.

If the friction disks are rings, as in the collar bearing of Fig. 6/10, the limits of integration are the inside and outside radii $R_i$ and $R_o$, respectively, and the frictional torque becomes

$$M = \tfrac{2}{3}\mu P \frac{R_o^{\,3} - R_i^{\,3}}{R_o^{\,2} - R_i^{\,2}} \tag{6/5a}$$

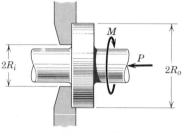

**Figure 6/10**

After the initial wearing-in period is over, the surfaces retain their new relative shape and further wear is therefore constant over the surface. This wear depends both on the circumferential distance traveled and the pressure $p$. Since the distance traveled is proportional to $r$, the expression $rp = K$ may be written, where $K$ is a constant. The value of $K$ is determined from equilibrium of the axial forces or

$$P = \int p \, dA = K \int_0^{2\pi} \int_0^R dr \, d\theta = 2\pi K R$$

With $pr = K = P/(2\pi R)$, we may write the expression for $M$ as

$$M = \int \mu pr \, dA = \frac{\mu P}{2\pi R} \int_0^{2\pi} \int_0^R r \, dr \, d\theta$$

which becomes

$$M = \tfrac{1}{2}\mu P R \tag{6/6}$$

The frictional moment for worn-in plates is, therefore, only $(\tfrac{1}{2})/(\tfrac{2}{3})$, or $\tfrac{3}{4}$ as much as for new surfaces. If the friction disks are rings of inside radius $R_i$ and outside radius $R_o$, substitution of these limits gives for the frictional torque for worn-in surfaces

$$M = \tfrac{1}{2}\mu P (R_o + R_i) \tag{6/6a}$$

The student should be prepared to deal with other disk-friction problems, where the pressure $p$ is some other function of $r$.

## Sample Problem 6/8

The bell crank fits over a 100-mm-diameter shaft which is fixed and cannot rotate. The horizontal force $T$ is applied to maintain equilibrium of the crank under the action of the vertical force $P = 100$ N. Determine the maximum and minimum values that $T$ may have without causing the crank to rotate in either direction. The coefficient of static friction $\mu$ between the shaft and the bearing surface of the crank is 0.20.

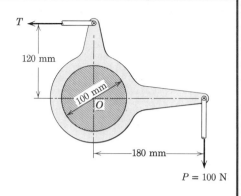

**Solution.** Impending rotation occurs when the reaction $R$ of the fixed shaft on the bell crank makes an angle $\phi = \tan^{-1}\mu$ with the normal to the bearing surface and is, therefore, tangent to the friction circle. Also, equilibrium requires that the three forces acting on the crank be concurrent at point $C$. These facts are shown in the free-body diagrams for the two cases of impending motion.

The following calculations are needed:

Friction angle $\phi = \tan^{-1}\mu = \tan^{-1} 0.20 = 11.31°$

Radius of friction circle $r_f = r \sin\phi = 50 \sin 11.31° = 9.81$ mm

Angle $\theta = \tan^{-1}\dfrac{120}{180} = 33.7°$

Angle $\beta = \sin^{-1}\dfrac{r_f}{OC} = \sin^{-1}\dfrac{9.81}{\sqrt{(120)^2 + (180)^2}} = 2.60°$

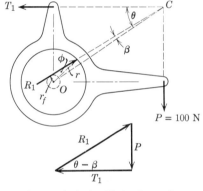

(a) Counterclockwise Motion Impends

**(a) Impending counterclockwise motion.** The equilibrium triangle of forces is drawn and gives

$$T_1 = P \cot(\theta - \beta) = 100 \cot(33.7° - 2.60°)$$

$$T_1 = T_{max} = 165.8 \text{ N} \qquad\qquad Ans.$$

**(b) Impending clockwise motion.** The equilibrium triangle of forces for this case gives

$$T_2 = P \cot(\theta + \beta) = 100 \cot(33.7° + 2.60°)$$

$$T_2 = T_{min} = 136.2 \text{ N} \qquad\qquad Ans.$$

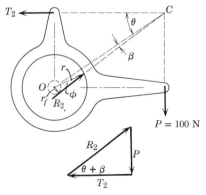

(b) Clockwise Motion Impends

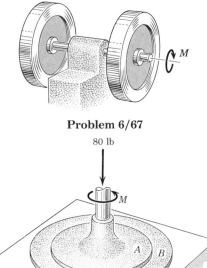

**Problem 6/67**

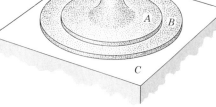

80 lb

**Problem 6/68**

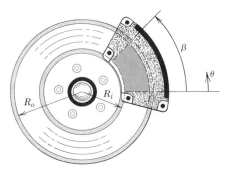

**Problem 6/69**

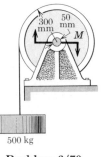

500 kg

**Problem 6/70**

## PROBLEMS

### Introductory problems

**6/67** The two flywheels are mounted on a common shaft which is supported by a journal bearing between them. Each flywheel has a mass of 40 kg, and the diameter of the shaft is 40 mm. If a 3-N·m couple $M$ on the shaft is required to maintain rotation of the flywheels and shaft at a constant low speed, compute (a) the coefficient of friction in the bearing and (b) the radius $r_f$ of the friction circle.

*Ans.* (a) $\mu = 0.1947$, (b) $r_f = 3.82$ mm

**6/68** Circular disk $A$ is placed on top of disk $B$ and is subjected to a compressive force of 80 lb. The diameters of $A$ and $B$ are 9 in. and 12 in., respectively, and the pressure under each disk is constant over its surface. If the coefficient of friction between $A$ and $B$ is 0.40, determine the couple $M$ which will cause $A$ to slip on $B$. Also, what is the minimum coefficient of friction $\mu$ between $B$ and the supporting surface $C$ which will prevent $B$ from rotating?

**6/69** An automobile disk brake consists of a flat-faced rotor and a caliper which contains a disk pad on each side of the rotor. For equal forces $P$ behind the two pads with the pressure $p$ uniform over the pad, show that the moment applied to the hub is independent of the angular span $\beta$ of the pads. Would pressure variation with $\theta$ change the moment?

*Ans.* $M = \dfrac{4\mu P}{3}\dfrac{R_o^3 - R_i^3}{R_o^2 - R_i^2}$

**6/70** A torque $M$ of 1510 N·m must be applied to the 50-mm-diameter shaft of the hoisting drum to raise the 500-kg load at constant speed. The drum and shaft together have a mass of 100 kg. Calculate the coefficient of friction $\mu$ for the bearing.

**6/71** Calculate the torque $M$ on the shaft of the hoisting drum of Prob. 6/70 that is required to lower the 500-kg load at constant speed. Use the value $\mu = 0.271$ calculated in Prob. 6/70 for the coefficient of friction.

*Ans.* $M = 1.433$ kN·m

*Representative problems*

**6/72** The pulley of radius $r$ is used to hoist the load $W$. Derive a relation for the tension $T$ required (*a*) to raise the load and (*b*) to lower the load, both at constant speed. The coefficient of friction for the pulley bearing of radius $r_0$ is $\mu$, which is quite small, thus permitting the substitution of $\mu$ for $\sin \phi$, where $\phi$ is the friction angle.

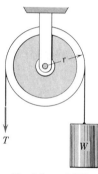

**Problem 6/72**

**6/73** The shaft $A$ fits loosely in the wrist-pin bearing of the connecting rod with center of gravity at $G$ as shown. With the rod initially in the vertical position the shaft is rotated slowly until the rod slips at the angle $\alpha$. Write an exact expression for the coefficient of friction $\mu$.

$$Ans. \ \mu = \frac{1}{\sqrt{\left(\dfrac{d/2}{\bar{r} \sin \alpha}\right)^2 - 1}}$$

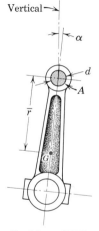

**Problem 6/73**

**6/74** The 20-kg steel ring $A$ with inside and outside radii of 50 mm and 60 mm, respectively, rests on a fixed horizontal shaft of 40-mm radius. If a downward force $P = 150$ N applied to the periphery of the ring is just sufficient to cause the ring to slip, calculate the coefficient of friction $\mu$ and the angle $\theta$.

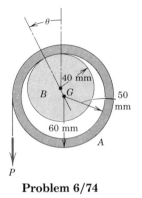

**Problem 6/74**

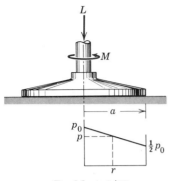

Problem 6/75

**6/75** For the flat sanding disk of radius $a$ the pressure $p$ developed between the disk and the sanded surface decreases linearly with $r$ from a value $p_0$ at the center to $p_0/2$ at $r = a$. If the coefficient of friction is $\mu$, derive the expression for the torque $M$ required to turn the shaft under an axial force $L$.

*Ans.* $M = \frac{5}{8}\mu La$

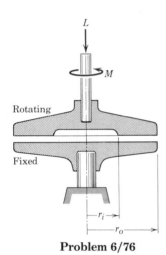

Problem 6/76

**6/76** The axial section of the two mating circular disks is shown. Derive the expression for the torque $M$ required to turn the upper disk on the fixed lower one if the pressure $p$ between the disks follows the relation $p = k/r^2$, where $k$ is a constant to be determined. The coefficient of friction $\mu$ is constant over the entire surface.

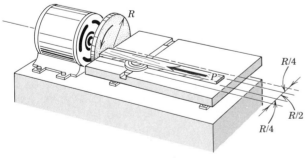

Problem 6/77

**6/77** An end of the thin board is being sanded by the disk sander under application of the force $P$. If the effective coefficient of kinetic friction is $\mu$ and if the pressure is essentially constant over the board end, determine the moment $M$ which must be applied by the motor in order to rotate the disk at a constant angular speed. The board end is centered along a radius of the disk.

*Ans.* $M = \frac{1}{2}\mu PR$

**6/78** Equations 6/5a and 6/6a describe the frictional moment for the disk friction of Fig. 6/9 where the pressure, respectively, is constant and varies inversely with $r$. For both conditions, express the frictional moments in dimensionless form as $M' = M/(\mu P R_o)$ for given values of $\mu$, $P$, and $R_o$ and plot $M'$ as functions of $R_i/R_o$ from 0 (solid disk) to 1 (infinitely thin ring).

**6/79** In the figure is shown a multiple-disk clutch for marine use. The driving disks $A$ are splined to the driving shaft $B$ so that they are free to slip along the shaft but must rotate with it. The disks $C$ drive the housing $D$ by means of the bolts $E$, along which they are free to slide. In the clutch shown there are five pairs of friction surfaces. Assume the pressure is uniformly distributed over the area of the disks and determine the maximum torque $M$ which can be transmitted if the coefficient of friction is 0.15 and $P = 500$ N. *Ans.* $M = 335$ N·m

**6/80** Calculate the torque $M$ required to rotate the 540-lb reel of telephone cable clockwise against the 400-lb tension in the cable. The diameter of the bearing is 2.50 in., and the coefficient of friction for the bearing is 0.30.

**6/81** The pulley system shown is used to hoist the 200-kg block. The diameter of the bearing for the upper pulley is 20 mm, and that for the lower pulley is 12 mm. For a coefficient of friction $\mu = 0.25$ for both bearings, calculate the tensions $T$, $T_1$, and $T_2$ in the three cables if the block is being raised slowly. *Ans.* $T = 1069$ N, $T_1 = 1013$ N, $T_2 = 949$ N

**6/82** Calculate the tensions $T$, $T_1$, and $T_2$ for Prob. 6/81 if the block is being lowered slowly.

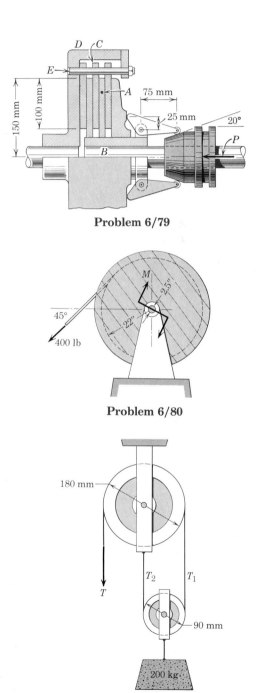

**Problem 6/79**

**Problem 6/80**

**Problem 6/81**

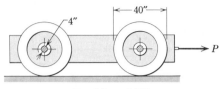

Problem 6/83

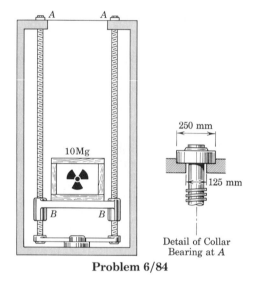

250 mm

125 mm

Detail of Collar
Bearing at $A$

Problem 6/84

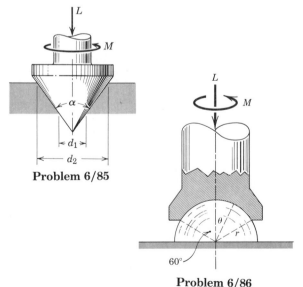

Problem 6/85

Problem 6/86

**6/83** Each of the four wheels of the vehicle weighs 40 lb and is mounted on a 4-in.-diameter journal (shaft). The total weight of the vehicle is 960 lb, including wheels, and is distributed equally on all four wheels. If a force $P = 16$ lb is required to keep the vehicle rolling at a constant low speed on a horizontal surface, calculate the coefficient of friction which exists in the wheel bearings. (*Hint:* Draw a complete free-body diagram of one wheel.)    *Ans.* $\mu = 0.204$

**6/84** The 10-Mg crate is lowered into an underground storage facility on a two-screw elevator as shown. Each screw has a mass of 0.9 Mg, is 120 mm in mean diameter, and has a single square thread with a lead of 11 mm. The screws are turned in synchronism by a motor unit in the base of the facility. The entire mass of the crate, screws, and 3-Mg elevator platform is supported equally by flat collar bearings at $A$, each of which has an outside diameter of 250 mm and an inside diameter of 125 mm. The pressure on the bearings is assumed to be uniform over the bearing surface. If the coefficient of friction for the collar bearing and the screws at $B$ is 0.15, calculate the torque $M$ which must be applied to each screw (*a*) to raise the elevator and (*b*) to lower the elevator.

**6/85** Determine the expression for the torque $M$ required to turn the shaft whose thrust $L$ is supported by a conical pivot bearing. The coefficient of friction is $\mu$, and the bearing pressure is constant.

$$Ans.\ M = \frac{\mu L}{3 \sin \frac{\alpha}{2}} \frac{d_2{}^3 - d_1{}^3}{d_2{}^2 - d_1{}^2}$$

**6/86** A thrust bearing for a shaft under an axial load $L$ consists of a partial spherical cup of radius $r$. If the pressure between the bearing surfaces at any point varies according to $p = p_0 \cos \theta$, derive the expression for the torque $M$ required to maintain constant rotational speed. The coefficient of friction is $\mu$.

## 6/8 FLEXIBLE BELTS

The impending slippage of flexible cables, belts, and ropes over sheaves and drums is of importance in the design of belt drives of all types, band brakes, and hoisting rigs. In Fig. 6/11a is shown a drum subjected to the two belt tensions $T_1$ and $T_2$, the torque $M$ necessary to prevent rotation, and a bearing reaction $R$. With $M$ in the direction shown, $T_2$ is greater than $T_1$. The free-body diagram of an element of the belt of length $r\,d\theta$ is shown in the b-part of the figure. We proceed with the force analysis of this element in a manner similar to that which we have illustrated for other variable-force problems where the equilibrium of a differential element is established. The tension increases from $T$ at the angle $\theta$ to $T + dT$ at the angle $\theta + d\theta$. The normal force is a differential $dN$, since it acts on a differential element of area. Likewise the friction force, which must act on the belt in a direction to oppose slipping, is a differential and is $\mu\,dN$ for impending motion. Equilibrium in the $t$-direction gives

$$T \cos \frac{d\theta}{2} + \mu\,dN = (T + dT) \cos \frac{d\theta}{2}$$

or
$$\mu\,dN = dT$$

since the cosine of a differential quantity is unity in the limit. Equilibrium in the $n$-direction requires that

$$dN = (T + dT) \sin \frac{d\theta}{2} + T \sin \frac{d\theta}{2}$$

or
$$dN = T\,d\theta$$

In this reduction we recall that the sine of a differential angle in the limit equals the angle and that the product of two differentials must be neglected in the limit compared with the first-order differentials remaining. Combining the two equilibrium relations gives

$$\frac{dT}{T} = \mu\,d\theta$$

Integrating between corresponding limits yields

$$\int_{T_1}^{T_2} \frac{dT}{T} = \int_0^\beta \mu\,d\theta$$

or
$$\ln \frac{T_2}{T_1} = \mu\beta$$

where the $\ln (T_2/T_1)$ is a natural logarithm (base $e$). Solving for $T_2$ gives

$$T_2 = T_1 e^{\mu\beta} \qquad\qquad (6/7)$$

We note that $\beta$ is the total angle of belt contact and must be expressed in radians. If a rope were wrapped around a drum $n$ times, the angle

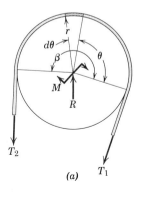

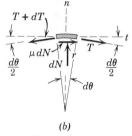

(b)

**Figure 6/11**

$\beta$ would be $2\pi n$ radians. Equation 6/7 holds equally well for a noncircular section where the total angle of contact is $\beta$. This conclusion is evident from the fact that the radius $r$ of the circular drum of Fig. 6/11 does not enter into the equations for the equilibrium of the differential element of the belt.

The relation expressed by Eq. 6/7 also applies to belt drives where both the belt and the pulley are rotating at constant speed. In this case the equation describes the ratio of belt tensions for slipping or impending slipping. When the speed of rotation becomes large, there is a tendency for the belt to leave the rim, so that Eq. 6/7 will involve some error.

### 6/9  ROLLING RESISTANCE

Deformation at the point of contact between a rolling wheel and its supporting surface introduces a resistance to rolling which we will mention only briefly. This resistance is not due to tangential friction forces and therefore is an entirely different phenomenon from that of dry friction.

To describe rolling resistance, we consider the wheel of Fig. 6/12 under the action of a load $L$ on the axle and a force $P$ applied at its center to produce rolling. The deformation of the wheel and supporting surfaces as shown is greatly exaggerated. The distribution of pressure $p$ over the area of contact is similar to that indicated, and the resultant $R$ of this distribution will act at some point $A$ and will pass through the center of the wheel for equilibrium. We find the force $P$ necessary to initiate and maintain rolling at constant velocity by equating the moments of all forces about $A$ to zero. This gives us

$$P = \frac{a}{r} L = \mu_r L$$

where the moment arm of $P$ is taken to be $r$. The ratio $\mu_r = a/r$ is referred to as the coefficient of rolling resistance. The coefficient as defined is the ratio of resisting force to normal force and in this respect is analogous to the coefficient of static or kinetic friction. On the other hand, there is no slipping or impending slipping in the interpretation of $\mu_r$.

The dimension $a$ depends on many factors which are difficult to quantify, so that a comprehensive theory of rolling resistance is not available. The distance $a$ is a function of the elastic and plastic properties of the mating materials, the radius of the wheel, the speed of travel, and the roughness of the surfaces. Some tests indicate only a small variation with wheel radius, and $a$ is often taken to be independent of the rolling radius. Unfortunately, the quantity $a$ has also been referred to as the coefficient of rolling friction in some references. However, $a$ has the dimension of length and therefore is not a dimensionless coefficient in the usual sense.

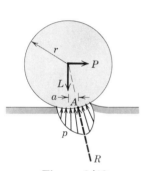

**Figure 6/12**

## Sample Problem 6/9

A flexible cable which supports the 100-kg load is passed over a fixed circular drum and subjected to a force $P$ to maintain equilibrium. The coefficient of static friction $\mu$ between the cable and the fixed drum is 0.30. (*a*) For $\alpha = 0$, determine the maximum and minimum values which $P$ may have in order not to raise or lower the load. (*b*) For $P = 500$ N, determine the minimum value which the angle $\alpha$ may have before the load begins to slip.

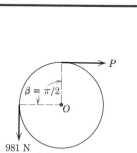

**Solution.** Impending slipping of the cable over the fixed drum is given by Eq. 6/7, which is $T_2/T_1 = e^{\mu\beta}$.

① (*a*) With $\alpha = 0$ the angle of contact is $\beta = \pi/2$ rad. For impending upward motion of the load, $T_2 = P_{max}$, $T_1 = 981$ N, and we have

② $$P_{max}/981 = e^{0.30(\pi/2)} \qquad P_{max} = 981(1.602) = 1572 \text{ N} \qquad Ans.$$

For impending downward motion of the load, $T_2 = 981$ N, $T_1 = P_{min}$. Thus,

$$981/P_{min} = e^{0.30(\pi/2)} \qquad P_{min} = 981/1.602 = 612 \text{ N} \qquad Ans.$$

(*b*) With $T_2 = 981$ N and $T_1 = P = 500$ N, Eq. 6/7 gives us

$$981/500 = e^{0.30\beta} \qquad 0.30\beta = \ln(981/500) = 0.674$$

$$\beta = 2.247 \text{ rad} \qquad or \qquad \beta = 2.247\left(\frac{360}{2\pi}\right) = 128.7°$$

③ $$\alpha = 128.7° - 90° = 38.7°$$

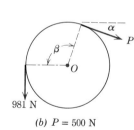

(*a*) $\alpha = 0$

(*b*) $P = 500$ N

① We are careful to note that $\beta$ must be expressed in radians.

② In our derivation of Eq. 6/7 be certain to note that $T_2 > T_1$.

③ As was noted in the derivation of Eq. 6/7, the radius of the drum does not enter into the calculations. It is only the angle of contact and the coefficient of friction that determine the limiting conditions for impending motion of the flexible cable over the curved surface.

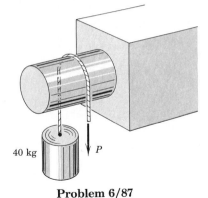

**Problem 6/87**

**Problem 6/88**

**Problem 6/89**

# PROBLEMS

## Introductory problems

**6/87** Determine the force $P$ required to $(a)$ raise and $(b)$ lower the 40-kg cylinder at a slow steady speed. The coefficient of friction between the cord and its supporting surface is 0.30.

*Ans.* $(a)$ $P = 1007$ N, $(b)$ $P = 152.9$ N

**6/88** A force $P = mg/6$ is required to lower the cylinder at a constant slow speed with the cord making $1\frac{1}{4}$ turns around the fixed shaft. Calculate the coefficient of friction $\mu$ between the cord and the shaft.

**6/89** A dockworker adjusts a spring line (rope) which keeps a ship from drifting alongside a wharf. If he exerts a pull of 200 N on the rope, which has $1\frac{1}{4}$ turns around the mooring bit, what force $T$ can he support? The coefficient of friction between the rope and the cast-steel mooring bit is 0.30.

*Ans.* $T = 2.11$ kN

**6/90** If the dockworker of Prob. 6/89 is to support a tension of 18 kN in the rope leading to the ship, how many turns around the mooring bit are necessary if he exerts a pull of 240 N on the free end? The coefficient of friction between the rope and the bit is 0.30.

**6/91** In western movies, cowboys are frequently observed hitching their horses by casually winding a few turns of the reins around a horizontal pole and letting the end hang free as shown—no knots! If the freely hanging length of rein weighs 2 oz and the number of turns is as shown, what tension $T$ does the horse have to produce in the direction shown in order to gain freedom? The coefficient of friction between the reins and wooden pole is 0.70.    *Ans.* $T = 1720$ lb

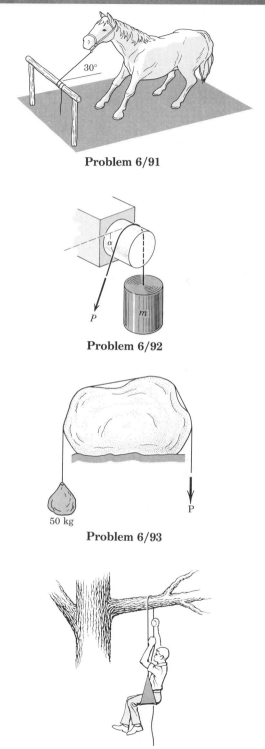

**Problem 6/91**

### Representative problems

**6/92** For a certain coefficient of friction $\mu$ and a certain angle $\alpha$, the force $P$ required to raise $m$ is 4 kN and that required to lower $m$ at a constant slow speed is 1.6 kN. Calculate the mass $m$.

**Problem 6/92**

**6/93** A 50-kg package is attached to a rope which passes over an irregularly shaped boulder with uniform surface texture. If a downward force $P = 70$ N is required to lower the package at a constant rate, (*a*) determine the coefficient of friction $\mu$ between the rope and the boulder. (*b*) What force $P'$ would be required to raise the package at a constant rate?
    *Ans.* (*a*) $\mu = 0.620$, (*b*) $P' = 3.44$ kN

**Problem 6/93**

**6/94** The 180-lb tree surgeon lowers himself with the rope over a horizontal limb of the tree. If the coefficient of friction between the rope and the limb is 0.60, compute the force which the man must exert on the rope to let himself down slowly.

**Problem 6/94**

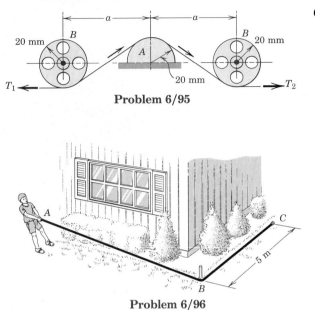

20 mm

*a*    *a*

B    B

A    20 mm

20 mm

$T_1$    $T_2$

**Problem 6/95**

**6/95** Magnetic tape passes around the light idler pulleys *B* and over the fixed circular recording head *A* with a constant speed. The tape tension is unchanged as it passes around the idler pulleys. Calculate the minimum spacing *a* for which the ratio of the tensions $T_1$ and $T_2$ will not exceed 1.15. The coefficient of friction between the tape and the head is 0.10.

*Ans. a = 62.2 mm*

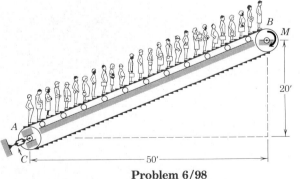

A

C

5 m

B

**Problem 6/96**

**6/96** A garden hose with a mass of 1.2 kg/m is in full contact with the ground from *B* to *C*. What is the horizontal component $P_x$ of the force which the gardener must exert in order to pull the hose around the small cylindrical guard at *B*? The coefficient of friction between the hose and ground is 0.50 and that between the hose and the cylinder is 0.40. Assume that the hose does not touch the ground between *A* and *B*.

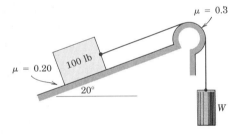

$\mu = 0.3$

$\mu = 0.20$    100 lb

20°

W

**Problem 6/97**

**6/97** Determine the range of cylinder weights *W* for which the system is in equilibrium. The coefficient of friction between the 100-lb block and the incline is 0.20 and that between the cord and cylindrical support surface is 0.30.    *Ans. $8.66 \le W \le 94.3$ lb*

B

M

20'

A

C

50'

**Problem 6/98**

**6/98** The endless belt of an escalator passes around idler drum *A* and is driven by a torque *M* applied to drum *B*. Belt tension is adjusted by a turnbuckle at *C*, which produces an initial tension of 1000 lb in each side of the belt when the escalator is unloaded. Calculate the minimum coefficient of friction $\mu$ between drum *B* and the belt to prevent slipping if the escalator handles 30 people uniformly distributed along the belt and averaging 150 lb each. (*Note:* It can be shown that the increase in belt tension on the upper side of drum *B* and the decrease in belt tension at the lower drum *A* are each equal to half the component of the combined passenger weight along the incline.)

**6/99** Calculate the horizontal force $P$ required to raise the 100-kg load. The coefficient of friction between the rope and the fixed bars is 0.40.   *Ans. $P = 3.30$ kN*

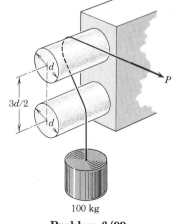

100 kg

**Problem 6/99**

**6/100** The uniform 3-m beam is suspended by the cable that passes over the large pulley. A locking pin at $A$ prevents rotation of the pulley. If the coefficient of friction between the cable and the pulley is 0.25, determine the minimum value of $x$ for which the cable will not slip on the pulley.

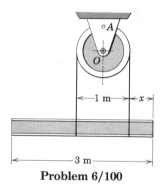

**Problem 6/100**

**6/101** A device for lowering a person in a sling down a rope at a constant controlled rate is shown in the figure. The rope passes around a central shaft fixed to the frame and leads freely out of the lower collar. The number of turns is adjusted by turning the lower collar, which winds or unwinds the rope around the shaft. Entrance of the rope into the upper collar at $A$ is equivalent to $\frac{1}{4}$ of a turn, and passage around the corner at $B$ is also equivalent to $\frac{1}{4}$ of a turn. Friction of the rope through the straight portions of the collars averages 10 N for each collar. If three complete turns around the shaft, in addition to the corner turns, are required for a 75-kg man to lower himself at a constant rate without exerting a pull on the free end of the rope, calculate the coefficient of friction $\mu$ between the rope and the contact surfaces of the device. Neglect the small helix angle of the rope around the shaft.   *Ans. $\mu = 0.195$*

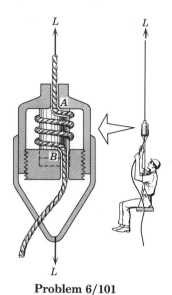

**Problem 6/101**

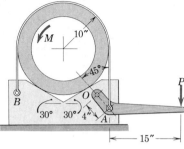

**Problem 6/102**

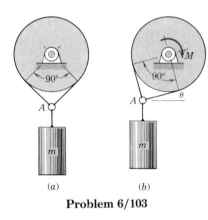

(a)          (b)

**Problem 6/103**

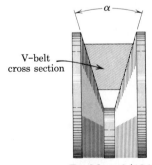

V-belt
cross section

**Problem 6/104**

**6/102** Find the couple $M$ required to turn the pipe in the V-block against the action of the flexible band. A force $P = 25$ lb is applied to the lever which is pivoted about $O$. The coefficient of friction between the band and the pipe is 0.30, and that between the pipe and the block is 0.40. The weights of the parts are negligible.

**6/103** The cylinder of mass $m$ is attached to the ring $A$, which is suspended by the cable that passes over the pulley, as shown in the $a$-part of the figure. A couple $M$ applied to the pulley turns it until slipping of the cable on the pulley occurs at the position $\theta = 20°$, shown in the $b$-part of the figure. Calculate the coefficient of friction $\mu$ between the cable and the pulley. *Ans.* $\mu = 0.214$

**6/104** Replace the flat belt and pulley of Fig. 6/11 by a V-belt and matching grooved pulley as indicated by the cross-sectional view accompanying this problem. Derive the relation among the belt tensions, the angle of contact, and the coefficient of friction for the V-belt when slipping impends. Use of a V-belt with $\alpha = 35°$ would be equivalent to increasing the coefficient of friction for a flat belt of the same material by what factor $n$?

**6/105** Shown in the figure is a band-type oil filter wrench. If the coefficient of friction between the band and the fixed filter is 0.25, determine the minimum value of $h$ which ensures that the wrench will not slip on the filter, regardless of the magnitude of the force $P$. Neglect the mass of the wrench and assume that the effect of the small part at $A$ is equivalent to that of a band wrap which begins at the three-o'clock position and runs clockwise. *Ans.* $h = 27.8$ mm

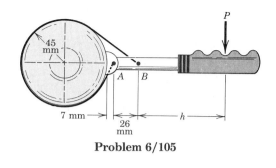

**Problem 6/105**

**6/106** The chain has a mass $\rho$ per unit length. Determine the overhang $h$ below the fixed cylindrical guide for which the chain will be on the verge of slipping. The coefficient of friction is $\mu$. (*Hint:* The resulting differential equation involving the variable chain tension $T$ at the corresponding angle $\theta$ is of the form $dT/d\theta + KT = f(\theta)$, a first-order, linear, nonhomogeneous equation with constant coefficient. The solution is

$$T = Ce^{-K\theta} + e^{-K\theta} \int e^{K\theta} f(\theta) \, d\theta$$

where $C$ and $K$ are constants.)

$$Ans. \ h = \frac{2\mu r}{1 + \mu^2} (1 + e^{\mu\pi})$$

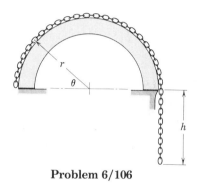

**Problem 6/106**

## *6/10  PROBLEM FORMULATION AND REVIEW*

In our study of friction we have concentrated our attention on dry or Coulomb friction where a simple mechanical model of surface irregularities between the contacting bodies, Fig. 6/1, suffices to explain the phenomenon adequately for most engineering purposes. By having this model clearly in mind, we can easily visualize the three types of dry-friction problems which are encountered. These categories are:

1. Static friction less than the maximum possible value and determined by the equations of equilibrium. (This usually requires a check to see that $F < \mu_s N$.)

2. Limiting static friction with impending motion $(F = \mu_s N)$.

3. Kinetic friction where sliding motion occurs between contacting surfaces $(F = \mu_k N)$.

A coefficient of friction applies to a given pair of mating surfaces. It is meaningless to speak of a coefficient of friction for a single surface. The static coefficient of friction $\mu_s$ for a given pair of surfaces is usually slightly greater than the kinetic coefficient $\mu_k$. The friction force which acts on a body is always in the direction to oppose the slipping of the body which takes place or the slipping which would take place in the absence of friction.

When we encounter friction forces distributed in some prescribed manner over a surface or along a line, we select a representative element of the surface or line and evaluate the force and moment effects of the elemental friction force acting on the element. We then integrate these effects over the entire surface or line.

Friction coefficients are subject to considerable variation, depending on the exact condition of the mating surfaces. Computation of coefficients of friction to three significant figures represents an accuracy which cannot easily be duplicated by experiment, and when cited, such values are included for purposes of computational check only. For design computations in engineering practice the use of a handbook value for a coefficient of static of kinetic friction must be viewed only as an approximation.

In reviewing the foregoing introduction to frictional problems, the reader should bear in mind the existence of the other forms of friction mentioned in the introductory article of the chapter. Problems which involve fluid friction, for example, are among the most important of the friction problems encountered in engineering, and a study of this phenomenon is included in the subject of fluid mechanics.

# REVIEW PROBLEMS

**6/107** A 100-lb block is placed on a 30° incline and released from rest. The coefficient of static friction between the block and the incline is 0.30. (*a*) Determine the maximum and minimum values of the initial tension $T$ in the spring for which the block will not slip when released. (*b*) Calculate the friction force $F$ on the block if $T$ = 40 lb.

> Ans. (*a*) $T_{max}$ = 76.0 lb, $T_{min}$ = 24.0 lb
>
> (*b*) $F$ = 10 lb

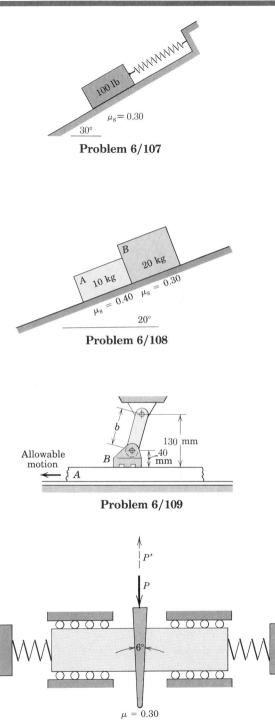

**Problem 6/107**

**6/108** Two boxes are placed on an incline in contact with each other and released from rest. The coefficients of static friction are 0.40 and 0.30 under boxes $A$ and $B$, respectively. Describe what happens.

**Problem 6/108**

**6/109** A frictional locking device allows bar $A$ to move to the left but prevents movement to the right. If the coefficient of friction between the shoe $B$ and the bar $A$ is 0.40, specify the maximum length $b$ of the link which will permit the device to work as described.

> Ans. $b$ = 96.9 mm

**Problem 6/109**

**6/110** The 6° wedge is used to separate the spring-loaded blocks. For a coefficient of friction of 0.30 between the wedge and the blocks, determine the force $P$ required to produce a 300-lb compression in each spring. After force $P$ is removed, find the force $P'$ required to extract the wedge.

**Problem 6/110**

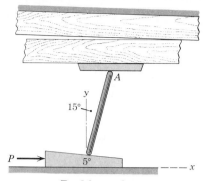

**Problem 6/111**

**6/111** The toggle-wedge is an effective device to close the gap between two planks during construction of a wooden boat. For the combination shown, if a force $P$ of 300 lb is required to move the wedge, determine the friction force $F$ acting on the upper end $A$ of the toggle. The coefficients of static and kinetic friction for all pairs of mating surfaces are taken to be 0.40.                         *Ans.* $F = 120.3$ lb

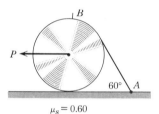

**Problem 6/112**

**6/112** The circular cylinder weighs 50 lb and is held by a cord fixed to its periphery at $B$ and to the ground at $A$. If the coefficient of static friction is 0.60, calculate the force $P$ required to cause the cylinder to slip.

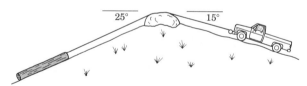

**Problem 6/113**

**6/113** A 500-kg log is being steadily pulled up the incline by means of the cable attached to the winch on the truck. If the coefficient of kinetic friction is 0.80 between the log and the incline and 0.50 between the cable and rock, determine the tension $T$ which must be developed by the winch.

*Ans.* $T = 7980$ N

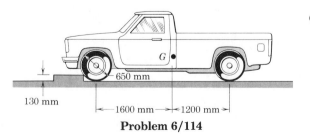

**Problem 6/114**

**6/114** Calculate the torque $M$ which the engine of the pickup truck must supply to the rear axle to roll the front wheels over the curbing from a rest position if the rear wheels do not slip. Determine the minimum effective coefficient of friction at the rear wheels to prevent slipping. The mass of the loaded truck with mass center at $G$ is 1900 kg.

**6/115** A compressive force of 600 N is to be applied to the two boards in the grip of the C-clamp. The threaded screw has a mean diameter of 10 mm and advances 2.5 mm per turn. The coefficient of static friction is 0.20. Determine the force $F$ which must be applied normal to the handle at $C$ in order to ($a$) tighten and ($b$) loosen the clamp. Neglect friction at point $A$.

*Ans.* ($a$) $F = 8.52$ N, ($b$) $F = 3.56$ N

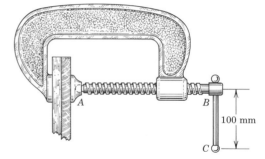

**Problem 6/115**

**6/116** A 1600-kg car with 3000-mm wheelbase has a center of mass 600 mm from the road and midway between the front and rear axles. If the effective coefficient of friction between the tires and the road is 0.80, find the angle $\theta$ with the horizontal made by the steepest grade that the car can climb at constant speed before the rear driving wheels slip. What torque $M$ is applied to each of the 660-mm-diameter rear wheels by the engine under these conditions? Neglect any friction under the front wheels.

**6/117** The wedge-mass system of Prob. 6/54 is repeated here where the mass of the wedge is negligible. If the coefficient of friction for all contacting surfaces is 0.20, determine the magnitude and direction of the horizontal force $P$ required to lower the 200-kg mass. *Ans.* $P = 266$ N to the right

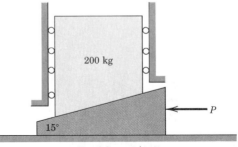

**Problem 6/117**

**6/118** The screw of the small press has a mean diameter of 25 mm and has a double square thread with a lead of 8 mm. The flat thrust bearing at $A$ is shown in the enlarged view and has surfaces which are well worn. If the coefficient of friction for both the threads and the bearing at $A$ is 0.25, calculate the torque $M$ on the handwheel required ($a$) to produce a compressive force of 4 kN and ($b$) to loosen the press from the 4-kN compression.

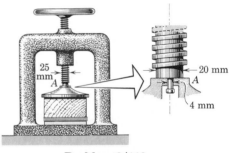

**Problem 6/118**

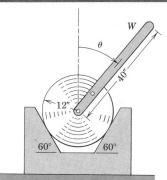

**Problem 6/119**

**6/119** The cylinder weighs 80 lb and the attached uniform slender bar has an unknown weight $W$. The unit remains in static equilibrium for values of the angle $\theta$ ranging up to 45° but slips if $\theta$ exceeds 45°. If the coefficient of static friction is known to be 0.30, determine $W$. *Ans.* $W = 70.1$ lb

**6/120** For what range of weights $W$ is the system in static equilibrium? The coefficient of static friction is $\mu_s = 0.20$ everywhere.

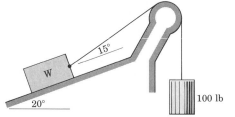

**Problem 6/120**

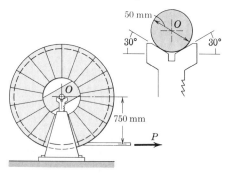

**Problem 6/121**

**6/121** The reel of telephone cable has a mass of 3 Mg and is supported on its shaft in the V-notched blocks on both sides of the reel. The reel is raised off the ground by jacking up the supports so that cable may be pulled off in the horizontal direction as shown. The shaft is fastened to the reel and turns with it. If the coefficient of friction between the shaft and the V-surfaces is 0.30, calculate the pull $P$ in the cable required to turn the reel. *Ans.* $P = 313$ N

**6/122** The small roller of the uniform slender rod rests against the vertical surface at $A$ while the rounded end at $B$ rests on the platform which is slowly pivoted downward beginning from the horizontal position shown. If the bar begins to slip when $\theta = 25°$, determine the coefficient of static friction $\mu_s$ between the bar and the platform. Neglect friction in the roller and the small thickness of the platform.

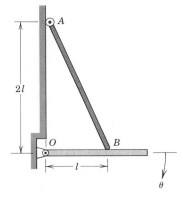

**Problem 6/122**

**6/123**  Under the action of the applied couple $M$ the 25-kg cylinder bears against the roller $A$, which is free to turn. If the coefficients of static and kinetic friction between the cylinder and the horizontal surface are 0.50 and 0.40, respectively, determine the friction force $F$ acting on the cylinder if $(a)$ $M = 20$ N·m and $(b)$ $M = 40$ N·m.

*Ans.* $(a)$ $F = 133.3$ N, $(b)$ $F = 127.6$ N

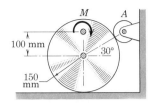

**Problem 6/123**

**6/124**  The truck unloads its cargo box by sliding it off the elevated rack, as the truck rolls slowly forward with its brakes applied for control. The box has a total weight of 10,000 lb with center of mass at $G$ in the center of the box. The coefficient of static friction between the box and the rack is 0.30. Calculate the braking force $F$ between the tires and the level road as the box is on the verge of slipping down the rack from the position shown and the truck is on the verge of rolling forward. No slipping occurs at the lower corner of the box.

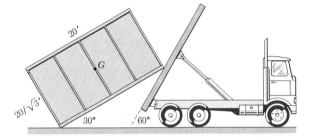

**Problem 6/124**

**6/125**  The two brake shoes and their lining pivot about the points $O$ and are expanded against the brake drum through the action of the hydraulic cylinder $C$. The pressure $p$ between the drum and the lining may be shown to vary directly as the sine of the angle $\theta$ measured from the pin $O$ for each shoe and has a value $p_0$ at $\theta = \beta$. The width of the lining in contact with the drum is $b$. Write the expression for the braking torque $M_f$ on the wheel if the coefficient of friction between the drum and the lining is $\mu$.

*Ans.* $M_f = 2\mu b r^2 p_0 \left(\dfrac{1 - \cos \beta}{\sin \beta}\right)$

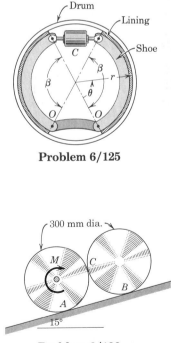

**Problem 6/125**

**6/126**  Calculate the couple $M$ applied to the lower of the two 20-kg cylinders which will allow them to roll slowly *down* the incline. The coefficients of static and kinetic friction for all contacting surfaces are $\mu_s = 0.60$ and $\mu_k = 0.50$.

**Problem 6/126**

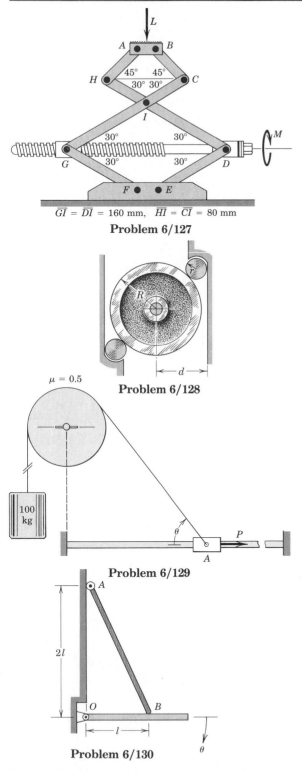

$\overline{GI} = \overline{DI} = 160$ mm,   $\overline{HI} = \overline{CI} = 80$ mm

**Problem 6/127**

**Problem 6/128**

$\mu = 0.5$

**Problem 6/129**

**Problem 6/130**

▶**6/127** A scissors-type jack has a single square thread which engages the threaded collar $G$ and turns in a ball thrust bearing at $D$. The thread has a mean diameter of 10 mm and a lead (advancement per revolution) of 3 mm. With a coefficient of friction of 0.20 for the greased threads, (*a*) calculate the torque $M$ on the screw required to raise a load of 1000 kg from the position shown and (*b*) calculate the torque $M$ required to lower the load from the same position. Note that links $AH$, $BC$, $FG$, and $ED$ have meshed gears integral with the links at points $A$, $B$, $E$, and $F$ which prevent rotation of the top and bottom platforms. Neglect friction in the bearing at $D$.

*Ans.* (*a*) $M = 35.7$ N·m, (*b*) $M = 12.15$ N·m

▶**6/128** The device shown prevents clockwise rotation in the horizontal plane of the central wheel by means of frictional locking of the two small rollers. For given values of $R$ and $r$ and for a common coefficient of friction $\mu$ at all contact surfaces, determine the range of values of $d$ for which the device will operate as described.   *Ans.* $\dfrac{2r + (1 - \mu^2)R}{1 + \mu^2} < d < (R + 2r)$

### *Computer-oriented problems*

\***6/129** The 100-kg load is elevated by the cable which slides over the fixed drum with a coefficient of friction of 0.50. The cable is secured to the slider $A$ which is pulled slowly along its smooth horizontal guide bar by the force $P$. Plot $P$ as a function of $\theta$ from $\theta = 90°$ to $\theta = 10°$ and determine its maximum value along with the corresponding angle $\theta$. Check your plotted value of $P_{\text{max}}$ analytically.

*Ans.* $P_{\text{max}} = 2430$ N, $\theta = 26.6°$

\***6/130** The system of Prob. 6/122 is repeated here. For a coefficient of static friction $\mu_s = 0.50$, determine the angle $\theta$ of the platform at which slipping will occur. Neglect friction in the roller and the small thickness of the platform.

**6/131** The uniform slender pole rests against a small roller at $B$. End $A$ will not slip on the horizontal surface if the coefficient of static friction $\mu_s$ is sufficiently large. (*a*) Determine the required minimum value of $\mu_s$ to prevent slipping for any value of $\theta$ from $\theta = 0$ to $\theta = 60°$ and plot $\mu_s$ versus $\theta$. From these results find the range of $\theta$ for which the pole will be unstable if $\mu_s = 0.4$. (*b*) At what angle $\theta$ is the pole most unstable and what is the least coefficient of static friction $\mu_s$ which would be required to prevent slipping for this angle?

*Ans.* (*a*) With $\mu_s = 0.4$ unstable for
$$24.0° < \theta < 57.2°$$
(*b*) $\theta = 41.8°$ requires $\mu_s = 0.554$

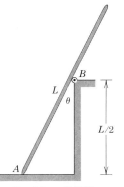

**Problem 6/131**

**6/132** The figure for Prob. 6/29 is shown again here. If $R = 1.2$ m, $L = 2$ m, and the rod has a mass of 3 kg, determine the maximum equilibrium angle $\theta$ which the rod can reach before slipping takes place. The coefficient of static friction between the rod and its support is 0.15.

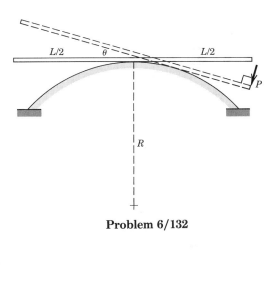

**Problem 6/132**

**6/133** Refer to Prob. 6/132, where the mass of the uniform rod is 3 kg, $L = 2$ m, and $R = 1.2$ m. As the force $P$ is increased, the equilibrium angle $\theta$ also increases, provided no slippage occurs. The limiting value of $\theta$ is reached when $R\theta = L/2$, where $\theta$ is in radians. For values of $\theta$ between zero and $L/(2R)$, plot the corresponding equilibrium value of $P$, the normal contact force $N$, and the minimum value $\mu_s$ of the static friction coefficient which prevents slipping at that angle. Identify the angle $\theta_{cr}$ at which $\mu_s$ reaches a maximum value.

*Ans.* $P = \dfrac{35.3\theta \cos \theta}{1 - 1.2\theta}$, $N = \dfrac{29.4 \cos \theta}{1 - 1.2\theta}$
$$\mu_s = (1 - 1.2\theta) \tan \theta$$
$$\mu_{s_{max}} = 0.222 \text{ at } \theta = 25.5°$$

**6/134** A heavy cable with a mass of 12 kg per meter of length passes over the two fixed pipes 300 m apart on the same level. One end supports a 1600-kg cylinder. By experiment it is found that a downward force $P$ of 60 kN is required to induce slipping of the cable over both pipes at a constant rate. Determine the coefficient of kinetic friction $\mu_k$ between the cable and the pipes, the maximum tension $T$ in the cable between the pipes, and the sag $h$ in the cable.

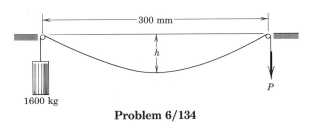

**Problem 6/134**

The analysis of structures that change their configuration under load is generally best handled by applying the principle of virtual work. This aircraft-loading platform is a typical example.

# VIRTUAL WORK

7

## 7/1 INTRODUCTION

In the previous chapters we have analyzed the equilibrium of a body by isolating it with a free-body diagram and writing the zero-force and zero-moment summation equations. For the most part this approach is employed for a body whose equilibrium position is known or specified and where one or more of the external forces is an unknown to be determined.

There is a separate class of problems in which bodies are composed of interconnected members that allow relative motion between the parts, thus permitting various possible equilibrium configurations to be examined. For problems of this type, the force- and moment-equilibrium equations, although valid and adequate, are often not the most direct and convenient approach. Here we find that a method based on the concept of the work done by a force is more useful and direct. Also, the method provides a deeper insight into the behavior of mechanical systems and allows us to examine the question of the stability of systems in equilibrium. We will now develop this approach, called the *method of virtual work*.

## 7/2 WORK

We must first define the term *work*, which is used in a quantitative sense as contrasted to its common nontechnical usage.

*(a) Work of a force.* Consider the constant force $\mathbf{F}$ acting on the body, Fig. 7/1a, whose movement along the plane from $A$ to $A'$ is represented by the vector $\Delta\mathbf{s}$, called the *displacement* of the body. By definition the work $U$ done by the force $\mathbf{F}$ on the body during this displacement is the component of the force in the direction of the displacement times the displacement or

$$U = (F \cos \alpha)\, \Delta s$$

(a)

(b)

**Figure 7/1**

From Fig. 7/1$b$ we see that the same result is obtained if we multiply the magnitude of the force by the component of the displacement in the direction of the force, which is

$$U = F(\Delta s \cos \alpha)$$

Because we obtain the same result regardless of the direction in which we resolve the vectors, we observe immediately that work $U$ is a scalar quantity.

Work is a positive quantity when the working component of the force is in the same direction as the displacement. When the working component is in the direction opposite to the displacement, Fig. 7/2, the work done is negative. Thus,

$$U = (F \cos \alpha)\, \Delta s = -(F \cos \theta)\, \Delta s$$

We now generalize the definition of work to account for conditions under which the direction of the displacement and the magnitude and direction of the force may be variable. Figure 7/3$a$ shows a force $\mathbf{F}$ acting on a body at a point $A$ which moves along the path shown from $A_1$ to $A_2$. Point $A$ is located by its position vector $\mathbf{r}$ measured from some arbitrary but convenient origin $O$, and the infinitesimal displacement in the motion from $A$ to $A'$ is given by the differential change $d\mathbf{r}$ of the position vector. The work done by the force $\mathbf{F}$ during the displacement $d\mathbf{r}$ is defined as

$$\boxed{dU = \mathbf{F} \cdot d\mathbf{r}} \qquad (7/1)$$

If $F$ denotes the magnitude of the force $\mathbf{F}$ and $ds$ denotes the magnitude of the differential displacement $d\mathbf{r}$, then by using the definition of the dot product, we may write

$$dU = F\, ds \cos \alpha$$

Again, we may interpret this expression as the force component $F \cos \alpha$ in the direction of the displacement times the displacement, or as the displacement component $ds \cos \alpha$ in the direction of the force times the force, as represented in Fig. 7/3$b$. If we express $\mathbf{F}$ and $d\mathbf{r}$ in terms of their rectangular components, we have

$$dU = (\mathbf{i}F_x + \mathbf{j}F_y + \mathbf{k}F_z) \cdot (\mathbf{i}\, dx + \mathbf{j}\, dy + \mathbf{k}\, dz)$$

$$= F_x\, dx + F_y\, dy + F_z\, dz$$

To obtain the total work $U$ done by $\mathbf{F}$ during a finite movement of point $A$ from $A_1$ to $A_2$, Fig. 7/3$a$, we must integrate $dU$ between these positions. Thus,

$$U = \int \mathbf{F} \cdot d\mathbf{r} = \int (F_x\, dx + F_y\, dy + F_z\, dz)$$

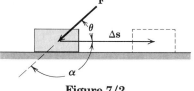

**Figure 7/2**

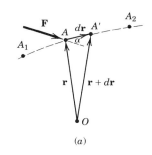

(a)

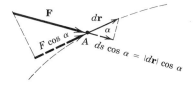

(b)

**Figure 7/3**

or

$$U = \int F \cos \alpha \, ds$$

To carry out this integration, we must know the relation between the force components and their respective coordinates or the relations between $F$ and $s$ and between $\cos \alpha$ and $s$.

In the case of concurrent forces that are applied at any particular point on a body, the work done by their resultant equals the total work done by the several forces. This we conclude from the fact that the component of the resultant in the direction of the displacement equals the sum of the components of the several forces in the same direction.

*(b) Work of a couple.* In addition to the work done by forces, couples also may do work. In Fig. 7/4a we have a couple $M$ acting on a body that changes its angular position by an amount $d\theta$. The work done by the couple is easily determined from the combined work of the two forces which constitute the couple. In the $b$-part of the figure we represent the couple by two equal and opposite forces $\mathbf{F}$ and $-\mathbf{F}$ acting at two arbitrary points $A$ and $B$ such that $F = M/b$. During the infinitesimal movement in the plane of the figure, line $AB$ moves to $A''B'$. We now take the displacement of $A$ in two steps, first, a displacement $d\mathbf{r}_B$ equal to that of $B$ and, second, a displacement $d\mathbf{r}_{A/B}$ (read as the displacement of $A$ with respect to $B$) due to the rotation about $B$. We see that the work done by $\mathbf{F}$ during the displacement from $A$ to $A'$ is equal and opposite to that due to $-\mathbf{F}$ acting through the equal displacement from $B$ to $B'$. Thus, we conclude that no work is done by a couple during a translation (movement without rotation). During the rotation, however, $\mathbf{F}$ does work equal to $\mathbf{F} \cdot d\mathbf{r}_{A/B} = Fb \, d\theta$, where $dr_{A/B} = b \, d\theta$ and where $d\theta$ is the infinitesimal angle of rotation in radians. Since $M = Fb$, we have

$$\boxed{dU = M \, d\theta} \qquad\qquad (7/2)$$

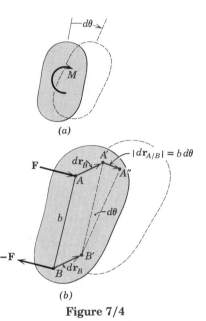

(a)

(b)

**Figure 7/4**

The work of the couple is positive if $M$ has the same sense as $d\theta$ (clockwise in this illustration) and negative if $M$ has a sense opposite to that of the rotation. The total work of a couple during a finite rotation in its plane becomes

$$U = \int M \, d\theta$$

Work has the dimensions of (force) × (distance). In SI units the unit of work is the joule (J), which is the work done by a force of one newton moving through a distance of one meter in the direction of the force (J = N·m). In the U.S. customary system the unit of work is the foot-pound (ft-lb), which is the work done by a one-pound force moving through a distance of one foot in the direction of the force. Dimensionally, the work of a force and the moment of a force are the same although they are entirely different physical quantities. We observe carefully that work is a scalar given by the dot product and involves the product of a force and a distance, both measured along the same line. Moment, on the other hand, is a vector given by the cross product and involves the product of force and distance measured at right angles to the force. To distinguish between these two quantities when we write their units, in SI units we shall use the joule (J) for work and reserve the combined units newton-meter (N·m) for moment. In the U.S. customary system we normally use the sequence foot-pound (ft-lb) for work and pound-foot (lb-ft) for moment.

*(c) Virtual work.* We consider now a particle whose static equilibrium position is determined by the forces which act on it. Any assumed and arbitrary small displacement $\delta\mathbf{r}$ away from this natural position and consistent with the system constraints is called a *virtual displacement*. The term *virtual* is used to indicate that the displacement does not exist in reality but is only assumed so that we may compare various possible equilibrium positions in the process of selecting the correct one. The work done by any force $\mathbf{F}$ acting on the particle during the virtual displacement $\delta\mathbf{r}$ is called *virtual work* and is

$$\delta U = \mathbf{F} \cdot \delta\mathbf{r} \qquad \text{or} \qquad \delta U = F\, \delta s \cos\alpha$$

where $\alpha$ is the angle between $\mathbf{F}$ and $\delta\mathbf{r}$ and $\delta s$ is the magnitude of $\delta\mathbf{r}$. The difference between $d\mathbf{r}$ and $\delta\mathbf{r}$ is that $d\mathbf{r}$ refers to an infinitesimal change in an actual movement and can be integrated, whereas $\delta\mathbf{r}$ refers to an infinitesimal virtual or assumed movement and cannot be integrated. Mathematically both quantities are first-order differentials.

A virtual displacement may also be a rotation $\delta\theta$ of a body. By Eq. 7/2 the virtual work done by a couple $M$ during a virtual angular displacement $\delta\theta$ is, then, $\delta U = M\, \delta\theta$.

We may regard the force $\mathbf{F}$ or couple $M$ as remaining constant during any infinitesimal virtual displacement. If we account for any change in $\mathbf{F}$ or $M$ during the infinitesimal motion, higher-order terms will result which disappear in the limit. This consideration is the same mathematically as that which permits us to neglect the product $dx\, dy$ when writing $dA = y\, dx$ for the element of area under the curve $y = f(x)$.

## 7/3 EQUILIBRIUM

We now express the equilibrium conditions in terms of virtual work, first for a particle, second for a single rigid body, and third for a system of connected rigid bodies.

***(a) Particle.*** Consider the particle or small body in Fig. 7/5 which finds its equilibrium position as a result of the forces in the springs to which it is attached. If the mass of the particle is significant, then the weight $mg$ would also be included as one of the forces. For an assumed virtual displacement $\delta\mathbf{r}$ of the particle away from its equilibrium position, the total virtual work done on the particle will be

$$\delta U = \mathbf{F}_1 \cdot \delta\mathbf{r} + \mathbf{F}_2 \cdot \delta\mathbf{r} + \mathbf{F}_3 \cdot \delta\mathbf{r} + \cdots = \Sigma\mathbf{F} \cdot \delta\mathbf{r}$$

We now express $\Sigma\mathbf{F}$ in terms of its scalar sums and $\delta\mathbf{r}$ in terms of its component virtual displacements in the coordinate directions and write

$$\delta U = \Sigma\mathbf{F} \cdot \delta\mathbf{r} = (\mathbf{i}\,\Sigma F_x + \mathbf{j}\,\Sigma F_y + \mathbf{k}\,\Sigma F_z) \cdot (\mathbf{i}\,\delta x + \mathbf{j}\,\delta y + \mathbf{k}\,\delta z)$$

$$= \Sigma F_x\,\delta x + \Sigma F_y\,\delta y + \Sigma F_z\,\delta z = 0$$

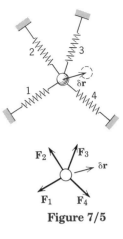

**Figure 7/5**

The sum is zero, since $\Sigma\mathbf{F} = \mathbf{0}$ and also $\Sigma F_x = 0$, $\Sigma F_y = 0$, and $\Sigma F_z = 0$. We see therefore that the equation $\delta U = 0$ is an alternative statement of the equilibrium conditions for a particle. This condition of zero virtual work for equilibrium is both necessary and sufficient, since we may apply it to virtual displacements taken one at a time in each of the three mutually perpendicular directions, in which case it becomes equivalent to the three known scalar requirements for equilibrium.

The principle of zero virtual work for the equilibrium of a single particle usually does not simplify this already simple problem since $\delta U = 0$ and $\Sigma\mathbf{F} = \mathbf{0}$ provide the same information. The concept of virtual work for a particle is introduced so that it may be applied to systems of particles in the development that follows.

***(b) Rigid body.*** We easily extend the principle of virtual work for a single particle to a rigid body which can be treated as a system of small elements or particles rigidly attached to one another. Since the virtual work done on each particle of the body in equilibrium is zero, it follows that the virtual work done on the entire rigid body is zero. Only the virtual work done by *external* forces appears in the evaluation of $\delta U = 0$ for the entire body, since all internal forces occur in pairs of equal, opposite, and collinear forces and the net work done by these forces during any movement is zero.

Again, as in the case of a particle, we find that the principle of virtual work offers no particular advantage to the solution for a single rigid body in equilibrium. Any assumed virtual displacement defined by a linear or angular movement will appear in each term in $\delta U = 0$ and when canceled will leave us with the same expression as we would have obtained by using one of the force or moment equations of equilibrium directly. This condition is illustrated in Fig. 7/6, where we are asked to determine the reaction $R$ under the roller for the hinged plate of negligible weight under the action of

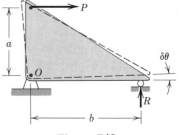

**Figure 7/6**

a given force $P$. A small assumed rotation $\delta\theta$ of the plate about $O$ is consistent with the hinge constraint at $O$ and is taken as the virtual displacement. The work done by $P$ is $-Pa\,\delta\theta$, and the work done by $R$ is $+Rb\,\delta\theta$. Therefore, the principle $\delta U = 0$ gives

$$-Pa\,\delta\theta + Rb\,\delta\theta = 0$$

Canceling out $\delta\theta$ leaves

$$Pa - Rb = 0$$

which is simply the equation of moment equilibrium about $O$. Therefore, nothing is gained by the use of the virtual-work principle for a single rigid body. Use of the principle will, however, provide us with a decided advantage for interconnected bodies, as described in the next section.

*(c) Ideal systems of rigid bodies.*  We now extend the principle of virtual work to describe the equilibrium of an interconnected system of rigid bodies. Our treatment here will be limited to so-called *ideal systems*, which are systems composed of two or more rigid members linked together by mechanical connections which are incapable of absorbing energy through elongation or compression and in which friction is sufficiently small to be neglected. Figure 7/7a shows a simple example of an ideal system where motion between its two parts is possible and where the equilibrium position is determined by the applied external forces **P** and **F**. For such an interconnected mechanical system we identify three types of forces which act. They are as follows:

(1)  *Active forces* are external forces capable of doing virtual work during possible virtual displacements. In Fig. 7/7a forces **P** and **F** are active forces since they would do work as the links move.

(2)  *Reactive forces* are forces which act at positions of fixed support where no virtual displacement in the direction of the force takes place. Reactive forces do no work during a virtual displacement. In Fig. 7/7b the horizontal force $\mathbf{F}_B$ exerted on the roller end of the member by the vertical guide can do no work since there can be no

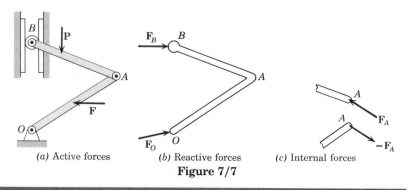

(a) Active forces          (b) Reactive forces          (c) Internal forces

**Figure 7/7**

horizontal displacement of the roller. The force $\mathbf{F}_O$ exerted on the system by the fixed support at $O$ is also a nonworking reactive force since no displacement of $O$ takes place.

(3) *Internal forces* are forces in the connections between members. During any possible movement of the system or its parts we see that the *net work done by the internal forces at the connections is zero*. This is so because the internal forces always exist in pairs of equal and opposite forces, as indicated for the internal forces $\mathbf{F}_A$ and $-\mathbf{F}_A$ at joint $A$ in Fig. 7/7c, and the work of one force necessarily cancels the work of the other force during their identical displacements.

With the observation that only the external active forces do work during any possible movement of the system, we may now state the principle of virtual work as follows:

**The virtual work done by external active forces on an ideal mechanical system in equilibrium is zero for any and all virtual displacements consistent with the constraints.**

By constraint we mean restriction of the motion by the supports. In this form the principle finds its greatest use for ideal systems. We state the principle mathematically by the equation

$$\boxed{\delta U = 0} \qquad\qquad (7/3)$$

where $\delta U$ stands for the total virtual work done on the system by all active forces during a virtual displacement.

Only now can we see the real advantages of the method of virtual work. There are essentially two. First, it is not necessary for us to dismember ideal systems in order to establish the relations between the active forces, as is generally the case with the equilibrium method based on forces and moment summations. Second, we may determine the relations between the active forces directly without reference to the reactive forces. These advantages make the method of virtual work particularly useful in determining the position of equilibrium of a system under known loads. This type of problem is in contrast to the problem of determining the forces acting on a body whose equilibrium position is fixed or specified.

The method of virtual work is especially useful for the purposes mentioned but requires that the internal friction forces do negligible work during any virtual displacement. Consequently, if internal friction in a mechanical system is appreciable, the method of virtual work will produce error when applied to the system as a whole unless the work done by internal friction is included.

In the method of virtual work a diagram which isolates the system under consideration should be drawn. Unlike the free-body diagram, where all forces are shown, the diagram for the method of work need show only the *active forces*, since the reactive forces do not enter into the application of $\delta U = 0$. Such a drawing will be termed an *active-force diagram*. Figure 7/7a is an active-force diagram for the system shown.

### (d) Degrees of freedom.
The number of independent coordinates needed to specify completely the configuration of a mechanical system is referred to as the number of *degrees of freedom* for that system. Figure 7/8a shows three examples of one-degree-of-freedom systems where only one coordinate is needed to determine the position of

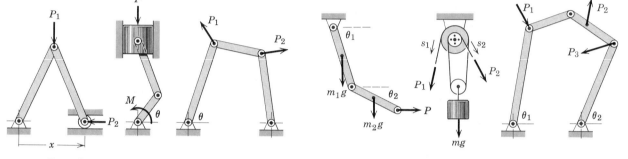

| (a) Examples of One–Degree–of–Freedom Systems | (b) Examples of Two–Degree–of–Freedom Systems |
|---|---|

**Figure 7/8**

every part of the system. The coordinate can be a distance or an angle. Figure 7/8b shows three examples of two-degree-of-freedom systems where two independent coordinates are needed to determine the configuration of the system. By adding more links to the mechanism in the right-hand figure, there is no limit to the number of degrees of freedom which can be introduced.

The principle of virtual work $\delta U = 0$ may be applied as many times as there are degrees of freedom. With each application we allow only one independent coordinate to change at a time while holding the others constant. In our treatment of virtual work in this chapter we shall restrict application to one-degree-of-freedom systems.*

### (e) Systems with friction; mechanical efficiency.
When sliding friction is present to any appreciable degree in a mechanical system, the system is said to be "real." In real systems some of the positive work done on the system by external active forces (input work) is dissipated in the form of heat generated by the negative work of the kinetic friction forces during movement of the system. When there is sliding between contacting surfaces, the friction force does negative work since its direction is always opposite to the movement of the body on which it acts. Thus, the kinetic friction force $\mu_k N$ acting on the sliding block in Fig. 7/9a does work on the block during the displacement $x$ in the amount of $-\mu_k N x$. During a virtual displacement $\delta x$ the friction force does work equal to $-\mu_k N \, \delta x$. The static

friction force acting on the rolling wheel in Fig. 7/9*b*, on the other hand, does no work if the wheel does not slip as it rolls. In Fig. 7/9*c* the moment $M_f$ about the center of the pinned joint due to the friction forces that act at the contacting surfaces will do negative work during any relative angular movement between the two parts. Thus, for a virtual displacement $\delta\theta$ between the two parts, which have the separate virtual displacements $\delta\theta_1$ and $\delta\theta_2$ as shown, the negative work done is $-M_f\,\delta\theta_1 - M_f\,\delta\theta_2 = -M_f(\delta\theta_1 + \delta\theta_2)$, or merely $-M_f\,\delta\theta$. For each part, $M_f$ is in the sense to oppose the relative motion of rotation. Negative work done by kinetic friction forces is dissipated in the form of heat and cannot be regained.

It was noted earlier in the article that a major advantage of the method of virtual work is the analysis of an entire system of connected members without taking them apart. If there is appreciable kinetic friction internal to the system, it becomes necessary to dismember the system to determine the friction forces. In such cases the method of virtual work finds only limited use.

Because of energy loss by friction, the output work of a machine is always less than the input work. The ratio of the two works is the *mechanical efficiency e.* Thus,

$$e = \frac{\text{output work}}{\text{input work}}$$

The mechanical efficiency of simple machines which have a single degree of freedom and which operate in a uniform manner may be determined by the method of work by evaluating the numerator and denominator of the expression for *e* during a virtual displacement. As an example, consider the block being moved up the inclined plane in Fig. 7/10. For the virtual displacement $\delta s$ shown, the output work is that necessary to elevate the block or $mg\,\delta s\,\sin\,\theta$, and the input work is $T\,\delta s = (mg\,\sin\,\theta + \mu_k mg\,\cos\,\theta)\,\delta s$. The efficiency of the inclined plane is, therefore,

$$e = \frac{mg\,\delta s\,\sin\,\theta}{mg(\sin\,\theta + \mu_k\,\cos\,\theta)\,\delta s} = \frac{1}{1 + \mu_k\,\cot\,\theta}$$

As a second example, consider the screw jack described in Art. 6/5 and shown in Fig. 6/5. Equation 6/3 gives the moment $M$ required to raise the load $W$, where the screw has a mean radius $r$ and a helix angle $\alpha$, and where the friction angle is $\phi = \tan^{-1}\mu_k$. During a small rotation $\delta\theta$ of the screw, the input work is $M\,\delta\theta = Wr\,\delta\theta\,\tan\,(\alpha + \phi)$. The output work is that required to elevate the load or $Wr\,\delta\theta\,\tan\,\alpha$, so that the efficiency of the jack becomes

$$e = \frac{Wr\,\delta\theta\,\tan\,\alpha}{Wr\,\delta\theta\,\tan\,(\alpha + \phi)} = \frac{\tan\,\alpha}{\tan\,(\alpha + \phi)}$$

As friction is decreased, $\phi$ becomes smaller, and the efficiency approaches unity.

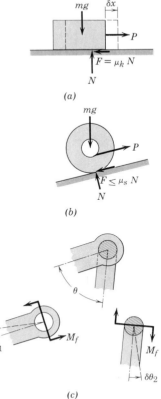

(a)

(b)

(c)

**Figure 7/9**

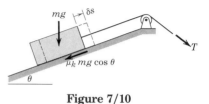

**Figure 7/10**

# Sample Problem 7/1

Each of the two uniform hinged bars has a mass $m$ and a length $l$, and is supported and loaded as shown. For a given force $P$ determine the angle $\theta$ for equilibrium.

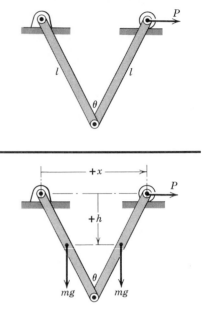

**Solution.** The active-force diagram for the system composed of the two members is shown separately and includes the weight $mg$ of each bar in addition to he force $P$. All other forces acting externally on the system are reactive forces which do no work during a virtual movement $\delta x$ and are therefore not shown.

The principle of virtual work requires that the total work of all external active forces be zero for any virtual displacement consistent with the constraints. Thus, for a movement $\delta x$ the virtual work becomes

① $[\delta U = 0]$ $\qquad\qquad P\,\delta x + 2mg\,\delta h = 0$

We now express each of these virtual displacements in terms of the variable $\theta$, the required quantity. Hence,

$$x = 2l \sin \frac{\theta}{2} \quad \text{and} \quad \delta x = l \cos \frac{\theta}{2}\,\delta\theta$$

② Similarly,

$$h = \frac{l}{2} \cos \frac{\theta}{2} \quad \text{and} \quad \delta h = -\frac{l}{4} \sin \frac{\theta}{2}\,\delta\theta$$

Substitution into the equation of virtual work gives us

$$Pl \cos \frac{\theta}{2}\,\delta\theta - 2mg\,\frac{l}{4} \sin \frac{\theta}{2}\,\delta\theta = 0$$

from which we get

$$\tan \frac{\theta}{2} = \frac{2P}{mg} \quad \text{or} \quad \theta = 2 \tan^{-1} \frac{2P}{mg} \qquad\qquad Ans.$$

To obtain this result by the principles of force and moment summation, it would be necessary to dismember the frame and take into account all forces acting on each member. Solution by the method of virtual work involves a simpler operation.

① Note carefully that with $x$ positive to the right $\delta x$ is also positive to the right in the direction of $P$, so that the virtual work is $P(+\delta x)$. With $h$ positive down $\delta h$ is also mathematically positive down in the direction of $mg$, so that the correct mathematical expression for the work is $mg(+\delta h)$. When we express $\delta h$ in terms of $\delta\theta$ in the next step, $\delta h$ will have a negative sign, thus bringing our mathematical expression into agreement with the physical observation that the weight $mg$ does negative work as each center of mass moves upward with an increase in $x$ and $\theta$.

② We obtain $\delta h$ and $\delta x$ with the same mathematical rules of differentiation with which we may obtain $dh$ and $dx$.

## Sample Problem   7/2

The mass $m$ is brought to an equilibrium position by the application of the couple $M$ to the end of one of the two parallel links that are hinged as shown. The links have negligible mass, and all friction is assumed to be absent. Determine the expression for the equilibrium angle $\theta$ assumed by the links with the vertical for a given value of $M$. Consider the alternative of a solution by force and moment equilibrium.

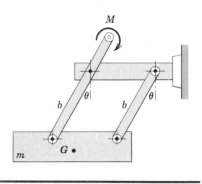

***Solution.***   The active-force diagram shows the weight $mg$ acting through the center of mass $G$ and the couple $M$ applied to the end of the link. There are no other external active forces or moments which do work on the system during a change in the angle $\theta$.

The vertical position of the center of mass $G$ is designated by the distance $h$ below the fixed horizontal reference line and is $h = b \cos \theta + c$. The work done by $mg$ during a movement $\delta h$ in the direction of $mg$ is

$$+mg\ \delta h = mg\ \delta(b \cos \theta + c)$$

$$= mg(-b \sin \theta\ \delta\theta + 0)$$

$$= -mgb \sin \theta\ \delta\theta$$

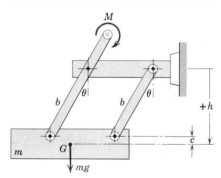

① The minus sign shows that the work is negative for a positive value of $\delta\theta$. The constant $c$ drops out since its variation is zero.

With $\theta$ measured positive in the clockwise sense, $\delta\theta$ is also positive clockwise. Thus, the work done by the clockwise couple $M$ is $+M\ \delta\theta$. Substitution into the virtual work equation gives us

$[\delta U = 0]$ $\qquad\qquad M\ \delta\theta + mg\ \delta h = 0$

which yields

$$M\ \delta\theta = mgb \sin \theta\ \delta\theta$$

$$\theta = \sin^{-1} \frac{M}{mgb} \qquad\qquad Ans.$$

Inasmuch as $\sin \theta$ cannot exceed unity, we see that for equilibrium $M$ is limited to values that do not exceed $mgb$.

The advantage of the virtual-work solution for this problem is readily seen when we observe what would be involved with a solution by force and moment equilibrium. For the latter approach, it would be necessary for us to draw separate free-body diagrams of all of the three moving parts and account for all of the internal reactions at the pin connections. To carry out these steps, it would be necessary for us to include in the analysis the horizontal position of $G$ with respect to the attachment points of the two links, even though reference to this position would finally drop out of the equations when solved. We conclude, then, that the virtual-work method in this problem deals directly with cause and effect and avoids reference to irrelevant quantities.

① Again, as in Sample Problem 7/1, we are consistent mathematically with our definition of work, and we see that the algebraic sign of the resulting expression agrees with the physical change.

## Sample Problem 7/3

For link $OA$ in the horizontal position shown, determine the force $P$ on the sliding collar which will prevent $OA$ from rotating under the action of the couple $M$. Neglect the mass of the moving parts.

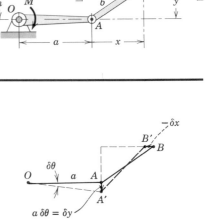

**Solution.** The given sketch serves as the active-force diagram for the system. All other forces are either internal or nonworking reactive forces due to the constraints.

We will give the crank $OA$ a small clockwise angular movement $\delta\theta$ as our virtual displacement and determine the resulting virtual work done by $M$ and $P$. From the horizontal position of the crank the angular movement gives a downward displacement of $A$ equal to

① $$\delta y = a\ \delta\theta$$

where $\delta\theta$ is, of course, expressed in radians.

From the right triangle for which link $AB$ is the constant hypotenuse we may write

$$b^2 = x^2 + y^2$$

We now take the differential of the equation and get

② $$0 = 2x\ \delta x + 2y\ \delta y \quad \text{or} \quad \delta x = -\frac{y}{x}\ \delta y$$

Thus,

$$\delta x = -\frac{y}{x}\ a\ \delta\theta$$

and the virtual-work equation becomes

③ $$[\delta U = 0] \quad M\ \delta\theta + P\ \delta x = 0 \quad M\ \delta\theta + P\left(-\frac{y}{x}\ a\ \delta\theta\right) = 0$$

$$P = \frac{Mx}{ya} = \frac{Mx}{ha} \qquad Ans.$$

Again, we observe that the virtual-work method produces a direct relationship between the active force $P$ and the couple $M$ without involving other forces which are irrelevant to this relationship. Solution by the force and moment equations of equilibrium, although fairly simple in this problem, would require accounting for all forces initially and then eliminating the irrelevant ones.

① Note that the displacement $a\ \delta\theta$ of point $A$ would no longer equal $\delta y$ if the crank $OA$ were not in a horizontal position.

② The length $b$ is constant so that $\delta b = 0$. Notice the negative sign, which merely tells us that if one change is positive, the other must be negative.

③ We could just as well use a counterclockwise virtual displacement for the crank, which would merely reverse the signs of all terms.

# PROBLEMS

(Assume that the negative work of friction is negligible in the following problems unless otherwise
indicated.)

## *Introductory problems*

**7/1** The mass of the uniform bar of length $l$ is $m$ while
that of the uniform bar of length $2l$ is $2m$. For a
given force $P$, determine the angle $\theta$ for equilibrium.

$$Ans. \ \ \theta = 2 \tan^{-1}\left(\frac{4P}{mg}\right)$$

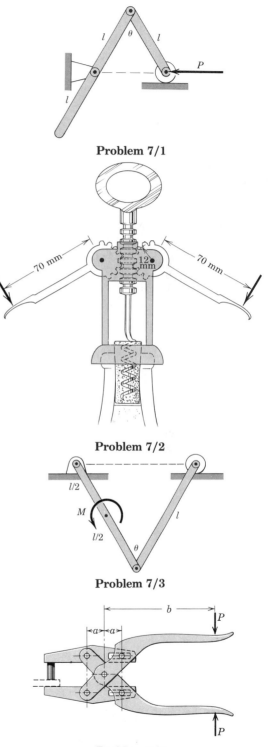

**Problem 7/1**

**7/2** By means of a rack-and-pinion mechanism, large
forces can be developed by the cork puller shown. If
the mean radius of the pinion gears is 12 mm,
determine the force $R$ which is exerted on the cork
for given forces $P$ on the handles.

**Problem 7/2**

**7/3** Determine the couple $M$ required to maintain equilibrium at an angle $\theta$. Each of the two uniform bars
has mass $m$ and length $l$.

$$Ans. \ \ M = mgl \sin\frac{\theta}{2}$$

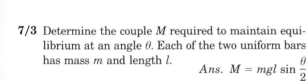

**Problem 7/3**

**7/4** Find the force $Q$ exerted on the paper by the paper
punch of Prob. 4/87, repeated here.

**Problem 7/4**

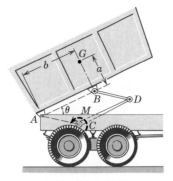

**Problem 7/5**

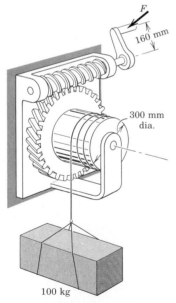

**Problem 7/6**

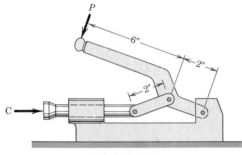

100 kg

**Problem 7/7**

**7/5** Determine the torque $M$ on the activating lever of the dump truck necessary to balance the load of mass $m$ with center of mass at $G$ when the dump angle is $\theta$. The polygon $ABDC$ is a parallelogram.

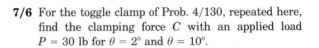

*Ans.* $M = mg(b \cos \theta - a \sin \theta)$

**7/6** For the toggle clamp of Prob. 4/130, repeated here, find the clamping force $C$ with an applied load $P = 30$ lb for $\theta = 2°$ and $\theta = 10°$.

**7/7** The hand-operated hoist is lifting a 100-kg load where 25 turns of the handle on the worm shaft produce one revolution of the drum. Assuming a 40 percent loss of energy due to friction in the mechanism, calculate the force $F$ normal to the handle arm required to lift the load.    *Ans.* $F = 61.3$ N

**7/8** What force $P$ is required to hold the segmented 3-panel door in the position shown where the center panel is in the 45° position? Each panel has a mass $m$, and friction in the roller guides (shown dotted) may be neglected.

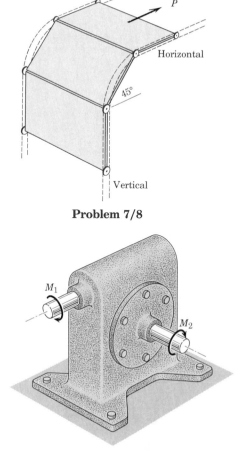

**Problem 7/8**

### Representative problems

**7/9** The speed reducer shown has a gear ratio of 40:1. With an input torque $M_1 = 30$ N·m, the measured output torque is $M_2 = 1180$ N·m. Determine the mechanical efficiency $e$ of the unit.

*Ans. e = 0.983*

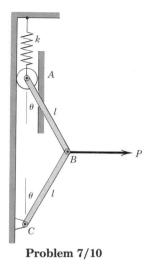

**Problem 7/9**

**7/10** The spring of constant $k$ is unstretched when $\theta = 0$. Derive an expression for the force $P$ required to deflect the system to an angle $\theta$. The mass of the bars is negligible.

**Problem 7/10**

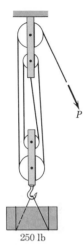

250 lb

**Problem 7/11**

·**7/11** For each unit of movement of the free end of the rope in the direction of the applied force $P$, the 250-lb load moves one-fourth of a unit. If the mechanical efficiency $e$ of the hoist is 0.75, calculate the force $P$ required to raise the load and the force $P'$ required to lower the load.

*Ans.* $P = 83.3$ lb, $P' = 46.9$ lb

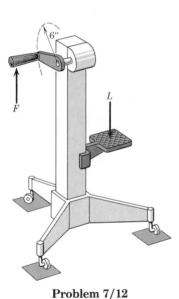

**Problem 7/12**

**7/12** For the screw-lift jack shown, 12 turns of the handle are required to elevate the lifting pad 1 in. If a force $F = 10$ lb applied normal to the crank is required to elevate a load $L = 2700$ lb, determine the efficiency $e$ of the screw in raising the load.

**Problem 7/13**

**7/13** The portable car hoist is operated by the hydraulic cylinder which controls the horizontal movement of end $A$ of the link in the horizontal slot. Determine the compression $C$ in the piston rod of the cylinder to support the load $P$ at a height $h$.

*Ans.* $C = P \sqrt{\left(\frac{2b}{h}\right)^2 - 1}$

**7/14** The torque $M$ applied to the light link $OA$ through its shaft at $O$ rotates $OA$ through an angle $\theta$ and raises end $A$ of the uniform bar $AB$ of mass $m$. End $B$ is supported by a small roller on the horizontal surface. When $\theta = 0$, the bar $AB$ is horizontal. Determine the equilibrium angle for a given value of $M$. What would happen if $M$ were greater than $mgr/2$?

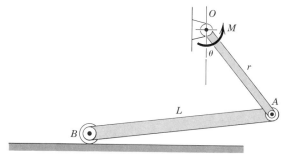

**Problem 7/14**

**7/15** Determine the force $F$ which the person must apply tangent to the rim of the handwheel of a wheelchair in order to roll up the incline of angle $\theta$. The combined mass of the chair and person is $m$. (If $s$ is the displacement of the center of the wheel measured along the incline and $\beta$ the corresponding angle in radians through which the wheel turns, it is easily shown that $s = R\beta$ if the wheel rolls without slipping.)

$$Ans. \ F = mg\frac{R}{r}\sin\theta$$

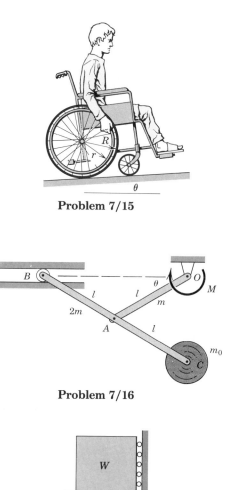

**Problem 7/15**

**7/16** Determine the couple $M$ which must be applied at $O$ in order to support the mechanism in the position $\theta = 30°$. The masses of the disk at $C$, bar $OA$, and bar $BC$ are $m_0$, $m$, and $2m$, respectively.

**Problem 7/16**

**7/17** Calculate the efficiency with which the 5° wedge elevates the weight $W$ under the action of the horizontal force $P$ on the wedge. The coefficient of friction between the wedge and block is 0.30.

$$Ans. \ e = 0.220$$

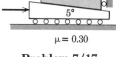

**Problem 7/17**

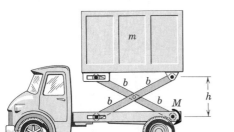

**Problem 7/18**

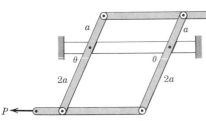

**Problem 7/19**

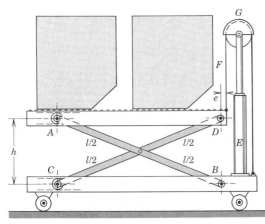

**Problem 7/20**

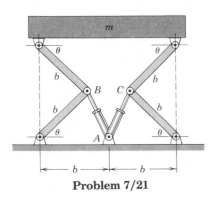

**Problem 7/21**

**7/18** The cargo box of the food-delivery truck for aircraft servicing has a loaded mass $m$ and is elevated by the application of a torque $M$ on the lower end of the link which is hinged to the truck frame. The horizontal slots allow the linkage to unfold as the cargo box is elevated. Express $M$ as a function of $h$.

**7/19** Each of the four uniform movable bars has a mass $m$, and their equilibrium position in the vertical plane is controlled by the force $P$ applied to the end of the lower bar. For a given value of $P$, determine the equilibrium angle $\theta$. Is it possible for the equilibrium position shown to be maintained by replacing the force $P$ by a couple $M$ applied to the end of the lower horizontal bar?

$$Ans. \quad \theta = \tan^{-1}\frac{mg}{P}, \text{ no}$$

**7/20** The platform $AD$ of an aircraft cargo loader is elevated to the proper height by the mechanism shown. There are two sets of linkages and hydraulic lifts, one set on each side. Cables $F$ which lift the platform are controlled by the hydraulic cylinders $E$ whose piston rods elevate the pulleys $G$. If the total weight of the platform and containers is $W$, determine the compressive force $P$ in each of the two piston rods. Does $P$ depend on the height $h$? What force $Q$ is supported by each link at its center joint when $W$ is centered between $A$ and $D$?

**7/21** The vertical position of the platform of mass $m$ supported by the four identical links is controlled by the hydraulic cylinders $AB$ and $AC$ which are pivoted at point $A$. Determine the compression $P$ in each of the cylinders required to support the platform for a specified angle $\theta$.

$$Ans. \quad P = mg\,\frac{\cos\theta}{\cos\dfrac{\theta}{2}}$$

**7/22** The postal scale consists of a sector of mass $m_0$ hinged at $O$ and with center of mass at $G$. The pan and vertical link $AB$ have a mass $m_1$ and are hinged to the sector at $B$. End $A$ is hinged to the uniform link $AC$ of mass $m_2$, which in turn is hinged to the fixed frame. The figure $OBAC$ forms a parallelogram, and the angle $GOB$ is a right angle. Determine the relation between the mass $m$ to be measured and the angle $\theta$, assuming that $\theta = \theta_0$ when $m = 0$.

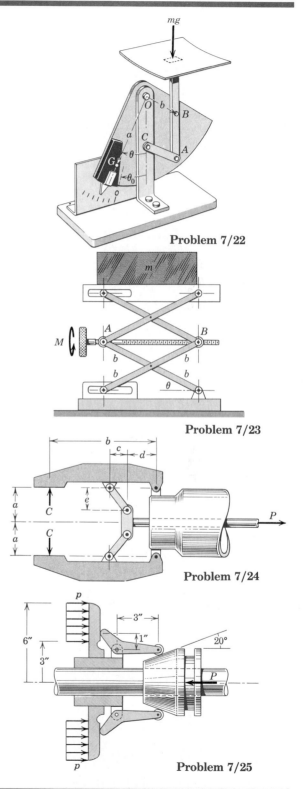

**Problem 7/22**

**7/23** The vertical position of the load of mass $m$ is controlled by the adjusting screw which connects joints $A$ and $B$. The change in the distance between $A$ and $B$ for one revolution of the screw equals the lead $L$ of the screw (advancement per revolution). If a moment $M_f$ is required to overcome friction in the threads and thrust bearing of the screw, determine the expression for the total moment $M$, applied to the adjusting screw, necessary to raise the load.

$$\textit{Ans.} \ \ M = M_f + \frac{mgL}{\pi} \cot \theta$$

**Problem 7/23**

**7/24** The claw of the remote-action actuator develops a clamping force $C$ as a result of the tension $P$ in the control rod. Express $C$ in terms of $P$ for the configuration shown where the jaws are parallel.

**Problem 7/24**

**7/25** The figure shows a sectional view of a mechanism for engaging a marine clutch. The position of the conical collar on the shaft is controlled by the engaging force $P$, and a slight movement to the left causes the two levers to bear against the back side of the clutch plate, which generates a uniform pressure $p$ over the contact area of the plate. This area is that of a circular ring with 6-in. outside and 3-in. inside radii. Determine by the method of virtual work the force $P$ required to generate a clutch-plate pressure of 30 lb/in.$^2$. The clutch plate is free to slide along the collar to which the levers are pivoted.

$$\textit{Ans.} \ \ P = 309 \text{ lb}$$

**Problem 7/25**

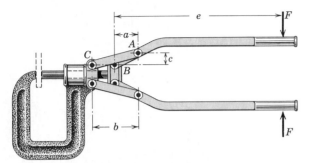

**Problem 7/26**

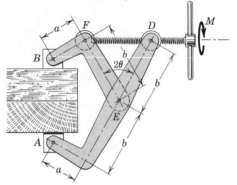

**Problem 7/27**

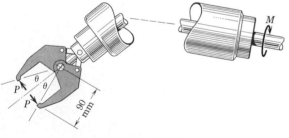

**Problem 7/28**

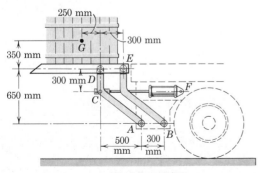

**Problem 7/29**

**7/26** Determine the force $P$ developed at the jaws of the rivet squeezer of Prob. 4/126, repeated here.

**7/27** Determine the force $F$ between the jaws of the clamp in terms of a torque $M$ exerted on the handle of the adjusting screw. The screw has a lead (advancement per revolution) $L$, and friction is to be neglected.

$$Ans. \ F = \frac{2\pi M}{L(\tan \theta + a/b)}$$

**7/28** The input end and the output claw of a remote mechanical actuator are shown. The internal connecting linkage is designed so that one complete revolution of the input shaft with applied torque $M$ creates a 2° change in the claw angle $\theta$. Calculate the compression $P$ developed by the claw for an input torque $M = 20$ N·m with a claw angle $\theta = 30°$ in the position shown. Neglect friction.

**7/29** A power-operated loading platform for the back of a truck is shown in the figure. The position of the platform is controlled by the hydraulic cylinder, which applies force at $C$. The links are pivoted to the truck frame at $A$, $B$, and $F$. Determine the force $P$ supplied by the cylinder in order to support the platform in the position shown. The mass of the platform and links may be neglected compared with that of the 250-kg crate with center of mass at $G$.

$$Ans. \ P = 3.5 \text{ kN}$$

▶**7/30** For the "cherry picker" maintenance vehicle, a motor-and-gear unit mounted in the joint of the arms at $A$ supplies a torque internal to the joint and maintains angle $CAB$ at exactly twice $\theta$ as the load $W$ is being raised. If $b_2 > b_1$, determine the external moment $M_1$ which must be applied at $B$ to $BA$ in order to begin raising the load. Neglect the weights of the two members. Discuss the work done on the system and justify the inclusion of the internal moment at $A$ in applying the virtual-work equation.

*Ans.* $M = W(b_2 - b_1)\cos\theta$

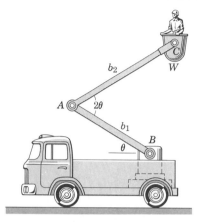

**Problem 7/30**

▶**7/31** In the screw-activated clamp, a torque $M$ applied to the handle of the screw tightens the clamp by increasing the distance $\overline{BD}$, thereby producing a clamping force $C$. The screw is threaded through the pivoted collar at $D$ and has a lead (advancement per revolution) of $\frac{1}{6}$ in. Assume no friction and express the clamping force $C$ in pounds in terms of the torque $M$ measured in lb-in. for the position $\theta = 30°$.

*Ans.* $C = 26.9M$

**Problem 7/31**

▶**7/32** Determine the force $Q$ at the jaw of the shear of Prob. 4/102, repeated here, for the 400-N force applied with $\theta = 30°$. (*Hint:* Replace the 400-N force by a force and a couple at the center of the small gear. The absolute angular displacement of the gear must be carefully determined.)

*Ans.* $Q = 13.18$ kN

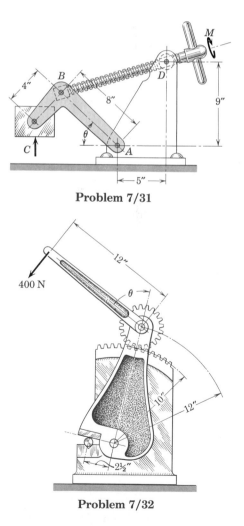

**Problem 7/32**

## 7/4 POTENTIAL ENERGY AND STABILITY

In the previous article we have dealt with the equilibrium configuration of mechanical systems composed of individual members that were assumed to be perfectly rigid. We now extend our method to account for mechanical systems that include elastic elements in the form of springs. For this purpose we find it helpful to introduce the concept of potential energy, which leads us directly to the important problem of determining the stability of equilibrium.

*(a) Elastic potential energy.* The work done on an elastic member is stored in the member and is called *elastic potential energy* $V_e$. This energy is potentially available by allowing the member to do work on some other body during the relief of its compression or extension. Consider a spring, Fig. 7/11, which is being compressed by a force $F$. The spring is assumed to be elastic and linear, that is, the force $F$ is directly proportional to the deflection $x$. We write this relation as $F = kx$, where $k$ is the *spring constant* or *stiffness* of the spring. The work done on the spring by $F$ during a movement $dx$ is $dU = F\,dx$, so that the elastic potential energy of the spring for a compression $x$ is the total work done on the spring

$$V_e = \int_0^x F\,dx = \int_0^x kx\,dx$$

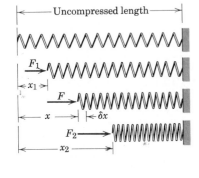

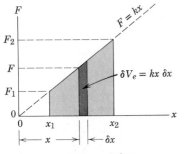

**Figure 7/11**

or

$$\boxed{V_e = \tfrac{1}{2}kx^2} \qquad\qquad (7/4)$$

Thus, we see that the potential energy of the spring equals the triangular area in the diagram of $F$ versus $x$ from 0 to $x$.

During an increase in the compression of the spring from $x_1$ to $x_2$, the work done on the spring equals its *change* in elastic potential energy or

$$\Delta V_e = \int_{x_1}^{x_2} kx\,dx = \tfrac{1}{2}k(x_2{}^2 - x_1{}^2)$$

which equals the trapezoidal area from $x_1$ to $x_2$.

During a virtual displacement $\delta x$ of the spring the virtual work done on the spring is the virtual change in elastic potential energy

$$\delta V_e = F\,\delta x = kx\,\delta x$$

During a decrease in the compression of the spring as it is relaxed from $x = x_2$ to $x = x_1$, the *change* (final minus initial) in the potential energy of the spring would be negative. Correspondingly if $\delta x$ is negative, $\delta V_e$ will also be negative.

When we have a spring in tension rather than compression, the work and energy relations are the same as those for compression, where $x$ now represents the elongation of the spring rather than its compression. While the spring is being stretched, we note that the force again acts in the direction of the displacement doing positive work on the spring and increasing its potential energy.

Since the force acting on the movable end of a spring is the negative of the force exerted by the spring on the body to which its movable end is attached, we conclude that the *work done on the body is the negative of the potential energy change of the spring.*

In the case of a torsional spring, which resists the rotation of a shaft or another element, potential energy can also be stored and released. If the torsional stiffness, expressed in torque per radian of twist, is a constant $K$, and if $\theta$ is the angle of twist in radians, then the resisting torque is $M = K\theta$. The potential energy becomes $V_e = \int_0^\theta K\theta \, d\theta$ or

$$V_e = \tfrac{1}{2}K\theta^2 \qquad (7/4a)$$

which is analogous to the expression for the linear extension spring.

The units of elastic potential energy are the same as those of work and are expressed in joules (J) in SI units and in foot-pounds (ft-lb) in U.S. customary units.

**(b) Gravitational potential energy.** In the previous article we treated the work of a gravitational force or weight acting on a body in the same way as the work of any other active force. Thus, for an upward displacement $\delta h$ of the body in Fig. 7/12 the weight $W = mg$ does negative work $\delta U = -mg\,\delta h$. Or, if the body has a downward displacement $\delta h$, with $h$ measured positive downward, the weight does positive work $\delta U = +mg\,\delta h$.

We now adopt an alternative to the foregoing treatment by expressing the work done by gravity in terms of a change in potential energy of the body. This alternative treatment is a useful representation when we describe a mechanical system in terms of its total energy. The *gravitational potential energy* $V_g$ of a body is defined simply as the work done on the body by a force equal and opposite to the weight in bringing the body to the position under consideration from some arbitrary datum plane where the potential energy is defined to be zero. The potential energy, then, is the negative of the work done by the weight. When the body is raised, for example, the work done is converted into energy that is potentially available, since the body is capable of doing work on some other body as it returns to its original lower position. If we take $V_g$ to be zero at $h = 0$, Fig. 7/12, then at a height $h$ above the datum plane the gravitational potential energy of the body is

$$V_g = mgh \qquad (7/5)$$

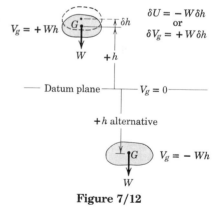

Figure 7/12

If the body is a distance $h$ below the datum plane, its gravitational potential energy is $-mgh$.

It is important to observe that the datum plane for zero potential energy is perfectly arbitrary inasmuch as it is only the *change* in energy with which we are concerned, and this change is the same no matter where we take the datum plane. Also, the gravitational potential energy is independent of the path followed in arriving at the particular level $h$ under consideration. Thus, the body of mass $m$ in Fig. 7/13 has the same potential-energy change no matter which path it follows in going from datum plane 1 to datum plane 2 since $\Delta h$ is the same for both. Our measurement of $h$, of course, is to the center of mass of the body.

The virtual change in gravitational potential energy is simply

$$\delta V_g = mg\,\delta h$$

where $\delta h$ is the upward virtual displacement of the mass center of the body. If the mass center should have a downward virtual displacement, then $\delta V_g$ is negative.

The units of gravitational potential energy are the same as those for work and elastic potential energy, joules (J) in SI units and foot-pounds (ft-lb) in U.S. customary units.

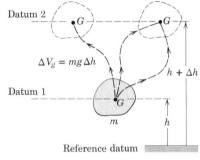

**Figure 7/13**

*(c) Energy equation.* In the previous two sections we have noted that the work done *by* a linear spring *on* the body to which its movable end is attached is the negative of the change in the elastic potential energy of the spring. Also, the work done by the gravitational force or weight $mg$ is the negative of the change in gravitational potential energy. Therefore, when we apply the virtual-work equation to systems with springs and with changes in the vertical position of its members, we may replace the work of the springs and the work of the weights by the negative of the respective potential energy changes.

With these substitutions in mind, we may write the total virtual work $\delta U$ in Eq. 7/3 as the sum of the work $\delta U'$ done by all active forces *other than spring forces and weight forces* and the work $-(\delta V_e + \delta V_g)$ done by the spring and weight forces. Equation 7/3 then becomes

$$\boxed{\delta U' - (\delta V_e + \delta V_g) = 0} \quad \text{or} \quad \boxed{\delta U' = \delta V} \qquad \textbf{(7/6)}$$

where $V = V_e + V_g$ stands for the total potential energy of the system. With this formulation a spring becomes *internal* to the system, and the work of spring and gravitational forces is accounted for in the $\delta V$ term.

With the method of virtual work it is useful to construct the active-force diagram of the system that is being analyzed. The boundary of the system must clearly distinguish those members that

are a part of the system from other bodies which are not a part of
the system. When an elastic member is included within the boundary
of our system, we see that the forces of interaction between it and
the movable members to which it is attached are *internal* to the
system and need not be shown because their effects are accounted
for in the $V_e$ term. Similarly, weight forces are not shown because
their work is accounted for in the $V_g$ term.

Figure 7/14 illustrates the difference between the use of Eqs.
7/3 and 7/6. The body under consideration in the *a*-part of the figure
is taken to be a particle for simplicity, and the virtual displacement
is taken to be along the fixed path. The particle is in equilibrium
under the action of the applied forces $F_1$ and $F_2$, the gravitational
force $mg$, the spring force $kx$, and a normal reaction force. In Fig.
7/14*b*, where the particle alone is isolated, $\delta U$ includes the virtual
work of all forces shown on the active-force diagram of the particle.
(The normal reaction exerted on the particle by the smooth guide
does no work and is omitted.) In Fig. 7/14*c* the spring is *included*
in the system, and $\delta U'$ is the virtual work only of $F_1$ and $F_2$, which
are the only external forces which do work. The work of the weight

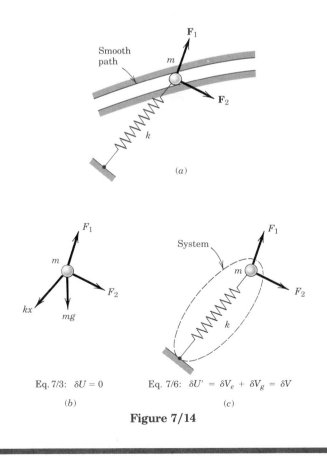

Smooth path

$\mathbf{F}_1$

$m$

$\mathbf{F}_2$

$k$

(*a*)

$F_1$

$m$

$F_2$

$kx$

$mg$

Eq. 7/3:  $\delta U = 0$

(*b*)

System

$F_1$

$m$

$F_2$

$k$

Eq. 7/6:  $\delta U' = \delta V_e + \delta V_g = \delta V$

(*c*)

**Figure 7/14**

$mg$ is accounted for in the $\delta V_g$ term, and the work of the spring force is included in the $\delta V_e$ term.

Thus, for a mechanical system with elastic members and members that undergo changes in position, we may restate the principle of virtual work as follows:

> *The virtual work done by all external active forces (other than the gravitational and spring forces accounted for in the potential energy terms) on a mechanical system in equilibrium equals the corresponding change in the total elastic and gravitational potential energy of the system for any and all virtual displacements consistent with the constraints.*

**(d) Stability of equilibrium.** Consider now the case of mechanical systems where movement is accompanied by changes in gravitational and elastic potential energies and where no work is done on the system by external active forces. The mechanism of Sample Problem 7/6 is an example of such a system. With $\delta U' = 0$ the virtual-work relation, Eq. 7/6, becomes

$$\delta(V_e + V_g) = 0 \quad \text{or} \quad \delta V = 0 \qquad (7/7)$$

Equation 7/7 expresses the requirement that the equilibrium configuration of a mechanical system is one for which the total potential energy $V$ of the system has a stationary value. For a system of one degree of freedom where the potential energy and its derivatives are continuous functions of the single variable, say, $x$, which describes the configuration, the equilibrium condition $\delta V = 0$ is equivalent mathematically to the requirement

$$\frac{dV}{dx} = 0 \qquad (7/8)$$

Equation 7/8 states that a mechanical system is in equilibrium when the derivative of its total potential energy is zero. For systems with several degrees of freedom the partial derivative of $V$ with respect to each coordinate in turn must be zero for equilibrium.*

There are three conditions under which Eq. 7/8 applies, namely, when the total potential energy is a minimum (stable equilibrium), a maximum (unstable equilibrium), or a constant (neutral equilibrium). We see a simple example of these three conditions in Fig. 7/15 where the potential energy of the roller is clearly a minimum

Stable        Unstable        Neutral

**Figure 7/15**

---

* For examples of two-degree-of-freedom systems see Art. 43, Chapter 7, of the senior author's *Statics, 2nd Edition SI Version* 1975.

in the stable position, a maximum in the unstable position, and a constant in the neutral position.

We may also characterize the stability of a mechanical system by noting that a small displacement away from the stable position results in an increase in potential energy and a tendency to return to the position of lower energy. On the other hand, a small displacement away from the unstable position results in a decrease in potential energy and a tendency to move farther away from the equilibrium position to one of still lower energy. For the neutral position a small displacement one way or the other results in no change in potential energy and no tendency to move either way.

When a function and its derivatives are continuous, the second derivative is positive at a point of minimum value of the function and negative at a point of maximum value of the function. Thus, the mathematical conditions for equilibrium and stability of a system with a single degree of freedom $x$ are:

$$
\begin{array}{lll}
\text{Equilibrium} & \dfrac{dV}{dx} = 0 & \\[3mm]
\text{Stable} & \dfrac{d^2V}{dx^2} > 0 & \qquad\textbf{(7/9)} \\[3mm]
\text{Unstable} & \dfrac{d^2V}{dx^2} < 0 &
\end{array}
$$

Occasionally we may have a situation where the second derivative of $V$ is also zero at the equilibrium position, in which case we must examine the sign of a higher derivative to ascertain the type of equilibrium. When the order of the lowest remaining nonzero derivative is even, the equilibrium will be stable or unstable according to whether the sign of this lowest even-order derivative is positive or negative. If the order of the lowest remaining nonzero derivative is odd, the equilibrium is classified as unstable, and the plot of $V$ versus $x$ for this case appears as an inflection point in the curve with zero slope at the equilibrium value.

Stability criteria for multiple degrees of freedom require more advanced treatment. For two degrees of freedom, for example, we use a Taylor-series expansion for two variables.

## Sample Problem 7/4

The 10-kg cylinder is suspended by the spring which has a stiffness of 2 kN/m. Plot the potential energy $V$ of the system and show that it is minimum at the equilibrium position.

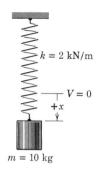

$k = 2$ kN/m

$V = 0$

$+x$

$m = 10$ kg

---

**Solution.** (Although the equilibrium position in this simple problem is clearly where the force in the spring equals the weight $mg$, we will proceed as though this fact were unknown in order to illustrate the energy relationships in the simplest way.) We choose the datum plane for zero
①  potential energy at the position for which the spring is unextended.

The elastic potential energy for an arbitrary position $x$ is $V_e = \frac{1}{2}kx^2$ and the gravitational potential energy is $-mgx$, so that the total potential energy is

$$[V = V_e + V_g] \qquad\qquad V = \tfrac{1}{2}kx^2 - mgx$$

Equilibrium occurs where

$$\left[\frac{dV}{dx} = 0\right] \qquad \frac{dV}{dx} = kx - mg = 0 \qquad x = mg/k$$

Although we know in this simple case that the equilibrium is stable, we prove it by evaluating the sign of the second derivative of $V$ at the equilibrium position. Thus, $d^2V/dx^2 = k$, which is positive, proving that the equilibrium is stable.

Substituting numerical values gives

$$V = \tfrac{1}{2}(2000)x^2 - 10(9.81)x$$

expressed in joules, and the equilibrium value of $x$ is

$$x = 10(9.81)/2000 = 0.049 \text{ m} \qquad \text{or} \qquad 49 \text{ mm} \qquad Ans.$$

We calculate $V$ for various values of $x$ and plot $V$ versus $x$ as shown.
②  The minimum value of $V$ occurs at $x = 0.049$ m where $dV/dx = 0$ and $d^2V/dx^2$ is positive.

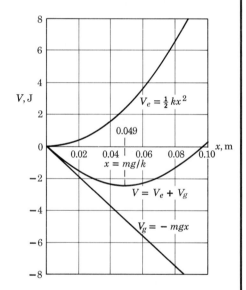

① The choice is arbitrary but simplifies the algebra.

② We could have chosen different datum planes for $V_e$ and $V_g$ without affecting our conclusions. Such a change would merely shift the separate curves for $V_e$ and $V_g$ up or down but would not affect the position of the minimum value of $V$.

# Sample Problem 7/5

The two uniform links, each of mass $m$, are in the vertical plane and are connected and constrained as shown. As the angle $\theta$ between the links increases with the application of the horizontal force $P$, the light rod, which is connected at $A$ and passes through a pivoted collar at $B$, compresses the spring of stiffness $k$. If the spring is uncompressed in the position equivalent to that for which $\theta = 0$, determine the force $P$ which will produce equilibrium at the angle $\theta$.

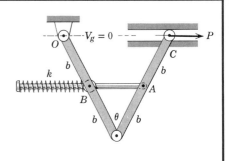

**Solution.** The given sketch serves as the active-force diagram of the system. The compression $x$ of the spring is the distance that $A$ has moved away from $B$, which is $x = 2b \sin \theta/2$. Thus, the elastic potential energy of the spring is

$$[V_e = \tfrac{1}{2}kx^2] \qquad V_e = \tfrac{1}{2}k\left(2b \sin \frac{\theta}{2}\right)^2 = 2kb^2 \sin^2 \frac{\theta}{2}$$

With the datum for zero gravitational potential energy taken through the support at $O$ for convenience, the expression for $V_g$ becomes

$$[V_g = mgh] \qquad V_g = 2mg\left(-b \cos \frac{\theta}{2}\right)$$

The distance between $O$ and $C$ is $4b \sin \theta/2$, so that the virtual work done by $P$ is

$$\delta U' = P \, \delta\left(4b \sin \frac{\theta}{2}\right) = 2Pb \cos \frac{\theta}{2} \, \delta\theta$$

The virtual-work equation now gives

$$[\delta U' = \delta V_e + \delta V_g]$$

$$2Pb \cos \frac{\theta}{2} \, \delta\theta = \delta\left(2kb^2 \sin^2 \frac{\theta}{2}\right) + \delta\left(-2mgb \cos \frac{\theta}{2}\right)$$

$$= 2kb^2 \sin \frac{\theta}{2} \cos \frac{\theta}{2} \, \delta\theta + mgb \sin \frac{\theta}{2} \, \delta\theta$$

Simplifying gives finally

$$P = kb \sin \frac{\theta}{2} + \tfrac{1}{2}mg \tan \frac{\theta}{2} \qquad \qquad Ans.$$

If we had been asked to express the equilibrium value of $\theta$ corresponding to a given force $P$, we would have difficulty solving explicitly for $\theta$ in this particular case. But for a numerical problem we could resort to a computer solution and graphical plot of numerical values of the sum of the two functions of $\theta$ to determine the value of $\theta$ for which the sum equals $P$.

## Sample Problem 7/6

The ends of the uniform bar of mass $m$ slide freely in the horizontal and vertical guides. Examine the stability conditions for the positions of equilibrium. The spring of stiffness $k$ is undeformed when $x = 0$.

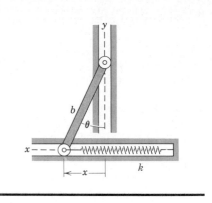

**Solution.** The system consists of the spring and the bar. Since there ① are no external active forces, the given sketch serves as the active-force diagram. We shall take the $x$-axis as the datum for zero gravitational potential energy. In the displaced position the elastic and gravitational potential energies are

$$V_e = \tfrac{1}{2}kx^2 = \tfrac{1}{2}kb^2 \sin^2 \theta \quad \text{and} \quad V_g = mg \frac{b}{2} \cos \theta$$

The total potential energy is then

$$V = V_e + V_g = \tfrac{1}{2}kb^2 \sin^2 \theta + \tfrac{1}{2}mgb \cos \theta$$

Equilibrium occurs for $dV/d\theta = 0$ so that

$$\frac{dV}{d\theta} = kb^2 \sin \theta \cos \theta - \tfrac{1}{2}mgb \sin \theta = (kb^2 \cos \theta - \tfrac{1}{2}mgb) \sin \theta = 0$$

① With no external active forces there is no $\delta U'$ term, and $\delta V = 0$ is equivalent to $dV/d\theta = 0$.

The two solutions to this equation are given by

$$\sin \theta = 0 \quad \text{and} \quad \cos \theta = \frac{mg}{2kb}$$

② ② Be careful not to overlook the solution $\theta = 0$ given by $\sin \theta = 0$.

We now determine the stability by examining the sign of the second derivative of $V$ for each of the two equilibrium positions. The second derivative is

$$\frac{d^2V}{d\theta^2} = kb^2(\cos^2 \theta - \sin^2 \theta) - \tfrac{1}{2}mgb \cos \theta$$

$$= kb^2(2 \cos^2 \theta - 1) - \tfrac{1}{2}mgb \cos \theta$$

**Solution I.** $\sin \theta = 0, \theta = 0$

$$\frac{d^2V}{d\theta^2} = kb^2(2 - 1) - \tfrac{1}{2}mgb = kb^2 \left(1 - \frac{mg}{2kb}\right)$$

$$= \text{positive (stable)} \quad \text{if } k > mg/2b$$

$$= \text{negative (unstable)} \quad \text{if } k < mg/2b \qquad Ans.$$

Thus, if the spring is sufficiently stiff, the bar will return to the vertical ③ position even though there is no force in the spring at that position.

③ This result is one that we might not have anticipated without the mathematical analysis of the stability.

**Solution II.** $\cos \theta = \frac{mg}{2kb}, \theta = \cos^{-1} \frac{mg}{2kb}$

$$\frac{d^2V}{d\theta^2} = kb^2 \left(2 \left[\frac{mg}{2kb}\right]^2 - 1\right) - \tfrac{1}{2}mgb \left(\frac{mg}{2kb}\right) = kb^2 \left(\left[\frac{mg}{2kb}\right]^2 - 1\right)$$

$$Ans.$$

④ Again, without the benefit of the mathematical analysis of the stability we might have supposed erroneously that the bar could come to rest in a stable equilibrium position for some value of $\theta$ between 0 and 90°.

Since the cosine must be less than unity, we see that this solution is limited to the case where $k > mg/2b$, which makes the second derivative ④ of $V$ negative. Hence, equilibrium for Solution II is never stable. If $k < mg/2b$, we no longer have Solution II since the spring will be too weak to maintain equilibrium at a value of $\theta$ between 0 and 90°.

## PROBLEMS

(Assume that the negative work of friction is negligible in the following problems.)

### *Introductory problems*

**7/33** The small cylinder of mass $m$ and radius $r$ is confined to roll on the circular surface of radius $R$. By the methods of this article, prove that the cylinder is unstable in case $(a)$ and stable in case $(b)$.

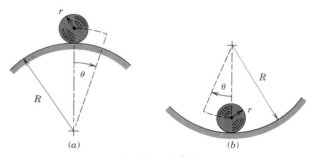

**Problem 7/33**

**7/34** The uniform rectangular plate is hinged about a horizontal axis through point $O$. Prove the stability conditions for the two equilibrium positions.

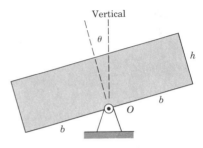

**Problem 7/34**

**7/35** The potential energy of a mechanical system is given by $V = 6x^4 - 3x^2 + 5$, where $x$ is the position coordinate which defines the configuration of the single-degree-of-freedom system. Determine the equilibrium values of $x$ and the stability condition of each.

*Ans.* $x = 0$, unstable; $x = \frac{1}{2}$, stable; $x = -\frac{1}{2}$, stable

**7/36** The uniform bar of mass $m$ and length $l$ is hinged about a horizontal axis through its end $O$ and is attached to a torsional spring which exerts a torque $M = K\theta$ on the rod, where $K$ is the torsional stiffness of the spring in units of torque per radian and $\theta$ is the angular deflection from the vertical in radians. Determine the maximum value of $l$ for which equilibrium at the position $\theta = 0$ is stable.

**Problem 7/36**

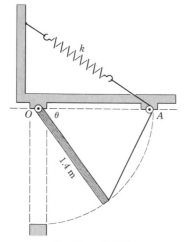

**Problem 7/37**

**7/37** The figure shows the cross section of a uniform 60-kg ventilator door hinged along its upper horizontal edge at $O$. The door is controlled by the spring-loaded cable which passes over the small pulley at $A$. The spring has a stiffness of 160 N per meter of stretch and is undeformed when $\theta = 0$. Determine the angle $\theta$ for equilibrium. *Ans.* $\theta = 52.7°$

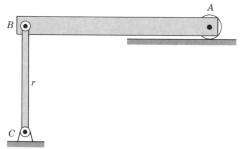

**Problem 7/38**

**7/38** The uniform bar $AB$ of mass $m$ is supported by the roller at $A$ and the light link $CB$ at $B$. Prove that the system is unstable in the equilibrium position shown, where $CB$ is vertical.

*Representative problems*

**7/39** When $u = 0$, the spring of stiffness $k$ is uncompressed. As $u$ increases, the rod slides through the pivoted collar at $A$ and compresses the spring between the collar and the end of the rod. Determine the force $P$ required to produce a given displacement $u$. Assume the absence of friction and neglect the mass of the rod.
$$Ans.\ P = \left(1 - \frac{b}{\sqrt{b^2 + u^2}}\right)ku$$

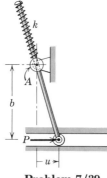

**Problem 7/39**

**7/40** The cylinder of mass $M$ and radius $R$ rolls without slipping on the circular surface of radius $3R$. Attached to the cylinder is a small body of mass $m$. Determine the required relationship between $M$ and $m$ if the body is to be stable in the equilibrium position shown.

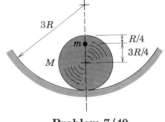

**Problem 7/40**

**7/41** The cross section of a spring-loaded ventilator door hinged about a horizontal axis at $O$ and having a mass $m$ with mass center at $G$ is shown in the figure. As the door opens, the spring is compressed by the hinged rod, which slides through the swivel collar at $A$. The spring is uncompressed when the door is closed. Show that the door will remain in equilibrium for any angle $\theta$ if the spring has the proper stiffness $k$.      *Ans.* $k = mg/a$

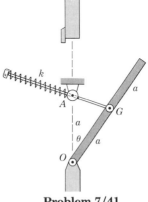

**Problem 7/41**

**7/42** The system of pivoted light bar, end mass $m$, and spring of stiffness $k$ is shown in its equilibrium configuration. Show that the system is stable in that position.

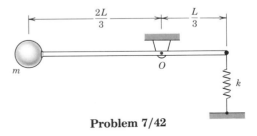

**Problem 7/42**

**7/43** For the device shown the spring would be unstretched in the position $\theta = 0$. Specify the stiffness $k$ of the spring which will establish an equilibrium position $\theta$ in the vertical plane. The mass of the links is negligible compared with $m$.

$$Ans. \ k = \frac{mg}{2b} \frac{\cot \theta}{1 - \cos \theta}$$

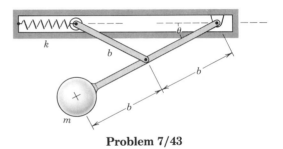

**Problem 7/43**

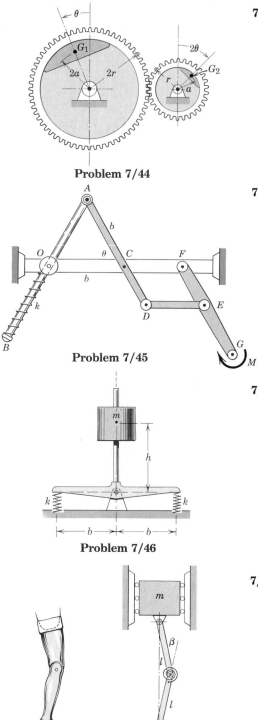

**Problem 7/44**

**7/44** Each of the two gears carries an eccentric mass $m$ and is free to rotate in the vertical plane about its bearing. Determine the values of $\theta$ for equilibrium and identify the type of equilibrium for each position.

**Problem 7/45**

**7/45** In the mechanism shown the rod $AB$ slides through the pivoted collar at $O$ and compresses the spring when the couple $M$ is applied to link $GF$ to increase the angle $\theta$. The spring has a stiffness $k$ and would be uncompressed in the position $\theta = 0$. Determine the angle $\theta$ for equilibrium. The weights of the links are negligible. Figure $CDEF$ is a parallelogram.

$$Ans. \quad \theta = \sin^{-1}\frac{M}{kb^2}$$

**Problem 7/46**

**7/46** Determine the maximum height $h$ of the mass $m$ for which the inverted pendulum will be stable in the vertical position shown. Each of the springs has a stiffness $k$, and they have equal precompressions in this position. Neglect the mass of the remainder of the mechanism.

**Problem 7/47**

**7/47** One of the critical requirements in the design of an artificial leg for an amputee is to prevent the knee joint from buckling under load when the leg is straight. As a first approximation, simulate the artificial leg by the two light links with a torsion spring at their common joint. The spring develops a torque $M = K\beta$, which is proportional to the angle of bend $\beta$ at the joint. Determine the minimum value of $K$ that will ensure stability of the knee joint for $\beta = 0$. $\qquad Ans. \; K_{min} = \frac{1}{2}mgl$

**7/48** For the mechanism of Prob. 4/84, repeated here, the spring of stiffness $k$ has an unstretched length of essentially zero, and the larger link has a mass $m$ with mass center at $B$. The mass of the smaller link is negligible. Determine the equilibrium angle $\theta$ for a given downward force $P$.

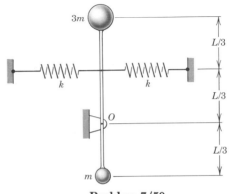

**Problem 7/48**

**7/49** In the figure is shown a small industrial lift with a foot release. There are four identical springs, two on each side of the central shaft. The stiffness of each pair of springs is $2k$. Specify the value of $k$ that will ensure stable equilibrium when the lift supports a load (weight) $L$ in the position where $\theta = 0$ with no force $P$ on the pedal. The springs have an equal initial precompression and may be assumed to act in the horizontal direction at all times.

$$Ans. \;\; k > \frac{L}{2l}$$

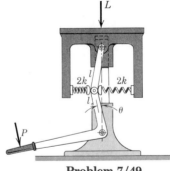

**Problem 7/49**

**7/50** Consider the system of the light bar pivoted freely at $O$, two end masses, and two springs. If the springs are equally stretched when the bar is in the vertical position shown, determine the positions of static equilibrium and investigate the stability of the system at each position. Take the springs to be very long so that they remain horizontal.

**Problem 7/50**

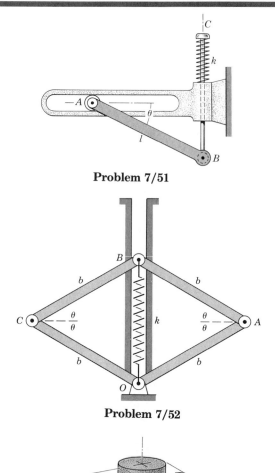

**Problem 7/51**

**Problem 7/52**

**Problem 7/53**

**Problem 7/54**

**7/51** The uniform link $AB$ has a mass $m$, and its left end $A$ travels freely in the fixed horizontal slot. End $B$ is attached to the vertical plunger, which compresses the spring as $B$ falls. The spring would be uncompressed at the position $\theta = 0$. Determine the angle $\theta$ for equilibrium (other than the impossible position corresponding to $\theta = 90°$) and designate the condition which will ensure stability.

$$Ans. \quad \sin\theta = \frac{mg}{2kl}, \; k > \frac{mg}{2l}$$

**7/52** Each of the four uniform links has a mass $m$. The joint $B$ slides freely in the vertical guide and compresses the spring of stiffness $k$ and uncompressed length $2b$. Determine the angle $\theta < 90°$ for equilibrium and prove that the system is stable in that position.

**7/53** The solid hemisphere of diameter $2b$ and concentric cylindrical knob of diameter $b$ are resting on a horizontal surface. Determine the maximum height $h$ which the knob may have without causing the unit to be unstable in the upright position shown. Both parts are made from the same material.

$$Ans. \quad h < b\sqrt{2}$$

**7/54** Predict through calculation whether the homogeneous semicylinder and the half-cylindrical shell will remain in the positions shown or whether they will roll off the lower cylinder.

**7/55** In the mechanism shown, the spring of stiffness $k$ would be uncompressed for the position equivalent to $\theta = 0$. Determine the equilibrium positions and specify their stability. Neglect the mass of the links compared with the mass $m$ of the cylinder.
*Ans.*

$$\theta = \pi, \text{ stable if } k < \frac{3mg}{4b} \text{ and unstable if } k > \frac{3mg}{4b}$$

$$\theta = 2 \sin^{-1}\frac{3mg}{4kb}, \text{ requires } k > \frac{3mg}{4b}, \text{ stable}$$

**7/56** The figure shows a tilting desk chair together with a detail of the spring-loaded tilting mechanism. The frame of the seat is pivoted about the fixed point $O$ on the base. The increase in distance between $A$ and $B$ as the chair tilts back about $O$ is the increase in compression of the spring. The spring, which has a stiffness of 96 kN/m, is uncompressed when $\theta = 0$. For small angles of tilt it may be assumed with negligible error that the axis of the spring remains parallel to the seat. The center of mass of an 80-kg person who sits in the chair is at $G$ on a line through $O$ perpendicular to the seat. Determine the angle of tilt $\theta$ for equilibrium. (*Hint:* The deformation of the spring may be visualized by allowing the base to tilt through the required angle $\theta$ about $O$ while the seat is held in a fixed position.)

**7/57** The front-end suspension of Prob. 4/98 is repeated here. The frame $F$ must be jacked up so that $h = 350$ mm in order to relieve the compression in the coil springs. Determine the value of $h$ when the jack is removed. Each spring has a stiffness of 120 kN/m. The load $L$ is 12 kN, and the central frame $F$ has a mass of 40 kg. Each wheel and attached link has a mass of 35 kg with a center of mass 680 mm from the vertical centerline.      *Ans.* $h = 265$ mm

▶**7/58** The uniform garage door $AB$ shown in section has a mass $m$ and is equipped with two of the spring-loaded mechanisms shown, one on each side of the door. The arm $OB$ has negligible mass, and the upper corner $A$ of the door is free to move horizontally on a roller. The unstretched length of the spring is $r - a$, so that in the top position with $\theta = \pi$ the spring force is zero. To ensure smooth action of the door as it reaches the vertical closed position $\theta = 0$, it is desirable that the door be insensitive to movement in this position. Determine the required spring stiffness $k$.      *Ans.* $k = \dfrac{mg(r + a)}{8a^2}$

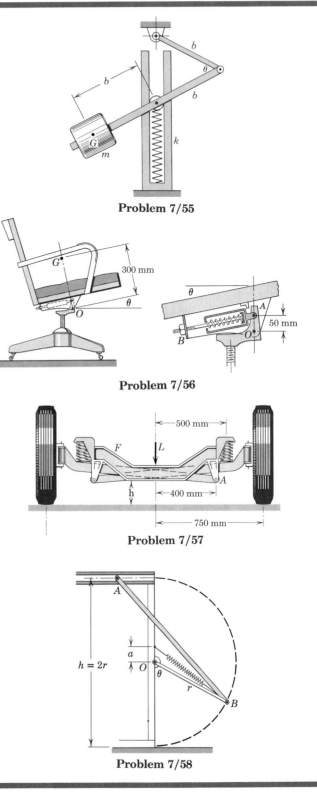

**Problem 7/55**

**Problem 7/56**

**Problem 7/57**

**Problem 7/58**

## 7/5  *PROBLEM FORMULATION AND REVIEW*

When various configurations are possible for a body or a system of interconnected bodies as a result of applied forces, the equilibrium position is found by applying the principle of virtual work. The only forces which need to be considered when determining the equilibrium position by this method are those which do work (active forces) during the assumed differential movement of the body or system away from its equilibrium position. Those external forces which do no work (reactive forces) need not be involved. For this reason we have constructed the active-force diagram of the body or system (rather than the free-body diagram) to focus attention on only those external forces which do work during the virtual displacements.

Relating the corresponding virtual displacements, linear and angular, of the parts of a mechanical system during a virtual movement consistent with the constraints is often the most difficult part of the analysis. First, the geometric relationships which describe the configuration of the system must be written. Next, the differential changes in the positions of parts of the system are established by the process of differentiation of the geometrical relationship to obtain expressions for the differential virtual movements.

In the method of virtual work we take special note of the fact that a virtual displacement is a first-order differential change in a length or an angle. This change is fictitious in that it is an assumed movement which need not take place in reality. Mathematically, a virtual displacement is treated the same as a differential change in an actual movement. We use the symbol $\delta$ for the differential virtual change and the usual symbol $d$ for the differential change in a real movement.

In Chapter 7 we have restricted our attention to mechanical systems for which the positions of the members can be specified by a single variable (single-degree-of-freedom systems). For two or more degrees of freedom, we would apply the virtual-work equation as many times as there are degrees of freedom, allowing one variable to change at a time while holding the remaining ones constant.

We have found that the concept of potential energy, both gravitational ($V_g$) and elastic ($V_e$), is useful in solving equilibrium problems where changes in the vertical position of the mass centers of the bodies occur and where corresponding changes in the length of elastic members (springs) also result during the virtual displacement. Here we obtain an expression for the total potential energy $V$ of the system in terms of the variable that specifies the possible position of the system. The first and second derivatives of $V$ are used to establish, respectively, the position of equilibrium and the type of stability which exists.

# REVIEW PROBLEMS

**7/59** A "black box" contains a series of interconnected racks and pinions, gears, and other internal mechanical elements which transform the linear motion of the pushrod $A$ to produce the linear motion of the pushrod $B$. For every unit of inward movement of $A$ under the action of force $P_1$, rod $B$ moves outward from the box one-fourth of a unit against the action of force $P_2$. If $P_1 = 100$ N, calculate $P_2$ for equilibrium. Neglect all friction and assume all mechanical components are ideally connected light rigid bodies.               *Ans. $P_2 = 400$ N*

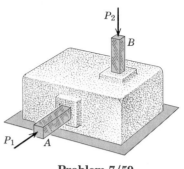

**Problem 7/59**

**7/60** Identify which of the problems ($a$) through ($f$) are best solved ($A$) by the force and moment equilibrium equations and ($B$) by virtual work. Outline briefly the procedure for each solution.

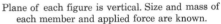

Plane of each figure is vertical. Size and mass of each member and applied force are known.

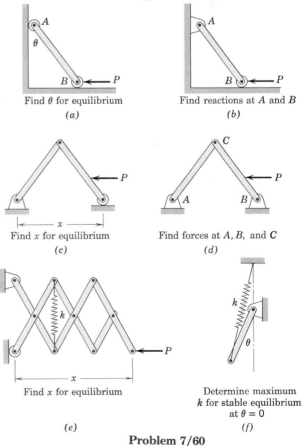

Find $\theta$ for equilibrium
(a)

Find reactions at $A$ and $B$
(b)

Find $x$ for equilibrium
(c)

Find forces at $A, B,$ and $C$
(d)

Find $x$ for equilibrium
(e)

Determine maximum $k$ for stable equilibrium at $\theta = 0$
(f)

**Problem 7/60**

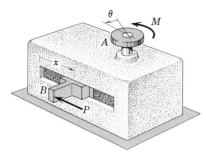

**Problem 7/61**

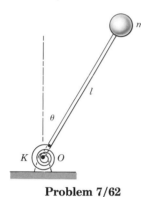

**Problem 7/62**

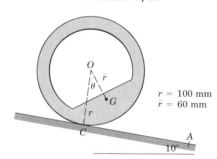

**Problem 7/63**

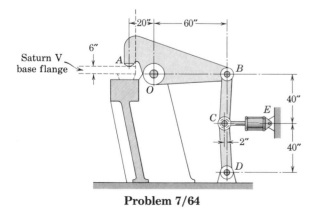

**Problem 7/64**

**7/61** A control mechanism consists of an input shaft at $A$ which is turned by applying a couple $M$ and an output slider $B$ which moves in the $x$-direction against the action of force $P$. The mechanism is arranged so that the linear movement of $B$ is proportional to the angular movement of $A$, with $x$ increasing 60 mm for every complete turn of $A$. If $M = 10$ N·m, determine $P$ for equilibrium. Neglect internal friction and assume all mechanical components are ideally connected rigid bodies.          *Ans. P = 1047 N*

**7/62** The system of light rod, torsional spring $K$, and end mass $m$ is in static equilibrium in the position $\theta = 30°$. Determine the necessary value of the spring constant $K$ for stability in this position and the corresponding static angular deflection $\theta_{st}$ of the spring.

**7/63** Determine the equilibrium values of $\theta$ and the stability of equilibrium at each position for the unbalanced wheel on the 10° incline. Static friction is sufficient to prevent slipping. The mass center is at $G$.          *Ans. $\theta = -6.82°$ or $\theta = 207°$*

**7/64** The sketch shows the approximate configuration of one of the four toggle-action hold-down assemblies that clamp the base flange of the Saturn V rocket vehicle to the pedestal of its platform prior to launching. Calculate the preset clamping force $F$ at $A$ if the link $CE$ is under tension produced by a fluid pressure of 2000 lb/in.² acting on the left side of the piston in the hydraulic cylinder. The piston has a net area of 16 in.² The weight of the assembly is considerable, but it is small compared with the clamping force produced and is therefore neglected here.

**7/65** The figure shows the cross section of a container composed of a hemispherical shell of radius $r$ and a cylindrical shell of height $h$, both made from the same material. Specify the limitation of $h$ for stability in the upright position when the container is placed on the horizontal surface.        *Ans.* $h < r$

**7/66** The potential energy of a mechanical system with negligible friction is $V = a \cos \theta + b \sin^2 \theta$, where $a$ and $b$ are positive constants and $\theta$ is the angular coordinate that defines the position of the system. Determine the position or positions of equilibrium and the stability condition at each position.

**7/67** The uniform aluminum disk of radius $R$ and mass $m$ rolls without slipping on the fixed circular surface of radius $2R$. Fastened to the disk is a lead cylinder also of mass $m$ with its center located a distance $b$ from the center $C$ of the disk. Determine the minimum value of $b$ for which the disk will remain in stable equilibrium on the cylindrical surface in the top position shown.        *Ans.* $b_{\min} = \frac{2}{3}R$

**7/68** The figure shows the cross section of a 135-kg overhead industrial door with hinged segments which pass over the rollers on the cylindrical guide. When the door is closed, the end $A$ is at position $B$. On each side of the door there is a control cable fastened to the bottom of the door and wound around a drum $C$. Each of the two drums is connected to a torsion spring, which offers a torsional resistance that increases by 10 N·m for each revolution of the drum, starting from an unwound position of the spring where $x = 0$. Determine the equilibrium value of $x$ and prove that equilibrium in this position is stable.

**7/69** The unbalanced circular disk with center of mass $G$, a distance $\bar{r}$ from its center $O$, is placed on a concave circular path of radius $R$. Determine the maximum value that $\bar{r}$ may have and still ensure that the disk remains stable in the bottom position shown. (*Hint:* With the aid of the geometry shown in the displaced position, write an expression for the potential energy $V_g$ in terms of $\theta$.)

$$Ans. \quad \bar{r}_{\max} = \frac{r}{\dfrac{R}{r} - 1}$$

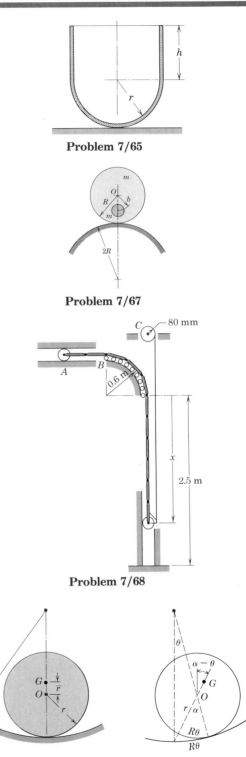

**Problem 7/65**

**Problem 7/67**

**Problem 7/68**

**Problem 7/69**

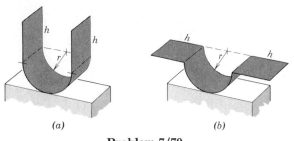

**Problem 7/70**

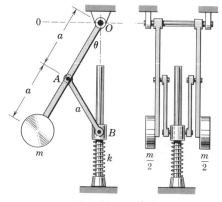

**Problem 7/71**

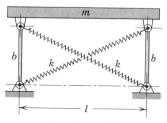

**Problem 7/72**

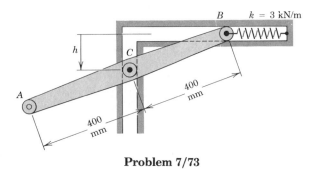

**Problem 7/73**

**7/70** Two semicylindrical shells with equal projecting rectangles are formed from sheet metal, one with configuration (*a*) and the other with configuration (*b*). Both shells rest on a horizontal surface. For case (*a*) determine the maximum value of *h* for which the shell will remain stable in the position shown. For case (*b*) prove that stability in the position shown is not affected by the dimension *h*.

▶**7/71** In the mechanism shown the spring of stiffness *k* is uncompressed when $\theta = 60°$. Also the masses of the parts are small compared with the sum *m* of the masses of the two cylinders. The mechanism is constructed so that the arms may swing past the vertical, as seen in the right-hand side view. Determine the values of $\theta$ for equilibrium and investigate the stability of the mechanism in each position. Neglect friction.

*Ans.* $\theta = 0$; stable if $k < mg/a$,
unstable if $k > mg/a$

$$\theta = \cos^{-1}\frac{1}{2}\left(1 + \frac{mg}{ka}\right)$$
only if $k > mg/a$; stable

▶**7/72** The platform of mass *m* is supported by equal legs and braced by the two springs as shown. If the masses of the legs and springs are negligible, determine the minimum stiffness *k* of each spring that will ensure stability of the platform in the position shown. Each spring has a tensile preset deflection equal to $\Delta$.

*Ans.* $k_{\min} = \dfrac{mg}{2b}\left(1 + \dfrac{b^2}{l^2}\right)$

***Computer-oriented problems***

***7/73** The small roller *C* at the mass center of the 10-kg link *AB* is confined to move in the vertical guide. The roller at end *B* is attached to the spring of stiffness $k = 3$ kN/m and is confined to move in the horizontal guide. When $h = 0$, the spring is unstretched. Determine the equilibrium value of *h*.

*Ans.* $h = 203$ mm

**\*7/74** The bar $OA$, which weighs 50 lb with center of gravity at $G$, is pivoted about its end $O$ and swings in the vertical plane under the constraint of the 20-lb counterweight. Write the expression for the total potential energy of the system, taking $V_g = 0$ when $\theta = 0$, and compute $V_g$ as a function of $\theta$ from $\theta = 0$ to $\theta = 360°$. From your plot of the results, determine the position or positions of equilibrium and the stability of equilibrium at each position.

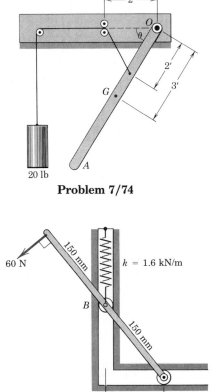

**Problem 7/74**

**\*7/75** Determine the equilibrium value of the coordinte $x$ for the mechanism under the action of the 60-N force applied normal to the light bar. The spring has a stiffness of 1600 N/m and is unstretched when $x = 0$. (*Hint:* Replace the applied force by a force–couple system at point $B$.)      *Ans.* $x = 130.3$ mm

**Problem 7/75**

**\*7/76** The uniform link $OA$ has a mass of 20 kg and is supported in the vertical plane by the spring $AB$ whose unstretched length is 400 mm. Plot the total potential energy $V$ and its derivative $dV/d\theta$ as functions of $\theta$ from $\theta = 0$ to $\theta = 120°$. From the plots identify the equilibrium values of $\theta$ and the corresponding stability of equilibrium. Take $V_g = 0$ on a level through $O$.

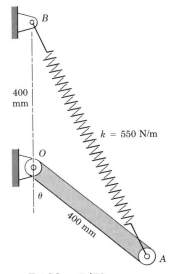

**Problem 7/76**

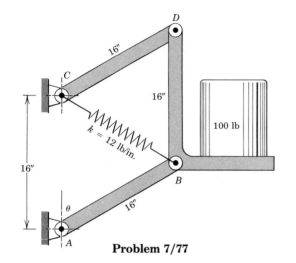

**Problem 7/77**

*7/77  Determine the equilibrium angle $\theta$ for the mechanism shown. The spring of stiffness $k = 12$ lb/in. has an unstretched length of 8 in. Each of the uniform links *AB* and *CD* has a weight of 10 lb, and member *BD* with its load weighs 100 lb. Motion is in the vertical plane.                                      *Ans.* $\theta = 71.7°$

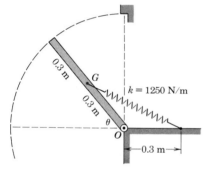

**Problem 7/78**

*7/78  The figure shows the edge view of a 10-kg ventilator door with mass center at *G* and hinged about a horizontal axis through its edge at *O*. When $\theta = 90°$, the restraining spring of stiffness $k = 1250$ N/m is unstretched. Compute and plot the total potential energy of the door and spring as a function of $\theta$ from $\theta = 0$ to $\theta = 90°$. Take $V_g = 0$ at $\theta = 0$. From the plot, specify the equilibrium position, other than $\theta = \pm 90°$, and its stability condition. Can you justify the fact that the equilibrium at $\theta = +90°$ is stable with the spring attached and unstable without the spring?

# AREA MOMENTS OF INERTIA

# A

## A/1 INTRODUCTION

When forces are distributed continuously over an area on which they act, it is often necessary to calculate the moment of these forces about some axis either in or perpendicular to the plane of the area. Frequently the intensity of the force (pressure or stress) is proportional to the distance of the force from the moment axis. The elemental force acting on an element of area, then, is proportional to distance times differential area, and the elemental moment is proportional to distance squared times differential area. We see, therefore, that the total moment involves an integral that has the form $\int (\text{distance})^2 \, d(\text{area})$. This integral is known as the *moment of inertia* or the *second moment* of the area. The integral is a function of the geometry of the area and occurs so frequently in the applications of mechanics that we find it useful to develop its properties in some detail and to have these properties available for ready use when the integral arises.

Figure A/1 illustrates the physical origin of these integrals. In the *a*-part of the figure the surface area *ABCD* is subjected to a distributed pressure *p* whose intensity is proportional to the distance *y* from the axis *AB*. This situation was treated in Art. 5/8 of Chapter 5, where we described the action of liquid pressure on a plane surface. The moment about *AB* that is due to the pressure on the element of area $dA$ is $py \, dA = ky^2 \, dA$. Thus, the integral in question appears when the total moment $M = k \int y^2 \, dA$ is evaluated.

In Fig. A/1*b* we show the distribution of stress acting on a transverse section of a simple elastic beam bent by equal and opposite couples applied to its ends. At any section of the beam a linear distribution of force intensity or stress $\sigma$, given by $\sigma = ky$, is present, the stress being positive (tensile) below the axis *O–O* and negative (compressive) above the axis. We see that the elemental moment about the axis *O–O* is $dM = y(\sigma \, dA) = ky^2 \, dA$. Thus, the same integral appears when the total moment $M = k \int y^2 \, dA$ is evaluated.

A third example is given in Fig. A/1*c*, which shows a circular shaft subjected to a twist or torsional moment. Within the elastic limit of the material this moment is resisted at each cross section

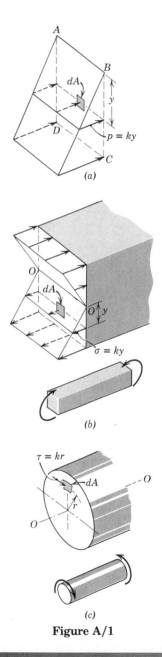

(a)

(b)

(c)

**Figure A/1**

of the shaft by a distribution of tangential or shear stress $\tau$, which is proportional to the radial distance $r$ from the center. Thus, $\tau = kr$, and the total moment about the central axis is $M = \int r(\tau \, dA) = k \int r^2 \, dA$. Here the integral differs from that in the preceding two examples in that the area is normal instead of parallel to the moment axis and in that $r$ is a radial coordinate instead of a rectangular one.

Although the integral illustrated in the preceding examples is generally called the *moment of inertia* of the area about the axis in question, a more fitting term is the *second moment of area*, since the first moment $y \, dA$ is multiplied by the moment arm $y$ to obtain the second moment for the element $dA$. The word *inertia* appears in the terminology by reason of the similarity between the mathematical form of the integrals for second moments of areas and those for the resultant moments of the so-called inertia forces in the case of rotating bodies. The moment of inertia of an area is a purely mathematical property of the area and in itself has no physical significance.

## A/2 DEFINITIONS

The following definitions of terms form the basis for the analysis of area moments of inertia.

**(a) Rectangular and polar moments of inertia.** Consider the area $A$ in the $x$-$y$ plane, Fig. A/2. The moments of inertia of the element $dA$ about the $x$- and $y$-axes are, by definition, $dI_x = y^2 \, dA$ and $dI_y = x^2 \, dA$, respectively. Therefore, the moments of inertia of $A$ about the same axes are

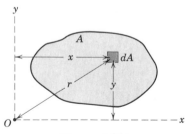

Figure A/2

$$I_x = \int y^2 \, dA$$
$$I_y = \int x^2 \, dA$$
(A/1)

where we carry out the integration over the entire area.

The moment of inertia of $dA$ about the pole $O$ ($z$-axis) is, by similar definition, $dI_z = r^2 \, dA$, and the moment of inertia of the entire area about $O$ is

$$I_z = \int r^2 \, dA$$
(A/2)

The expressions defined by Eqs. A/1 are known as *rectangular* moments of inertia, whereas the expression of Eq. A/2 is known as

the *polar* moment of inertia.* Since $x^2 + y^2 = r^2$, it is clear that

$$I_z = I_x + I_y \qquad\qquad (A/3)$$

A polar moment of inertia for an area whose boundaries are more simply described in rectangular coordinates than in polar coordinates is easily calculated with the aid of Eq. A/3.

We note that the moment of inertia of an element involves the square of the distance from the inertia axis to the element. An element whose coordinate is negative contributes as much to the moment of inertia as does an equal element with a positive coordinate of the same magnitude. Consequently we see that the area moment of inertia about any axis is always a positive quantity. In contrast, the first moment of the area, which was involved in the computations of centroids, could be either positive, negative, or zero.

The dimensions of moments of inertia of areas are clearly $L^4$, where $L$ stands for the dimension of length. Thus, the SI units for area moments of inertia are expressed as quartic meters $(m^4)$ or quartic millimeters $(mm^4)$. The U.S. customary units for area moments of inertia are quartic feet $(ft^4)$ or quartic inches $(in.^4)$.

The choice of the coordinates to use for the calculation of moments of inertia is important. Rectangular coordinates should be used for shapes whose boundaries are most easily expressed in these coordinates. Polar coordinates will usually simplify problems involving boundaries that are easily described in $r$ and $\theta$. The choice of an element of area which simplifies the integration as much as possible is also important. These considerations are quite analogous to those we discussed and illustrated in Chapter 5 for the calculation of centroids.

*(b) Radius of gyration.*    Consider an area $A$, Fig. A/3a, which has rectangular moments of inertia $I_x$ and $I_y$ and a polar moment of inertia $I_z$ about $O$. We now visualize this area to be concentrated into a long narrow strip of area $A$ a distance $k_x$ from the $x$-axis, Fig.

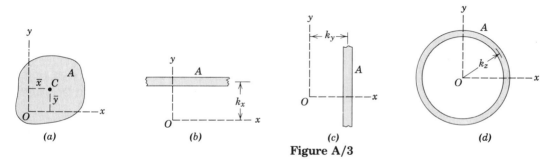

(a)          (b)          (c)          (d)

**Figure A/3**

---

* The polar moment of inertia of an area is sometimes denoted in mechanics literature by the symbol $J$.

A/3*b*. By definition the moment of inertia of the strip about the *x*-axis will be the same as that of the original area if $k_x^2 A = I_x$. The distance $k_x$ is known as the *radius of gyration* of the area about the *x*-axis. A similar relation for the *y*-axis is written by considering the area to be concentrated into a narrow strip parallel to the *y*-axis as shown in Fig. A/3*c*. Also, if we visualize the area to be concentrated into a narrow ring of radius $k_z$ as shown in Fig. A/3*d*, we may express the polar moment of inertia as $k_z^2 A = I_z$. In summary we write

$$\boxed{\begin{aligned} I_x &= k_x^2 A \\ I_y &= k_y^2 A \\ I_z &= k_z^2 A \end{aligned}} \quad \text{or} \quad \boxed{\begin{aligned} k_x &= \sqrt{I_x/A} \\ k_y &= \sqrt{I_y/A} \\ k_z &= \sqrt{I_z/A} \end{aligned}} \quad \textbf{(A/4)}$$

The radius of gyration, then, is a measure of the distribution of the area from the axis in question. A rectangular or polar moment of inertia may be expressed by specifying the radius of gyration and the area.

When we substitute Eqs. A/4 into Eq. A/3, we have

$$\boxed{k_z^2 = k_x^2 + k_y^2} \qquad \textbf{(A/5)}$$

Thus, the square of the radius of gyration about a polar axis equals the sum of the squares of the radii of gyration about the two corresponding rectangular axes.

It is imperative that there be no confusion between the coordinate to the centroid $C$ of an area and the radius of gyration. In Fig. A/3*a* the square of the centroidal distance from the *x*-axis, for example, is $\bar{y}^2$, which is the square of the mean value of the distances from the elements of the area to the *x*-axis. The quantity $k_x^2$, on the other hand, is the mean of the squares of these distances. The moment of inertia is *not* equal to $A\bar{y}^2$, since the square of the mean is less than the mean of the squares.

**(c) Transfer of axes.** The moment of inertia of an area about a noncentroidal axis may be easily expressed in terms of the moment of inertia about a parallel centroidal axis. In Fig. A/4 the $x_0$-$y_0$ axes pass through the centroid $C$ of the area. Let us now determine the moments of inertia of the area about the parallel *x*-*y* axes. By definition the moment of inertia of the element $dA$ about the *x*-axis is

$$dI_x = (y_0 + d_x)^2 \, dA$$

Expanding and integrating give us

$$I_x = \int y_0^2 \, dA + 2d_x \int y_0 \, dA + d_x^2 \int dA$$

We see that the first integral is by definition the moment of inertia $\bar{I}_x$ about the centroidal $x_0$-axis. The second integral is zero, since $\int y_0 \, dA = A\bar{y}_0$ and $\bar{y}_0$ is automatically zero with the centroid on the

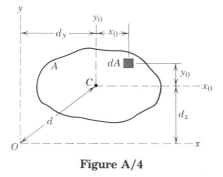

**Figure A/4**

$x_0$-axis. The third term is simply $Ad_x{}^2$. Thus, the expression for $I_x$ and the similar expression for $I_y$ become

$$
\begin{aligned}
I_x &= \bar{I}_x + Ad_x{}^2 \\
I_y &= \bar{I}_y + Ad_y{}^2
\end{aligned}
\tag{A/6}
$$

By Eq. A/3 the sum of these two equations gives

$$
I_z = \bar{I}_z + Ad^2
\tag{A/6a}
$$

Equations A/6 and A/6a are the so-called *parallel-axis theorems*. Two points in particular should be noted. First, the axes between which the transfer is made *must be parallel*, and second, one of the axes *must pass through the centroid* of the area.

If a transfer is desired between two parallel axes neither one of which passes through the centroid, it is first necessary for us to transfer from one axis to the parallel centroidal axis and then to transfer from the centroidal axis to the second axis.

The parallel-axis theorems also hold for radii of gyration. With substitution of the definition of $k$ into Eqs. A/6, the transfer relation becomes

$$
k^2 = \bar{k}^2 + d^2
\tag{A/6b}
$$

where $\bar{k}$ is the radius of gyration about a centroidal axis parallel to the axis about which $k$ applies and $d$ is the distance between the two axes. The axes may be either in the plane or normal to the plane of the area.

A summary of the moment-of-inertia relations for some of the common plane figures is given in Table D/3, Appendix D.

## Sample Problem A/1

Determine the moments of inertia of the rectangular area about the centroidal $x_0$- and $y_0$-axes, the centroidal polar axis $z_0$ through $C$, the $x$-axis, and the polar axis $z$ through $O$.

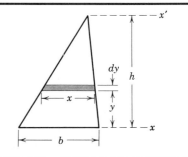

**Solution.** For the calculation of the moment of inertia $\bar{I}_x$ about the $x_0$-axis, a horizontal strip of area $b\,dy$ is chosen so that all elements of the strip have the same $y$-coordinate. Thus,

$$[I_x = \int y^2\,dA] \qquad \bar{I}_x = \int_{-h/2}^{h/2} y^2 b\,dy = \tfrac{1}{12}bh^3 \qquad Ans.$$

By interchanging symbols the moment of inertia about the centroidal $y_0$-axis is

$$\bar{I}_y = \tfrac{1}{12}hb^3 \qquad Ans.$$

The centroidal polar moment of inertia is

$$[I_z = I_x + I_y] \qquad \bar{I}_z = \tfrac{1}{12}(bh^3 + hb^3) = \tfrac{1}{12}A(b^2 + h^2) \qquad Ans.$$

By the parallel-axis theorem the moment of inertia about the $x$-axis is

$$[I_x = \bar{I}_x + Ad_x^2] \qquad I_x = \tfrac{1}{12}bh^3 + bh\left(\frac{h}{2}\right)^2 = \tfrac{1}{3}bh^3 = \tfrac{1}{3}Ah^2 \qquad Ans.$$

We also obtain the polar moment of inertia about $O$ by the parallel-axis theorem, which gives us

$$[I_z = \bar{I}_z + Ad^2] \qquad I_z = \tfrac{1}{12}A(b^2 + h^2) + A\left[\left(\frac{b}{2}\right)^2 + \left(\frac{h}{2}\right)^2\right]$$

$$I_z = \tfrac{1}{3}A(b^2 + h^2) \qquad Ans.$$

① If we had started with the second-order element $dA = dx\,dy$, integration with respect to $x$ holding $y$ constant amounts simply to multiplication by $b$ and gives us the expression $y^2 b\,dy$, which we chose at the outset.

## Sample Problem A/2

Determine the moments of inertia of the triangular area about its base and about parallel axes through its centroid and vertex.

①②  **Solution.** A strip of area parallel to the base is selected as shown in the figure, and it has the area $dA = x\,dy = [(h - y)b/h]\,dy$. By definition

$$[I_x = \int y^2\,dA] \qquad I_x = \int_0^h y^2 \frac{h - y}{h} b\,dy = b\left[\frac{y^3}{3} - \frac{y^4}{4h}\right]_0^h = \frac{bh^3}{12} \qquad Ans.$$

By the parallel-axis theorem the moment of inertia $\bar{I}$ about an axis through the centroid, a distance $h/3$ above the $x$-axis, is

$$[\bar{I} = I - Ad^2] \qquad \bar{I} = \frac{bh^3}{12} - \left(\frac{bh}{2}\right)\left(\frac{h}{3}\right)^2 = \frac{bh^3}{36} \qquad Ans.$$

A transfer from the centroidal axis to the $x'$-axis through the vertex gives

$$[I = \bar{I} + Ad^2] \qquad I_{x'} = \frac{bh^3}{36} + \left(\frac{bh}{2}\right)\left(\frac{2h}{3}\right)^2 = \frac{bh^3}{4} \qquad Ans.$$

① Here again we choose the simplest possible element. If we had chosen $dA = dx\,dy$, we would have to integrate $y^2\,dx\,dy$ with respect to $x$ first. This gives us $y^2 x\,dy$, which is the expression we chose at the outset.

② Expressing $x$ in terms of $y$ should cause no difficulty if we observe the proportional relationship between the similar triangles.

## Sample Problem   A/3

Calculate the moments of inertia of the area of a circle about a diametral axis and about the polar axis through the center. Specify the radii of gyration.

**Solution.** A differential element of area in the form of a circular ring may be used for the calculation of the moment of inertia about the polar $z$-axis through $O$ since all elements of the ring are equidistant from $O$. The elemental area is $dA = 2\pi r_0\, dr_0$, and thus,

$$[I_z = \int r^2\, dA] \qquad I_z = \int_0^r r_0{}^2(2\pi r_0\, dr_0) = \frac{\pi r^4}{2} = \tfrac{1}{2}Ar^2 \qquad Ans.$$

The polar radius of gyration is

$$\left[k = \sqrt{\frac{I}{A}}\right] \qquad\qquad k_z = \frac{r}{\sqrt{2}} \qquad\qquad Ans.$$

By symmetry $I_x = I_y$, so that from Eq. A/3

$$[I_z = I_x + I_y] \qquad I_x = \tfrac{1}{2}I_z = \frac{\pi r^4}{4} = \tfrac{1}{4}Ar^2 \qquad Ans.$$

The radius of gyration about the diametral axis is

$$\left[k = \sqrt{\frac{I}{A}}\right] \qquad\qquad k_x = \frac{r}{2} \qquad\qquad Ans.$$

The foregoing determination of $I_x$ is the simplest possible. The result may also be obtained by direct integration, using the element of area $dA = r_0\, dr_0\, d\theta$ shown in the lower figure. By definition

$$[I_x = \int y^2\, dA] \qquad I_x = \int_0^{2\pi}\int_0^r (r_0 \sin\theta)^2 r_0\, dr_0\, d\theta$$

$$= \int_0^{2\pi} \frac{r^4 \sin^2\theta}{4}\, d\theta$$

$$= \frac{r^4}{4}\frac{1}{2}\left[\theta - \frac{\sin 2\theta}{2}\right]_0^{2\pi} = \frac{\pi r^4}{4} \qquad Ans.$$

① Polar coordinates are certainly indicated here. Also, as before, we choose the simplest and lowest-order element possible, which is the differential ring. It should be evident immediately from the definition that the polar moment of inertia of the ring is its area $2\pi r_0\, dr_0$ times $r_0{}^2$.

② This integration is straightforward, but the use of Eq. A/3 along with the result for $I_z$ is certainly simpler.

458

## Sample Problem A/4

Determine the moment of inertia of the area under the parabola about the x-axis. Solve by using (a) a horizontal strip of area and (b) a vertical strip of area.

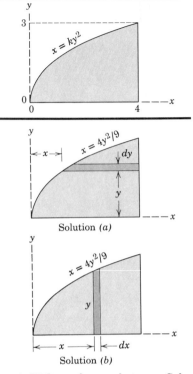

**Solution.** The constant $k = \frac{4}{9}$ is obtained first by substituting $x = 4$ and $y = 3$ into the equation for the parabola.

**(a) Horizontal strip.** Since all parts of the horizontal strip are the same distance from the x-axis, the moment of inertia of the strip about the x-axis is $y^2\, dA$ where $dA = (4 - x)\, dy = 4(1 - y^2/9)\, dy$. Integrating with respect to $y$ gives us

$$[I_x = \int y^2\, dA] \quad I_x = \int_0^3 4y^2\left(1 - \frac{y^2}{9}\right) dy = \frac{72}{5} = 14.4 \text{ (units)}^4 \quad Ans.$$

**(b) Vertical strip.** Here all parts of the element are at different distances from the x-axis, so we must use the correct expression for the moment of inertia of the elemental rectangle about its base, which, from Sample Problem A/1, is $bh^3/3$. For the width $dx$ and the height $y$ the expression becomes

$$dI_x = \tfrac{1}{3}(dx)y^3$$

To integrate with respect to $x$, we must express $y$ in terms of $x$, which gives $y = 3\sqrt{x}/2$, and the integral becomes

$$I_x = \tfrac{1}{3}\int_0^4 \left(\frac{3\sqrt{x}}{2}\right)^3 dx = \frac{72}{5} = 14.4 \text{ (units)}^4 \quad Ans.$$

① There is little preference between Solutions (a) and (b). Solution (b) requires knowing the moment of inertia for a rectangular area about its base.

Solution (a) / Solution (b)

## Sample Problem A/5

Find the moment of inertia about the x-axis of the semicircular area.

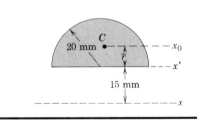

**Solution.** The moment of inertia of the semicircular area about the x'-axis is one-half of that for a complete circle about the same axis. Thus, from the results of Sample Problem A/3

$$I_{x'} = \frac{1}{2}\frac{\pi r^4}{4} = \frac{20^4\pi}{8} = 2\pi(10^4) \text{ mm}^4$$

We obtain the moment of inertia $\bar{I}$ about the parallel centroidal axis $x_0$ next. Transfer is made through the distance $\bar{r} = 4r/(3\pi) = (4)(20)/(3\pi) = 80/(3\pi)$ mm by the parallel-axis theorem. Hence,

$$[\bar{I} = I - Ad^2] \quad \bar{I} = 2(10^4)\pi - \left(\frac{20^2\pi}{2}\right)\left(\frac{80}{3\pi}\right)^2 = 1.755(10^4) \text{ mm}^4$$

Finally, we transfer from the centroidal $x_0$-axis to the x-axis. Thus,

$$[I = \bar{I} + Ad^2] \quad I_x = 1.755(10^4) + \left(\frac{20^2\pi}{2}\right)\left(15 + \frac{80}{3\pi}\right)^2$$

$$= 1.755(10^4) + 34.66(10^4) = 36.4(10^4) \text{ mm}^4 \quad Ans.$$

① This problem illustrates the caution we should observe in using a double transfer of axes since neither the x'- nor the x-axis passes through the centroid C of the area. If the circle were complete with the centroid on the x'-axis, only one transfer would be needed.

## Sample Problem   A/6

Calculate the moment of inertia about the $x$-axis of the area enclosed between the $y$-axis and the circular arcs of radius $a$ whose centers are at $O$ and $A$.

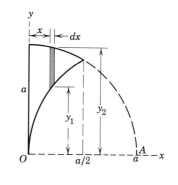

*Solution.*   The choice of a vertical differential strip of area permits one integration to cover the entire area. A horizontal strip would require two integrations with respect to $y$ by virtue of the discontinuity. The moment of inertia of the strip about the $x$-axis is that of a strip of height $y_2$ minus that of a strip of height $y_1$. Thus, from the results of Sample Problem A/1 we write

$$dI_x = \tfrac{1}{3}(y_2\,dx)y_2{}^2 - \tfrac{1}{3}(y_1\,dx)y_1{}^2 = \tfrac{1}{3}(y_2{}^3 - y_1{}^3)\,dx$$

The values of $y_2$ and $y_1$ are obtained from the equations of the two curves, which are $x^2 + y_2{}^2 = a^2$ and $(x - a)^2 + y_1{}^2 = a^2$, and which give ① $y_2 = \sqrt{a^2 - x^2}$ and $y_1 = \sqrt{a^2 - (x - a)^2}$. Thus,

① We choose the positive signs for the radicals here since both $y_1$ and $y_2$ lie above the $x$-axis.

$$I_x = \tfrac{1}{3}\int_0^{a/2}\left\{(a^2 - x^2)\sqrt{a^2 - x^2} - [a^2 - (x - a)^2]\sqrt{a^2 - (x - a)^2}\right\}dx$$

Simultaneous solution of the two equations which define the two circles gives the $x$-coordinate of the intersection of the two curves, which, by inspection, is $a/2$. Evaluation of the integrals gives

$$\int_0^{a/2} a^2\sqrt{a^2 - x^2}\,dx = \frac{a^4}{4}\left(\frac{\sqrt{3}}{2} + \frac{\pi}{3}\right)$$

$$-\int_0^{a/2} x^2\sqrt{a^2 - x^2}\,dx = \frac{a^4}{16}\left(\frac{\sqrt{3}}{4} - \frac{\pi}{3}\right)$$

$$-\int_0^{a/2} a^2\sqrt{a^2 - (x - a)^2}\,dx = \frac{a^4}{4}\left(\frac{\sqrt{3}}{2} - \frac{2\pi}{3}\right)$$

$$\int_0^{a/2} (x - a)^2\sqrt{a^2 - (x - a)^2}\,dx = \frac{a^4}{8}\left(\frac{\sqrt{3}}{8} + \frac{\pi}{3}\right)$$

Collection of the integrals with the factor of $\tfrac{1}{3}$ gives

$$I_x = \frac{a^4}{96}(9\sqrt{3} - 2\pi) = 0.0969a^4 \qquad\qquad Ans.$$

If we had started from a second-order element $dA = dx\,dy$, we would write $y^2\,dx\,dy$ for the moment of inertia of the element about the $x$-axis. Integrating from $y_1$ to $y_2$ holding $x$ constant produces for the vertical strip

$$dI_x = \left[\int_{y_1}^{y_2} y^2\,dy\right]dx = \tfrac{1}{3}(y_2{}^3 - y_1{}^3)\,dx$$

which is the expression we started with by having the moment-of-inertia result for a rectangle in mind.

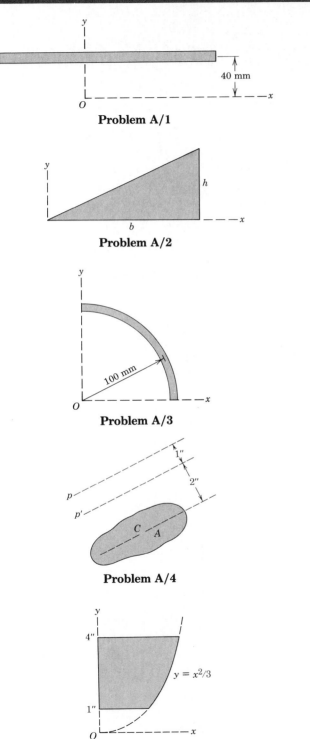

**Problem A/1**

**Problem A/2**

**Problem A/3**

**Problem A/4**

**Problem A/5**

# PROBLEMS

## Introductory problems

**A/1** If the moment of inertia of the thin strip of area about the $x$-axis is $2.56(10^6)$ mm$^4$, determine the area $A$ of the strip to within a close approximation.

*Ans.* $A = 1600$ mm$^2$

**A/2** Determine by direct integration the moment of inertia of the triangular area about the $y$-axis.

**A/3** The thin quarter-circular ring has an area of 1600 mm$^2$. Determine the moment of inertia of the ring about the $x$-axis to a close approximation.

*Ans.* $I_x = 8(10^6)$ mm$^4$

**A/4** The moments of inertia of the area $A$ about the parallel $p$- and $p'$-axes differ by 50 in.$^4$ Compute the area $A$, which has its centroid at $C$.

**A/5** Calculate the moment of inertia of the shaded area about the $y$-axis. *Ans.* $I_y = 21.5$ in.$^4$

**A/6** Show that the moment of inertia of the rectangular area about the $x$-axis through one end may be used for its polar moment of inertia about point $O$ when $b$ is small compared with $a$. What is the percentage error $n$ when $b/a = 1/10$?

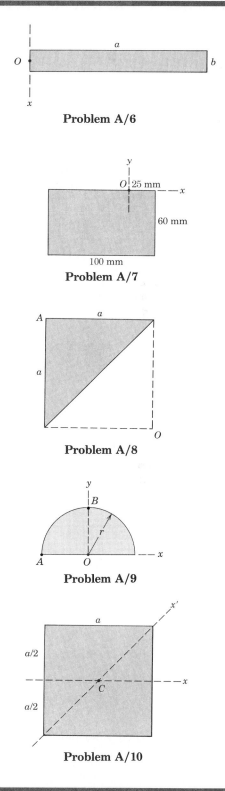

**Problem A/6**

**A/7** Calculate the moment of inertia of the rectangular area about the $x$-axis and find the polar moment of inertia about point $O$.

*Ans.* $I_x = 7.2(10^6)$ mm$^4$, $I_O = 15.95(10^6)$ mm$^4$

**Problem A/7**

### Representative problems

**A/8** Determine the polar radii of gyration of the triangular area about points $O$ and $A$.

**Problem A/8**

**A/9** Determine the polar moments of inertia of the semicircular area about points $A$ and $B$.

*Ans.* $I_A = \frac{3}{4}\pi r^4$, $I_B = r^4\left(\frac{3\pi}{4} - \frac{4}{3}\right)$

**Problem A/9**

**A/10** In two different ways show that the moments of inertia of the square area about the $x$- and $x'$-axes are the same.

**Problem A/10**

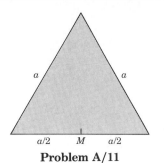

**Problem A/11**

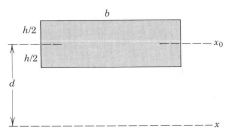

**Problem A/12**

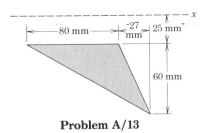

**Problem A/13**

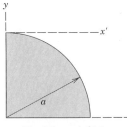

**Problem A/14**

**A/11** Determine the polar radius of gyration of the area of the equilateral triangle about the midpoint $M$ of its base.

$$Ans. \ k_M = \frac{a}{\sqrt{6}}$$

**A/12** The moment of inertia about the $x$-axis of the rectangle of area $A$ is approximately equal to $Ad^2$ if $h$ is small compared with $d$. Determine and plot the percentage error $n$ of the approximate value for $h/d$ ratios from 0.1 to 1. What is the percentage error for $h = d/4$?

**A/13** Calculate the moment of inertia of the triangular area about the $x$-axis.    $Ans. \ I_x = 534(10^4) \ mm^4$

**A/14** Determine the moment of inertia of the quarter-circular area about the tangent $x'$-axis.

**A/15** Calculate the moment of inertia of the shaded area about the *x*-axis.                          *Ans.* $I_x = 9(10^4)$ mm$^4$

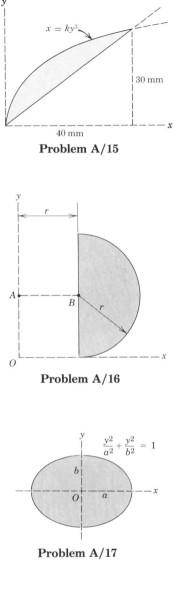

**Problem A/15**

**A/16** Determine the moment of inertia about the *x*-axis and the polar radius of gyration about point *O* for the semicircular area shown.

**Problem A/16**

**A/17** Determine the moment of inertia of the elliptical area about the *y*-axis and find the polar radius of gyration about the origin *O* of the coordinates.

$$Ans. \ I_y = \frac{\pi a^3 b}{4}, \ k_O = \tfrac{1}{2}\sqrt{a^2 + b^2}$$

**Problem A/17**

**A/18** Calculate by direct integration the moment of inertia of the shaded area about the *x*-axis. Solve, first, by using a horizontal strip having differential area and, second, by using a vertical strip of differential area.

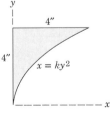

**Problem A/18**

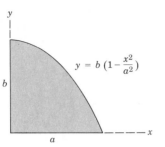

**Problem A/19**

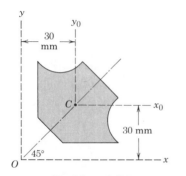

**Problem A/20**

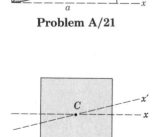

**Problem A/21**

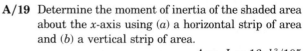

**Problem A/22**

**A/19** Determine the moment of inertia of the shaded area about the $x$-axis using ($a$) a horizontal strip of area and ($b$) a vertical strip of area.

*Ans.* $I_x = 16ab^3/105$

**A/20** The plane figure is symmetrical with respect to the 45° line and has an area of 1600 mm². Its polar moment of inertia about its centroid $C$ is $40(10^4)$ mm⁴. Compute ($a$) the polar radius of gyration about $O$ and ($b$) the radius of gyration about the $x_0$-axis.

**A/21** Determine the moments of inertia of the shaded area about the $x$- and $y$-axes. Use the same differential element for both calculations.

*Ans.* $I_x = a^4/28, I_y = a^4/20$

**A/22** Show that the moment of inertia of the area of the square about any axis $x'$ through its center is the same as that about a central axis $x$ parallel to a side.

**A/23** Determine the moment of inertia of the shaded area about the $x$- and $y$-axes.

$$Ans. \ I_x = \frac{4}{9}\frac{ab^3}{\pi}, \ I_y = \frac{ba^3}{\pi}\left(1 - \frac{4}{\pi^2}\right)$$

**A/24** From considerations of symmetry show that $I_{x'} = I_{y'} = I_x = I_y$ for the semicircular area regardless of the angle $\alpha$.

**A/25** Determine the moment of inertia of the shaded area about the $y$- and $y'$-axes.

$$Ans. \ I_y = 30{,}000 \text{ in.}^4, \ I_{y'} = 16{,}000 \text{ in.}^4$$

**A/26** Determine the radius of gyration about the $y$-axis of the shaded area shown.

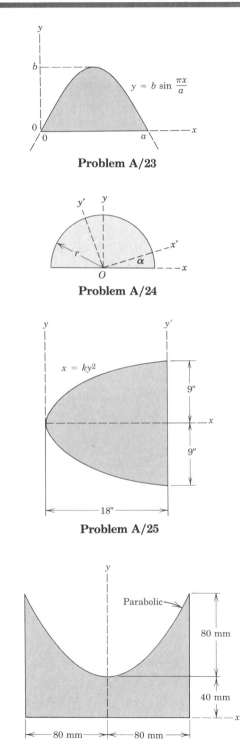

$$y = b \sin \frac{\pi x}{a}$$

**Problem A/23**

**Problem A/24**

$$x = ky^2$$

9"

9"

18"

**Problem A/25**

Parabolic

80 mm

40 mm

80 mm

80 mm

**Problem A/26**

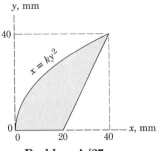

**Problem A/27**

**A/27** Determine the moment of inertia of the shaded area about the $y$-axis.      *Ans.* $I_y = 27.8(10^4)$ mm$^4$

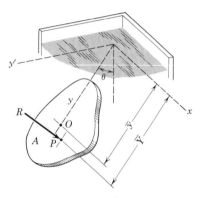

**Problem A/28**

**A/28** A flat plate of area $A$ and centroid $O$ is submerged in a liquid of density $\rho$. Show that the resultant $R$ of the hydrostatic forces exerted on one face of the plate by the liquid acts at a point $P$ (called the *center of pressure*) located by $\overline{Y} = I_x/(A\bar{y})$, where $I_x$ is the moment of inertia of the area about the $x$-axis at the surface and $\bar{y}$ is the $y$-coordinate of the centroid $O$ measured in the plane of the plate.

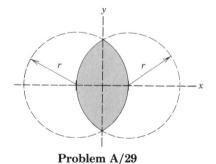

**Problem A/29**

►**A/29** Calculate the moment of inertia of the overlapping shaded area of the two circles about the $x$-axis.
      *Ans.* $I_x = 0.1988r^4$

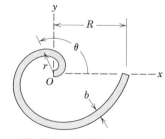

**Problem A/30**

►**A/30** A narrow strip of area of constant width $b$ has the form of a spiral $r = k\theta$. After one complete turn from $\theta = 0$ to $\theta = 2\pi$, the end radius of the spiral is $R$. Determine the polar moment of inertia and the radius of gyration of the area about $O$.
      *Ans.* $I_O = 1.609R^3b$, $k_O = 0.690R$

## A/3 COMPOSITE AREAS

It is frequently necessary to calculate the moment of inertia of an area composed of a number of distinct parts of simple and calculable geometric shape. Since a moment of inertia is the integral or sum of the products of distance squared times element of area, it follows that the moment of inertia of a positive area is always a positive quantity. Therefore, the moment of inertia of a composite area about a particular axis is simply the sum of the moments of inertia of its component parts about the same axis. It is often convenient to regard a composite area as being composed of positive and negative parts. We may then treat the moment of inertia of a negative area as a negative quantity.

When a composite area is composed of a large number of parts, it is convenient to tabulate the results for each of the parts in terms of its area $A$, its centroidal moment of inertia $\bar{I}$, the distance $d$ from its centroidal axis to the axis about which the moment of inertia of the entire section is being computed, and the product $Ad^2$. For any one of the parts the moment of inertia about the desired axis by the transfer-of-axis theorem is $\bar{I} + Ad^2$. Thus, for the entire section the desired moment of inertia becomes $I = \Sigma\bar{I} + \Sigma Ad^2$.

For such an area in the $x$-$y$ plane, for example, and with the notation of Fig. A/4, where $\bar{I}_x$ is the same as $I_{x_0}$ and $\bar{I}_y$ is the same as $I_{y_0}$, the tabulation would include

| Part | Area, $A$ | $d_x$ | $d_y$ | $Ad_x^2$ | $Ad_y^2$ | $\bar{I}_x$ | $\bar{I}_y$ |
|------|-----------|-------|-------|----------|----------|-------------|-------------|
|      |           |       |       |          |          |             |             |
|      |      | Sums | | $\Sigma Ad_x^2$ | $\Sigma Ad_y^2$ | $\Sigma\bar{I}_x$ | $\Sigma\bar{I}_y$ |

From the sums of the four columns, then, the moments of inertia for the composite area about the $x$- and $y$-axes become

$$I_x = \Sigma\bar{I}_x + \Sigma Ad_x^2$$

$$I_y = \Sigma\bar{I}_y + \Sigma Ad_y^2$$

Although we may add the moments of inertia of the individual parts of a composite area about a given axis, we may not add their radii of gyration. The radius of gyration for the composite area about the axis in question is given by $k = \sqrt{I/A}$, where $I$ is the total moment of inertia and $A$ is the total area of the composite figure. Similarly, the radius of gyration $k$ about a polar axis through some point equals $\sqrt{I_z/A}$, where $I_z = I_x + I_y$ for $x$-$y$ axes through that point.

## Sample Problem A/7

Calculate the moment of inertia and radius of gyration about the $x$-axis for the shaded area shown.

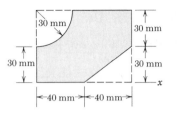

**Solution.** The composite area is composed of the positive area of the rectangle (1) and the negative areas of the quarter circle (2) and triangle (3). For the rectangle the moment of inertia about the $x$-axis, from Sample Problem A/1 (or Table D/3), is

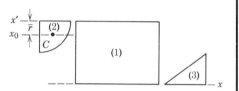

$$I_x = \tfrac{1}{3}Ah^2 = \tfrac{1}{3}(80)(60)(60)^2 = 5.76(10^6) \text{ mm}^4$$

From Sample Problem A/3 (or Table D/3), the moment of inertia of the negative quarter-circular area about its base axis $x'$ is

$$I_{x'} = -\frac{1}{4}\left(\frac{\pi r^4}{4}\right) = -\frac{\pi}{16}(30)^4 = -0.1590(10^6) \text{ mm}^4$$

We now transfer this result through the distance $\bar{r} = 4r/(3\pi) = 4(30)/(3\pi) = 12.73$ mm by the transfer-of-axis theorem to get the centroidal moment of inertia of part (2) (or use Table D/3 directly).

① $[\bar{I} = I - Ad^2]$    $\bar{I}_x = -0.1590(10^6) - \left[ -\frac{\pi(30)^2}{4}(12.73)^2 \right]$

$$= -0.0445(10^6) \text{ mm}^4$$

The moment of inertia of the quarter-circular part about the $x$-axis is now

② $[I = \bar{I} + Ad^2]$    $I_x = -0.0445(10^6) + \left[ -\frac{\pi(30)^2}{4} \right](60 - 12.73)^2$

$$= -1.624(10^6) \text{ mm}^4$$

Finally, the moment of inertia of the negative triangular area (3) about its base, from Sample Problem A/2 (or Table D/3), is

$$I_x = -\tfrac{1}{12}bh^3 = -\tfrac{1}{12}(40)(30)^3 = -0.09(10^6) \text{ mm}^4$$

The total moment of inertia about the $x$-axis of the composite area is, consequently,

③    $I_x = 5.76(10^6) - 1.624(10^6) - 0.09(10^6) = 4.046(10^6) \text{ mm}^4$    *Ans.*

The net area of the figure is $A = 60(80) - \tfrac{1}{4}\pi(30)^2 - \tfrac{1}{2}(40)(30) = 3493$ mm$^2$ so that the radius of gyration about the $x$-axis is

$$k_x = \sqrt{I_x/A} = \sqrt{4.046(10^6)/3493} = 34.0 \text{ mm} \qquad Ans.$$

① Note that we must transfer the moment of inertia for the quarter-circular area to its centroidal axis $x_0$ before we can transfer it to the $x$-axis, as was done in Sample Problem A/5.

② We watch our signs carefully here. Since the area is negative, both $\bar{I}$ and $A$ carry negative signs.

③ If there had been more than the three parts to the composite area, we would have arranged a tabulation of the $\bar{I}$ terms and the $Ad^2$ terms so as to keep a systematic account of the terms and obtain $I = \Sigma\bar{I} + \Sigma Ad^2$.

# PROBLEMS

### *Introductory problems*

**A/31**  Determine the moment of inertia about the $x$-axis of the square area without and with the central circular hole.                $Ans.$  $I_x = 21.3R^4, I_x = 20.6R^4$

**A/32**  Determine the polar moment of inertia of the circular area without and with the central square hole.

**A/33**  Calculate the polar radius of gyration of the area of the angle section about point $A$. Note that the width of the legs is small compared to the length of each leg.                $Ans.$  $k_A = 10.4$ in.

**A/34**  The cross-sectional area of a wide-flange I-beam has the dimensions shown. Obtain a close approximation to the handbook value of $\bar{I}_x = 657$ in.$^4$ by treating the section as being composed of three rectangles.

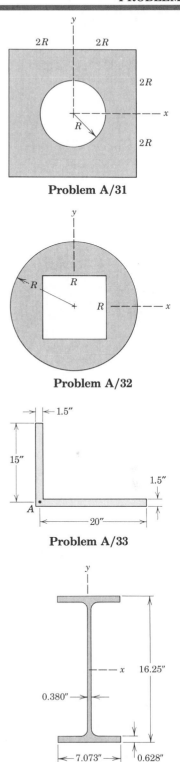

**Problem A/31**

**Problem A/32**

**Problem A/33**

**Problem A/34**

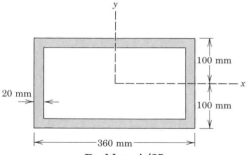

**Problem A/35**

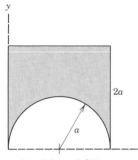

**Problem A/36**

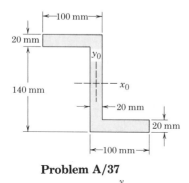

**Problem A/37**

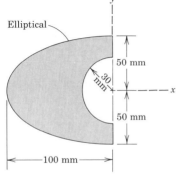

**Problem A/38**

**A/35** Determine the moment of inertia of the shaded area about the $x$-axis in two ways. The wall thickness is 20 mm on all four sides of the rectangle.

*Ans.* $I_x = 130.8(10^6)$ mm$^4$

**A/36** Determine the moments of inertia of the shaded area about the $x$- and $y$-axes.

### Representative problems

**A/37** Determine the moments of inertia of the Z-section about its centroidal $x_0$- and $y_0$-axes.

*Ans.* $\bar{I}_x = 22.6(10^6)$ mm$^4$
$\bar{I}_y = 9.81(10^6)$ mm$^4$

**A/38** Determine the moment of inertia of the shaded area about the $y$-axis.

**A/39** Determine the moment of inertia of the shaded area about the *x*-axis in two different ways.

Ans. $I = \frac{58}{3}a^4$

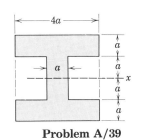

**Problem A/39**

**A/40** Calculate the moment of inertia of the cross section of the beam about its centroidal $x_0$-axis.

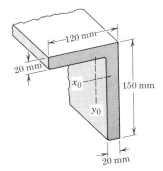

**Problem A/40**

**A/41** A floor joist which measures a full 2 in. by 8 in. has a 1-in. hole drilled through it for a water-pipe installation. Determine the percent reduction *n* in the moment of inertia of the cross-sectional area about the *x*-axis (compared with the undrilled joist) for hole locations in the range $0 \le y \le 3.5$ in. Evaluate your expression for $y = 2$ in.

Ans. $n = 0.1953 + 2.34y^2$, $n = 9.57\%$

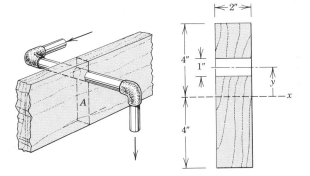

**Problem A/41**

**A/42** Make use of the results of Prob. A/29 and determine the moment of inertia $I_y$ of the shaded area shown.

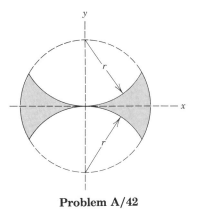

**Problem A/42**

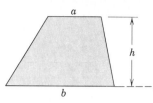

**Problem A/43**

**A/43** Derive the expression for the moment of inertia of the trapezoidal area about the $x$-axis through its base.    *Ans.* $I_x = \frac{1}{12}(b + 3a)h^3$

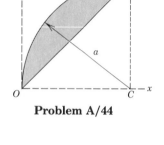

**Problem A/44**

**A/44** For the shaded area shown determine the polar moments of inertia about point $C$.

**Problem A/45**

**A/45** Approximate the moment of inertia about the $x$-axis of the semicircular area by dividing it into five horizontal strips of equal width. Treat the moment of inertia of each strip as its area (width times length of its horizontal midline) times the square of the distance from its midline to the $x$-axis. Compare your result with the exact value.
*Ans.* $I_x \cong 254$ in.$^4$, $I_{x_{\text{exact}}} = 245.4$ in.$^4$

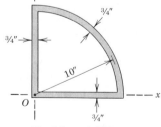

**Problem A/46**

**A/46** Calculate the polar radius of gyration about point $O$ of the area shown. Note that the widths of the elements are small compared with their lengths.

**A/47** A hollow mast of circular section as shown is to be stiffened by bonding two strips of the same material and of rectangular section to the mast throughout its length. Determine the proper dimension $h$ of each near rectangle which will exactly double the stiffness of the mast to bending in the $y$-$z$ plane. (Stiffness in the $y$-$z$ plane is proportional to the area moment of inertia about the $x$-axis.) Take the inner boundary of each strip to be a straight line.

*Ans.* $h = 1.90$ in.

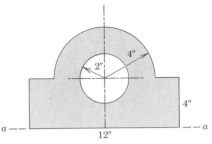

**Problem A/47**

**A/48** The cross section of a bearing block is shown in the figure by the shaded area. Calculate the moment of inertia of the section about its base $a$-$a$.

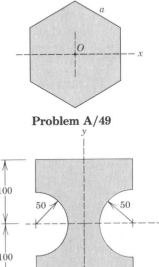

**Problem A/48**

**A/49** Develop a formula for the moment of inertia of the regular hexagonal area of side $a$ about its central $x$-axis.

*Ans.* $I_x = \dfrac{5\sqrt{3}}{16} a^4$

**Problem A/49**

**A/50** Calculate the moments of inertia of the shaded area about the $x$- and $y$-axes.

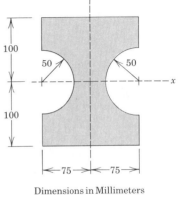

Dimensions in Millimeters

**Problem A/50**

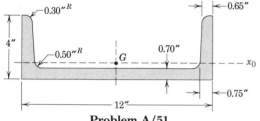

**Problem A/51**

**A/51** Calculate the moment of inertia of the standard $12 \times 4$ in. channel section about the centroidal $x_0$-axis. Neglect the fillets and radii and compare with the handbook value of $\bar{I}_x = 16.0$ in.[4]

*Ans.* $\bar{I}_x = 16.00$ in.[4]

▶**A/52** For the 2 in. by 8 in. floor joist with the circular hole of Prob. A/41, determine an expression for the percent $n$ by which the moment of inertia of the shaded area about its centroidal $x'$-axis (parallel to $x$) is reduced from the moment of inertia of the complete undrilled section about the $x$-axis. Express $n$ in terms of $y$ for the range $0 \le y \le 3.5$ in. Evaluate your expression for $y = 2$ in.

*Ans.* $n = 0.1953 + 2.68y^2$, $n = 10.91\%$

$y$

$10 \mid 10$

$h$

$10$

$30 \mid 30$

$x$

Dimensions in Millimeters

**Problem A/53**

▶**A/53** Calculate the value of $h$ for which $I_x = I_y$ for the shaded area shown. (*Hint:* The solution of a cubic equation is required here. Refer to Art. C/4 or C/11 of Appendix C for solving a cubic equation.)

*Ans.* $h = 20.0$ mm

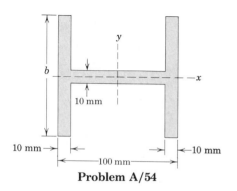

**Problem A/54**

▶**A/54** For the H-beam section, determine the flange width $b$ that will make the moments of inertia about the central $x$- and $y$-axes equal. (*Hint:* Read the hint for the previous problem.)

*Ans.* $b = 161$ mm

## A/4 PRODUCTS OF INERTIA AND ROTATION OF AXES

***(a) Definition.*** In certain problems involving unsymmetrical cross sections and in the calculation of moments of inertia about rotated axes, an expression $dI_{xy} = xy \, dA$ occurs, which has the integrated form

$$I_{xy} = \int xy \, dA \qquad \text{(A/7)}$$

where $x$ and $y$ are the coordinates of the element of area $dA = dx \, dy$. The quantity $I_{xy}$ is called the *product of inertia* of the area $A$ with respect to the $x$-$y$ axes. Unlike moments of inertia, which are always positive for positive areas, the product of inertia may be positive, negative, or zero.

The product of inertia is zero whenever either one of the reference axes is an axis of symmetry, such as the $x$-axis for the area of Fig. A/5. Here we see that the sum of the terms $x(-y) \, dA$ and $x(+y) \, dA$ due to symmetrically placed elements vanishes. Since the entire area may be considered to be composed of pairs of such elements, it follows that the product of inertia $I_{xy}$ for the entire area is zero.

***(b) Transfer of axes.*** By definition the product of inertia of the area $A$ in Fig. A/4 with respect to the $x$- and $y$-axes in terms of the coordinates $x_0$, $y_0$ to the centroidal axes is

$$I_{xy} = \int (x_0 + d_y)(y_0 + d_x) \, dA$$

$$= \int x_0 y_0 \, dA + d_x \int x_0 \, dA + d_y \int y_0 \, dA + d_x d_y \int dA$$

The first integral is by definition the product of inertia about the centroidal axes, which we write $\bar{I}_{xy}$. The middle two integrals are both zero since the first moment of the area about its own centroid is necessarily zero. The third integral is merely $d_x d_y A$. Thus, the transfer-of-axis theorem for products of inertia becomes

$$I_{xy} = \bar{I}_{xy} + d_x d_y A \qquad \text{(A/8)}$$

***(c) Rotation of axes.*** The product of inertia is useful when we need to calculate the moment of inertia of an area about inclined axes. This consideration leads directly to the important problem of determining the axes about which the moment of inertia is a maximum and a minimum.

In Fig. A/6 the moments of inertia of the area about the $x'$- and $y'$-axes are

$$I_{x'} = \int y'^2 \, dA = \int (y \cos \theta - x \sin \theta)^2 \, dA$$

$$I_{y'} = \int x'^2 \, dA = \int (y \sin \theta + x \cos \theta)^2 \, dA$$

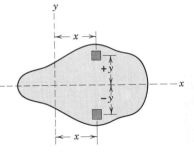

**Figure A/5**

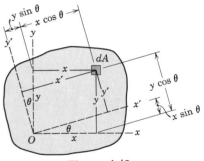

**Figure A/6**

where $x'$ and $y'$ have been replaced by their equivalent expressions as seen from the geometry of the figure.

Expanding and substituting the trigonometric identities,

$$\sin^2 \theta = \frac{1 - \cos 2\theta}{2} \qquad \cos^2 \theta = \frac{1 + \cos 2\theta}{2}$$

and the defining relations for $I_x, I_y, I_{xy}$ give us

$$I_{x'} = \frac{I_x + I_y}{2} + \frac{I_x - I_y}{2} \cos 2\theta - I_{xy} \sin 2\theta$$

$$I_{y'} = \frac{I_x + I_y}{2} - \frac{I_x - I_y}{2} \cos 2\theta + I_{xy} \sin 2\theta$$

$$\textbf{(A/9)}$$

In a similar manner we write the product of inertia about the inclined axes as

$$I_{x'y'} = \int x'y' \, dA = \int (y \sin \theta + x \cos \theta)(y \cos \theta - x \sin \theta) \, dA$$

Expanding and substituting the trigonometric identities

$$\sin \theta \cos \theta = \tfrac{1}{2} \sin 2\theta \qquad \cos^2 \theta - \sin^2 \theta = \cos 2\theta$$

and the defining relations for $I_x, I_y, I_{xy}$ give us

$$I_{x'y'} = \frac{I_x - I_y}{2} \sin 2\theta + I_{xy} \cos 2\theta \qquad \textbf{(A/9a)}$$

Adding Eqs. A/9 gives $I_{x'} + I_{y'} = I_x + I_y = I_z$, the polar moment of inertia about $O$, which checks the results of Eq. A/3.

The angle which makes $I_{x'}$ and $I_{y'}$ a maximum or a minimum may be determined by setting the derivative of either $I_{x'}$ or $I_{y'}$ with respect to $\theta$ equal to zero. Thus,

$$\frac{dI_{x'}}{d\theta} = (I_y - I_x) \sin 2\theta - 2I_{xy} \cos 2\theta = 0$$

Denoting this critical angle by $\alpha$ gives

$$\tan 2\alpha = \frac{2I_{xy}}{I_y - I_x} \qquad \textbf{(A/10)}$$

Equation A/10 gives two values for $2\alpha$ which differ by $\pi$ since $\tan 2\alpha = \tan (2\alpha + \pi)$. Consequently the two solutions for $\alpha$ will differ by $\pi/2$. One value defines the axis of maximum moment of inertia, and the other value defines the axis of minimum moment of inertia. These two rectangular axes are known as the *principal axes of inertia*.

When we substitute Eq. A/10 for the critical value of $2\theta$ in Eq. A/9a, we see that the product of inertia is zero for the principal axes of inertia. Substitution of $\sin 2\alpha$ and $\cos 2\alpha$, obtained from Eq. A/10, for $\sin 2\theta$ and $\cos 2\theta$ in Eqs. A/9 gives the magnitudes of the principal moments of inertia as

$$I_{max} = \frac{I_x + I_y}{2} + \frac{1}{2}\sqrt{(I_x - I_y)^2 + 4I_{xy}^2}$$

$$I_{min} = \frac{I_x + I_y}{2} - \frac{1}{2}\sqrt{(I_x - I_y)^2 + 4I_{xy}^2}$$

**(A/11)**

*(d) Mohr's circle of inertia.* We may represent the relations in Eqs. A/9, A/9a, A/10, and A/11 graphically by a diagram known as Mohr's circle. For given values of $I_x$, $I_y$, and $I_{xy}$ the corresponding values of $I_{x'}$, $I_{y'}$, and $I_{x'y'}$ may be determined from the diagram for any desired angle $\theta$. A horizontal axis for the measurement of moments of inertia and a vertical axis for the measurement of products of inertia are first selected, Fig. A/7. Next, point $A$, which has the coordinates $(I_x, I_{xy})$, and point $B$, which has the coordinates $(I_y, -I_{xy})$, are located. We now draw a circle with these two points as the extremities of a diameter. The angle from the radius $OA$ to the horizontal axis is $2\alpha$ or twice the angle from the $x$-axis of the area in question to the axis of maximum moment of inertia. The angle on the diagram and the angle on the area are both measured in the same sense as shown. The coordinates of any point $C$ are $(I_{x'}, I_{x'y'})$, and those of the corresponding point $D$ are $(I_{y'}, -I_{x'y'})$. Also the angle between $OA$ and $OC$ is $2\theta$ or twice the angle from the $x$-axis to the $x'$-axis. Again we measure both angles in the same sense as shown. We may verify from the trigonometry of the circle that Eqs. A/9, A/9a, and A/10 agree with the statements made.

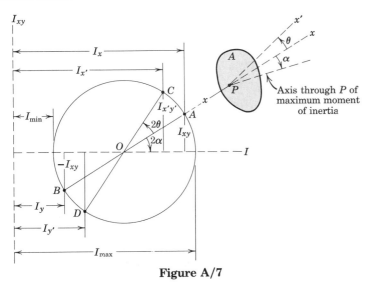

**Figure A/7**

## Sample Problem A/8

Determine the product of inertia of the rectangular area with centroid at $C$ with respect to the $x$-$y$ axes parallel to its sides.

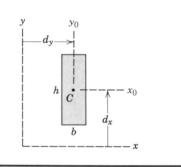

**Solution.** Since the product of inertia $\bar{I}_{xy}$ about the axes $x_0$-$y_0$ is zero by symmetry, the transfer-of-axis theorem gives us

$$[I_{xy} = \bar{I}_{xy} + d_x d_y A] \qquad I_{xy} = d_x d_y bh \qquad \qquad Ans.$$

In this example both $d_x$ and $d_y$ are shown positive. We must be careful to be consistent with the positive directions of $d_x$ and $d_y$ as defined so that their proper signs are observed.

## Sample Problem A/9

Determine the product of inertia about the $x$-$y$ axes for the area under the parabola.

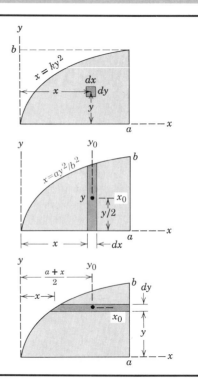

**Solution.** With the substitution of $x = a$ when $y = b$, the equation of the curve becomes $x = ay^2/b^2$.

**Solution I.** If we start with the second-order element $dA = dx\,dy$, we have $dI_{xy} = xy\,dx\,dy$. The integral over the entire area is

$$I_{xy} = \int_0^b \int_{ay^2/b^2}^a xy\,dx\,dy = \int_0^b \frac{1}{2}\left(a^2 - \frac{a^2 y^4}{b^4}\right) y\,dy = \tfrac{1}{6}a^2 b^2 \qquad Ans.$$

**Solution II.** Alternatively we can start with a first-order elemental strip and save one integration by using the results of Sample Problem ① A/8. Taking a vertical strip $dA = y\,dx$ gives $dI_{xy} = 0 + (\tfrac{1}{2}y)(x)(y\,dx)$, where the distances to the centroidal axes of the elemental rectangle are $d_x = y/2$ and $d_y = x$. Now we have

$$I_{xy} = \int_0^a \frac{y^2}{2} x\,dx = \int_0^a \frac{xb^2}{2a} x\,dx = \frac{b^2}{6a} x^3 \Big]_0^a = \tfrac{1}{6}a^2 b^2 \qquad Ans.$$

① If we had chosen a horizontal strip, our expression would have become $dI_{xy} = y\tfrac{1}{2}(a + x)[(a - x)\,dy]$, which when integrated, of course, gives us the same result as before.

## Sample Problem A/10

Determine the product of inertia of the semicircular area with respect to the $x$-$y$ axes.

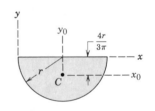

① **Solution.** We use the transfer-of-axis theorem, Eq. A/8, to write

$$[I_{xy} = \bar{I}_{xy} + d_x d_y A] \qquad I_{xy} = 0 + \left(-\frac{4r}{3\pi}\right)(r)\left(\frac{\pi r^2}{2}\right) = -\frac{2r^4}{3} \qquad Ans.$$

where the $x$- and $y$-coordinates of the centroid $C$ are $d_y = +r$ and $d_x = -4r/(3\pi)$. Since one of the centroidal axes is an axis of symmetry, $\bar{I}_{xy} = 0$.

① Proper use of the transfer-of-axis theorem saves a great deal of labor in computing products of inertia.

## Sample Problem A/11

Determine the orientation of the principal axes of inertia through the centroid of the angle section and determine the corresponding maximum and minimum moments of inertia.

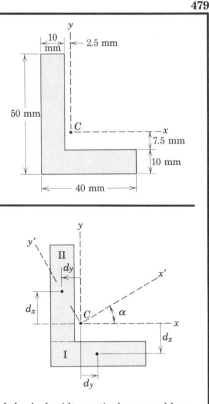

**Solution.** The location of the centroid $C$ is easily calculated, and its position is shown on the diagram.

**Products of inertia.** The product of inertia for each rectangle about its centroidal axes parallel to the $x$-$y$ axes is zero by symmetry. Thus, the product of inertia about the $x$-$y$ axes for part I is

$$[I_{xy} = \bar{I}_{xy} + d_x d_y A] \quad I_{xy} = 0 + (-12.5)(+7.5)(400) = -3.75(10^4) \text{ mm}^4$$

where

$$d_x = -(7.5 + 5) = -12.5 \text{ mm}$$

and

$$d_y = +(20 - 10 - 2.5) = 7.5 \text{ mm}$$

Likewise for part II,

$$[I_{xy} = \bar{I}_{xy} + d_x d_y A] \quad I_{xy} = 0 + (12.5)(-7.5)(400) = -3.75(10^4) \text{ mm}^4$$

where $\quad d_x = +(20 - 7.5) = 12.5 \text{ mm}, \quad d_y = -(5 + 2.5) = -7.5 \text{ mm}$

For the complete angle

$$I_{xy} = -3.75(10^4) - 3.75(10^4) = -7.50(10^4) \text{ mm}^4$$

**Moments of inertia.** The moments of inertia about the $x$- and $y$-axes for part I are

$$[I = \bar{I} + Ad^2] \quad I_x = \tfrac{1}{12}(40)(10)^3 + (400)(12.5)^2 = 6.583(10^4) \text{ mm}^4$$

$$I_y = \tfrac{1}{12}(10)(40)^3 + (400)(7.5)^2 = 7.583(10^4) \text{ mm}^4$$

and the moments of inertia for part II about these same axes are

$$[I = \bar{I} + Ad^2] \quad I_x = \tfrac{1}{12}(10)(40)^3 + (400)(12.5)^2 = 11.583(10^4) \text{ mm}^4$$

$$I_y = \tfrac{1}{12}(40)(10)^3 + (400)(7.5)^2 = 2.583(10^4) \text{ mm}^4$$

Thus, for the entire section we have

$$I_x = 6.583(10^4) + 11.583(10^4) = 18.167(10^4) \text{ mm}^4$$

$$I_y = 7.583(10^4) + 2.583(10^4) = 10.167(10^4) \text{ mm}^4$$

**Principal axes.** The inclination of the principal axes of inertia is given by Eq. A/10, so we have

$$\left[\tan 2\alpha = \frac{2I_{xy}}{I_y - I_x}\right] \quad \tan 2\alpha = \frac{2(-7.50)}{10.167 - 18.167} = 1.875$$

$$2\alpha = 61.9° \qquad \alpha = 31.0° \qquad\qquad Ans.$$

We now compute the principal moments of inertia from Eqs. A/9 using $\alpha$ for $\theta$ and get $I_{max}$ from $I_{x'}$ and $I_{min}$ from $I_{y'}$. Thus,

$$I_{max} = \left[\frac{18.167 + 10.167}{2} + \frac{18.167 - 10.167}{2}(0.4705) + (7.50)(0.8824)\right](10^4)$$

$$= 22.67(10^4) \text{ mm}^4 \qquad\qquad Ans.$$

$$I_{min} = \left[\frac{18.167 + 10.167}{2} - \frac{18.167 - 10.167}{2}(0.4705) - (7.50)(0.8824)\right](10^4)$$

$$= 5.67(10^4) \text{ mm}^4 \qquad\qquad Ans.$$

*Mohr's circle.* Alternatively we could use Eqs. A/11 to obtain the results for $I_{max}$ and $I_{min}$, or we could construct the Mohr circle from the calculated values of $I_x$, $I_y$, and $I_{xy}$. These values are spotted on the diagram to locate points $A$ and $B$, which are the extremities of the diameter of the circle. The angle $2\alpha$ and $I_{max}$ and $I_{min}$ are obtained from the figure as shown.

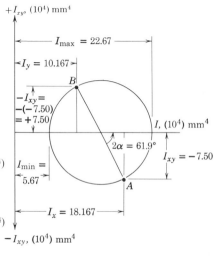

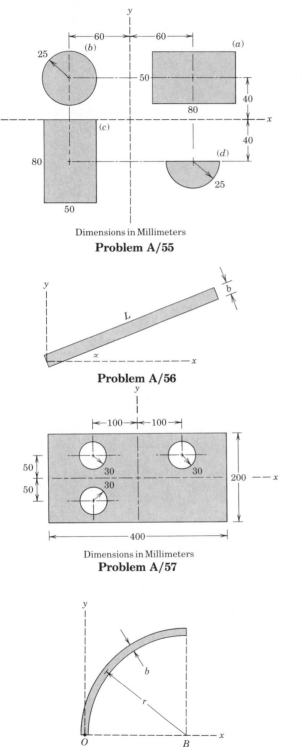

Dimensions in Millimeters
**Problem A/55**

**Problem A/56**

Dimensions in Millimeters
**Problem A/57**

**Problem A/58**

# PROBLEMS

### *Introductory problems*

**A/55** Determine the product of inertia of each of the four areas about the $x$-$y$ axes.

> *Ans.* (a) and (c) $I_{xy} = 9.60(10^6)$ mm$^4$
> (b) $I_{xy} = -4.71(10^6)$ mm$^4$
> (d) $I_{xy} = -2.98(10^6)$ mm$^4$

**A/56** Determine the product of inertia of the rectangular area about the $x$-$y$ axes. Treat the case where $b$ is small compared with $L$.

**A/57** Determine $I_x$, $I_y$, and $I_{xy}$ for the rectangular plate with three equal circular holes.

> *Ans.* $I_x = 2.44(10^8)$ mm$^4$, $I_y = 9.80(10^8)$ mm$^4$
> $I_{xy} = -14.14(10^6)$ mm$^4$

**A/58** Determine the product of inertia of the area of the quarter-circular ring about the $x$-$y$ axes. Treat the case where $b$ is small compared with $r$.

**A/59** Determine the product of inertia of the elliptical area about the $x$-$y$ axes.  *Ans.* $I_{xy} = a^2b^2/8$

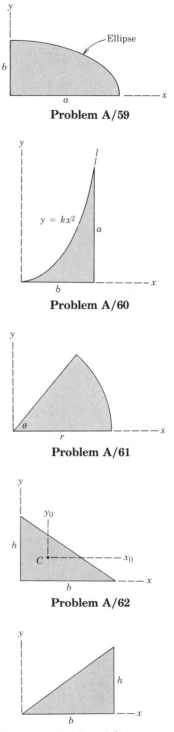

**Problem A/59**

**A/60** Determine the product of inertia of the shaded area about the $x$-$y$ axes.

**Problem A/60**

**A/61** Obtain the product of inertia of the area of the circular sector with respect to the $x$-$y$ axes. Evaluate your general expression for the specific cases of ($a$) quarter-circular and ($b$) semicircular areas.

$$Ans.\ I_{xy} = \frac{r^4}{16}\,(1 - \cos 2\theta)$$

$$(a)\ I_{xy} = r^4/8,\ (b)\ I_{xy} = 0$$

**Problem A/61**

**A/62** Derive the expression for the product of inertia of the right-triangular area about the $x$-$y$ axes and about the centroidal $x_0$-$y_0$ axes.

**Problem A/62**

**A/63** Derive the expression for the product of inertia of the right-triangular area about the $x$-$y$ axes. Solve, first, by double integration and, second, by single integration starting with a vertical strip as the element.

$$Ans.\ I_{xy} = \frac{b^2h^2}{8}$$

**Problem A/63**

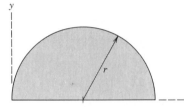

**Problem A/64**

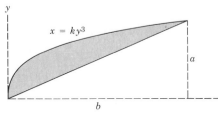

**Problem A/65**

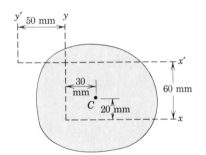

**Problem A/66**

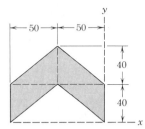

Dimensions in Millimeters

**Problem A/67**

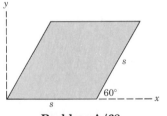

**Problem A/68**

**A/64** Solve for the product of inertia of the semicircular area about the *x*-*y* axes in two different ways.

### *Representative problems*

**A/65** Determine the product of inertia of the shaded area about the *x*-*y* axes.        *Ans.* $I_{xy} = a^2b^2/16$

**A/66** The products of inertia of the shaded area with respect to the *x*-*y* and *x'*-*y'* axes are $8(10^6)$ mm$^4$ and $-42(10^6)$ mm$^4$, respectively. Compute the area of the figure, whose centroid is *C*.

**A/67** Determine the product of inertia of the shaded area with respect to the assigned axes. (*Hint:* Locate the centroid of the symmetrical area.)
                                        *Ans.* $I_{xy} = -8(10^6)$ mm$^4$

**A/68** Determine the product of inertia of the rhombic area about the *x*-*y* axes. (*Hint:* Regard the area as a combination of a rectangle and triangles and use the results of Prob. A/63.)

**A/69** Calculate the product of inertia of the shaded area about the $x$-$y$ axes. (*Hint:* Take advantage of the transfer-of-axes relations.)

$Ans. \; I_{xy} = -1968 \; in.^4$

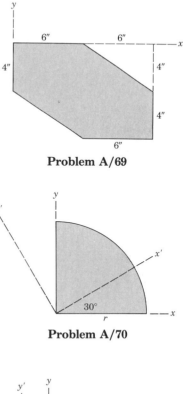

**Problem A/69**

**A/70** Determine the moments and product of inertia of the quarter-circular area with respect to the $x'$-$y'$ axes.

**Problem A/70**

**A/71** Determine the moments and product of inertia of the area of the equilateral triangle with respect to the $x'$-$y'$ axes.

$Ans. \; I_{x'} = 0.0277b^4, \; I_{y'} = 0.1527b^4, \; I_{x'y'} = 0.0361b^4$

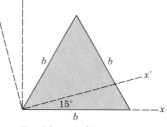

**A/72** Where $I_x = I_y$ for an area which is symmetrical about either the $x$- or the $y$-axis, prove that, for all axes through the origin, the moment of inertia is constant and the product of inertia is zero.

**Problem A/71**

**A/73** Determine the proportions of the rectangular area for which the moment of inertia about an $x'$-axis through the center point $C$ of the base is a constant value regardless of $\theta$. (See Prob. A/72.)

$Ans. \; a = 2b$

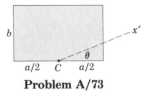

**Problem A/73**

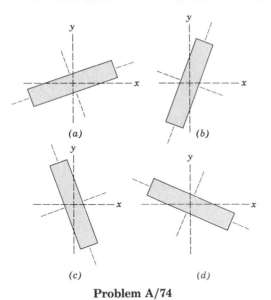

(a)          (b)

(c)          (d)

**Problem A/74**

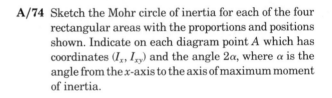

**A/74** Sketch the Mohr circle of inertia for each of the four rectangular areas with the proportions and positions shown. Indicate on each diagram point $A$ which has coordinates $(I_x, I_{xy})$ and the angle $2\alpha$, where $\alpha$ is the angle from the $x$-axis to the axis of maximum moment of inertia.

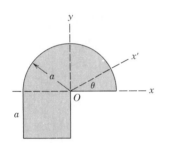

**Problem A/75**

**A/75** Determine the maximum and minimum moments of inertia for the shaded area about axes through point $O$ and identify the angle $\theta$ to the axis of minimum moment of inertia.

   *Ans.* $I_{max} = 0.976a^4$, $I_{min} = 0.476a^4$ with $\theta = \pi/4$

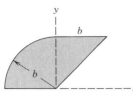

**Problem A/76**

**A/76** Find $I_x$ and $I_y$ for the shaded area and show that the $x$-$y$ axes are principal axes of inertia.

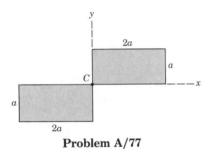

**Problem A/77**

**A/77** Determine the minimum and maximum moments of inertia with respect to centroidal axes through $C$ for the composite of the two rectangular areas shown. Find the angle $\alpha$ measured from the $x$-axis to the axis of maximum moment of inertia.

   *Ans.* $I_{min} = 0.505a^4$, $I_{max} = 6.16a^4$, $\alpha = 112.5°$

**A/78** Prove that the magnitude of the product of inertia can be computed from the relation

$$I_{xy} = \sqrt{I_x I_y - I_{max} I_{min}}$$

**A/79** Determine the maximum moment of inertia about an axis through $O$ and the angle $\alpha$ to this axis for the triangular area shown. Also construct the Mohr circle of inertia.

*Ans.* $I_{max} = 183.6$ in.$^4$, $\alpha = -16.85°$

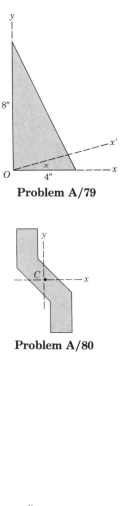

**Problem A/79**

**A/80** The maximum and minimum moments of inertia of the shaded area are $12(10^6)$ mm$^4$ and $2(10^6)$ mm$^4$, respectively, about axes passing through the centroid $C$, and the product of inertia with respect to the $x$-$y$ axes has a magnitude of $4(10^6)$ mm$^4$. Use the proper sign for the product of inertia and calculate $I_x$ and the angle $\alpha$ measured counterclockwise from the $x$-axis to the axis of maximum moment of inertia.

**Problem A/80**

**A/81** The moments and product of inertia of an area with respect to $x$-$y$ axes are $I_x = 14$ in.$^4$, $I_y = 24$ in.$^4$, and $I_{xy} = 12$ in.$^4$ Construct the Mohr circle of inertia and use it to determine the principal moments of inertia and the angle $\alpha$ from the $x$-axis to the axis of maximum moment of inertia.

*Ans.* $I_{max} = 32$ in.$^4$, $I_{min} = 6$ in.$^4$

$\alpha = 56.3°$ clockwise

**A/82** Calculate the maximum and minimum moments of inertia of the structural angle about axes through its corner $A$ and find the angle $\alpha$ measured counterclockwise from the $x$-axis to the axis of maximum inertia. Neglect the small radii and fillet.

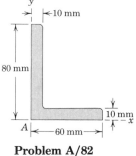

**Problem A/82**

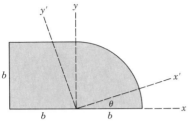

**Problem A/83**

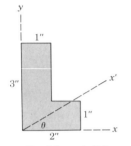

**Problem A/84**

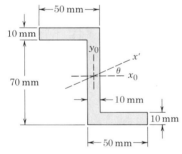

**Problem A/85**

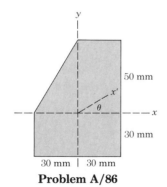

**Problem A/86**

## *Computer-oriented problems

**\*A/83** Plot the moment of inertia of the shaded area about the $x'$-axis as a function of $\theta$ from $\theta = 0$ to $\theta = 180°$. Determine the maximum and minimum values of $I_{x'}$ and the corresponding values of $\theta$ from the graph. Check your results by applying Eqs. A/10 and A/11.

$Ans.\ I_{max} = 0.655b^4$ at $\theta = 45°$
$I_{min} = 0.405b^4$ at $\theta = 135°$

**\*A/84** Plot the moment of inertia of the shaded area about the $x'$-axis as a function of $\theta$ from $\theta = 0$ to $\theta = 90°$ and determine the minimum value of $I_{x'}$ and the corresponding value of $\theta$.

**\*A/85** Plot the moment of inertia of the Z-section area about the $x'$-axis as a function of $\theta$ from $\theta = 0$ to $\theta = 90°$. Determine the maximum value of $I_{x'}$ and the corresponding value of $\theta$ from your plot, then verify these results by using Eqs. A/10 and A/11.

$Ans.\ I_{max} = 1.820(10^6)$ mm$^4$ at $\theta = 30.1°$

**\*A/86** Plot the moment of inertia about the $x'$-axis as a function of $\theta$ from $\theta = 0$ to $\theta = 90°$ and determine the minimum value of $I_{x'}$ and the corresponding value of $\theta$.

**\*A/87** Plot the moment of inertia of the shaded area about the $x'$-axis as a function of $\theta$ from $\theta = 0$ to $\theta = 180°$. Determine the maximum and minimum values of $I_{x'}$ and the corresponding values of $\theta$.

$$Ans. \ I_{max} = 0.2860b^4 \text{ with } \theta = 131.1°$$
$$I_{min} = 0.0547b^4 \text{ with } \theta = 41.1°$$

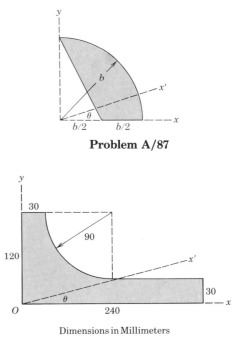

**Problem A/87**

**\*A/88** Determine the moment of inertia of the shaded area about the $x'$-axis through $O$ in terms of $\theta$ and plot it for the range $\theta = 0$ to $\theta = 180°$. Find the maximum and minimum values and their corresponding angles $\theta$.

Dimensions in Millimeters

**Problem A/88**

# MASS MOMENTS OF INERTIA

# B

See *Vol. 2 Dynamics* for Appendix B, which fully treats the concept and calculation of mass moment of inertia. Because this quantity is an important element in the study of rigid-body dynamics and is not a factor in statics, we present only a brief definition in this *Statics* volume so that the student can appreciate the basic differences between area and mass moments of inertia.

Consider a three-dimensional body of mass $m$ as shown in Fig. B/1. The mass moment of inertia $I$ about the axis $O$–$O$ is defined as

$$I = \int r^2 \, dm$$

where $r$ is the perpendicular distance of the mass element $dm$ from the axis $O$–$O$ and where the integration is over the entire body. For

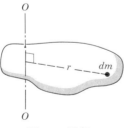

**Figure B/1**

a given rigid body the mass moment of inertia is a measure of the distribution of its mass relative to the axis in question, and for that axis is a constant property of the body. Note that the dimensions are (mass)(length)$^2$, which are kg·m$^2$ in SI units and lb-ft-sec$^2$ in U.S. customary units. Contrast these dimensions with those of area moment of inertia, which are (length)$^4$, m$^4$ in SI units and ft$^4$ in U.S. customary units.

# SELECTED TOPICS OF MATHEMATICS

# C

## C/1 INTRODUCTION

Appendix C contains an abbreviated summary and reminder of selected topics in basic mathematics which find frequent use in mechanics. The relationships are cited without proof. The student of mechanics will have frequent occasion to use many of these relations, and he or she will be handicapped if they are not well in hand. Other topics not listed will also be needed from time to time.

As the reader reviews and applies mathematics, he or she should bear in mind that mechanics is an applied science descriptive of real bodies and actual motions. Therefore, the geometric and physical interpretation of the applicable mathematics should be kept clearly in mind during the development of theory and the formulation and solution of problems.

## C/2 PLANE GEOMETRY

1. When two intersecting lines are, respectively, perpendicular to two other lines, the angles formed by each pair are equal.

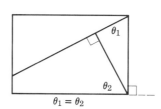

$\theta_1 = \theta_2$

2. Similar triangles

$$\frac{x}{b} = \frac{h - y}{h}$$

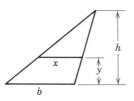

3. Any triangle

Area $= \frac{1}{2}bh$

4. Circle

Circumference $= 2\pi r$
Area $= \pi r^2$
Arc length $s = r\theta$
Sector area $= \frac{1}{2}r^2\theta$

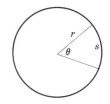

5. Every triangle inscribed within a semicircle is a right triangle.

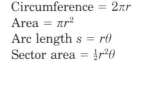

$\theta_1 + \theta_2 = \pi/2$

6. Angles of a triangle

$$\theta_1 + \theta_2 + \theta_3 = 180°$$
$$\theta_4 = \theta_1 + \theta_2$$

## C/3 SOLID GEOMETRY

1. Sphere

   Volume = $\frac{4}{3}\pi r^3$
   Surface area = $4\pi r^2$

3. Right-circular cone

   Volume = $\frac{1}{3}\pi r^2 h$
   Lateral area = $\pi r L$
   $L = \sqrt{r^2 + h^2}$

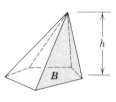

2. Spherical wedge

   Volume = $\frac{2}{3}r^3\theta$

4. Any pyramid or cone

   Volume = $\frac{1}{3}Bh$
   where $B$ = area of base

## C/4 ALGEBRA

1. Quadratic equation

   $ax^2 + bx + c = 0$

   $x = \dfrac{-b \pm \sqrt{b^2 - 4ac}}{2a}$, $b^2 \geq 4ac$ for real roots

2. Logarithms

   $b^x = y$, $x = \log_b y$

   Natural logarithms

   $b = e = 2.718\ 282$
   $e^x = y$, $x = \log_e y = \ln y$
   $\log(ab) = \log a + \log b$
   $\log(a/b) = \log a - \log b$
   $\log(1/n) = -\log n$
   $\log a^n = n \log a$
   $\log 1 = 0$
   $\log_{10} x = 0.4343 \ln x$

3. Determinants
   2nd order

   $\begin{vmatrix} a_1 & b_1 \\ a_2 & b_2 \end{vmatrix} = a_1 b_2 - a_2 b_1$

   3rd order

   $\begin{vmatrix} a_1 & b_1 & c_1 \\ a_2 & b_2 & c_2 \\ a_3 & b_3 & c_3 \end{vmatrix} = \begin{array}{l} +a_1 b_2 c_3 + a_2 b_3 c_1 + a_3 b_1 c_2 \\ -a_3 b_2 c_1 - a_2 b_1 c_3 - a_1 b_3 c_2 \end{array}$

4. Cubic equation

   $x^3 = Ax + B$

   Let $p = A/3$, $q = B/2$.

   Case I:  $q^2 - p^3$ negative (three roots real and distinct)

   $\cos u = q/(p\sqrt{p})$, $0 < u < 180°$

   $x_1 = 2\sqrt{p} \cos(u/3)$

   $x_2 = 2\sqrt{p} \cos(u/3 + 120°)$

   $x_3 = 2\sqrt{p} \cos(u/3 + 240°)$

   Case II:  $q^2 - p^3$ positive (one root real, two roots imaginary)

   $x_1 = (q + \sqrt{q^2 - p^3})^{1/3} + (q - \sqrt{q^2 - p^3})^{1/3}$

   Case III:  $q^2 - p^3 = 0$ (three roots real, two roots equal)

   $x_1 = 2q^{1/3}$, $x_2 = x_3 = -q^{1/3}$

   For general cubic equation

   $$x^3 + ax^2 + bx + c = 0$$

   Substitute $x = x_0 - a/3$ and get $x_0{}^3 = Ax_0 + B$. Then proceed as above to find values of $x_0$ from which $x = x_0 - a/3$.

## C/5 ANALYTIC GEOMETRY

### 1. Straight line

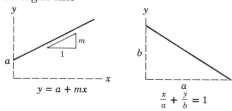

$$y = a + mx$$

$$\frac{x}{a} + \frac{y}{b} = 1$$

### 2. Circle

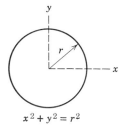 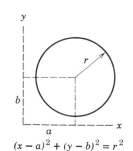

$$x^2 + y^2 = r^2$$

$$(x - a)^2 + (y - b)^2 = r^2$$

### 3. Parabola

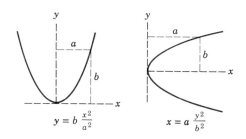

$$y = b\,\frac{x^2}{a^2}$$

$$x = a\,\frac{y^2}{b^2}$$

### 4. Ellipse

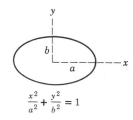

$$\frac{x^2}{a^2} + \frac{y^2}{b^2} = 1$$

### 5. Hyperbola

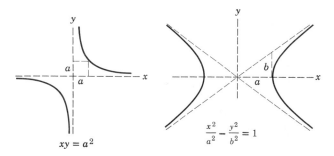

$$xy = a^2$$

$$\frac{x^2}{a^2} - \frac{y^2}{b^2} = 1$$

## C/6 TRIGONOMETRY

### 1. Definitions

$\sin \theta = a/c \quad \csc \theta = c/a$
$\cos \theta = b/c \quad \sec \theta = c/b$
$\tan \theta = a/b \quad \cot \theta = b/a$

### 2. Signs in the four quadrants

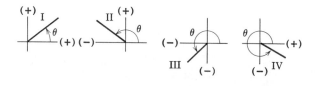

| | I | II | III | IV |
|---|---|---|---|---|
| $\sin \theta$ | + | + | − | − |
| $\cos \theta$ | + | − | − | + |
| $\tan \theta$ | + | − | + | − |
| $\csc \theta$ | + | + | − | − |
| $\sec \theta$ | + | − | − | + |
| $\cot \theta$ | + | − | + | − |

3. Miscellaneous relations

$\sin^2 \theta + \cos^2 \theta = 1$
$1 + \tan^2 \theta = \sec^2 \theta$
$1 + \cot^2 \theta = \csc^2 \theta$

$\sin \dfrac{\theta}{2} = \sqrt{\tfrac{1}{2}(1 - \cos \theta)}$

$\cos \dfrac{\theta}{2} = \sqrt{\tfrac{1}{2}(1 + \cos \theta)}$

$\sin 2\theta = 2 \sin \theta \cos \theta$
$\cos 2\theta = \cos^2 \theta - \sin^2 \theta$
$\sin (a \pm b) = \sin a \cos b \pm \cos a \sin b$
$\cos (a \pm b) = \cos a \cos b \mp \sin a \sin b$

4. Law of sines

$\dfrac{a}{b} = \dfrac{\sin A}{\sin B}$

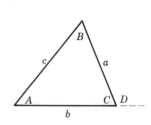

5. Law of cosines

$c^2 = a^2 + b^2 - 2ab \cos C$
$c^2 = a^2 + b^2 + 2ab \cos D$

---

## C/7  VECTOR OPERATIONS

1. *Notation.*   Vector quantities are printed in boldface type, and scalar quantities appear in lightface italic type. Thus, the vector quantity **V** has a scalar magnitude $V$. In longhand work vector quantities should always be consistently indicated by a symbol such as $\underline{V}$ or $\vec{V}$ to distinguish them from scalar quantities.

2. *Addition*

   Triangle addition   **P** + **Q** = **R**
   Parallelogram addition   **P** + **Q** = **R**
   Commutative law   **P** + **Q** = **Q** + **P**
   Associative law   **P** + (**Q** + **R**) = (**P** + **Q**) + **R**.

3. *Subtraction*

$$\mathbf{P} - \mathbf{Q} = \mathbf{P} + (-\mathbf{Q})$$

4. *Unit vectors* **i**, **j**, **k**

$$\mathbf{V} = V_x \mathbf{i} + V_y \mathbf{j} + V_z \mathbf{k}$$

   where

$$|\mathbf{V}| = V = \sqrt{V_x^2 + V_y^2 + V_z^2}$$

5. *Direction cosines* $l$, $m$, $n$ are the cosines of the angles between **V** and the $x$-, $y$-, $z$-axes. Thus,

$$l = V_x/V \qquad m = V_y/V \qquad n = V_z/V$$

   so that

$$\mathbf{V} = V(l\mathbf{i} + m\mathbf{j} + n\mathbf{k})$$

   and

$$l^2 + m^2 + n^2 = 1$$

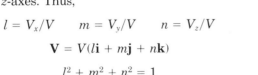

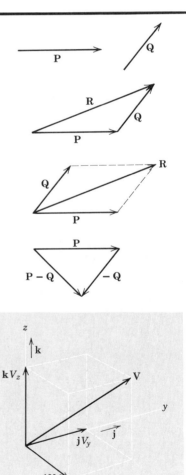

6. *Dot or scalar product*

$$\mathbf{P} \cdot \mathbf{Q} = PQ \cos \theta$$

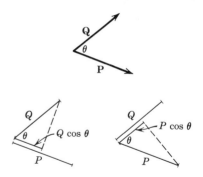

This product may be viewed as the magnitude of $\mathbf{P}$ multiplied by the component $Q \cos \theta$ of $\mathbf{Q}$ in the direction of $\mathbf{P}$, or as the magnitude of $\mathbf{Q}$ multiplied by the component $P \cos \theta$ of $\mathbf{P}$ in the direction of $\mathbf{Q}$.

Commutative law    $\mathbf{P} \cdot \mathbf{Q} = \mathbf{Q} \cdot \mathbf{P}$

From the definition of the dot product

$$\mathbf{i} \cdot \mathbf{i} = \mathbf{j} \cdot \mathbf{j} = \mathbf{k} \cdot \mathbf{k} = 1$$

$$\mathbf{i} \cdot \mathbf{j} = \mathbf{j} \cdot \mathbf{i} = \mathbf{i} \cdot \mathbf{k} = \mathbf{k} \cdot \mathbf{i} = \mathbf{j} \cdot \mathbf{k} = \mathbf{k} \cdot \mathbf{j} = 0$$

$$\mathbf{P} \cdot \mathbf{Q} = (P_x\mathbf{i} + P_y\mathbf{j} + P_z\mathbf{k}) \cdot (Q_x\mathbf{i} + Q_y\mathbf{j} + Q_z\mathbf{k})$$
$$= P_xQ_x + P_yQ_y + P_zQ_z$$

$$\mathbf{P} \cdot \mathbf{P} = P_x^2 + P_y^2 + P_z^2$$

It follows from the definition of the dot product that two vectors $\mathbf{P}$ and $\mathbf{Q}$ are perpendicular when their dot product vanishes, $\mathbf{P} \cdot \mathbf{Q} = 0$.

The angle $\theta$ between two vectors $\mathbf{P}_1$ and $\mathbf{P}_2$ may be found from their dot product expression $\mathbf{P}_1 \cdot \mathbf{P}_2 = P_1P_2 \cos \theta$, which gives

$$\cos \theta = \frac{\mathbf{P}_1 \cdot \mathbf{P}_2}{P_1P_2} = \frac{P_{1_x}P_{2_x} + P_{1_y}P_{2_y} + P_{1_z}P_{2_z}}{P_1P_2} = l_1l_2 + m_1m_2 + n_1n_2$$

where $l$, $m$, $n$ stand for the respective direction cosines of the vectors. It is also observed that two vectors are perpendicular to each other when their direction cosines obey the relation $l_1l_2 + m_1m_2 + n_1n_2 = 0$.

Distributive law    $\mathbf{P} \cdot (\mathbf{Q} + \mathbf{R}) = \mathbf{P} \cdot \mathbf{Q} + \mathbf{P} \cdot \mathbf{R}$

7. *Cross or vector product.*    The cross product $\mathbf{P} \times \mathbf{Q}$ of the two vectors $\mathbf{P}$ and $\mathbf{Q}$ is defined as a vector with a magnitude

$$|\mathbf{P} \times \mathbf{Q}| = PQ \sin \theta$$

and a direction specified by the right-hand rule as shown. Reversing the vector order and using the right-hand rule give $\mathbf{Q} \times \mathbf{P} = -\mathbf{P} \times \mathbf{Q}$.

Distributive law    $\mathbf{P} \times (\mathbf{Q} + \mathbf{R}) = \mathbf{P} \times \mathbf{Q} + \mathbf{P} \times \mathbf{R}$

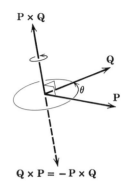

From the definition of the cross product, using a *right-handed coordinate system*, we get

$$\mathbf{i} \times \mathbf{j} = \mathbf{k} \qquad \mathbf{j} \times \mathbf{k} = \mathbf{i} \qquad \mathbf{k} \times \mathbf{i} = \mathbf{j}$$

$$\mathbf{j} \times \mathbf{i} = -\mathbf{k} \qquad \mathbf{k} \times \mathbf{j} = -\mathbf{i} \qquad \mathbf{i} \times \mathbf{k} = -\mathbf{j}$$

$$\mathbf{i} \times \mathbf{i} = \mathbf{j} \times \mathbf{j} = \mathbf{k} \times \mathbf{k} = 0$$

With the aid of these identities and the distributive law, the vector product may be written

$$\mathbf{P} \times \mathbf{Q} = (P_x\mathbf{i} + P_y\mathbf{j} + P_z\mathbf{k}) \times (Q_x\mathbf{i} + Q_y\mathbf{j} + Q_z\mathbf{k})$$
$$= (P_yQ_z - P_zQ_y)\mathbf{i} + (P_zQ_x - P_xQ_z)\mathbf{j} + (P_xQ_y - P_yQ_x)\mathbf{k}$$

The cross product may also be expressed by the determinant

$$\mathbf{P} \times \mathbf{Q} = \begin{vmatrix} \mathbf{i} & \mathbf{j} & \mathbf{k} \\ P_x & P_y & P_z \\ Q_x & Q_y & Q_z \end{vmatrix}$$

8. *Additional relations*

*Triple scalar product* $(\mathbf{P} \times \mathbf{Q}) \cdot \mathbf{R} = \mathbf{R} \cdot (\mathbf{P} \times \mathbf{Q})$. The dot and cross may be interchanged as long as the order of the vectors is maintained. Parentheses are unnecessary since $\mathbf{P} \times (\mathbf{Q} \cdot \mathbf{R})$ is meaningless because a vector $\mathbf{P}$ cannot be crossed into a scalar $\mathbf{Q} \cdot \mathbf{R}$. Thus, the expression may be written

$$\mathbf{P} \times \mathbf{Q} \cdot \mathbf{R} = \mathbf{P} \cdot \mathbf{Q} \times \mathbf{R}$$

The triple scalar product has the determinant expansion

$$\mathbf{P} \times \mathbf{Q} \cdot \mathbf{R} = \begin{vmatrix} P_x & P_y & P_z \\ Q_x & Q_y & Q_z \\ R_x & R_y & R_z \end{vmatrix}$$

*Triple vector product* $(\mathbf{P} \times \mathbf{Q}) \times \mathbf{R} = -\mathbf{R} \times (\mathbf{P} \times \mathbf{Q})$
$$= \mathbf{R} \times (\mathbf{Q} \times \mathbf{P})$$

Here the parentheses must be used since an expression $\mathbf{P} \times \mathbf{Q} \times \mathbf{R}$ would be ambiguous because it would not identify the vector to be crossed. It may be shown that the triple vector product is equivalent to

$$(\mathbf{P} \times \mathbf{Q}) \times \mathbf{R} = \mathbf{R} \cdot \mathbf{P}\mathbf{Q} - \mathbf{R} \cdot \mathbf{Q}\mathbf{P}$$

or
$$\mathbf{P} \times (\mathbf{Q} \times \mathbf{R}) = \mathbf{P} \cdot \mathbf{R}\mathbf{Q} - \mathbf{P} \cdot \mathbf{Q}\mathbf{R}$$

The first term in the first expression, for example, is the dot product $\mathbf{R} \cdot \mathbf{P}$, a scalar, multiplied by the vector $\mathbf{Q}$.

9. *Derivatives of vectors* obey the same rules as they do for scalars.

$$\frac{d\mathbf{P}}{dt} = \dot{\mathbf{P}} = \dot{P}_x\mathbf{i} + \dot{P}_y\mathbf{j} + \dot{P}_z\mathbf{k}$$

$$\frac{d(\mathbf{P}u)}{dt} = \mathbf{P}\dot{u} + \dot{\mathbf{P}}u$$

$$\frac{d(\mathbf{P} \cdot \mathbf{Q})}{dt} = \mathbf{P} \cdot \dot{\mathbf{Q}} + \dot{\mathbf{P}} \cdot \mathbf{Q}$$

$$\frac{d(\mathbf{P} \times \mathbf{Q})}{dt} = \mathbf{P} \times \dot{\mathbf{Q}} + \dot{\mathbf{P}} \times \mathbf{Q}$$

10. ***Integration of vectors.*** If **V** is a function of $x$, $y$, and $z$ and an element of volume is $d\tau = dx\ dy\ dz$, the integral of **V** over the volume may be written as the vector sum of the three integrals of its components. Thus,

$$\int \mathbf{V}\ d\tau = \mathbf{i} \int V_x\ d\tau + \mathbf{j} \int V_y\ d\tau + \mathbf{k} \int V_z\ d\tau$$

## C/8 SERIES

(Expression in brackets following series indicates range of convergence.)

$$(1 \pm x)^n = 1 \pm nx + \frac{n(n-1)}{2!}x^2 \pm \frac{n(n-1)(n-2)}{3!}x^3 + \cdots \quad [x^2 < 1]$$

$$\sin x = x - \frac{x^3}{3!} + \frac{x^5}{5!} - \frac{x^7}{7!} + \cdots \qquad\qquad [x^2 < \infty]$$

$$\cos x = 1 - \frac{x^2}{2!} + \frac{x^4}{4!} - \frac{x^6}{6!} + \cdots \qquad\qquad [x^2 < \infty]$$

$$\sinh x = \frac{e^x - e^{-x}}{2} = x + \frac{x^3}{3!} + \frac{x^5}{5!} + \frac{x^7}{7!} + \cdots \qquad [x^2 < \infty]$$

$$\cosh x = \frac{e^x + e^{-x}}{2} = 1 + \frac{x^2}{2!} + \frac{x^4}{4!} + \frac{x^6}{6!} + \cdots \qquad [x^2 < \infty]$$

$$f(x) = \frac{a_0}{2} + \sum_{n=1}^{\infty} a_n \cos \frac{n\pi x}{l} + \sum_{n=1}^{\infty} b_n \sin \frac{n\pi x}{l}$$

where $a_n = \dfrac{1}{l} \displaystyle\int_{-l}^{l} f(x) \cos \dfrac{n\pi x}{l}\ dx, \qquad b_n = \dfrac{1}{l} \displaystyle\int_{-l}^{l} f(x) \sin \dfrac{n\pi x}{l}\ dx$

[Fourier expansion for $-l < x < l$]

## C/9 DERIVATIVES

$$\frac{dx^n}{dx} = nx^{n-1}, \qquad \frac{d(uv)}{dx} = u\frac{dv}{dx} + v\frac{du}{dx}, \qquad \frac{d\left(\dfrac{u}{v}\right)}{dx} = \frac{v\dfrac{du}{dx} - u\dfrac{dv}{dx}}{v^2}$$

$$\lim_{\Delta x \to 0} \sin \Delta x = \sin dx = \tan dx = dx$$

$$\lim_{\Delta x \to 0} \cos \Delta x = \cos dx = 1$$

$$\frac{d \sin x}{dx} = \cos x, \qquad \frac{d \cos x}{dx} = -\sin x, \qquad \frac{d \tan x}{dx} = \sec^2 x$$

$$\frac{d \sinh x}{dx} = \cosh x, \qquad \frac{d \cosh x}{dx} = \sinh x, \qquad \frac{d \tanh x}{dx} = \operatorname{sech}^2 x$$

## C/10 INTEGRALS

$$\int x^n \, dx = \frac{x^{n+1}}{n+1}$$

$$\int \frac{dx}{x} = \ln x$$

$$\int \sqrt{a + bx} \, dx = \frac{2}{3b} \sqrt{(a + bx)^3}$$

$$\int x\sqrt{a + bx} \, dx = \frac{2}{15b^2} (3bx - 2a)\sqrt{(a + bx)^3}$$

$$\int x^2\sqrt{a + bx} \, dx = \frac{2}{105b^3} (8a^2 - 12abx + 15b^2x^2)\sqrt{(a + bx)^3}$$

$$\int \frac{dx}{\sqrt{a + bx}} = \frac{2\sqrt{a + bx}}{b}$$

$$\int \frac{\sqrt{a + x}}{\sqrt{b - x}} \, dx = -\sqrt{a + x}\sqrt{b - x} + (a + b)\sin^{-1}\sqrt{\frac{a + x}{a + b}}$$

$$\int \frac{x \, dx}{a + bx} = \frac{1}{b^2}[a + bx - a\ln(a + bx)]$$

$$\int \frac{x \, dx}{(a + bx)^n} = \frac{(a + bx)^{1-n}}{b^2}\left(\frac{a + bx}{2 - n} - \frac{a}{1 - n}\right)$$

$$\int \frac{dx}{a + bx^2} = \frac{1}{\sqrt{ab}}\tan^{-1}\frac{x\sqrt{ab}}{a} \quad \text{or} \quad \frac{1}{\sqrt{-ab}}\tanh^{-1}\frac{x\sqrt{-ab}}{a}$$

$$\int \frac{x \, dx}{a + bx^2} = \frac{1}{2b}\ln(a + bx^2)$$

$$\int \sqrt{x^2 \pm a^2} \, dx = \tfrac{1}{2}[x\sqrt{x^2 \pm a^2} \pm a^2\ln(x + \sqrt{x^2 \pm a^2})]$$

$$\int \sqrt{a^2 - x^2} \, dx = \tfrac{1}{2}\left(x\sqrt{a^2 - x^2} + a^2\sin^{-1}\frac{x}{a}\right)$$

$$\int x\sqrt{a^2 - x^2} \, dx = -\tfrac{1}{3}\sqrt{(a^2 - x^2)^3}$$

$$\int x^2\sqrt{a^2 - x^2} \, dx = -\frac{x}{4}\sqrt{(a^2 - x^2)^3} + \frac{a^2}{8}\left(x\sqrt{a^2 - x^2} + a^2\sin^{-1}\frac{x}{a}\right)$$

$$\int x^3\sqrt{a^2 - x^2} \, dx = -\tfrac{1}{5}(x^2 + \tfrac{2}{3}a^2)\sqrt{(a^2 - x^2)^3}$$

$$\int \frac{dx}{\sqrt{a + bx + cx^2}} = \frac{1}{\sqrt{c}}\ln\left(\sqrt{a + bx + cx^2} + x\sqrt{c} + \frac{b}{2\sqrt{c}}\right) \quad \text{or} \quad \frac{-1}{\sqrt{-c}}\sin^{-1}\left(\frac{b + 2cx}{\sqrt{b^2 - 4ac}}\right)$$

$$\int \frac{dx}{\sqrt{x^2 \pm a^2}} = \ln(x + \sqrt{x^2 \pm a^2})$$

$$\int \frac{dx}{\sqrt{a^2 - x^2}} = \sin^{-1} \frac{x}{a}$$

$$\int \frac{x \, dx}{\sqrt{x^2 - a^2}} = \sqrt{x^2 - a^2}$$

$$\int \frac{x \, dx}{\sqrt{a^2 \pm x^2}} = \pm \sqrt{a^2 \pm x^2}$$

$$\int x\sqrt{x^2 \pm a^2} \, dx = \frac{1}{3}\sqrt{(x^2 \pm a^2)^3}$$

$$\int x^2\sqrt{x^2 \pm a^2} \, dx = \frac{x}{4} \sqrt{(x^2 \pm a^2)^3} \mp \frac{a^2}{8} x\sqrt{x^2 \pm a^2} - \frac{a^4}{8} \ln (x + \sqrt{x^2 \pm a^2})$$

$$\int \sin x \, dx = -\cos x$$

$$\int \cos x \, dx = \sin x$$

$$\int \sec x \, dx = \frac{1}{2} \ln \frac{1 + \sin x}{1 - \sin x}$$

$$\int \sin^2 x \, dx = \frac{x}{2} - \frac{\sin 2x}{4}$$

$$\int \cos^2 x \, dx = \frac{x}{2} + \frac{\sin 2x}{4}$$

$$\int \sin x \cos x \, dx = \frac{\sin^2 x}{2}$$

$$\int \sinh x \, dx = \cosh x$$

$$\int \cosh x \, dx = \sinh x$$

$$\int \tanh x \, dx = \ln \cosh x$$

$$\int \ln x \, dx = x \ln x - x$$

$$\int e^{ax} \, dx = \frac{e^{ax}}{a}$$

$$\int x e^{ax} \, dx = \frac{e^{ax}}{a^2} (ax - 1)$$

$$\int e^{ax} \sin px \, dx = \frac{e^{ax}(a \sin px - p \cos px)}{a^2 + p^2}$$

$$\int e^{ax} \cos px \, dx = \frac{e^{ax}(a \cos px + p \sin px)}{a^2 + p^2}$$

$$\int e^{ax} \sin^2 x \, dx = \frac{e^{ax}}{4 + a^2} \left( a \sin^2 x - \sin 2x + \frac{2}{a} \right)$$

$$\int e^{ax} \cos^2 x \, dx = \frac{e^{ax}}{4 + a^2} \left( a \cos^2 x + \sin 2x + \frac{2}{a} \right)$$

$$\int e^{ax} \sin x \cos x \, dx = \frac{e^{ax}}{4 + a^2} \left( \frac{a}{2} \sin 2x - \cos 2x \right)$$

$$\int \sin^3 x \, dx = -\frac{\cos x}{3} (2 + \sin^2 x)$$

$$\int \cos^3 x \, dx = \frac{\sin x}{3} (2 + \cos^2 x)$$

$$\int \cos^5 x \, dx = \sin x - \tfrac{2}{3} \sin^3 x + \tfrac{1}{5} \sin^5 x$$

$$\int x \sin x \, dx = \sin x - x \cos x$$

$$\int x \cos x \, dx = \cos x + x \sin x$$

$$\int x^2 \sin x \, dx = 2x \sin x - (x^2 - 2) \cos x$$

$$\int x^2 \cos x \, dx = 2x \cos x + (x^2 - 2) \sin x$$

Radius of curvature
$$\begin{cases} \rho_{xy} = \dfrac{\left[ 1 + \left( \dfrac{dy}{dx} \right)^2 \right]^{3/2}}{\dfrac{d^2 y}{dx^2}} \\[4em] \rho_{r\theta} = \dfrac{\left[ r^2 + \left( \dfrac{dr}{d\theta} \right)^2 \right]^{3/2}}{r^2 + 2 \left( \dfrac{dr}{d\theta} \right)^2 - r \dfrac{d^2 r}{d\theta^2}} \end{cases}$$

## C/11  NEWTON'S METHOD FOR SOLVING INTRACTABLE EQUATIONS

Frequently, the application of the fundamental principles of mechanics leads to an algebraic or transcendental equation which is not solvable (or easily solvable) in closed form. In such cases, an iterative technique, such as Newton's method, can be a powerful tool for obtaining a good estimate to the root or roots of the equation.

Let us place the equation to be solved in the form $f(x) = 0$. The *a*-part of the accompanying figure depicts an arbitrary function $f(x)$ for values of $x$ in the vicinity of the desired root $x_r$. Note that $x_r$ is merely the value of $x$ at which the function crosses the x-axis. Suppose that we have available (perhaps via a hand-drawn plot) a rough estimate $x_1$ of this root. Provided that $x_1$ does not closely correspond to a maximum or minimum value of the function $f(x)$, we may obtain a better estimate of the root $x_r$ by projecting the tangent to $f(x)$ at

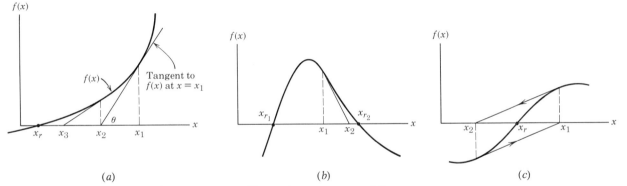

(a)　　　　　　　　(b)　　　　　　　　(c)

$x_1$ such that it intersects the x-axis at $x_2$. From the geometry of the figure, we may write

$$\tan \theta = f'(x_1) = \frac{f(x_1)}{x_1 - x_2}$$

where $f'(x_1)$ denotes the derivative of $f(x)$ with respect to $x$ evaluated at $x = x_1$. Solving the above equation for $x_2$ results in

$$x_2 = x_1 - \frac{f(x_1)}{f'(x_1)}$$

The term $-f(x_1)/f'(x_1)$ is the correction to the initial root estimate $x_1$. Once $x_2$ is calculated, we may repeat the process to obtain $x_3$, and so forth.

Thus, we generalize the above equation to

$$x_{k+1} = x_k - \frac{f(x_k)}{f'(x_k)}$$

where

$$x_{k+1} = \text{the } (k + 1)\text{th estimate to the desired root } x_r$$

$$x_k = \text{the } k\text{th estimate to the desired root } x_r$$

$$f(x_k) = \text{the function } f(x) \text{ evaluated at } x = x_k$$

$$f'(x_k) = \text{the function derivative evaluated at } x = x_k$$

This equation is repetitively applied until $f(x_{k+1})$ is sufficiently close to zero and $x_{k+1} \cong x_k$. The student should verify that the equation is valid for all possible sign combinations of $x_k$, $f(x_k)$, and $f'(x_k)$.

Several cautionary notes are in order:

1. Clearly, $f'(x_k)$ must not be zero or close to zero. This would mean, as restricted above, that $x_k$ exactly or approximately corresponds to a minimum or maximum of $f(x)$. If the slope $f'(x_k)$ is zero, then the slope projection never intersects the $x$-axis. If the slope $f'(x_k)$ is small, then the correction to $x_k$ may be so large that $x_{k+1}$ is a worse root estimate than $x_k$. For this reason, experienced engineers usually limit the size of the correction term; that is, if the absolute value of $f(x_k)/f'(x_k)$ is larger than a preselected maximum value, the maximum value is used.

2. If there are several roots of the equation $f(x) = 0$, we must be in the vicinity of the desired root $x_r$ in order that the algorithm actually converges to that root. The $b$-part of the figure depicts the condition that the initial estimate $x_1$ will result in convergence to $x_{r_2}$ rather than $x_{r_1}$.

3. Oscillation from one side of the root to the other can occur if, for example, the function is antisymmetric about a root which is an inflection point. The use of one-half of the correction will usually prevent this behavior, which is depicted in the $c$-part of the accompanying figure.

*Example:* Beginning with an initial estimate of $x_1 = 5$, estimate the single root of the equation $e^x - 10 \cos x - 100 = 0$.

The table below summarizes the application of Newton's method to the given equation. The iterative process was terminated when the absolute value of the correction $-f(x_k)/f'(x_k)$ became less than $10^{-6}$.

| $k$ | $x_k$ | $f(x_k)$ | $f'(x_k)$ | $x_{k+1} - x_k = -\dfrac{f(x_k)}{f'(x_k)}$ |
|---|---|---|---|---|
| 1 | 5.000 000 | 45.576 537 | 138.823 916 | $-0.328\ 305$ |
| 2 | 4.671 695 | 7.285 610 | 96.887 065 | $-0.075\ 197$ |
| 3 | 4.596 498 | 0.292 886 | 89.203 650 | $-0.003\ 283$ |
| 4 | 4.593 215 | 0.000 527 | 88.882 536 | $-0.000\ 006$ |
| 5 | 4.593 209 | $-2(10^{-8})$ | 88.881 956 | $2.25\ (10^{-10})$ |

## C/12  *SELECTED TECHNIQUES FOR NUMERICAL INTEGRATION*

### *1. Area determination.*

Consider the problem of determining the shaded area under the curve $y = f(x)$ from $x = a$ to $x = b$ as depicted in the $a$-part of the figure and suppose that analytical integration is not feasible. The function may be known in tabular form from experimental measurements or it may be known in analytical form. The function is taken to be continuous within the interval $a < x < b$. We may divide the area into $n$ vertical strips, each of width $\Delta x = (b - a)/n$, and then add the areas of all strips to obtain $A = \int y\, dx$. A representative strip of area $A_i$ is shown with darker shading in the figure. Three useful numerical approximations are cited. In each case the greater the number of strips, the more accurate becomes the approximation geometrically. As a general rule, one can begin with a relatively small number of strips and increase the number until the resulting changes in the area approximation no longer improve the desired accuracy.

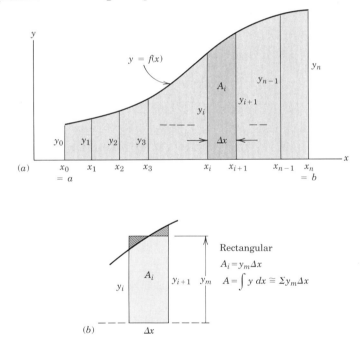

I. *Rectangular* [Figure (b)]   The areas of the strips are taken to be rectangles, as shown by the representative strip whose height $y_m$ is chosen visually so that the small cross-hatched areas are as nearly equal as possible. Thus, we form the sum $\Sigma y_m$ of the effective heights and multiply by $\Delta x$. For a function known in analytical form, a value for $y_m$ equal to that of the function at the midpoint $x_i + \Delta x/2$ may be calculated and used in the summation.

II. *Trapezoidal* [Figure (*c*)]   The areas of the strips are taken to be trapezoids, as shown by the representative strip. The area $A_i$ is the average height $(y_i + y_{i+1})/2$ times $\Delta x$. Adding the areas gives the area approximation as tabulated. For the example with the curvature shown, clearly the approximation will be on the low side. For the reverse curvature, the approximation will be on the high side.

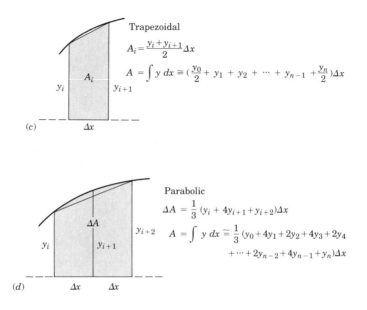

Trapezoidal

$$A_i = \frac{y_i + y_{i+1}}{2} \Delta x$$

$$A = \int y\, dx \cong (\frac{y_0}{2} + y_1 + y_2 + \cdots + y_{n-1} + \frac{y_n}{2}) \Delta x$$

(*c*)

Parabolic

$$\Delta A = \frac{1}{3}(y_i + 4y_{i+1} + y_{i+2}) \Delta x$$

$$A = \int y\, dx \cong \frac{1}{3}(y_0 + 4y_1 + 2y_2 + 4y_3 + 2y_4$$
$$+ \cdots + 2y_{n-2} + 4y_{n-1} + y_n) \Delta x$$

(*d*)

III. *Parabolic* [Figure (*d*)]   The area between the chord and the curve (neglected in the trapezoidal solution) may be accounted for by approximating the function by a parabola passing through the points defined by three successive values of *y*. This area may be calculated from the geometry of the parabola and added to the trapezoidal area of the pair of strips to give the area $\Delta A$ of the pair as cited. Adding all of the $\Delta A$'s produces the tabulation shown, which is known as Simpson's rule. To use Simpson's rule, the number *n* of strips must be even.

*Example:* Determine the area under the curve $y = x\sqrt{1 + x^2}$ from $x = 0$ to $x = 2$. (An integrable function is chosen here so that the three approximations can be compared with the exact value, which is $A = \int_0^2 x\sqrt{1 + x^2}\, dx = \frac{1}{3}(1 + x^2)^{3/2}\big|_0^2 = \frac{1}{3}(5\sqrt{5} - 1) = 3.393\ 447$.)

| NUMBER OF SUBINTERVALS | AREA APPROXIMATIONS | | |
|:---:|:---:|:---:|:---:|
| | RECTANGULAR | TRAPEZOIDAL | PARABOLIC |
| 4 | 3.361 704 | 3.456 731 | 3.392 214 |
| 10 | 3.388 399 | 3.403 536 | 3.393 420 |
| 50 | 3.393 245 | 3.393 850 | 3.393 447 |
| 100 | 3.393 396 | 3.393 547 | 3.393 447 |
| 1000 | 3.393 446 | 3.393 448 | 3.393 447 |
| 2500 | 3.393 447 | 3.393 447 | 3.393 447 |

Note that the worst approximation error is less than 2 percent, even with only four strips.

**2. Integration of first-order ordinary differential equations.** The application of the fundamental principles of mechanics frequently results in differential relationships. Let us consider the first-order form $\frac{dy}{dt} = f(t)$, where the function $f(t)$ may not be readily integrable or may be known only in tabular form. We may numerically integrate by means of a simple slope-projection technique, known as Euler integration, which is illustrated in the figure.

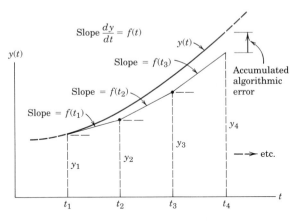

Beginning at $t_1$, at which the value $y_1$ is known, we project the slope over a horizontal subinterval or step $(t_2 - t_1)$ and see that $y_2 = y_1 + f(t_1)(t_2 - t_1)$. At $t_2$, the process may be repeated beginning at $y_2$, and so forth until the desired value of $t$ is reached. Hence, the general expression is

$$y_{k+1} = y_k + f(t_k)(t_{k+1} - t_k)$$

If $y$ versus $t$ were linear, i.e., if $f(t)$ were constant, the method would be exact, and there would be no need for a numerical approach in that case. Changes in the slope over the subinterval introduce error. For the case shown in the figure, the estimate $y_2$ is clearly less than the true value of the function $y(t)$ at $t_2$. More accurate integration techniques (such as Runge–Kutta methods) take into account changes in the slope over the subinterval and thus provide better results.

As with the area-determination techniques, experience is helpful in the selection of a subinterval or step size when dealing with analytical functions. As a rough rule, one begins with a relatively large step size and then steadily decreases the step size until the corresponding changes in the integrated result are much smaller than the desired accuracy. A step size which is too small, however, can result in increased error due to a very large number of computer operations. This type of error is generally known as "round-off error", while the error which results from a large step size is known as algorithm error.

*Example:* For the differential equation $\dfrac{dy}{dt} = 5t$ with the initial condition $y = 2$ when $t = 0$, determine the value of $y$ for $t = 4$.

Application of the Euler integration technique yields the following results:

| NUMBER OF SUBINTERVALS | STEP SIZE | $y$ at $t = 4$ | PERCENT ERROR |
|:---:|:---:|:---:|:---:|
| 10 | 0.4 | 38 | 9.5 |
| 100 | 0.04 | 41.6 | 0.95 |
| 500 | 0.008 | 41.92 | 0.19 |
| 1000 | 0.004 | 41.96 | 0.10 |

This simple example may be integrated analytically. The result is $y = 42$ (exactly).

# USEFUL TABLES

# D

## TABLE D/1   PHYSICAL PROPERTIES

*Density* (kg/m³) *and specific weight* (lb/ft³)

|  | kg/m³ | lb/ft³ |  | kg/m³ | lb/ft³ |
|---|---|---|---|---|---|
| Air* | 1.2062 | 0.07530 |  |  |  |
| Aluminum | 2 690 | 168 | Iron (cast) | 7 210 | 450 |
| Concrete (av.) | 2 400 | 150 | Lead | 11 370 | 710 |
| Copper | 8 910 | 556 | Mercury | 13 570 | 847 |
| Earth (wet, av.) | 1 760 | 110 | Oil (av.) | 900 | 56 |
| (dry, av.) | 1 280 | 80 | Steel | 7 830 | 489 |
| Glass | 2 590 | 162 | Titanium | 3 080 | 192 |
| Gold | 19 300 | 1205 | Water (fresh) | 1 000 | 62.4 |
| Ice | 900 | 56 | (salt) | 1 030 | 64 |
| * At 20°C (68°F) and atmospheric | | | Wood (soft pine) | 480 | 30 |
| pressure | | | (hard oak) | 800 | 50 |

*Coefficients of friction*

(The coefficients in the following table represent typical values under normal working conditions. Actual coefficients for a given situation will depend on the exact nature of the contacting surfaces. A variation of 25 to 100 percent or more from these values could be expected in an actual application, depending on prevailing conditions of cleanliness, surface finish, pressure, lubrication, and velocity.)

| CONTACTING SURFACE | TYPICAL VALUES OF COEFFICIENT OF FRICTION | |
|---|---|---|
| | STATIC, $\mu_s$ | KINETIC, $\mu_k$ |
| Steel on steel (dry) | 0.6 | 0.4 |
| Steel on steel (greasy) | 0.1 | 0.05 |
| Teflon on steel | 0.04 | 0.04 |
| Steel on babbitt (dry) | 0.4 | 0.3 |
| Steel on babbitt (greasy) | 0.1 | 0.07 |
| Brass on steel (dry) | 0.5 | 0.4 |
| Brake lining on cast iron | 0.4 | 0.3 |
| Rubber tires on smooth pavement (dry) | 0.9 | 0.8 |
| Wire rope on iron pulley (dry) | 0.2 | 0.15 |
| Hemp rope on metal | 0.3 | 0.2 |
| Metal on ice | | 0.02 |

## TABLE D/2   SOLAR SYSTEM CONSTANTS

Universal gravitational constant     $G = 6.673(10^{-11})$ m³/(kg·s²)
                                      $= 3.439(10^{-8})$ ft⁴/(lbf-s⁴)

Mass of Earth                         $m_e = 5.976(10^{24})$ kg
                                      $= 4.095(10^{23})$ lbf-s²/ft

Period of Earth's rotation (1 sidereal day)   = 23 h 56 min 4 s
                                              = 23.9344 h

Angular velocity of Earth             $\omega = 0.7292(10^{-4})$ rad/s
Mean angular velocity of Earth–Sun line   $\omega' = 0.1991(10^{-6})$ rad/s
Mean velocity of Earth's center about Sun   = 107 200 km/h
                                            = 66,610 mi/h

| BODY | MEAN DISTANCE TO SUN km (mi) | ECCENTRICITY OF ORBIT e | PERIOD OF ORBIT solar days | MEAN DIAMETER km (mi) | MASS RELATIVE TO EARTH | SURFACE GRAVITATIONAL ACCELERATION m/s² (ft/s²) | ESCAPE VELOCITY km/s (mi/s) |
|---|---|---|---|---|---|---|---|
| Sun | — | — | — | 1 392 000 (865 000) | 333 000 | 274 (898) | 616 (383) |
| Moon | 384 398* (238 854)* | 0.055 | 27.32 | 3 476 (2 160) | 0.0123 | 1.62 (5.32) | 2.37 (1.47) |
| Mercury | 57.3 × 10⁶ (35.6 × 10⁶) | 0.206 | 87.97 | 5 000 (3 100) | 0.054 | 3.47 (11.4) | 4.17 (2.59) |
| Venus | 108 × 10⁶ (67.2 × 10⁶) | 0.0068 | 224.70 | 12 400 (7 700) | 0.815 | 8.44 (27.7) | 10.24 (6.36) |
| Earth | 149.6 × 10⁶ (92.96 × 10⁶) | 0.0167 | 365.26 | 12 742† (7 918)† | 1.000 | 9.821‡ (32.22)‡ | 11.18 (6.95) |
| Mars | 227.9 × 10⁶ (141.6 × 10⁶) | 0.093 | 686.98 | 6 788 (4 218) | 0.107 | 3.73 (12.3) | 5.03 (3.13) |

\* Mean distance to Earth (center-to-center)

† Diameter of sphere of equal volume, based on a spheroid Earth with a polar diameter of 12 714 km (7900 mi) and an equatorial diameter of 12 756 km (7926 mi)

‡ For nonrotating spherical Earth, equivalent to absolute value at sea level and latitude 37.5°

## TABLE D/3   PROPERTIES OF PLANE FIGURES

| FIGURE | CENTROID | AREA MOMENTS OF INERTIA |
|---|---|---|
| Arc Segment | $\bar{r} = \dfrac{r \sin \alpha}{\alpha}$ | — |
| Quarter and Semicircular Arcs | $\bar{y} = \dfrac{2r}{\pi}$ | — |
| Circular Area | — | $I_x = I_y = \dfrac{\pi r^4}{4}$ <br><br> $I_z = \dfrac{\pi r^4}{2}$ |
| Semicircular Area | $\bar{y} = \dfrac{4r}{3\pi}$ | $I_x = I_y = \dfrac{\pi r^4}{8}$ <br><br> $\bar{I}_x = \left(\dfrac{\pi}{8} - \dfrac{8}{9\pi}\right) r^4$ <br><br> $I_z = \dfrac{\pi r^4}{4}$ |
| Quarter-Circular Area | $\bar{x} = \bar{y} = \dfrac{4r}{3\pi}$ | $I_x = I_y = \dfrac{\pi r^4}{16}$ <br><br> $\bar{I}_x = \bar{I}_y = \left(\dfrac{\pi}{16} - \dfrac{4}{9\pi}\right) r^4$ <br><br> $I_z = \dfrac{\pi r^4}{8}$ |
| Area of Circular Sector | $\bar{x} = \dfrac{2}{3} \dfrac{r \sin \alpha}{\alpha}$ | $I_x = \dfrac{r^4}{4}\left(\alpha - \tfrac{1}{2} \sin 2\alpha\right)$ <br><br> $I_y = \dfrac{r^4}{4}\left(\alpha + \tfrac{1}{2} \sin 2\alpha\right)$ <br><br> $I_z = \tfrac{1}{2} r^4 \alpha$ |

**TABLE D/3    PROPERTIES OF PLANE FIGURES**  *Continued*

| FIGURE | CENTROID | AREA MOMENTS OF INERTIA |
|--------|----------|-------------------------|
| Rectangular Area <br><br> | — | $I_x = \dfrac{bh^3}{3}$ <br><br> $\bar{I}_x = \dfrac{bh^3}{12}$ <br><br> $\bar{I}_z = \dfrac{bh}{12}(b^2 + h^2)$ |
| Triangular Area <br><br> | $\bar{x} = \dfrac{a+b}{3}$ <br><br> $\bar{y} = \dfrac{h}{3}$ | $I_x = \dfrac{bh^3}{12}$ <br><br> $\bar{I}_x = \dfrac{bh^3}{36}$ <br><br> $I_{x_1} = \dfrac{bh^3}{4}$ |
| Area of Elliptical Quadrant <br><br> | $\bar{x} = \dfrac{4a}{3\pi}$ <br><br> $\bar{y} = \dfrac{4b}{3\pi}$ | $I_x = \dfrac{\pi ab^3}{16}, \quad \bar{I}_x = \left(\dfrac{\pi}{16} - \dfrac{4}{9\pi}\right)ab^3$ <br><br> $I_y = \dfrac{\pi a^3 b}{16}, \quad \bar{I}_y = \left(\dfrac{\pi}{16} - \dfrac{4}{9\pi}\right)a^3 b$ <br><br> $I_z = \dfrac{\pi ab}{16}(a^2 + b^2)$ |
| Subparabolic Area <br><br> $y = kx^2 = \dfrac{b}{a^2}x^2$ <br> Area $A = \dfrac{ab}{3}$ <br> | $\bar{x} = \dfrac{3a}{4}$ <br><br> $\bar{y} = \dfrac{3b}{10}$ | $I_x = \dfrac{ab^3}{21}$ <br><br> $I_y = \dfrac{a^3 b}{5}$ <br><br> $I_z = ab\left(\dfrac{a^2}{5} + \dfrac{b^2}{21}\right)$ |
| Parabolic Area <br><br> $y = kx^2 = \dfrac{b}{a^2}x^2$ <br> Area $A = \dfrac{2ab}{3}$ <br> | $\bar{x} = \dfrac{3a}{8}$ <br><br> $\bar{y} = \dfrac{3b}{5}$ | $I_x = \dfrac{2ab^3}{7}$ <br><br> $I_y = \dfrac{2a^3 b}{15}$ <br><br> $I_z = 2ab\left(\dfrac{a^2}{15} + \dfrac{b^2}{7}\right)$ |

## TABLE D/4   PROPERTIES OF HOMOGENEOUS SOLIDS
($m$ = mass of body shown)

| BODY | MASS CENTER | MASS MOMENTS OF INERTIA |
|---|---|---|
| Circular Cylindrical Shell | — | $I_{xx} = \frac{1}{2}mr^2 + \frac{1}{12}ml^2$ <br> $I_{x_1x_1} = \frac{1}{2}mr^2 + \frac{1}{3}ml^2$ <br> $I_{zz} = mr^2$ |
| Half Cylindrical Shell | $\bar{x} = \dfrac{2r}{\pi}$ | $I_{xx} = I_{yy}$ <br> $\quad = \frac{1}{2}mr^2 + \frac{1}{12}ml^2$ <br> $I_{x_1x_1} = I_{y_1y_1}$ <br> $\quad = \frac{1}{2}mr^2 + \frac{1}{3}ml^2$ <br> $I_{zz} = mr^2$ <br> $\bar{I}_{zz} = \left(1 - \dfrac{4}{\pi^2}\right)mr^2$ |
| Circular Cylinder | — | $I_{xx} = \frac{1}{4}mr^2 + \frac{1}{12}ml^2$ <br> $I_{x_1x_1} = \frac{1}{4}mr^2 + \frac{1}{3}ml^2$ <br> $I_{zz} = \frac{1}{2}mr^2$ |
| Semicylinder | $\bar{x} = \dfrac{4r}{3\pi}$ | $I_{xx} = I_{yy}$ <br> $\quad = \frac{1}{4}mr^2 + \frac{1}{12}ml^2$ <br> $I_{x_1x_1} = I_{y_1y_1}$ <br> $\quad = \frac{1}{4}mr^2 + \frac{1}{3}ml^2$ <br> $I_{zz} = \frac{1}{2}mr^2$ <br> $\bar{I}_{zz} = \left(\dfrac{1}{2} - \dfrac{16}{9\pi^2}\right)mr^2$ |
| Rectangular Parallelepiped | — | $I_{xx} = \frac{1}{12}m(a^2 + l^2)$ <br> $I_{yy} = \frac{1}{12}m(b^2 + l^2)$ <br> $I_{zz} = \frac{1}{12}m(a^2 + b^2)$ <br> $I_{y_1y_1} = \frac{1}{12}mb^2 + \frac{1}{3}ml^2$ <br> $I_{y_2y_2} = \frac{1}{3}m(b^2 + l^2)$ |

**TABLE D/4    PROPERTIES OF HOMOGENEOUS SOLIDS**    *Continued*

($m$ = mass of body shown)

| BODY | MASS CENTER | MASS MOMENTS OF INERTIA |
|---|---|---|
| Spherical Shell | — | $I_{zz} = \frac{2}{3}mr^2$ |
| Hemispherical Shell | $\bar{x} = \dfrac{r}{2}$ | $I_{xx} = I_{yy} = I_{zz} = \frac{2}{3}mr^2$ <br> $\bar{I}_{yy} = \bar{I}_{zz} = \frac{5}{12}mr^2$ |
| Sphere | — | $I_{zz} = \frac{2}{5}mr^2$ |
| Hemisphere | $\bar{x} = \dfrac{3r}{8}$ | $I_{xx} = I_{yy} = I_{zz} = \frac{2}{5}mr^2$ <br> $\bar{I}_{yy} = \bar{I}_{zz} = \frac{83}{320}mr^2$ |
| Uniform Slender Rod | — | $I_{yy} = \frac{1}{12}ml^2$ <br> $I_{y_1y_1} = \frac{1}{3}ml^2$ |

**TABLE D/4**    **PROPERTIES OF HOMOGENEOUS SOLIDS**  *Continued*

($m$ = mass of body shown)

| BODY | MASS CENTER | MASS MOMENTS OF INERTIA |
|---|---|---|
| Quarter-Circular Rod | $\bar{x} = \bar{y}$ $= \dfrac{2r}{\pi}$ | $I_{xx} = I_{yy} = \frac{1}{2}mr^2$ $I_{zz} = mr^2$ |
| Elliptical Cylinder | — | $I_{xx} = \frac{1}{4}ma^2 + \frac{1}{12}ml^2$ $I_{yy} = \frac{1}{4}mb^2 + \frac{1}{12}ml^2$ $I_{zz} = \frac{1}{4}m(a^2 + b^2)$ $I_{y_1y_1} = \frac{1}{4}mb^2 + \frac{1}{3}ml^2$ |
| Conical Shell | $\bar{z} = \dfrac{2h}{3}$ | $I_{yy} = \frac{1}{4}mr^2 + \frac{1}{2}mh^2$ $I_{y_1y_1} = \frac{1}{4}mr^2 + \frac{1}{6}mh^2$ $I_{zz} = \frac{1}{2}mr^2$ $\bar{I}_{yy} = \frac{1}{4}mr^2 + \frac{1}{18}mh^2$ |
| Half Conical Shell | $\bar{x} = \dfrac{4r}{3\pi}$ $\bar{z} = \dfrac{2h}{3}$ | $I_{xx} = I_{yy}$ $\quad = \frac{1}{4}mr^2 + \frac{1}{2}mh^2$ $I_{x_1x_1} = I_{y_1y_1}$ $\quad = \frac{1}{4}mr^2 + \frac{1}{6}mh^2$ $I_{zz} = \frac{1}{2}mr^2$ $\bar{I}_{zz} = \left(\dfrac{1}{2} - \dfrac{16}{9\pi^2}\right)mr^2$ |
| Right-Circular Cone | $\bar{z} = \dfrac{3h}{4}$ | $I_{yy} = \frac{3}{20}mr^2 + \frac{3}{5}mh^2$ $I_{y_1y_1} = \frac{3}{20}mr^2 + \frac{1}{10}mh^2$ $I_{zz} = \frac{3}{10}mr^2$ $\bar{I}_{yy} = \frac{3}{20}mr^2 + \frac{3}{80}mh^2$ |

**TABLE D/4   PROPERTIES OF HOMOGENEOUS SOLIDS** *Continued*
($m$ = mass of body shown)

| BODY | MASS CENTER | MASS MOMENTS OF INERTIA |
|---|---|---|
| Half Cone | $\bar{x} = \dfrac{r}{\pi}$  $\bar{z} = \dfrac{3h}{4}$ | $I_{xx} = I_{yy}$ $= \frac{3}{20}mr^2 + \frac{3}{5}mh^2$ $I_{x_1x_1} = I_{y_1y_1}$ $= \frac{3}{20}mr^2 + \frac{1}{10}mh^2$ $I_{zz} = \frac{3}{10}mr^2$ $\bar{I}_{zz} = \left(\dfrac{3}{10} - \dfrac{1}{\pi^2}\right)mr^2$ |
| Semiellipsoid $\dfrac{x^2}{a^2} + \dfrac{y^2}{b^2} + \dfrac{z^2}{c^2} = 1$ | $\bar{z} = \dfrac{3c}{8}$ | $I_{xx} = \frac{1}{5}m(b^2 + c^2)$ $I_{yy} = \frac{1}{5}m(a^2 + c^2)$ $I_{zz} = \frac{1}{5}m(a^2 + b^2)$ $\bar{I}_{xx} = \frac{1}{5}m(b^2 + \frac{19}{64}c^2)$ $\bar{I}_{yy} = \frac{1}{5}m(a^2 + \frac{19}{64}c^2)$ |
| Elliptic Paraboloid $\dfrac{x^2}{a^2} + \dfrac{y^2}{b^2} = \dfrac{z}{c}$ | $\bar{z} = \dfrac{2c}{3}$ | $I_{xx} = \frac{1}{6}mb^2 + \frac{1}{2}mc^2$ $I_{yy} = \frac{1}{6}ma^2 + \frac{1}{2}mc^2$ $I_{zz} = \frac{1}{6}m(a^2 + b^2)$ $\bar{I}_{xx} = \frac{1}{6}m(b^2 + \frac{1}{3}c^2)$ $\bar{I}_{yy} = \frac{1}{6}m(a^2 + \frac{1}{3}c^2)$ |
| Rectangular Tetrahedron | $\bar{x} = \dfrac{a}{4}$  $\bar{y} = \dfrac{b}{4}$  $\bar{z} = \dfrac{c}{4}$ | $I_{xx} = \frac{1}{10}m(b^2 + c^2)$ $I_{yy} = \frac{1}{10}m(a^2 + c^2)$ $I_{zz} = \frac{1}{10}m(a^2 + b^2)$ $\bar{I}_{xx} = \frac{3}{80}m(b^2 + c^2)$ $\bar{I}_{yy} = \frac{3}{80}m(a^2 + c^2)$ $\bar{I}_{zz} = \frac{3}{80}m(a^2 + b^2)$ |
| Half Torus | $\bar{x} = \dfrac{a^2 + 4R^2}{2\pi R}$ | $I_{xx} = I_{yy} = \frac{1}{2}mR^2 + \frac{5}{8}ma^2$ $I_{zz} = mR^2 + \frac{3}{4}ma^2$ |

## PHOTO CREDITS

**Chapter Openers**

**Chapter 1:**  Michele Burgess/The Stock Market.

**Chapter 2:**  Robert Isaacs/Photo Researchers.

**Chapter 3:**  Peeter Vilms/Jeroboam, Inc.

**Chapter 4:**  Dan Chidester/The Image Works.

**Chapter 5:**  Jeff Albertson/Stock, Boston.

**Chapter 6:**  Grafton Marshall Smith/The Image Bank.

**Chapter 7:**  Chris Sorensen/The Stock Market.

# INDEX

Absolute system of units, 8
Acceleration, of a body, 6, 113
  due to gravity, 9
Accuracy, 11
Action and reaction, principle of, 6, 20, 103, 208, 229, 289
Active force, 412
Active-force diagram, 414
Addition of vectors, 4, 21, 23, 494
Aerostatics, 315
Angle, of friction, 350
  of repose, 352
Approximations, 12, 264
Archimedes, 1
Area, first moment of, 244
  second moment of, 244
Area moments of inertia, *see* Moments of inertia of areas
Atmospheric pressure, 316
Axes, choice of, 23, 62, 106, 179, 243, 246
  rotation of, 475
Axis, moment, 34, 71

Beams, concentrated loads on, 283
  definition of, 282
  distributed loads on, 283
  external effects, 282
  internal effects, 289
  loading-shear relation for, 290, 291
  resultant of forces on cross section of, 289
  shear-moment relation for, 290, 291, 292
  statically determinate and indeterminate, 282
  types of, 282
Bearing friction, 380
Belt friction, 389
Bending moment, 289
Bending-moment diagram, 290
Bodies, interconnected, 208, 412
Body, deformable, 3
  rigid, 2

Body force, 20, 240
Boundary conditions, 303
British system of units, 7
Buoyancy, center of, 322
  force of, 321
  principle of, 321

Cables, catenary, 305
  flexible, 301
  length of, 304, 306
  parabolic, 303
  tension in, 304, 307
Cajori, F., 6
Center, of buoyancy, 322
  of gravity, 20, 241
  of mass, 241, 243
  of pressure, 317
Centroids, 243
  of composite figures, 263
  by integration, 244
  of irregular volumes, 264
  table of, 509
  by theorems of Pappus, 275
Coefficient, of friction, 348, 349, 507
  of rolling resistance, 390
Collinear forces, equilibrium of, 113
Components, of a force, 21, 22
  rectangular, 5, 22, 23, 61
  scalar, 22
  of a vector, 4, 21, 22, 61
Composite areas, moment of inertia of, 467
Composite bodies, center of mass of, 263
Composite figures, centroid of, 263
Compression in truss members, 175, 176
Computer-oriented problems, 15, 170, 235, 342, 404, 448, 486
Concentrated forces, 20, 239
  on beams, 283
Concurrent forces, equilibrium of, 113, 144
  resultant of, 21, 85

Cone of friction, 350
Constant of gravitation, 10, 508
Constraint, 116, 146
    adequacy of, 117, 146
    partial, 146
    proper and improper, 118, 146
    redundant, 118, 146
Coordinates, choice of, 23, 62, 106, 246, 336, 453
Coplanar forces, equilibrium of, 113, 114
    resultant of, 23, 52
Coulomb, 346
Couple, 44, 72
    equivalent, 45
    moment of, 44, 72
    resolution of, 45, 86
    resultant, 52, 84
    vector representation of, 44, 72
    work of, 409
Cross or vector product, 35, 70, 495

D'Alembert, J., 1
da Vinci, 1
Deformable body, 2
Degrees of freedom, 414, 433, 444
Density, 242
Derivative of vector, 496
Derivatives, table of, 497
Diagram, active-force, 414
    bending-moment, 290
    free-body, 15, 102, 105, 143
    shear-force, 290
Differential element, choice of, 245
Differentials, order of, 12, 245, 336
Dimensions, homogeneity of, 14
Direction cosines, 5, 61
Disk friction, 381
Displacement, 407
    virtual, 410
Distributed forces, 20, 239, 240, 336
    on beams, 283
Distributive law, 36, 495
Dot or scalar product, 62, 408, 495
Dynamics, 2, 6

Efficiency, mechanical, 415
Elastic potential energy, 428
Energy, criterion for equilibrium, 432
    criterion for stability, 433
    elastic, 428
    potential, 428, 432
Equilibrium, alternative equations of, 115
    categories of, 113, 144
    of collinear forces, 113
    of concurrent forces, 113, 144
    condition of, 52, 113, 142, 410, 413
    of coplanar forces, 113, 114

energy criterion for, 432, 433
equations of, 113, 142
of interconnected rigid bodies, 208, 413
of machines, 208
necessary and sufficient conditions for, 113, 142
neutral, 433
of parallel forces, 114, 145
of a particle, 411
of a rigid body, 411
stability of, 117, 433
with two degrees of freedom, 414
by virtual work, 410, 411, 413
Euler, 1
External effects of force, 19

First moment of area, 245
Fixed vector, 3, 19
Flexible cables, 301
    differential equation for, 303
Fluids, 315
    friction in, 346
    incompressible, 316
    pressure in, 315
Force, action of, 19, 104, 144
    active, 412
    body, 20, 240
    buoyancy, 321
    components of, 21, 22, 61
    concentrated, 20, 239
    concept of, 2
    contact, 20
    coplanar system of, 52
    distributed, 20, 239, 240, 336
    effects of, 19
    friction, 103, 345
    gravitational, 9, 20, 105, 240
    inertia, 452
    intensity of, 240
    internal, 19, 239, 413
    kinds of, 20
    magnetic, 20, 105
    measurement of, 20
    mechanical action of, 104, 144
    moment of, 34, 70
    polygon, 52, 115
    reactive, 19, 412
    remote action of, 20, 105
    resolution of, 21, 22, 61, 62
    resultant, 52, 84, 86, 241, 336
    shear, 289, 315
    specifications of, 19
    unit of, 7
    work of, 407
Force-couple system, 45, 52, 73
Force system, concurrent, 53, 71, 85
    coplanar, 52

general, 19, 84
parallel, 21, 53, 85
Formulation of problems, 13
Frames, defined, 208, 229
equilibrium of, 208
Frames and machines, rigidity of, 208
Free-body diagram, 15, 102, 105, 143
Freedom, degrees of, 414, 432, 444
Free vector, 3, 4, 44, 72
Friction, angle of, 350
bearing, 380, 381
belt, 389
circle of, 380
coefficients of, 348, 349, 507
cone of, 350
disk, 381
dry or Coulomb, 346, 347
fluid, 346
internal, 346
journal bearing, 380
kinetic, 348
limiting, 348
in machines, 368
mechanism of, 347
pivot, 381
problems in dry friction, 351, 398
rolling, 390
screw thread, 370
static, 348
types of, 346
wedge, 368
work of, 414

Gage pressure, 316
Galileo, 1
Gas, 315
Graphical representation, 14, 21, 22, 52
Gravitation, constant of, 10, 508
law of, 10
Gravitational force, 9, 20, 105, 240
Gravitational potential energy, 429
Gravitational system of units, 8
Gravity, acceleration due to, 9
center of, 20, 241
Guldin, Paul, 275
Gyration, radius of, 453

Homogeneity, dimensional, 14
Hydrostatic pressure, 316, 319
Hydrostatics, 315
Hyperbolic functions, 306

Ideal systems, 412
Impending motion, 348, 350
Inclined axes, area moments of inertia about, 475

Inertia, 2, 452
area moments of, *see* Moments of inertia of areas
principal axes of, 476
products of, 475
Inertia force, 452
Integrals, table of selected, 498
Integration, choice of element for, 245, 336
numerical techniques for, 503, 505
of vectors, 497
Interconnected bodies, 208, 412
Internal effects of force, 19, 239, 413
Internal friction, 346
International System of units, 7

Joints, method of, 176, 201, 228
Joule, 409
Journal bearings, friction in, 380

Kilogram, 7, 8, 11
Kilopound, 8
Kinetic friction, 348
coefficient of, 349, 507

Lagrange, 1
Laplace, 1
Law, associative, 494
commutative, 494
of cosines, 494
distributive, 36, 495
of gravitation, 10
parallelogram, 4, 21, 52
of sines, 494
Pascal's, 315
triangle, 4, 21
Laws of motion, Newton's, 6
Length, standard unit of, 9
Limit, mathematical, 12
Line of action, 20
Liquids, 316
Loading-shear relation for beams, 290

Mach, Ernst, 36
Machines, defined, 208, 229
equilibrium of, 208
friction in, 368
ideal or real, 345
Mass, 2, 8
center of, 241, 243
unit of, 7, 8
Mathematical limit, 13
Mathematical model, 13
Mathematics, selected topics in, 491
Mechanical efficiency, 415
Mechanical system, 102
Mechanics, 1
Metacenter, 322

Metacentric height, 322
Meter, 9
Method, of joints, 176, 201, 228
   of problem solution, 14
   of sections, 189, 201, 228
   of virtual work, 407
Metric units, 7
Minimum energy, principle of, 433
Mohr's circle, 477
Moment, bending, 289
   components of, 71
   of a couple, 44, 72
   of a force, 34, 70
   torsional, 289, 451
   units of, 34
   vector representation of, 35, 70
Moment arm, 34
Moment axis, 34, 71
Moments, principle of, 53, 84, 241, 247, 336
Moments of inertia of areas, 451
   for composite areas, 467
   dimensions and units of, 453
   about inclined axes, 475
   by integration, 452
   maximum and minimum, 476, 477
   Mohr's circle representation of, 477
   polar, 453
   principal axes of, 476
   radius of gyration for, 453
   rectangular, 452
   table of, 509
   tabular computation of, 467
   transfer of axes for, 454, 475
Morin, 346
Motion, impending, 348, 350
Multi-force members, 208

Neutral equilibrium, 433
Newton, Isaac, 1
Newton's laws, 6
Newton (unit), 7
Newton's method, 501
Numerical integration, 503, 505

Order of differentials, 12, 245

Pappus, 275
   theorems of, 275
Parallel-axis theorems, for area moments of inertia,
      455
Parallel forces, equilibrium of, 114, 145
   resultant of, 21, 53, 85
Parallelogram law, 4, 21, 52
Particle, 2
Particles, equilibrium of, 411
Pascal (unit), 240

Pascal's law, 315
Pivot friction, 381
Polar moment of inertia, 453
Polygon, of forces, 52, 115
Potential energy, 428, 430
   datum for, 429, 430
   units of, 429, 430
Pound, standard, 8
Pound force, 8
Pound mass, 8
Pressure, 240, 315
   atmospheric, 316
   center of, 317
   fluid, 315
   gage, 316
   hydrostatic, 316, 319
   on submerged surfaces, 316, 319
Principal axes of inertia, 476
*Principia*, 6
Principle, of action and reaction, 6, 20, 103, 208, 229,
      289
   of buoyancy, 321
   of concurrency of forces, 115
   of minimum energy, 432
   of moments, 53, 84, 241, 247, 336
   of transmissibility, 3, 20, 52
   of virtual work, 411, 413, 432
Products of inertia, 475
   about inclined axes, 476
Products of vectors, 35, 62, 71, 408, 495

Radius of gyration, 453
Reactive forces, 19, 412
Rectangular components, 5, 22, 23, 61
Rectangular moments of inertia, 452
Redundancy, external and internal, 178, 200
Redundant supports, 117, 146
Repose, angle of, 352
Resolution, force, 21, 22, 61
   force and couple, 45, 52, 73
Resultant, of concurrent forces, 21, 53, 85
   of coplanar forces, 23, 24, 52
   couple, 84, 86
   of fluid pressure, 316, 319
   force, 52, 84, 86, 336
   of forces on beam cross section, 289
   of general force system, 84
   of parallel forces, 21, 53, 85
Right-hand rule, 35, 62, 70
Rigid bodies, interconnected, 208, 412
Rigid body, 2
   equilibrium of, 411
Rolling resistance, coefficient of, 390

Scalar, 3
Scalar components, 22

Scalar or dot product, 62, 408, 495
Screw, friction in, 370
Second moment of area, 244, 451
Sections, method of, 189, 201, 228
Series, selected expansions, 497
Shear force, 289, 315
Shear-force diagram, 290
Shear-moment relation for beams, 290, 291, 292
Shear stress, 452
Singularity functions, 292
SI units, 7
Sliding vector, 3, 20, 34, 72
Slug, 8
Space, 2
Space trusses, 200, 228
Specific weight, 240
Spring, linear and nonlinear, 105
    potential energy of, 428
    stiffness of, 428
Stability, of equilibrium, 117, 433
    of floating bodies, 322
    for single degree-of-freedom system, 432
    of trusses, 178, 200
Statically determinate structures, 117, 146, 173, 178
Statically indeterminate structures, 117, 146, 178, 200, 208
Static friction, 348
    coefficient of, 348, 507
Statics, 2
Stevinus, 1
Stiffness of spring, 428
Stress, 240
    shear, 452
Structures, statical determinacy of, 117, 146, 173, 178 200, 208
    types of, 173
Submerged surfaces, pressure on, 316, 319
Subtraction of vectors, 4, 494
Symmetry, considerations of, 243, 475
System, with elastic members, 428
    force-couple, 45, 52, 73
    of forces, concurrent, 21, 53, 71, 85, 113, 144
        coplanar, 52
        general, 19, 84
        parallel, 53, 85, 114, 145
    ideal, 412
    of interconnected bodies, 208, 412
    mechanical, 102
    real, 414
    of units, 7

Table, of area moments of inertia, 509
    of centroids, 509
    of coefficients of friction, 507
    of densities, 507
    of derivatives, 497

of mathematical relations, 491
of solar system constants, 508
Tension in truss members, 175, 176
Theorem, of Pappus, 275
    of Varignon, 35, 36, 53, 71
Three-force member, 115
Thrust bearing, friction in, 381
Time, 2, 9
Ton, 8
Torque, *see* Moment, of force
Torsional moment, 289, 451
Transfer of axes, for moments of inertia, 454
    for products of inertia, 475
Transmissibility, principle of, 3, 20, 52
Triangle law, 4, 21
Triple scalar product, 71, 496
Triple vector product, 496
Trusses, definition, 174
    plane, 174
    simple, 175, 200
    space, 200, 228
    stability of, 178, 200
    statical determinacy of, 178, 200, 228
    types of, 174
Two-force members, 114, 175

U.S. customary units, 7
Units, 7, 34, 409
Unit vectors, 5, 22, 61, 63, 71
Unstable equilibrium, 433

Varignon, 1
Varignon's theorem, 35, 36, 53, 71
Vector equation, 6
Vectors, 3, 19
    addition of, 4, 21, 23, 494
    components of, 4, 21, 22, 61
    couple, 44, 72
    cross or vector product of, 35, 70, 495
    derivative of, 496
    dot or scalar product of, 62, 408, 495
    fixed, 3, 19
    free, 3, 4, 44, 72
    moment, 34, 70
    notation for, 4
    resolution of, 21, 22, 61, 62
    sliding, 3, 20, 34, 72
    subtraction of, 4, 494
    unit, 5, 22, 61, 63, 71
Vector sum, of couples, 73
    of forces, 21, 23, 52, 84
Virtual displacement, 410
Virtual work, 407, 410
    for elastic systems, 432
    for ideal systems, 412, 413
    for a particle, 411

Virtual Work (*Continued*)
  for a rigid body, 412
Viscosity, 346

Wear in bearings, 382
Wedges, friction in, 368

Weight, 11, 20, 105, 240
Work, of a couple, 409
  of a force, 407, 408
  units of, 409
  virtual, 407, 410
Wrench, 85, 86